Human and Social Biology for CSEC™

New edition

Ann Fullick
Alexcia Morris
Farishazad Nagir

The Publishers would like to thank the following for permission to reproduce copyright material.

Acknowledgements

Every effort has been made to trace all copyright holders, but if any have been inadvertently overlooked, the Publishers will be pleased to make the necessary arrangements at the first opportunity.

Although every effort has been made to ensure that website addresses are correct at time of going to press, Hodder Education cannot be held responsible for the content of any website mentioned in this book. It is sometimes possible to find a relocated web page by typing in the address of the home page for a website in the URL window of your browser.

Hachette UK's policy is to use papers that are natural, renewable and recyclable products and made from wood grown in well-managed forests and other controlled sources. The logging and manufacturing processes are expected to conform to the environmental regulations of the country of origin.

Orders: please contact Hachette UK Distribution, Hely Hutchinson Centre, Milton Road, Didcot, Oxfordshire, OX11 7HH. Telephone: +44 (0)1235 827827. Email education@hachette.co.uk Lines are open from 9 a.m. to 5 p.m., Monday to Friday. You can also order through our website: www.hoddereducation.com

ISBN: 978 1 3983 7915 2

© Ann Fullick, Alexcia Morris and Farishazad Nagir 2022

First published in 2009

Second edition published in 2015

This edition published in 2022 by

Hodder Education,

An Hachette UK Company

Carmelite House

50 Victoria Embankment

London EC4Y 0DZ

www.hoddereducation.com

Impression number   10 9 8 7 6 5 4 3 2 1

Year   2026 2025 2024 2023 2022

All rights reserved. Apart from any use permitted under UK copyright law, no part of this publication may be reproduced or transmitted in any form or by any means, electronic or mechanical, including photocopying and recording, or held within any information storage and retrieval system, without permission in writing from the publisher or under licence from the Copyright Licensing Agency Limited. Further details of such licences (for reprographic reproduction) may be obtained from the Copyright Licensing Agency Limited, www.cla.co.uk

Cover images: f*ront cover background* © Paulista – stock.adobe.com, *l-r circles* © sidekick – stock.adobe.com, © Monkey Business – stock.adobe.com, © Iakov Filimonov – stock.adobe.com, © michaeljung – stock.adobe.com, © steklo- stock.adobe.com, *back cover* © nik962 – stock.adobe.com

Illustrations by Vian Oelofsen, Stéphan Theron

Typeset in Myriad Pro 11/14 pt by IO Publishing CC

Produced by DZS Grafik, Printed in Bosnia & Herzegovina

A catalogue record for this title is available from the British Library.

Contents

A **Section A** Living organisms and the environment 4
    Chapter 1: The structure and function of cells 6
    Chapter 2: Passive and active transport 24
    Chapter 3: Photosynthesis 44
    Chapter 4: Feeding relationships and the carbon cycle in nature 59

B **Section B** Life processes 82
    Chapter 5: Nutrition 84
    Chapter 6: The human digestive system 114
    Chapter 7: The human respiratory system 135
    Chapter 8: The circulatory system 165
    Chapter 9: The skeletal system 189
    Chapter 10: Excretion and homeostasis 206
    Chapter 11: Coordination and control 227
    Chapter 12: The human reproductive system 252

C **Section C** Heredity and variation 282
    Chapter 13: Cell reproduction and variation 284
    Chapter 14: Genetics 300

D **Section D** Disease and their impact on humans 326
    Chapter 15: Health and diseases 328
    Chapter 16: Infectious diseases caused by viruses and bacteria 356
    Chapter 17: Infectious diseases in action 373
    Chapter 18: Lifestyle diseases 388
    Chapter 19: Parasites and vectors 413
    Chapter 20: Preventing disease 430

E **Section E** The impact of health practices on the environment 444
    Chapter 21: Clean water 446
    Chapter 22: Pollution and waste management 456

F **Section F** The School-Based Assessment (SBA) and the Alternative Paper 488
    Chapter 23: Planning, conducting and reportng on your School-Based Assessment (SBA) project 490
    Chapter 24: A Sample School-Based Assessment (SBA) report and a worked Case study 508
    Glossary 522
    Index 529
    Acknowledgements 535

Section A: Living organisms and the environment

Learning outcomes

At the end of Section A, you will:

- understand the processes that govern the interactions of organisms in the environment, and the processes by which life is perpetuated
- understand the nature of the interdependence of the processes, structures and functions of the major systems within an organism in the maintenance of health.

Cells are the building blocks of life – your own body cells have many structures and functions in common with the cells of a whistling-cracking frog or a breadfruit! You will discover the secrets of cells, which can only be revealed with the help of a **microscope**. You will compare the cells of plants with those of animals, and you will also explore the structure of tiny organisms called microbes, which are very important to life on Earth, yet which can cause many deadly diseases.

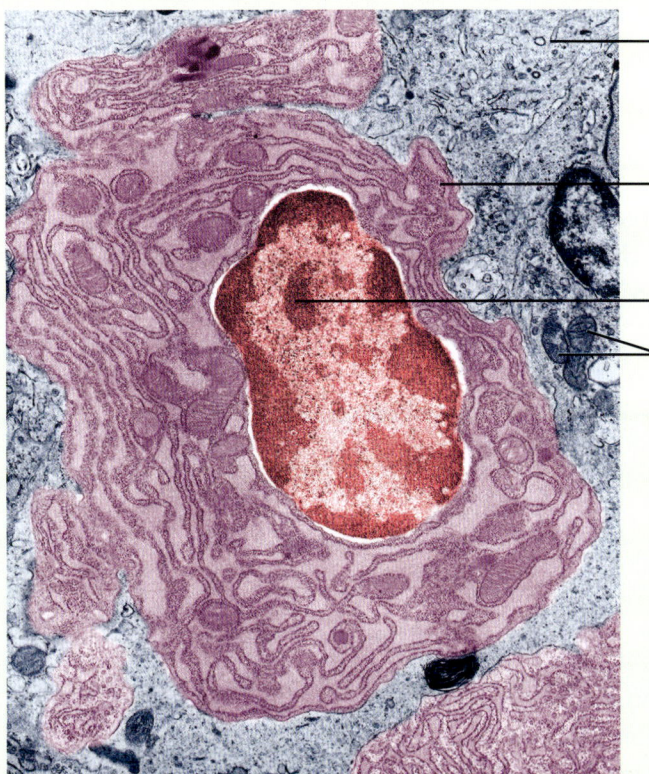

Figure Intro 1.1 The innermost workings of a cell are only visible when they are magnified many thousands of times, yet we have discovered a great deal about how our whole body works from studying the internal secrets of the cell.

Figure Intro 1.2 The cells of this whistling-cracking frog have a great deal in common with those of every other living creature, yet its place in the feeding relationships of the animals of the Caribbean is unique.

Living organisms are linked by their similarities at a cellular level, but they are interconnected at other levels as well. There are feeding relationships between all the different groups of living organisms, starting with plants, which have the ability to make their own food by photosynthesis. Almost all other types of living organisms depend on plants, and the plants rely on the actions of microbes to return nutrients to the soil. You will explore some of these feeding relationships and how they work. You can work out the energy flow right through a food chain until it reaches yourself!

Revision tip

Start early:
* **Early in the school year**
* **Early in the day – especially at weekends**

Chapter 1: The structure and function of cells

- identify selected specialised cells in the human body and explain how they are adapted to their function
- state the functions of cell structures including the cell membrane, cell wall, nucleus, mitochondria, vacuoles, ribosomes, chloroplasts and endoplasmic reticulum
- draw and label diagrams to show the structure of unspecialised animal and plant cells
- explain the processes of diffusion, osmosis and active transport, and why they are so important in living systems
- describe the characteristics of living organisms

When you have completed this chapter, you will be able to:

The planet we live on teems with a wide variety of living organisms including animals, plants and microbes. To understand these organisms, both as individuals and as part of an integrated **environment**, we need to consider the most basic facts of life. All living organisms are made up of units called cells. Some organisms, such as Amoeba, consist of single cells. Others, such as ourselves, are made up of many millions of cells all working together. Organisms that contain more than one cell are called multicellular, which means 'many cells'.

What the examiners say

- Candidates often demonstrate a basic understanding of cells, but are not as competent in explaining their relationship to other topics such as diseases or respiration.
- Candidates are sometimes unable to differentiate between fundamental concepts such as breathing and respiration, or discuss differences between living and non-living things.
- There are repeated misconceptions regarding the functions of cells. For example, one misconception is that one of the functions of red blood cells is to transport nutrients around the body, while white blood cells aid in coagulation.

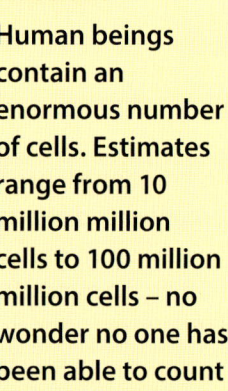

DID YOU KNOW?

Human beings contain an enormous number of cells. Estimates range from 10 million million cells to 100 million million cells – no wonder no one has been able to count them accurately!

The biologist's toolkit: Asking questions

Scientists ask a lot of questions! This is how we find out more about the living world around us.

There are many different types of questions. Some of them are answered by science. A scientific question is one we answer by collecting and thinking about **data**. Data is information. It may be numbers from measurements, or words or drawings describing observations.

Most science investigations start with a question. Here are some examples of questions that science can answer:

- Which flowers do copper-rumped hummingbirds like best?
- Which Caribbean island uses the most fossil fuels?
- Does being **obese** increase your risk of heart disease?

However, science cannot answer all questions, for example:

- What is my favourite food?
- Will I get into the school cricket team?

When you plan an investigation, choose the question you are going to answer carefully. Make sure it can be answered by science.

Copper-rumped hummingbird

The characteristics of living organisms

All living organisms have specific characteristics, which they display regardless of whether they have one cell or millions. In some cases, particularly in the larger multicellular animals like humans, it is easy to show that all these processes are taking place. In others, such as microscopic organisms and plants, we have to rely on technology to show us what is happening.

The seven life processes, described in Table 1.1, are common to most living organisms.

Table 1.1 Life processes

Life process	Illustration		Description
Movement		All these flowers have turned to face the Sun.	The process by which all organisms get nearer to things they need or away from problems. Animals move using muscles; plants move more slowly using growth (Figure 1.2).
Respiration		Respiration supplies energy to the cells of all organisms including this hummingbird.	The process by which living organisms get the energy from their food.

Sensitivity (irritability)		Eyes, ears and whiskers help this mouse detect its surroundings.	Living organisms are all sensitive to changes in their surroundings, in different ways.
Nutrition		Humans eat many different types of food.	All living organisms need food to provide the energy used by their cells. Plants make their own food by photosynthesis, while animals eat other organisms.
Excretion		Many animals excrete through their kidneys.	The process of removing the poisonous waste products produced by the cells.
Reproduction		Offspring often look similar to their parents.	The process of producing offspring is vital to the long-term survival of any type of living organism.
Growth		Growth is a permanent increase in size.	When organisms grow, they increase in both size and mass, using chemicals from their food to build new material and new cells.

The structure and function of cells

Looking at cells

Everything we know about the structure of cells has depended on the development of the microscope. We have been able to look at cells for more than 300 years, and as microscopes have improved, so has our knowledge and understanding of cell structure. Light microscopes let us magnify up to 1500 times. They are relatively cheap and widely used. **Electron microscopes** are big and very expensive, but they let us magnify things up to 1 million times or more, so they show us extra things inside our cells.

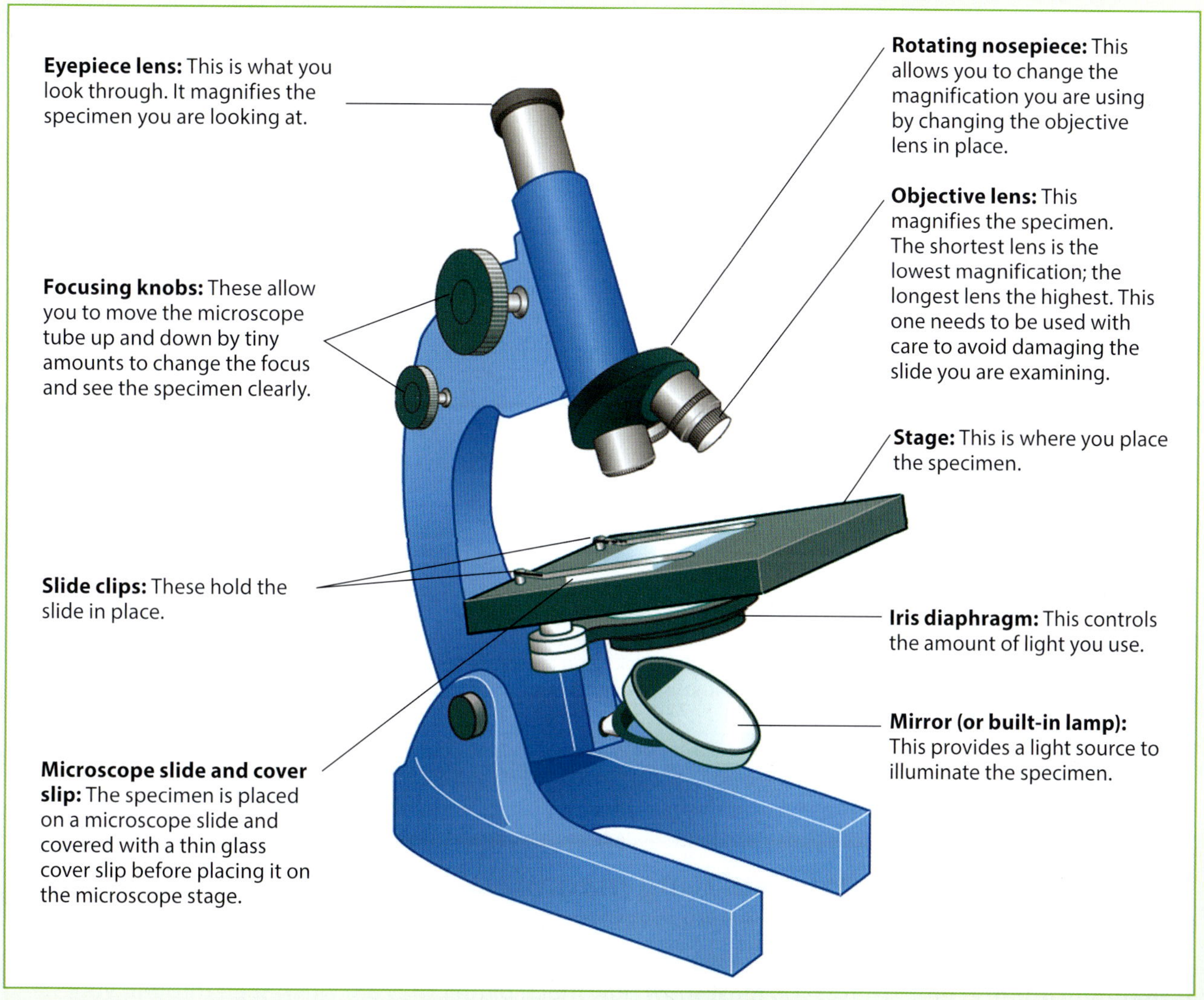

Eyepiece lens: This is what you look through. It magnifies the specimen you are looking at.

Focusing knobs: These allow you to move the microscope tube up and down by tiny amounts to change the focus and see the specimen clearly.

Slide clips: These hold the slide in place.

Microscope slide and cover slip: The specimen is placed on a microscope slide and covered with a thin glass cover slip before placing it on the microscope stage.

Rotating nosepiece: This allows you to change the magnification you are using by changing the objective lens in place.

Objective lens: This magnifies the specimen. The shortest lens is the lowest magnification; the longest lens the highest. This one needs to be used with care to avoid damaging the slide you are examining.

Stage: This is where you place the specimen.

Iris diaphragm: This controls the amount of light you use.

Mirror (or built-in lamp): This provides a light source to illuminate the specimen.

Figure 1.1 A compound light microscope has two sets of lenses that are used to magnify a specimen. These microscopes are widely used to look at cells.

There are some basic similarities between most animal and plant cells. For example, almost all cells have a **nucleus**, a **cell membrane**, **mitochondria**, **ribosomes**, **endoplasmic reticulum** and **cytoplasm**.

There are other features that are often seen in plant cells, particularly in the green parts of plants. This has led scientists to develop a **model** of the basic structure of an unspecialised animal cell and an unspecialised green plant cell. Although not many cells are this simple, the idea of unspecialised animal and plant cells gives us a very useful base point with which to compare other, more specialised cells.

DID YOU KNOW?

The biggest single cell is an unfertilised ostrich egg – it is around 180 mm long, 140 mm wide and weighs about 1.5 kg!

Structures and functions in unspecialised animal cells

Structures suspended within a cell membrane are called **organelles**. These organelles contain **enzymes** and chemicals to carry out specialised jobs within the cell. You will learn more about enzymes later in this book.

Mitochondria (singular: mitochondrion) are the powerhouses of a cell. They carry out most of the reactions of respiration, in which energy is released from food in a form that your cells can use. Cells that need a lot of energy – such as muscle cells and secreting cells – contain many mitochondria.

The **nucleus** controls all the activities of a cell. It also contains the instructions for making new cells or new organisms in the form of long threads called **chromosomes**. This is the **genetic material**. You will find out more about this in Chapter 14.

The **cell membrane** forms a barrier like a very thin 'skin' around the outside of the cell. The membrane controls the passage of substances – such as carbon dioxide, oxygen and water – into and out of the cell. Because it lets some substances through but not others, it is called a **partially permeable membrane**.

The **cytoplasm** is a liquid gel in which most of the chemical reactions needed for life take place. About 70% of the cytoplasm of a cell is actually water! The cytoplasm contains all the other organelles of the cell.

Ribosomes are found on the membrane stacks called endoplasmic reticulum that run through the cytoplasm of your cells. These stacks are involved in the manufacture, packaging and transport of many different substances. The ribosomes themselves are vital for protein synthesis, the process by which your body makes all the enzymes that control the reactions of your cells.

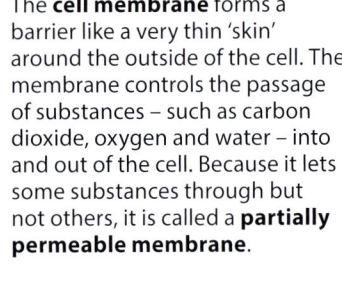

Figure 1.2 A simple animal cell like this shows the features that are common to almost all living cells.

The biologist's toolkit: Calculating magnification

When you look at a **specimen** using a magnifying lens or a microscope you need to record the **magnification**. In other words, you must show how much bigger they are when seen through your lenses than they are in real life.

When you use a light microscope you can calculate the magnification of your specimen easily. Multiply the magnification of your **eyepiece lens** (see Figure 1.1) by the magnification of the **objective lens** (see Figure 1.1).

Example:

If your eyepiece lens is ×4 and your objective lens is ×10: your total magnification = 4 × 10 = **40×**

Always give the magnification at which you looked at a specimen when you label drawings made using a microscope, for example: **human cells as viewed at 40×**

The structure and function of cells

Activity 1.1

Using the light microscope to look at animal cells

You will need:

- a microscope
- a lamp
- prepared microscope slides of human cheek cells/epidermal cells

Method

1. Set up your microscope with the lowest power lens (the smallest lens) in place.

2. Clip the prepared slide into place on the stage using the slide clips. Position the specimen over the hole in the stage.

3. If your microscope has a built-in lamp, switch it on. If it has a mirror, adjust the angle of the mirror until the specimen is illuminated.

4. Now look through the eyepiece lens and adjust the iris diaphragm until the light is bright but does not dazzle you. The illuminated area you can see is called the field of view.

5. Looking at your microscope from the side (not through the eyepiece lens), and using the coarse focusing knob, slowly move the objective lens down so it is as close as possible to the slide without touching it.

6. Now look through the eyepiece lens again. Turn the coarse focusing knob very gently in the opposite direction to move the objective lens away from the slide. Do this while you are looking through the eyepiece lens, and the specimen will gradually appear in focus. Once you can see the specimen clearly, use the fine focusing knob to get the focus as sharp as you can.

7. You may find that if you now shut the iris diaphragm down further, so that the hole for the light to pass through gets smaller, you will see the specimen better (the contrast is greater).

8. To use the higher magnifications, rotate the nosepiece so that the next lens clicks into place. Do not adjust the focusing knobs at this point as the specimen should still be in focus and, with the coarse focusing knob in particular, it is very easy to break the slide. If you do need to adjust the focus, use the fine focusing knob only with higher magnifications. Take great care to not touch the slide with the lens. You may want to adjust the iris diaphragm as well.

9. Human cheek cells and simple epithelial cells are very similar to the diagram of an unspecialised animal cell on page 10. Draw some of the cells you see and label them as fully as you can. Remember you will not see ribosomes or mitochondria under normal light microscopes.

Remember

Microscopes are expensive and delicate pieces of equipment, so always take care of them and handle them safely. You will not be able to see mitochondria and ribosomes with a light microscope.

The biologist's toolkit: Observations – biological drawings

A biologist must be very observant. We have lots of ways of making and recording our observations. One way is to draw biological specimens such as whole organisms, parts of organisms such as bones, flowers or feathers, and the inside of organisms or parts of organisms seen through a magnifying lens or a microscope.

Biological drawings are different from **diagrams**. A diagram represents a complicated structure and helps you understand it. A biological drawing is an accurate picture of what you actually observe. There are a few simple rules to follow when you make a biological drawing:

- Give your drawing a clear title to say what it is and how magnified it is, for example: drawn life size (x1) or drawn 400× magnification.
- Use plain, unlined paper if possible.
- Use a pencil – no colouring in.
- Put your drawing in the middle of the page so that there is room for labels
- Label your diagram. Label lines must be drawn with a ruler so they are straight and the label lines should not cross each other.
- Give a scale line to show how much bigger the drawing is than life size.

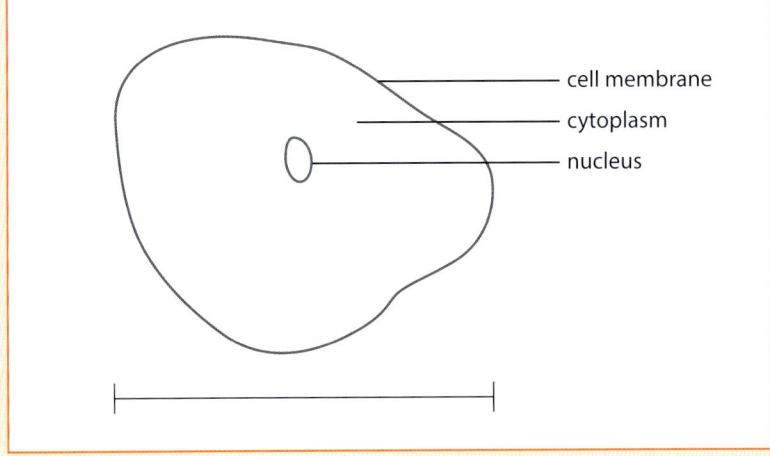

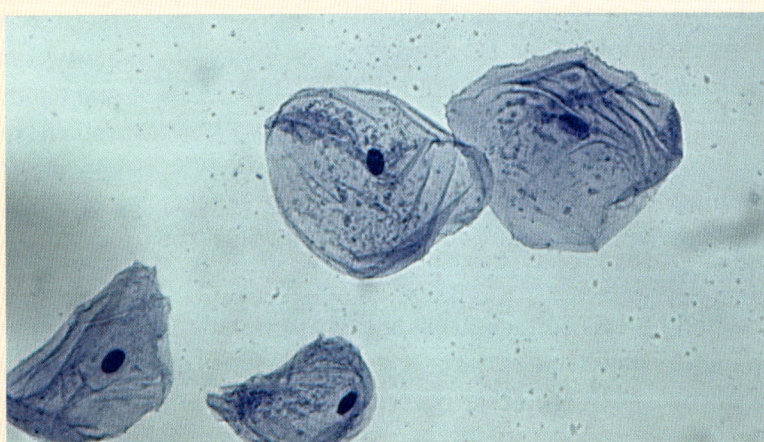

A micrograph of a human cheek cell

The structure and function of cells

Why do cells have organelles?

All the processes of life take place within a single cell. Imagine 100 different reactions going on in a laboratory test tube – chemical chaos and probably a few explosions would result! However, this is the level of chemical activity going on in a cell at any one time. Cell chemistry works because each reaction is controlled by an enzyme, a protein designed to control the rate of each specific reaction. Each enzyme makes sure that the reaction it controls takes place without becoming mixed up with any other reactions.

The enzymes involved in different chemical processes are usually found in different parts of a cell. For example, many of the enzymes:

- controlling the reactions of respiration are found in the mitochondria
- controlling the reactions of photosynthesis are found in the chloroplasts
- involved in protein synthesis are found on the surface of the ribosomes.

These cell compartments or organelles help to keep your cell chemistry well under control.

Structures and functions of unspecialised plant cells

Plants are very different from animals – they do not move their whole body about and they make their own food by photosynthesis. So, while plant cells have all the features of a typical animal cell – nucleus, cell membrane, cytoplasm, mitochondria, endoplasmic reticulum and ribosomes – they also have unique structures that they need for their own, very different way of life.

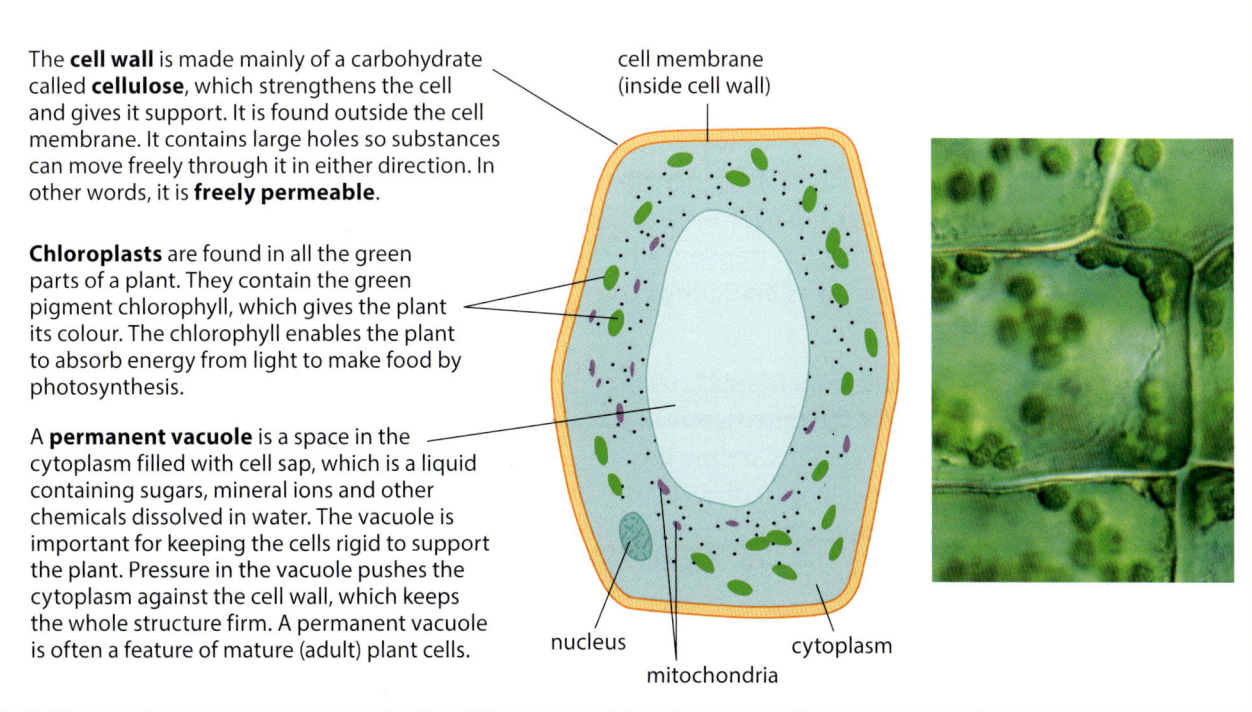

The **cell wall** is made mainly of a carbohydrate called **cellulose**, which strengthens the cell and gives it support. It is found outside the cell membrane. It contains large holes so substances can move freely through it in either direction. In other words, it is **freely permeable**.

Chloroplasts are found in all the green parts of a plant. They contain the green pigment chlorophyll, which gives the plant its colour. The chlorophyll enables the plant to absorb energy from light to make food by photosynthesis.

A **permanent vacuole** is a space in the cytoplasm filled with cell sap, which is a liquid containing sugars, mineral ions and other chemicals dissolved in water. The vacuole is important for keeping the cells rigid to support the plant. Pressure in the vacuole pushes the cytoplasm against the cell wall, which keeps the whole structure firm. A permanent vacuole is often a feature of mature (adult) plant cells.

Figure 1.3 A photosynthetic plant cell has many features in common with an animal cell, but it has others that are unique to plants.

Looking at cells

Activity 1.2

Making a slide of plant cells

The prepared slides you looked at in Activity 1.1 showed animal cells that were dead and stained to make them easier to see. In this activity, you are going to look at one of a number of different types of plant cells – onion, rhubarb or pondweed.

You will need:

- a microscope
- microscope slides
- cover slips
- forceps
- a mounted needle
- a pipette
- a lamp
- a piece of onion, rhubarb or pondweed, for example, Elodea (Canadian pondweed)

Remember
Microscopes are expensive and delicate pieces of equipment so always take care of them and handle them safely.

Method

Onion cells do not contain any chlorophyll, so they are not coloured. You can look at them as they are, or stain them using iodine, which reacts with the starch in the cells and turns blue-black.

1. Take your piece of onion and remove a small piece of the thin skin (inner epidermis) on the inside of the fleshy part using your forceps. It is very thin and quite tricky to handle.

2. Place the epidermis onto a microscope slide and very gently add either a drop of water or a drop of iodine from a pipette.

3. Using the mounted needle (or a sharp pencil), very gently lower the cover slip over the specimen. Take great care not to trap any air bubbles – these will show up as black rings under the microscope.

4. Remove any excess liquid from the slide using tissues and place it under the microscope. Starting with the low-power lens, follow the procedure for looking at cells described in Activity 1.1 on page 6. Use the higher power lenses to look at the cells in as much detail as possible.

5. Make a labelled drawing of several of the cells you can see.

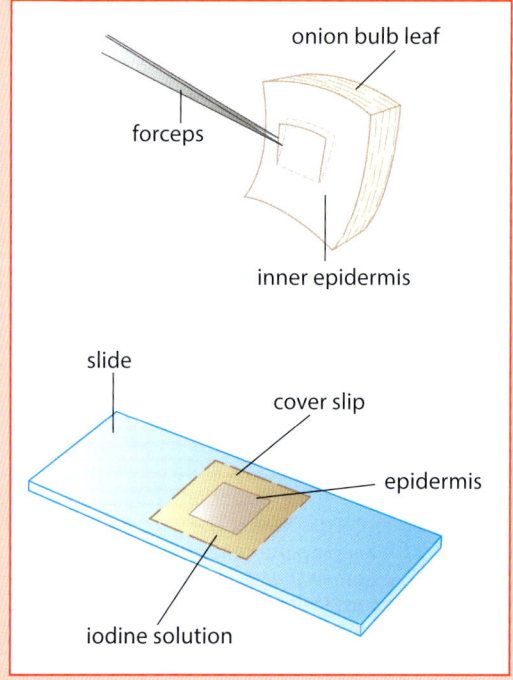

Figure 1.4 Making a slide on onion epidermis

Checkpoint questions

1. State the seven life processes that are common to all living organisms.
2. a Describe the cytoplasm of a cell.
   b State two of the main jobs of the cytoplasm of a cell
3. a What are enzymes?
   b Explain why they are so important in a cell
4. State what an organelle is and justify its importance in a cell.
5. Make a table to compare the similarities and differences in structure between unspecialised animal cells and unspecialised plant cells.

The structure of microbes

Animals and plants are not the only types of living organisms. Microbes (also called microorganisms) are tiny living organisms that are usually so small that we need to use a microscope to see them. Many of these microbes are very important, as shown in the diagram below.

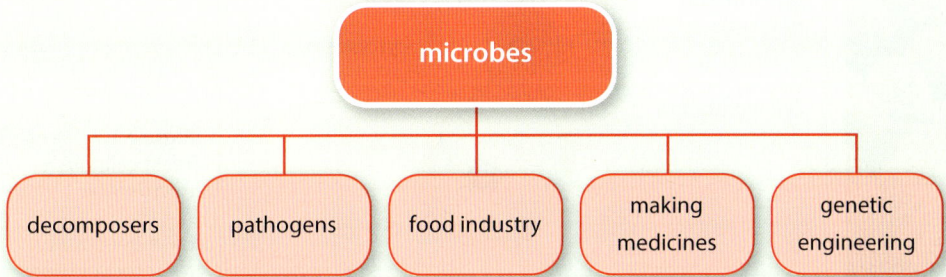

Examples of microbes are bacteria, viruses, fungi, yeasts and moulds.

Bacteria are single-celled organisms (Figure 1.5). They are much smaller than the smallest animal and plant cells. A bacterial cell has many similarities to animal and plant cells. It is made up of cytoplasm surrounded by a membrane and a **cell wall**. Inside the bacterial cell is the genetic material, but this is not contained in a nucleus. A bacterial cell wall differs from a plant cell wall because it is not made of **cellulose**. Some bacteria have other features, like flagella to help them move, or protective slime capsules. Bacteria also come in a variety of different shapes and sizes. Some are rod-shaped, some are round, some are comma-shaped and some are spirals. While some bacteria cause disease, many are harmless and many are actively useful to people.

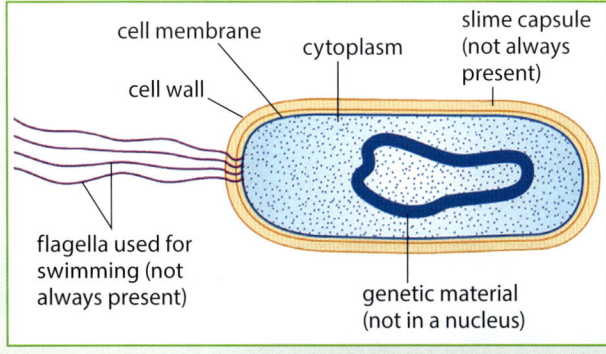

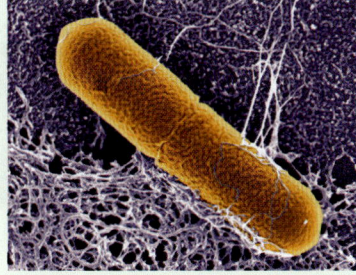

Figure 1.5 Bacteria come in many different shapes and sizes but they all have the same basic structure.

Viruses are even smaller than bacteria. They usually have regular geometric shapes, and they are made up of a protein coat surrounding genetic material that contains relatively few **genes**. They do not carry out any of the functions of normal living organisms except reproduction, and they can only reproduce by taking over another living cell. Most naturally occurring viruses cause disease.

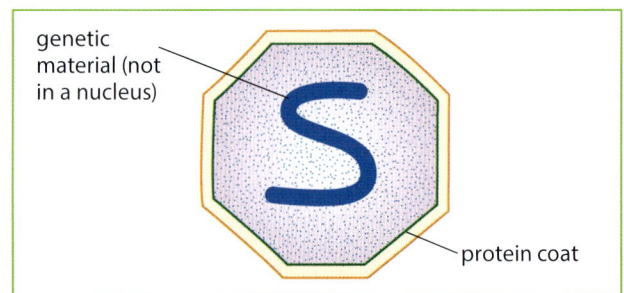

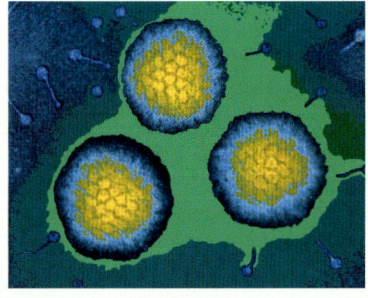

Figure 1.6 Viruses are very small with a very simple structure. There is some argument about whether or not they are living organisms.

Fungi are living organisms that obtain their food from other dead or living organisms. Some of the microorganisms that are most useful to people are moulds and yeasts, which are both types of fungi. Both moulds and yeasts are extremely important as decomposers, breaking down animal and plant material, and returning nutrients to the environment (see Chapter 4).

Yeasts are single-celled organisms. Each yeast cell has a nucleus, cytoplasm and a membrane surrounded by a cell wall. The main way in which yeasts reproduce is by asexual budding – splitting to form new yeast cells.

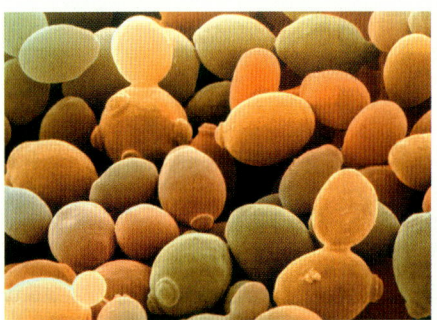

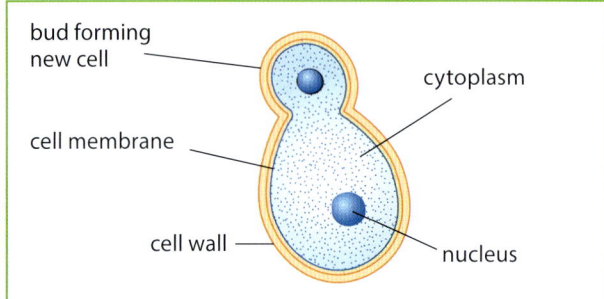

Figure 1.7 Yeast cells are microscopic organisms that have been useful to us for centuries.

Moulds are different. They are made up of tiny, thread-like structures called hyphae (Figure 1.8). The **hyphae** are not made up of individual cells – they are tubes that consist of a cell wall containing cytoplasm and lots of nuclei. Moulds, like yeasts, generally reproduce asexually, but they do it by producing fruiting bodies that contain spores. The mycelium of many fungi grow underground or through wood – and the fruiting bodies of these fungi are large. They appear as the toadstools, mushrooms and bracket fungi that we see around us.

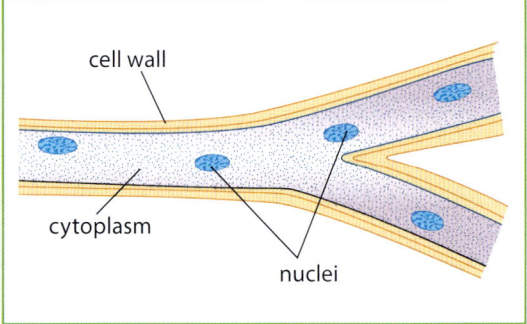

Figure 1.8 a) The mycelium of a fungus, made of hyphae b) The fruiting body of a fungus spreads the spores to make new fungi c) A diagram of fungal hyphae.

> **Checkpoint questions**
>
> 6 Make a table to compare the structure of the basic units of bacteria, viruses and fungi.
> 7 Identify features that are similar in all of the cells you have looked at – animals, plants and microbes.

Cell specialisation in humans

When an egg and a sperm combine to form an embryo, a single cell is formed. In multicellular organisms, like humans, this cell divides many times. As the embryo develops, the cells **differentiate** – they become **specialised**, which means they are adapted to carry out particular functions (jobs) in your body. For example, some cells differentiate to become red blood cells and carry oxygen, some become muscle cells, and others become **neurons** (nerve cells). This differentiation takes place as some of the genes (genetic material) in the nucleus of the cells are switched on and others are switched off. Scientists are still not quite sure what causes these changes to take place, but it seems to be at least partly due to the position of the cells in the embryo itself.

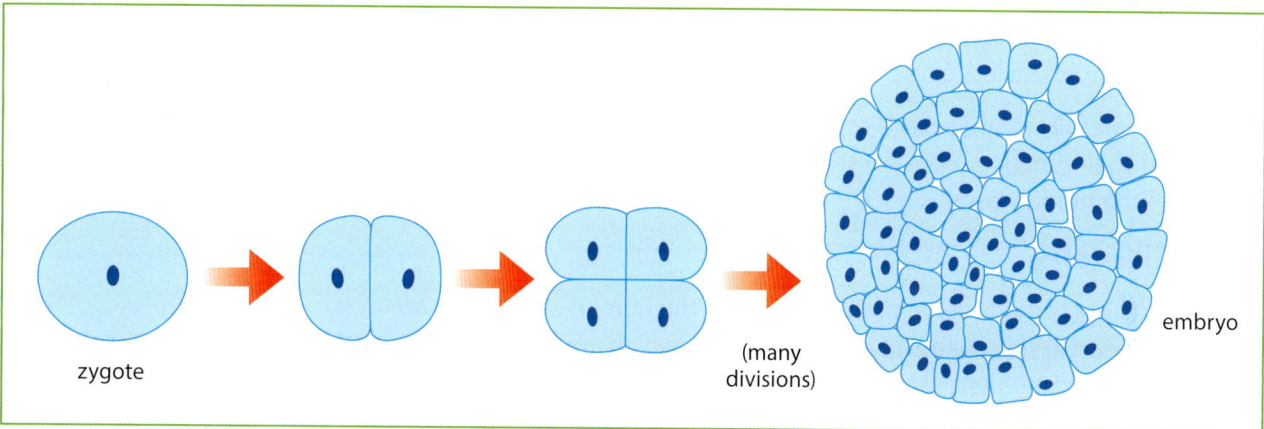

Figure 1.9 Animals and plants grow by cell division.

Specialised cells with similar functions are often grouped together to form a tissue, for example, in humans, **connective tissue** such as **tendons** and **cartilage** joins parts of the body together, while **nervous tissue** carries information around the body and **muscle tissue** contracts to move the body around.

In many living organisms, including people, there are more levels of organisation. Several different tissues work together to form an organ such as the heart, the kidneys or the lungs, and each organ carries out a particular function in your body. For example, the heart pumps blood around your body, your kidneys remove waste products and your lungs take in oxygen and remove waste **carbon dioxide**.

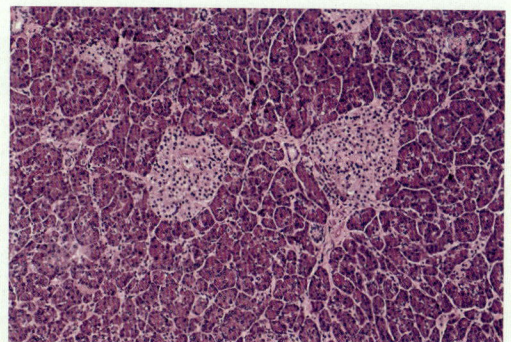

Figure 1.10 The pancreas controls your blood sugar and produces digestive enzymes to break down your food. Inside the organ, using a microscope, you can see two very different tissues made up of different types of cells that carry out these different jobs.

In turn, different organs work together in organ systems to carry out major functions in the body such as transporting the blood all around your body or reproduction. Examples include the cardiovascular system (the heart, lungs and blood vessels) and the digestive system (Figure 1.12).

These different levels or organisation are very important in large, multicellular organisms. Groups of cells or tissues working together to carry out particular functions in the body mean different parts of the animal or plant work together so the organism can carry out all of the functions of life efficiently. You will learn more about many of these organs and organ systems in the rest of your Human & Social Biology course.

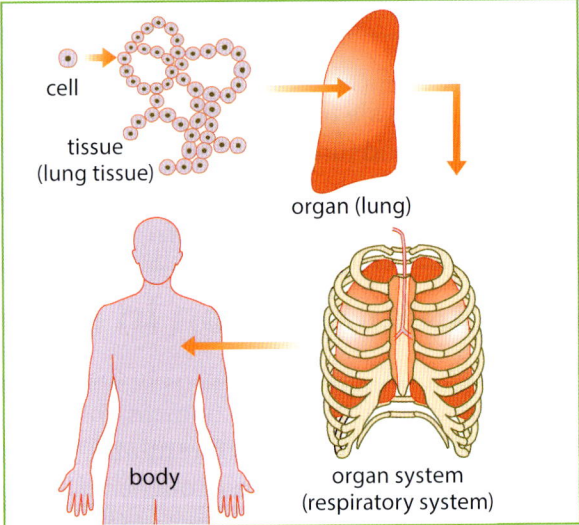

Figure 1.11 Large living organisms have many levels of organisation. As a result, each part of the body is perfectly adapted to carry out its functions.

Specialised cells

When cells become specialised, their structure is modified or adapted to suit the specialised job that the cell is doing. For example:

- cells that use a lot of energy have many mitochondria
- cells that are important for **diffusion** have a large surface area
- cells that produce lots of proteins have many ribosomes and mitochondria.

By looking carefully at specialised cells, you will see how their structure is adapted to their function. Here are some examples of the specialised cells you will find in the human body.

Epithelial cells

Epithelial cells play many very important roles in the human body. They are usually arranged in thin sheets of epithelial tissue (which are often only one cell thick), and they cover internal and external surfaces. So, your skin is made up of epithelial cells, and your gut, your respiratory system, your reproductive system and many other organ systems of your body are all lined with epithelial cells.

Table 1.2 The functions of epithelial cells

Type of epithelial cell		Function
	Squamous epithelium (flattened cells)	Found in the lungs and gut walls, where diffusion is very important.
	Cuboidal epithelium	Found in glands, and aids in secretion and **absorption** of chemicals.
	Ciliated columnar epithelium	Found in the respiratory and reproductive systems, with small, hair-like cilia that beat rhythmically to move substances through a tube. They move mucus and microbes away from the lungs, and help the ovum move towards the uterus.

The structure and function of cells

Reproductive cells – eggs and sperm

If you are female, your ovaries contain ova or egg cells. These female sex cells have only half the number of chromosomes in their nucleus compared to the number of chromosomes found in normal body cells (see Section C). They have a protective outer coat to make sure only one sperm gets through to fertilise the egg, and a store of food in the cytoplasm for the developing embryo. A small number of relatively large egg cells are released by the ovaries over a woman's reproductive life.

If you are male, once you have gone through **puberty**, your body produces millions of male sex cells called sperm. Like the egg cells, sperm have only half the chromosome number of normal body cells. Sperm cells are usually released a long way from the egg they are going to fertilise. They need to move through the female reproductive system to reach the egg. Then they need to be able to break into the egg. They have several adaptations to make all this possible. Sperm have a long tail that contains muscle-like proteins, which allows them to swim towards the egg. The middle section of a sperm is full of mitochondria, which provide the energy for the tail to work. They have a special sac called the acrosome, which stores digestive enzymes used to break down the outer layers of the egg. Finally, the sperm have a large nucleus containing the genetic information to be passed on to the offspring. Sperm cells are much smaller than egg cells, but they are produced in their millions every day.

Nerve cells (neurons)

Nerve cells or neurons are part of the communication and control system of your body. Electrical nerve impulses pass along these nerve cells at great speed, carrying information from one part of your body to another. Nerve cells have a cell body that contains:

- the nucleus
- **dendrites** that communicate with neighbouring nerve cells
- nerve fibres (called **axons**) that carry the nerve impulse long distances.

Nerve fibres are often covered by a protective myelin sheath, which insulates the axon and allows the nerve impulses to travel faster.

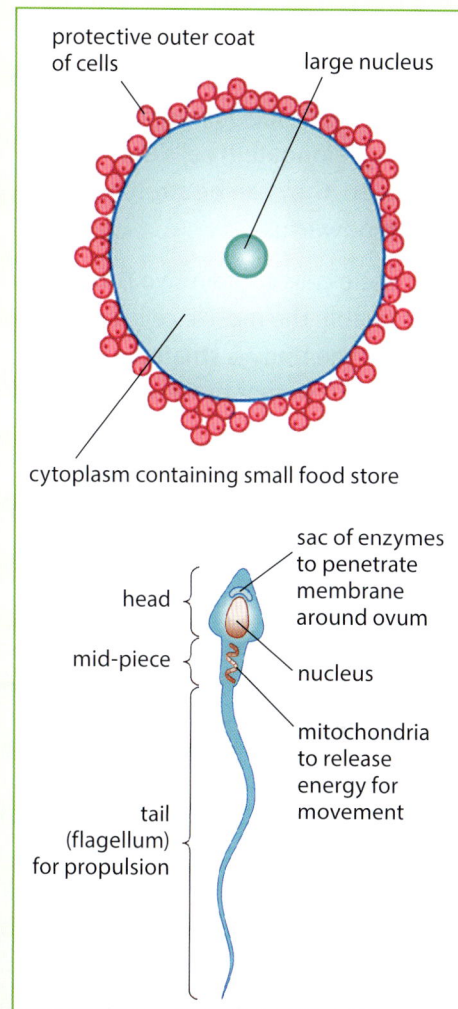

Figure 1.12 Egg and sperm cells show very clear adaptations to their specialised functions.

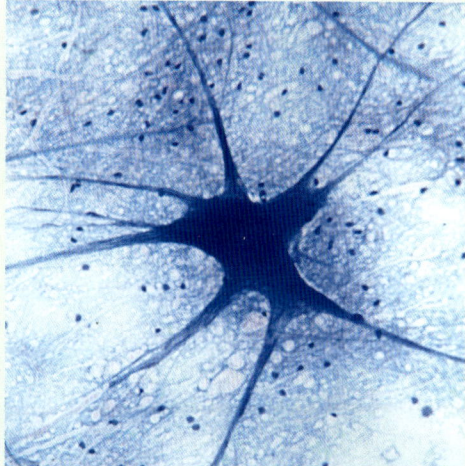

Figure 1.13 Nerve cells are very different from epidermal cells because they play a very different role in your body.

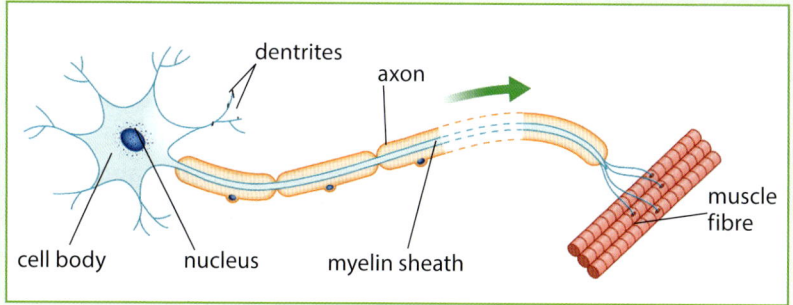

Cell specialisation in humans

Muscle cells

Your muscles are responsible for movement in your body. They are made up of specialised muscle cells or muscle fibres. These are very long (elongated) cells that can contract and relax. When they are relaxed, they can be stretched, and when they contract, they shorten powerfully.

The muscle cells contain two proteins called **actin** and **myosin**, which enable the muscle cells to contract. The most common muscle in the body is striated or striped muscle, and these proteins are arranged so that the muscle cell looks striped. Muscle cells also contain lots of mitochondria, which provide the energy for them to contract. Muscle cells are always found in bundles and they all contract together. You will find out more about muscle tissue and its role in your body in Chapter 9.

Connective tissue cells

There are lots of different types of connective tissue, from bone to the loose connective tissues that make up much of your skin and the adipose tissue where you store fat.

The main property of connective tissue cells is that the cells spread out in a matrix, made up of water and protein fibres produced by the connective tissue cells themselves. Some connective tissue cells make stretchy elastic fibres, and others make strong, inelastic fibres. The matrix may be very solid (for example, bone), stretchy (for example, **ligaments**) or liquid (for example, the liquid part of your blood). You will discover examples of different types of connective tissue as you work through this book.

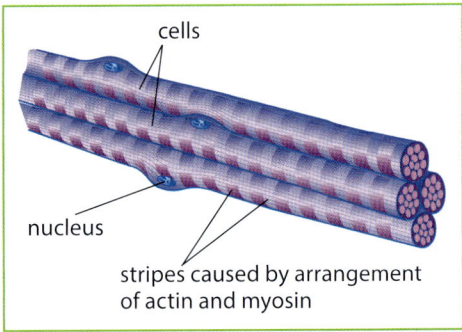

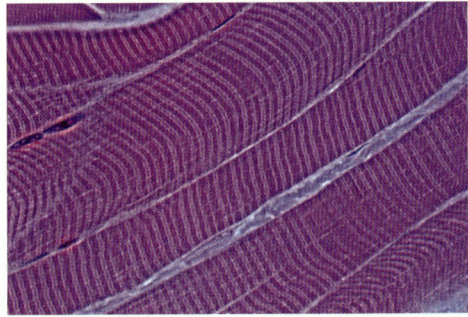

Figure 1.14 The arrangement of actin and myosin in your striated muscle cells makes them very distinctive, and allows them to contract and relax as they move the different parts of your body.

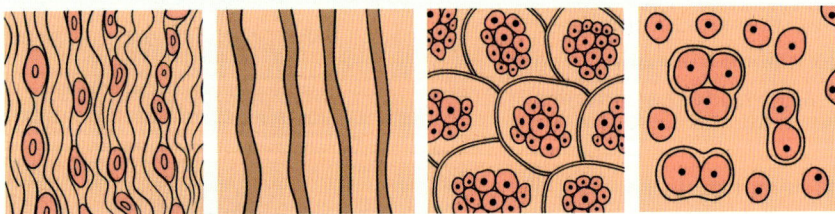

Figure 1.15 There are many types of connective tissue cells and they are all different. Learn some of the different types – and look out for widely spaced cells, a matrix and fibres.

All living cells carry out the characteristic functions of life. As a result, they all have some features in common. However, as you have seen, cells are specialised in many ways to carry out particular functions in your body. As you study more about the way the human body works in this book, you will discover more examples of specialised cells and their importance in the healthy functioning of your body.

DID YOU KNOW?

Some scientists work with unspecialised human **embryonic stem cells** to try and grow new adult tissues. They hope to use them to replace diseased tissues in people with serious illnesses. So far progress has been slow, partly because scientists are not sure how to make the undifferentiated cells become the specialised cells that the body needs, and partly because there are ethical issues about using cells from human embryos.

End-of-chapter summary

In this chapter, you have learnt that:

- all living organisms are based on units called cells
- there are seven life processes that are common to all living organisms: nutrition, respiration, excretion, growth, irritability(response to stimuli), movement and reproduction
- unspecialised animal cells all have the following structures and organelles: a cell membrane, cytoplasm, nucleus, mitochondria, ribosomes and endoplasmic reticulum; each of these parts has a characteristic structure and carries out clear functions in the working of the cell
- unspecialised plant cells all have the same basic structures and organelles as an animal cell; in addition, they have a cellulose cell wall and may have a permanent vacuole; in the green parts of a plant, all the cells contain chloroplasts, which in turn contain chlorophyll; each of these parts has a characteristic structure and carries out clear functions in the working of the cell
- microbes are very small organisms that can only be seen using a microscope; they include bacteria, viruses and fungi; each type of microbe has very distinctive cell structures
- in multicellular organisms like humans, the cells of the embryo differentiate to form specialised cells that carry out particular functions in the body
- cells specialised to carry out a particular function are grouped together to form a tissue, for example, muscle tissue and nervous tissue
- a group of different tissues that work together to carry out a particular function is called an organ, for example, the heart and the pancreas
- an organ system involves several different organs working together to carry out particular functions in your body, for example, the digestive system and the cardiovascular system
- there are many different specialised cells in the human body; they include epithelial cells, sperm cells, egg cells, nerve cells, muscle cells and connective tissue cells; a close look at their specialisations shows how they are adapted to their functions.

Possible SBAs using microscopes and knowledge of cells

→ How does the population of microscopic organisms in your local pond or stream vary over time?
→ How does the size of human cheek cells vary between different people?
→ Does the size of human cheek cells vary between students and adults?
→ How does the size and shape of the pollen produced by local plants vary depending on how the pollen is spread?

End-of-chapter questions

1. Which one of the following is not an organelle within a cell?
   A Nucleus
   B Chloroplast
   C Mitochondrion
   D Cytoplasm

2. Which one of the following is not one of the seven processes characteristic of living things?
   A Movement
   B Language
   C Reproduction
   D Respiration

3. Which one of the following is an example of a tissue in humans?
   A Heart
   B Stomach
   C Muscle
   D Uterus

4. a Why have microscopes been so important in developing our understanding of cells? (2)
   b Write a set of instructions that could be handed out with a microscope to make sure that students use it properly. (6)

5. This diagram shows an unspecialised plant cell from a blade of grass.

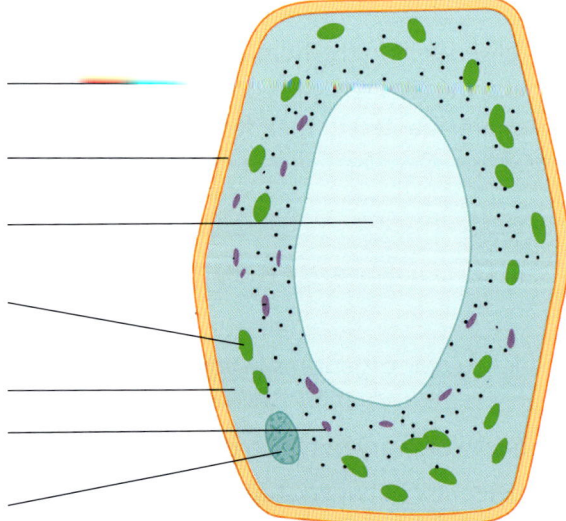

   a Copy the diagram and use words from the list given below to help you label it. (7)

   | cell membrane | cell wall | chloroplast | cytoplasm | nucleus |
   | mitochondrion | vacuole | | | |

   b Name **two** parts of this grass cell that you would not see in an animal cell. (2)

22 The structure and function of cells

6 a Copy and label this diagram of a yeast cell. (5)

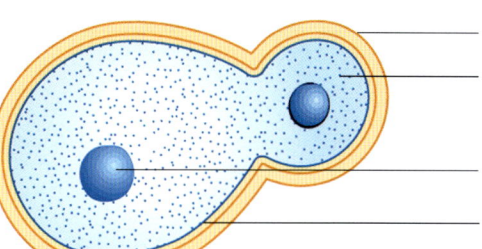

   b How are yeast cells similar to plant cells? (2)
   c How do yeast cells differ from plant cells? (2)

7 How is a sperm cell specialised for its role in reproduction? (4)

8 Read the following information about Chlamydomonas and then answer the questions below.

> Chlamydomonas is a single-celled organism that lives in water. It has an eyespot that is sensitive to light and it can move itself about. It 'swims' towards the light using long flagella. It has a large chloroplast and uses the light to photosynthesise. It stores any excess food as starch. When it is mature and has been in plenty of light, it will reproduce by splitting in two.

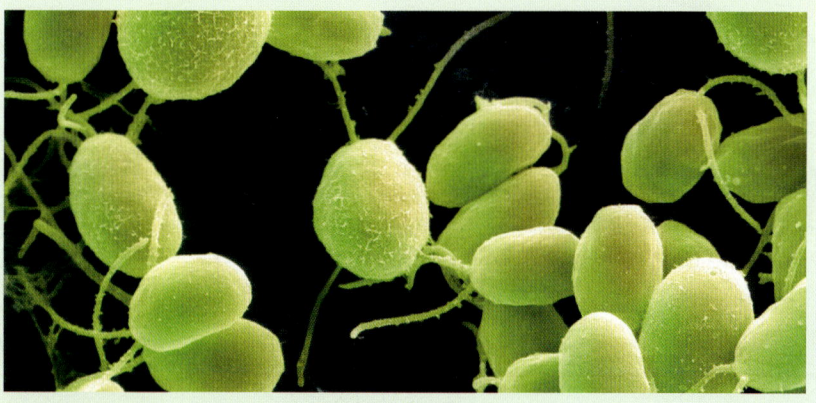

   a Chlamydomonas is a living organism. What features of Chlamydomonas show you that this is true? (4)
   b For many years, scientists were not sure whether to classify Chlamydomonas as an animal or a plant. Now it is put in a separate group altogether!
       i What features suggest that *Chlamydomonas* is an animal cell? (3)
       ii What features suggest that *Chlamydomonas* is a plant cell? (3)

9 a Why do cells become specialised in the human body? (2)
   b Choose **two** different types of cells and explain how they are adapted for the job they do in your body. (8)
   c Describe the different levels of organisation in the human body from cells to the whole body. (4)

Chapter 2 Passive and active transport

When you have completed this chapter, you will be able to:

- distinguish clearly between osmosis and diffusion
- explain the processes of simple diffusion and osmosis and their importance in living systems
- demonstrate simple diffusion experimentally
- discuss the advantages and disadvantages of simple diffusion, osmosis and active transport for moving substances into and out of cells
- demonstrate osmosis experimentally
- explain the process of active transport and its importance in living organisms

What the examiners say

→ Candidates often provide incomplete definitions of the key terms, suggesting unfamiliarity with the basic concepts covered in this topic. Among other things, candidates tend to omit mentioning 'water molecules' in their definition of osmosis.

Revision tip

Know your learning style – some people are visual, some learn by doing, others learn best by hearing, while others read and write. Go to our website for a brief description of each type of learner, as well as revision tips that each type of learner can use to maximise their study sessions.

Our cells need to take in substances such as oxygen and glucose, get rid of waste products, and move chemicals that are needed elsewhere in our body. In human beings, just like in all other living organisms, dissolved substances move into and out of cells across the cell membrane using both **passive transport** and **active transport**. Passive transport does not use any energy from the cells. It takes place as a result of passive processes such as **simple diffusion** and **osmosis**.

Active transport moves substances using energy produced during cell respiration. In this chapter, we look at each method of transport in turn.

Figure 2.1 The blood from an injured fish spreads through the water by simple diffusion, helped by the movement of the sea – and sharks like this reef shark will follow the gradient of blood to some easy prey!

Passive transport in living systems

Simple diffusion

When you get home from school, you probably know if there is a meal cooking before you get into the kitchen. How? Because the smell reaches you by simple diffusion. Simple diffusion happens when the particles of a gas, or any substance in solution, spread out.

Figure 2.2 An example of simple diffusion

Simple diffusion is the net (overall) movement of particles from an area where they are in a high concentration to an area where they are in a lower concentration. It takes place because of the random movements of the particles of a gas or of a substance in solution in water. All the particles are moving and bumping into one another, which moves them all around. Although the molecules are moving in both directions, more particles are moving into the area of high concentration. This means that the net (overall) movement is away from the area of high concentration towards the area of low concentration.

Imagine an empty room containing a group of boys and a group of girls. If everyone closes their eyes and moves around briskly but randomly, people will bump into one another and scatter until the room contains an even mixture of boys and girls. This example gives you a good working model of simple diffusion (see Figure 2.3).

> **DID YOU KNOW?**
>
> A reef shark like the one in Figure 2.1 senses 1 part of blood in 10 or even 100 million parts of water – so it's not a good idea to bleed in the sea!

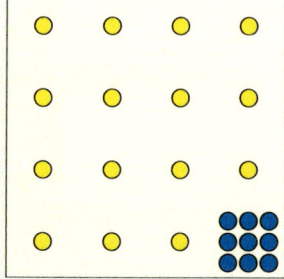

At the moment of adding blue particles to a yellow mixture, they are not mixed at all.

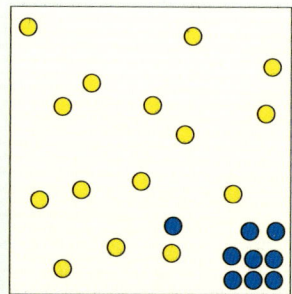

As the particles move randomly, the blue ones begin to mix with the yellow ones.

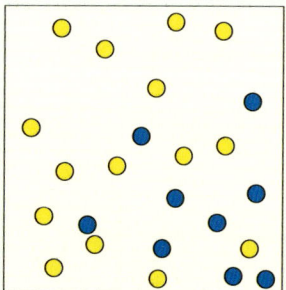

As the particles move and spread out, they bump into one another and keep spreading as a result of all the random movement.

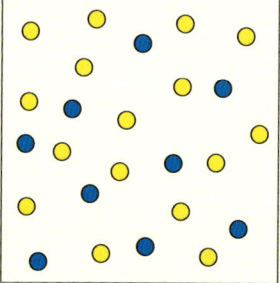

Eventually, the particles are completely mixed and simple diffusion is complete.

Figure 2.3 The random movement of particles results in substances spreading out or diffusing from an area where they are in a high concentration to an area where they are in a lower concentration.

Activity 2.1

Demonstrating diffusion

You will need:
- a stopwatch or timer

Method

If your classroom is suitable, try the idea described on page 25.

1. All the boys stand in one corner of the room (a high concentration of boys). All the girls stand in another corner of the room (a high concentration of girls).

2. Start a timer for 30 seconds and move around slowly with your eyes closed until the timer tells you all to stop.

3. Open your eyes and observe what is happening. Start the timer again and move slowly with your eyes closed. Repeat until the classroom contains an even mixture of boys and girls.

Rates of simple diffusion

Simple diffusion is a relatively slow process. A number of different factors affect the rate at which this process takes place.

- **Concentration:** If there is a big difference in concentration between two areas, simple diffusion takes place quickly. However, when a substance moves from an area of higher concentration to one that is just a bit lower, the rate of movement towards the less-concentrated area will appear to be quite slow. This is because although some particles move into the area of lower concentration by random movement, at the same time other identical particles are also leaving that area by random movement.

overall or net movement = particles moving in − particles moving out

In general, the bigger the difference in concentration, the faster the rate of diffusion. This difference between two areas of concentration is called the **concentration gradient**. The bigger the difference, the steeper the gradient.

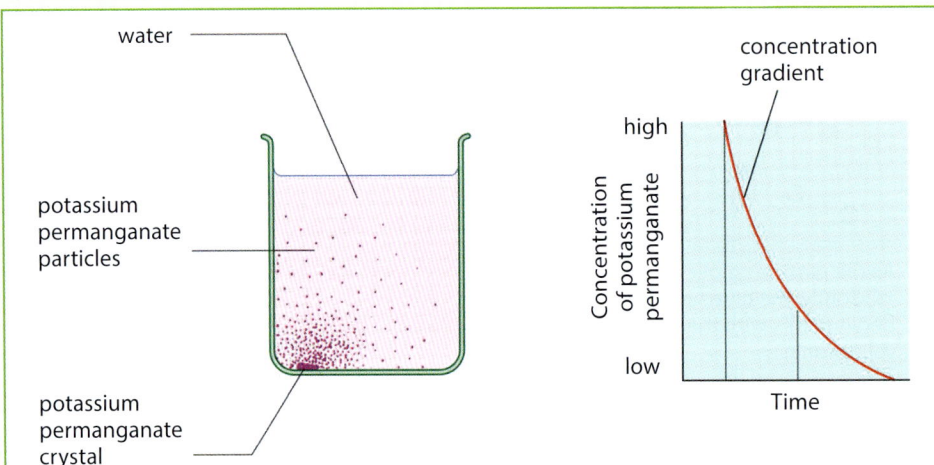

Figure 2.4 This diagram shows how the overall movement of particles in a particular direction is more effective if there is a big difference (a steep concentration gradient) between the two areas. This is why so many body systems are adapted to maintain steep concentration gradients.

Passive and active transport

- **Temperature:** Concentration isn't the only thing that affects the rate of simple diffusion. An increase in temperature means the particles in a gas or a solution move more quickly. This, in turn, means simple diffusion takes place more rapidly as the random movement of the particles speeds up.

The biologist's toolkit: Calculating rate

When we look at biological processes like simple diffusion or osmosis, we often want to measure the **rate** at which they happen. The rate is the speed at which something happens, or the number of times something happens within a particular period of time.

The units you use refer to the unit of time you are measuring, for example, molecules per second or breaths per minute.

> **For example:**
>
> If 100 grams of glucose moves by simple diffusion from the gut of an animal into its blood in 10 minutes, what is the rate of diffusion of the glucose?
>
> The rate of diffusion = 100 g/10 min = 10 g/min

Activity 2.2

Demonstrating simple diffusion in a liquid

Potassium manganate VII (potassium permanganate) are purple crystals that dissolve in water. This activity demonstrates both simple diffusion and the impact of temperature on the rate of diffusion.

You will need:
- two identical beakers (100, 200 or 250 cm^3)
- cold water
- two crystals of potassium manganate VII
- warm/hot water
- a stopwatch

> **Remember**
> Remember that diffusion is *passive* – it takes place along a concentration gradient from high to low concentration, and uses up no energy.

Method

1. Half fill one beaker with cold water. Put exactly the same amount of warm or hot water in the second beaker. (NB: If the water is hot, be careful how you handle it.)
2. Drop a crystal of potassium manganate VII into each beaker at the same time. Also, at the same time, start the stopwatch.
3. Time how long it takes the purple colour to reach different points in your beaker, and (if possible) the total time it takes for the liquid to become purple all the way through.
4. Write up the investigation. Explain your results in terms of diffusion and the effect of temperature on the movement of particles.

Passive transport in living systems

Simple diffusion in living organisms

Many important substances move across the cell membranes of living systems by simple diffusion. For example, the oxygen you need for respiration passes easily from the air into your lungs, into your blood and then into your body cells by simple diffusion. There is a steep concentration gradient between the oxygen in the air and your blood, so the oxygen moves into the blood quickly by diffusion.

Simple diffusion takes place over the surface of a cell and can be a relatively slow process. Some types of cells have adaptations to make diffusion easier and more rapid. The most common adaptation is to increase the surface area of the cell membrane, which allows more room for diffusion to occur.

Adaptations of organs for diffusion

As animals and plants get bigger, simple diffusion of substances from the air or water around them is no longer enough. The surface area compared to the volume of the organism is not big enough for simple diffusion to work.

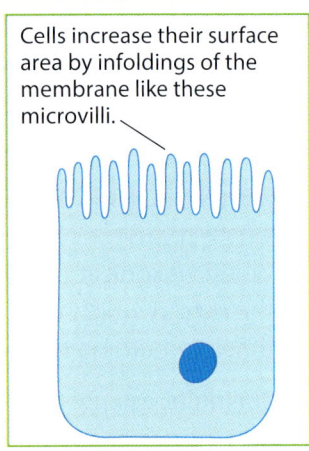

Cells increase their surface area by infoldings of the membrane like these microvilli.

Figure 2.5 An increase in the surface area of a cell membrane means more diffusion can take place.

The biologist's toolkit: Calculating surface area to volume ratio

The **surface area to volume ratio**, or **sa:vol ratio** is a very important idea in biology. It affects the size of animals and plants, and how many organs of our body work.

The surface area is the area of the outer surface of an organism or of an organ. The volume is the amount of space it contains.

The ratio of the surface area to the volume gets smaller as the object gets bigger. You can see this in the diagrams below:

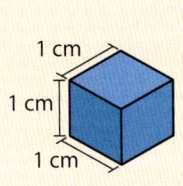 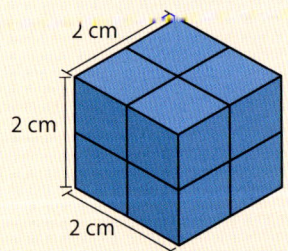 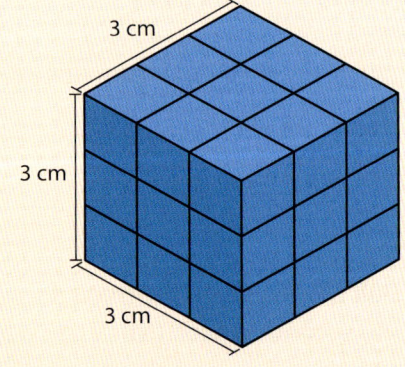

Surface area	6 cm²	24 cm²	54 cm²
Volume	1 cm³	8 cm³	27 cm³
Surface area : Volume	6:1	3:1	2:1

In small organisms, the sa:vol ratio is relatively big, so the diffusion distances are short. In these small organisms, simple diffusion works to exchange materials between the outside world and the cells.

In larger organisms, the sa:vol ratio gets smaller. The distances between the outside world and the cells in the middle get bigger, so simple diffusion is no longer enough. Specialised organs are needed.

Passive and active transport

In larger multicellular organisms, it isn't only the cells that are adapted to make diffusion more efficient. The tissues lining some organs and even whole organs are adapted as well. For example, the lungs of human beings and many other animals are specially adapted to make the movement of oxygen into the body and the removal of waste carbon dioxide more efficient. The lung tissue is arranged into clusters of tiny air sacs called **alveoli**, which have a huge surface area ideal for rapid gas exchange by simple diffusion. (See Figure 2.6 – you will learn more about the lungs in Chapter 7.) This makes it possible for enough simple diffusion to take place for us to exchange the gases we need for respiration.

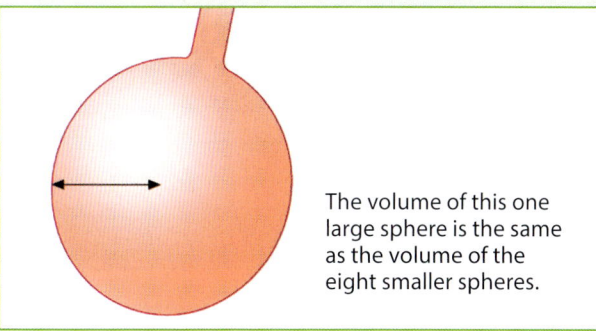
The volume of this one large sphere is the same as the volume of the eight smaller spheres.

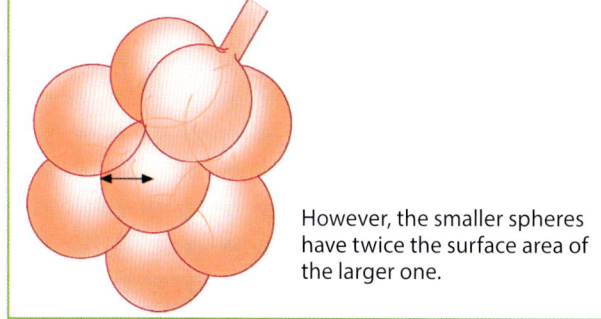
However, the smaller spheres have twice the surface area of the larger one.

Figure 2.6 The multitude of tiny air sacs in the lungs gives them a much larger surface area than if they were single balloon-like sacs, making increased rates of rapid diffusion possible.

Similarly, the lining of our gut is folded into thousands of tiny, finger-like projections, called villi, which greatly increase the uptake of digested food by diffusion.

Both your lungs and your digestive system also have an excellent blood supply that carries away the oxygen or glucose as soon as it has passed by simple diffusion from one side to the other. This maintains a steep concentration gradient so diffusion is as fast and efficient as possible.

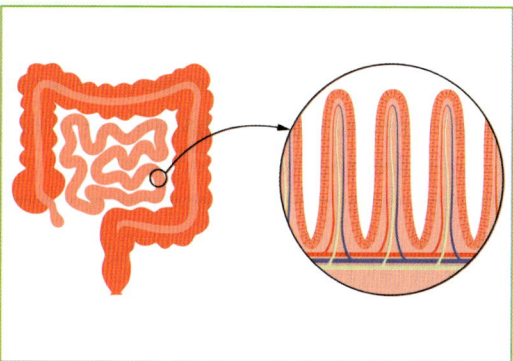

Figure 2.7 The villi lining your gut make it possible for your digested food to pass by simple diffusion into your blood to be carried to your cells.

Plants also rely heavily on simple diffusion for the carbon dioxide they need for photosynthesis and the minerals they need from the soil. The wide, flattened shape of the leaves increases the surface area for diffusion, and the many air spaces inside the leaf allow carbon dioxide to come into contact with lots of cells to aid diffusion (see Figure 2.8).

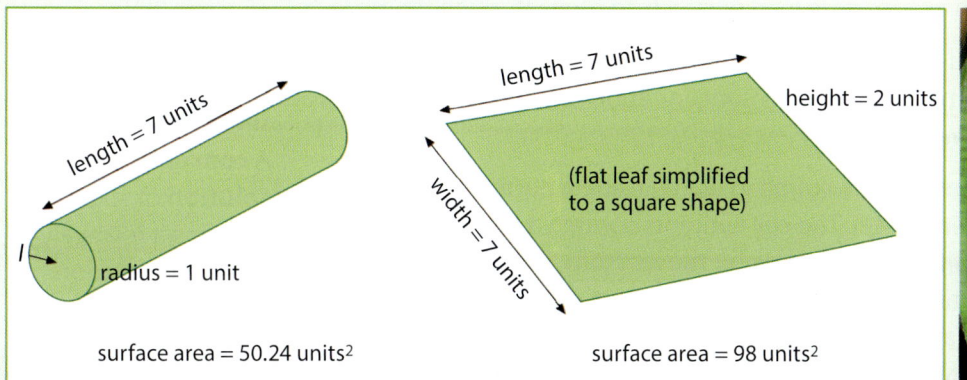

Figure 2.8 The wide, flat shape of most leaves increases the surface area for collecting light and exchanging gases.

Passive transport in living systems

DID YOU KNOW?
Many Caribbean moths – and others around the world – rely on diffusion to find a mate. The female moths produce a powerful chemical called a pheromone to attract males. Pheromone molecules spread through the air by diffusion, sometimes helped by breezes. Male moths can pick up these molecules from as far as five miles away from the female – and then fly up the concentration gradient until they reach their mate!

Checkpoint questions

1. Explain how sharks find an injured fish or person so easily.
2. Define the net movement of particles.
3. State two factors that affect the rate of diffusion and explain the effect they have.

Osmosis

Simple diffusion takes place where particles spread freely from one place to another. However, the solutions inside a cell are separated from those outside the cell by the cell membrane, which only lets the smallest particles through. Since it only lets some types of particles through, the membrane is called **partially permeable**.

Partially permeable cell membranes allow water to move across them. They do not allow larger molecules, such as sucrose (sugar) or starch, to move through them.

Osmosis is a special type of diffusion where only water moves across a partially permeable membrane, from an area of high concentration of water to an area of lower concentration of water.

A cell is basically some chemicals in solution in water inside a partially permeable bag (the cell membrane). The cell contents contain a fairly concentrated solution of salts and sugars. Water moves from a high concentration of water particles (a dilute solution) to a less-concentrated solution of water particles (a concentrated solution) across the membrane of the cell. In other words, osmosis takes place. Most of the other salts and sugars in solution are too large to pass through the partially permeable cell membrane.

Remember
A dilute solution of sugar contains a lot of water and a small amount of sugar. Put scientifically, a dilute sugar solution contains a **high** concentration of water (the **solvent**) and a **low** concentration of sugar (the **solute**). A **concentrated solution** of sugar contains a relatively **low** concentration of water and a **high** concentration of sugar.

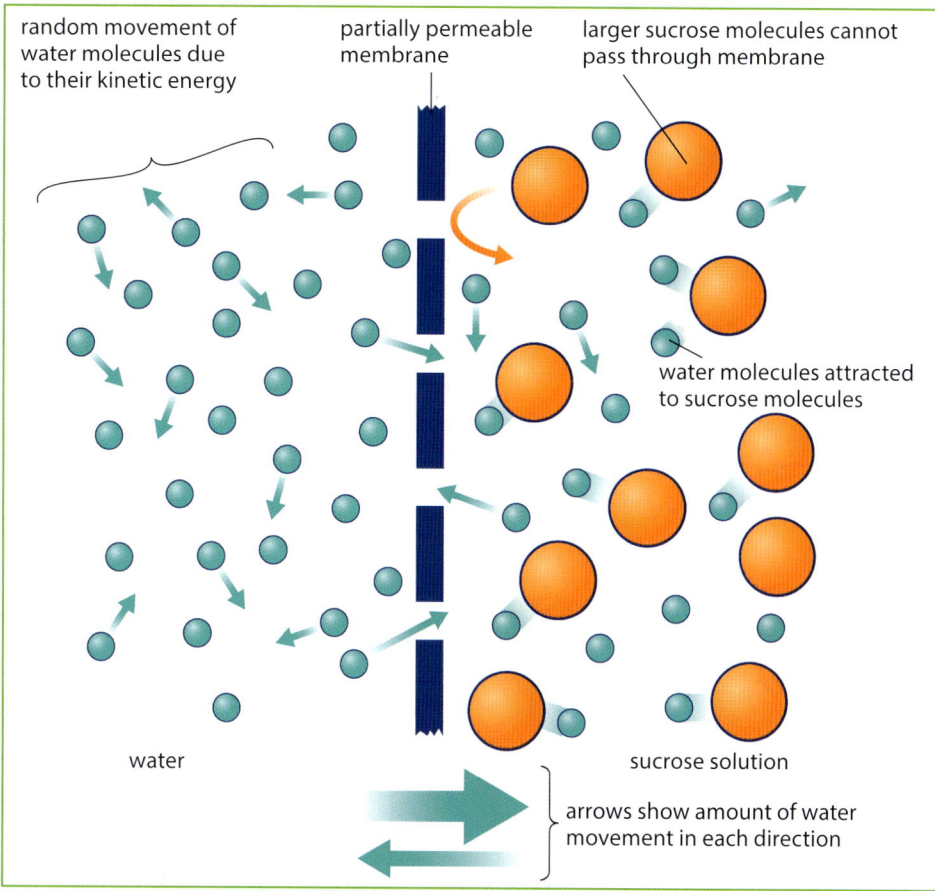

Figure 2.9 This model shows how osmosis works – with a net movement of water molecules from an area where they are in a high concentration to an area where they are in a lower concentration through a partially permeable membrane.

Cell membranes are not the only partially permeable membranes. There are artificial partially permeable membranes too, and these can be used to make a model cell (see Activity 2.3 on page 32). By changing the concentration of the solutions inside and outside your model cell, you can see exactly why osmosis is so important in living organisms – and what happens if things go wrong!

The biologist's toolkit: Using models

Sometimes what we want to investigate as scientists may be too big to work with – like the universe – or too small to see, like atoms and molecules. Sometimes it is just too dangerous, like some poisonous organisms or radioactive materials.

Scientists make models to help us understand what is going on and to answer scientific questions. We also use models to predict what might happen in a situation, or to explain what we have observed. There are different types of models:

- Physical models, for example our model cells in Figures 2.9, 2.10 and 2.11
- Mathematical models
- Computer models that rely on computer programs to find the answers for us.

Mathematical model

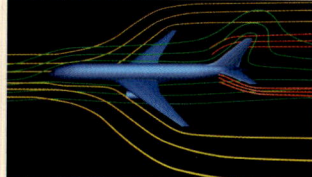
Computer model

Osmosis

Activity 2.3

Demonstrating osmosis

You will need:

- two sets of the equipment shown in Figure 2.10 to make model cells in different situations
- a chinagraph pencil or small stickers

Method

1. Set up two sets of the apparatus shown in Figure 2.10. Label them A and B.
2. In set A, put concentrated sucrose (sugar) solution in the Visking tubing bag and water surrounding it in the beaker.
3. In set B, put water in the Visking tubing bag and concentrated sucrose solution surrounding it in the beaker.
4. In both cases, mark the starting level of the liquid on the capillary tubing using the pencil or stickers, and observe the state of the Visking tubing bag.
5. Leave the model cells for 30 minutes or longer.
6. Observe the level of water in the capillary tubing and the state of the Visking tubing bag.
7. Write up your experiment with diagrams of your equipment at the beginning and the end of the experimental time.
8. Use Figure 2.11 below to help you explain your results in terms of osmosis and the water movements into and out of your model cells.

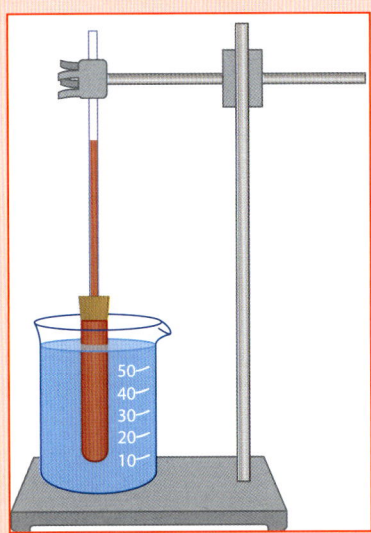

Figure 2.10 Apparatus to demonstrate osmosis: In this model, a coloured dye has been added to the solution in the model cell to make it easier for you to see the changes in the water levels.

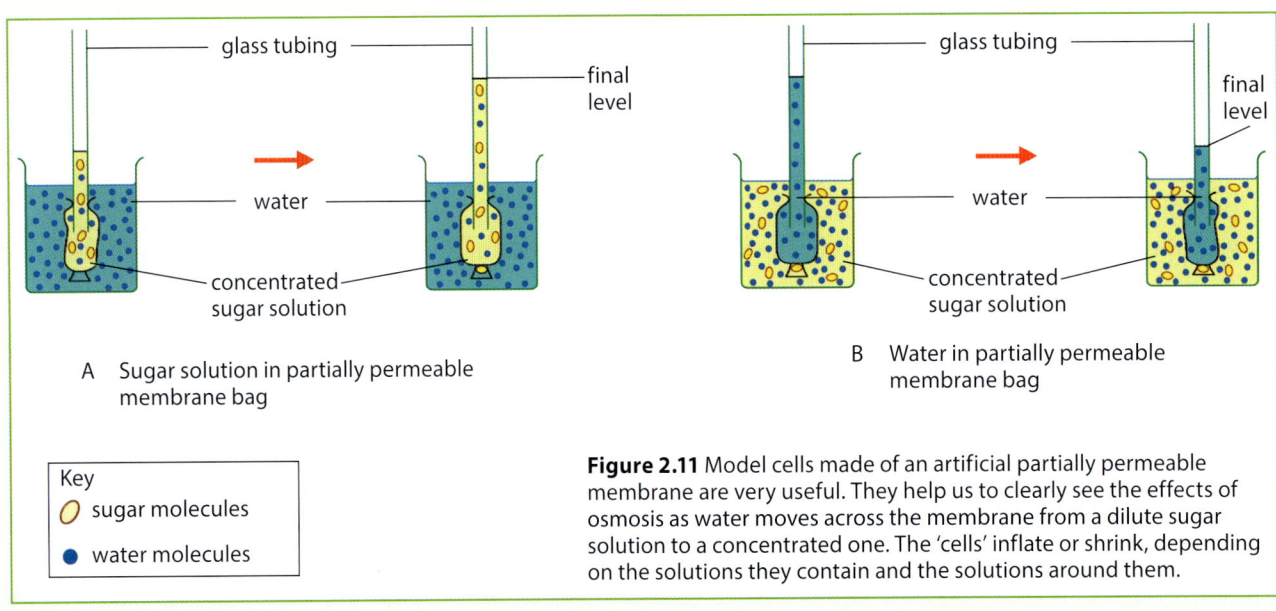

A Sugar solution in partially permeable membrane bag

B Water in partially permeable membrane bag

Key
- sugar molecules
- water molecules

Figure 2.11 Model cells made of an artificial partially permeable membrane are very useful. They help us to clearly see the effects of osmosis as water moves across the membrane from a dilute sugar solution to a concentrated one. The 'cells' inflate or shrink, depending on the solutions they contain and the solutions around them.

Passive and active transport

Osmosis in animals

Osmosis is an important way of moving water into and out of a cell when needed. If a cell uses up water in its chemical reactions, the cytoplasm becomes more concentrated and more water will immediately move in by osmosis. Similarly, if the cytoplasm becomes too dilute because water is produced during chemical reactions, water will leave the cell by osmosis, restoring the balance.

Osmosis can also cause some very serious problems in animal cells. If the solution outside the cell is more dilute than the cell contents, then water will move into the cell by osmosis, diluting the cytoplasm. The cell will swell and may eventually burst. On the other hand, if the solution outside the cell is more concentrated than the cell contents, then water will move out of the cell by osmosis, the cytoplasm will become too concentrated and the cell will shrivel up. Once you understand the effect that osmosis can have on cells, the importance of **homeostasis** and maintaining constant internal conditions will become clear. You will learn more about this in Section B.

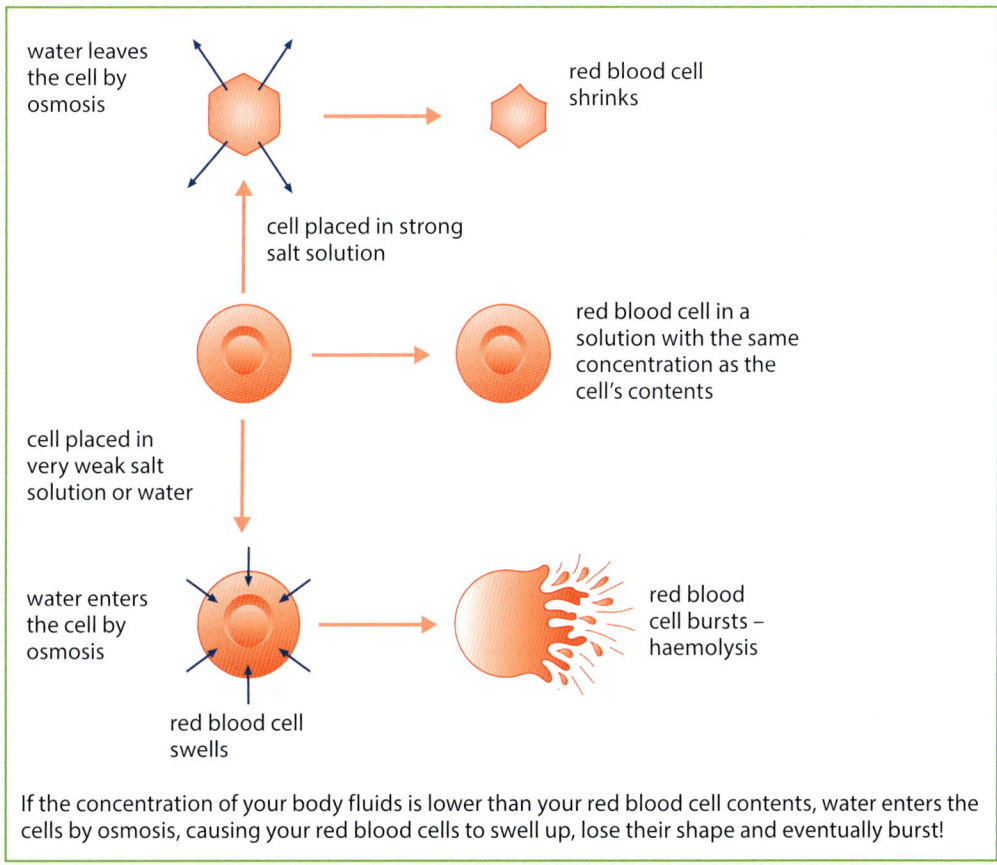

If the concentration of your body fluids is lower than your red blood cell contents, water enters the cells by osmosis, causing your red blood cells to swell up, lose their shape and eventually burst!

Figure 2.12 Keeping your body fluids at the right concentration is vital. When you see what happens to your red blood cells when things go wrong, you can understand why!

When the concentration of your body fluids is the same as your red blood cell contents, equal amounts of water enter and leave the cell by random movement and the cell keeps its shape.

If the concentration of the solution around the red blood cells is higher than the concentration of substances inside the cells, water will leave the cells by osmosis. This makes the cells shrivel and shrink so that they can no longer carry oxygen around your body.

If the concentration of your body fluids is lower than the contents of your red blood cells, water enters the cells by osmosis – your red blood cells swell up, lose their shape and eventually burst!

For many marine invertebrates like jellyfish, osmosis causes no problems because the concentration of solutes in the cells of their body is exactly the same as the seawater that surrounds them, so there is no net movement of water into or out of the cells.

Fish that live in fresh water have a real problem. They need a constant flow of water over their gills to get the oxygen they need for respiration. However, water moves into their gill cells and blood by osmosis at the same time. Like all vertebrates, fish have kidneys, which play a big part in controlling the concentration of their body fluids – a process known as **osmoregulation**. So, freshwater fish produce huge amounts of very dilute urine to get rid of the excess water that enters their body. They also have special salt-absorbing glands that use active transport to move salt against a concentration gradient from the water into the fish – rather like the situation in plant root cells (see active transport on page 39).

Osmosis in plants

Plants rely on well-regulated osmosis to provide the necessary pressure to support their stems and leaves. Water moves into plant cells by osmosis, making the cytoplasm swell and press against the plant cell walls. The pressure builds up until no more water can physically enter the cell, which is now hard and rigid. This swollen state (called **turgor**) keeps the leaves and stems of a plant rigid and firm. So, for plants, it is important that the fluid surrounding the cells always has a higher concentration of water (a more dilute solution of chemicals) than the cytoplasm of the cells, to keep osmosis working in the right direction.

To understand the difference between animal and plant cells when it comes to water moving in by osmosis, imagine blowing up a balloon. As more and more air moves in, the balloon gets bigger and bigger, and eventually bursts. This models an animal cell placed in pure water or a very dilute solution of salts.

Now imagine a balloon sealed into a cardboard shoe box. As the balloon inflates it fills the box and then presses out against the box walls. Eventually, you simply cannot force any more air into the balloon. The box feels very rigid and the balloon does not burst. This models a plant cell placed in pure water or a very dilute solution of salts (see Figure 2.13).

If the surrounding fluid becomes more concentrated than the contents of the plant cells, water will leave the cells by osmosis. The vacuole shrinks and the cell becomes much less rigid – it is **flaccid**. If water continues to leave the cell by osmosis, eventually the cytoplasm pulls away from the cell walls and the cell goes into a state called **plasmolysis** (see Figure 2.13).

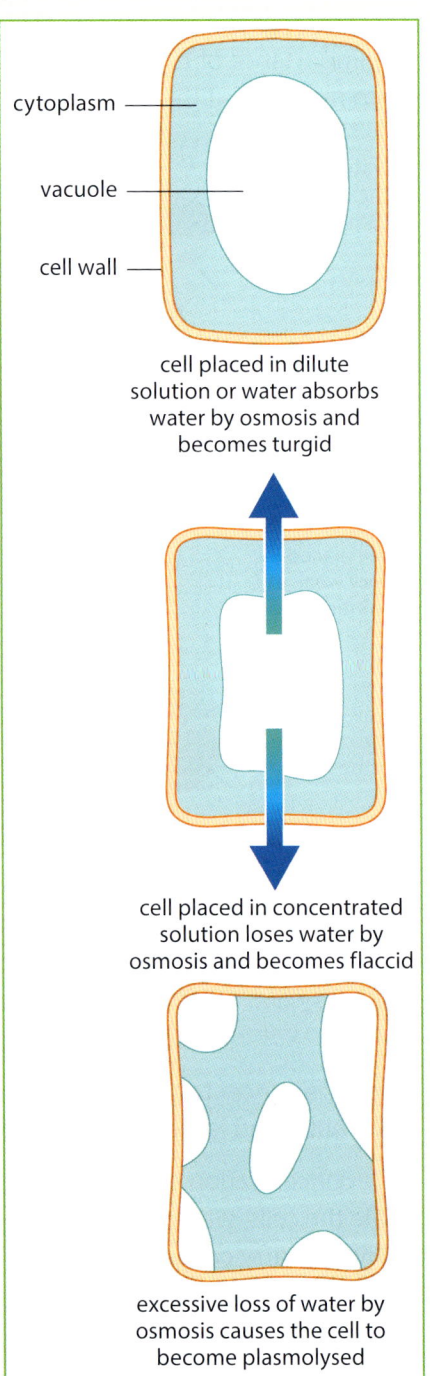

cell placed in dilute solution or water absorbs water by osmosis and becomes turgid

cell placed in concentrated solution loses water by osmosis and becomes flaccid

excessive loss of water by osmosis causes the cell to become plasmolysed

Figure 2.13 Osmosis plays an important role in maintaining the rigid structure of plants.

Passive and active transport

In normal conditions, water moves into plant cells by osmosis and keeps them rigid. This in turn helps to keep the plant upright. If conditions are very dry, the plant cannot take enough water from the soil up through its roots. The cells are no longer rigid and the plant wilts. When this happens, many of the chemical reactions slow down and so the plant is able to survive until more water is available. But this only lasts for so long – if the osmotic situation isn't put right fairly quickly, most plants will die. So, if your house plants wilt, take note – and give them some water!

Osmosis is also very important for moving water around within the plant itself. Water moves into the plant root cells by osmosis across the cell membrane. The roots are covered in special root hair cells. These have tiny, hair-like extensions that increase the surface area for osmosis to take place. The cytoplasm of these root hair cells is now more dilute than the cytoplasm of the adjacent cells. Water moves into these cells by osmosis (see Figure 2.14 on page 36). These cells now have more dilute cytoplasm than the cells next to them, and the water continues to move along by osmosis until it reaches the xylem and the **transpiration** stream.

DID YOU KNOW?

When you sprinkle salt on green mangoes or aubergines before using them in cooking, water moves out of the plant cells by osmosis and dissolves the salt crystals. This is why you are left with floppy fruit and a salty liquid!

Checkpoint questions

4 Compare diffusion and osmosis.
5 Explain why it so important for animals to keep the concentration of their body fluids constant.
6 Plants do not have rigid skeletons – instead, osmosis is an important part of the plant support system. Explain how osmosis keeps plant stems rigid.

Activity 2.4

Investigating the effects of osmosis in plant cells

You will need:

- red onion or rhubarb epidermis, or epidermis from the leaves of Rhoeo discolor; if nothing else is available, use white onion epidermis, but coloured cells show up the effects of osmosis much better
- two small beakers – one labelled 'water' and the other labelled 'sucrose solution'
- a microscope
- two microscope slides and cover slips
- a mounted needle
- 1 M sucrose solution
- two dropping pipettes
- tissue/filter paper

Method

1. Using one of your pipettes, place a drop of water on one of the microscope slides and then place the pipette in the beaker of water.

2. Using the other pipette, place one drop of sucrose solution on the other microscope slide and then place the pipette in the beaker of sucrose solution.

3. Either collect or prepare a piece of epidermis and place it on the drop of water on your microscope slide.

4. Add another drop of water on top of your piece of epidermis and, using the mounted needle, carefully lower the cover slip into place. Blot any excess water using tissues.

5. Now collect or prepare another piece of epidermis and place it on the drop of sucrose solution on your microscope slide.

6. Add another drop of sucrose solution on top of your piece of epidermis and, using the mounted needle, carefully lower the cover slip into place. Blot any excess liquid using tissues.

7. Examine both of your slides carefully under the microscope. Look for any differences between them. Draw and label a few representative cells from each slide.

8. Take the slide of the tissue in sucrose solution. Replace the sucrose solution with water and observe any changes in the cells. To do this, place some drops of water on one side of the slide beside the cover slip. Place some tissue or filter paper next to the cover slip on the other side (see Figure 2.14) and the sucrose solution will be drawn up into the absorbent paper, pulling the water under the cover slip. You may need to repeat this several times to make sure the cells are now in almost pure water.

9. Examine the cells again carefully and observe any changes in them.

10. Write up your investigation fully, including your drawings, and explain your observations in terms of osmosis.

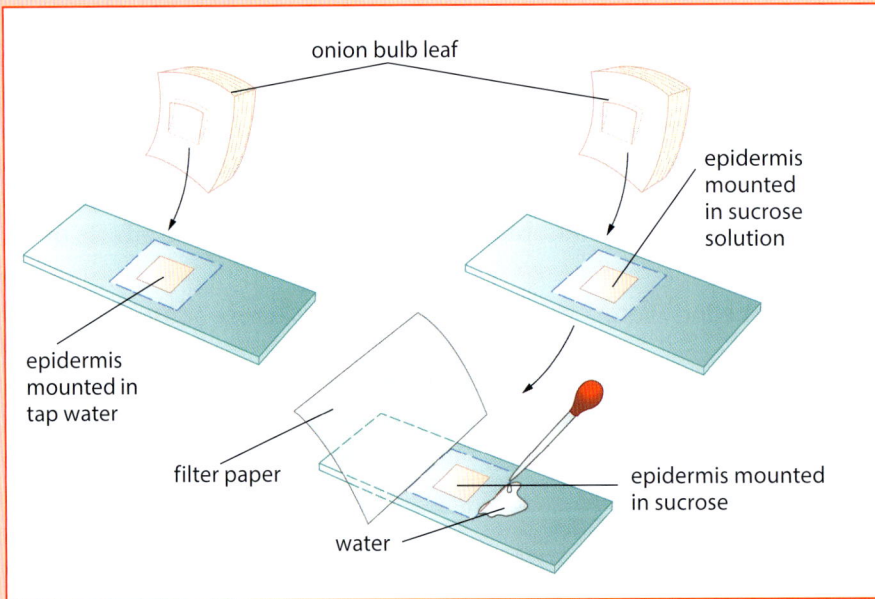

Figure 2.14 Using this simple technique, you can change the solution surrounding your plant epidermis cells as many times as you want to.

Passive and active transport

Activity 2.5

Investigating the effects of osmosis on potato tissue

There are two different ways of carrying out this experiment. Potato is the most common vegetable used, but you could use others and compare the results you obtain. The equipment is the same for both methods.

You will need:

- a potato or other starchy vegetable
- a cork borer or apple corer, and a sharp knife or scalpel
- a tile or chopping board
- tweezers
- a ruler
- 1 M sucrose solution
- three boiling tubes
- a balance (sensitive to 0.1 g)
- filter paper
- a marker pen

Method A

1. If you have a cork borer or apple corer, cut three cylinders out of your potato. Trim the skin off the top and bottom, and cut them all to approximately the same length. If not, cut three chip shapes from your potato (approximately 5 cm × 1 cm × 1 cm) and trim off any skin from the top and bottom.

2. Half fill one boiling tube with tap water and label it. Half fill another with 1 M sucrose solution and label it. Leave the third tube empty.

3. You are going to be measuring changes in your potato cylinders, so make sure that you know exactly which cylinder you are going to place in which boiling tube before you start measuring! Draw tables like those given below to record your observations.

Table 2.1 Investigating the effect of osmosis on potato cylinders: length

Tube	Starting length (mm)	Final length (mm)	Change in length (mm)	% change in length	Condition (flexible/stiff)
Water					
Sucrose solution					
Nothing (air)					

Table 2.2 Investigating the effect of osmosis on potato cylinders: mass

Tube	Starting mass (g)	Final mass (g)	Change in mass (g)	% change in mass	Condition (flexible/stiff)
Water					
Sucrose solution					
Nothing (air)					

4 Measure the length of each cylinder as accurately as you can and record the measurement.

5 Gently blot each potato cylinder with filter paper to remove excess moisture and then weigh each one. Record each mass carefully.

6 Place one potato cylinder in your tube of water, one in 1 M sucrose solution and one in the air. Leave them for a minimum of 30 minutes.

7 Using the tweezers, remove each cylinder of potato and blot it dry if necessary.

8 Measure each potato cylinder in turn and record the final length in your table.

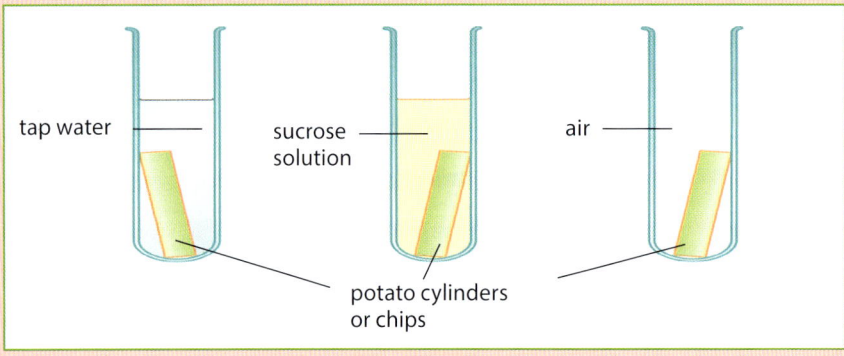

Figure 2.15 Apparatus for investigating the effects of osmosis on plant tissue

9 Observe the appearance of the potato cylinder compared to a freshly cut one and record it in your table.

10 Calculate the change in length from the start to the finish. This may be positive or negative, depending on whether the potato cylinder has gained or lost length.

11 Calculate the percentage change in length for each cylinder and record it in your table:

$$\% \text{ change} = \frac{\text{change in length}}{\text{starting length}} \times 100$$

12 Calculate the percentage change in mass for each cylinder and record it in your table:

$$\% \text{ change} = \frac{\text{change in mass}}{\text{starting mass}} \times 100$$

13 Write up your investigation, explaining your observations in terms of osmosis and the concentrations of the liquids surrounding the potato cylinders. Make suggestions for ways you feel the investigation might be improved and the results made more reliable. Do you think that measuring the length or finding the mass of the potato is the most reliable method to use?

Method B

1 Follow method A as far as step 3. In this second method, you are only going to investigate changes in mass, so you will only need one table for your results.

2 When you have cut and dried your three cylinders of potato, cut each one into a number of smaller discs.

3 Weigh each pile of disks and then place them into the different boiling tubes. Leave them for a minimum of 30 minutes.

4 Using the tweezers, remove all the disks from one tube, blot them dry if necessary and weigh them. Record your results in the table.

Passive and active transport

5   Repeat this for the discs in the other two tubes.

6   Calculate the percentage change in mass for each group of discs and enter it in your table:

% change = $\dfrac{\text{change in mass}}{\text{starting mass}} \times 100$

7   Write up your investigation as before.

Active transport

There are three main ways in which substances are moved into and out of cells.

- **Diffusion:**
  - Passive movement of substances – no energy is needed or used
  - Depends on a concentration gradient.
- **Osmosis:**
  - Passive movement of water – no energy is needed or used
  - Depends on the concentration gradient of water
  - Involves water moving across a partially permeable membrane.
- **Active transport:**
  - Active movement of substances – uses energy from respiration in the cell
  - Movement of substances against a concentration gradient or across a partially permeable membrane.

Active transport allows cells to move substances from an area of low concentration to an area of high concentration, against the concentration gradient, using energy. As a result, the cells can absorb ions from very dilute solutions. It also makes it possible for them to move substances like sugars and ions from one place to another through the cell membranes.

- **Absorption of mineral ions:** Mineral ions in the soil are usually found in very dilute solutions – more dilute than the solution in plant cells. By using active transport, plants can absorb them from the soil even though it is against the concentration gradient. These mineral ions are needed by plants for making proteins and other important chemicals.
- **Transporting glucose:** Glucose is always moved out of our gut and kidney tubules into our blood, even when it is against a steep gradient, so this movement relies on active transport (see Chapter 11).
- **Mangrove survival:** Mangrove trees can survive in the salty swamp water where they grow, because many species have glands in their leaves. They remove the excess salt that gets into their systems by active transport through these glands.

The cells of all living organisms contain sodium chloride and other chemicals in solution, so if they are surrounded by a solution with a lower concentration of salts than their cell contents, they will gain water by osmosis. If they are in a more concentrated solution, water is lost. Either way may be a disaster.

Caption Figure 2.16 Mangrove trees are a very important part of the swamp ecosystem. They depend on active transport of salt to survive.

Here are some of the different ways in which living organisms attempt – largely successfully – to beat osmosis. Active transport is usually an important part of the solution.

As we saw earlier in the chapter, fish that live in fresh water have a real problem with water moving into their body by osmosis and rely on active transport to help them solve it (see page 34).

Some marine birds and reptiles, such as the Caribbean flamingo, green turtles and the crocodiles found in the Dominican Republic, take in a great deal of salt in the seawater they drink, and their kidneys cannot get rid of it all. Their solution is to have special salt glands, usually found near the eyes and nostrils. Sodium ions are moved out of the body into the salt glands, which produce a very strong salt solution – up to six times more salty than their urine. The sodium ions have to be moved against a very steep concentration gradient, and so active transport is involved in the survival of these marine creatures.

> ### DID YOU KNOW?
> Cyanide is a deadly poison loved by crime writers. It smells faintly of almonds. Cyanide kills because it stops the reactions of respiration in mitochondria. If you give individual cells cyanide, all active transport stops as their energy supply dries up. But if you supply the cells with energy, even though the mitochondria are still poisoned, active transport starts up again!

Figure 2.17 Turtles, like flamingos and other marine animals, live in very extreme conditions. They rely on active transport in special salt glands to move excess salt out of their body.

The biologist's toolkit: Predictions and hypotheses

When a scientist is trying to answer a scientific question, they often produce a **hypothesis**. What does this mean? A hypothesis is theory or a possible explanation for an observation that can be tested out. Scientists will use their hypothesis to make a **prediction** – what they think will happen. They then do an investigation or take more observations to collect the data that will tell them if their prediction was right. If so, this is evidence that their hypothesis was probably correct – although they usually need a lot of evidence to convince everyone.

You must be able to test your hypothesis by:
- writing a prediction based on the hypothesis
- collecting data that shows you whether your original prediction is correct.

For example:
I have observed that white-throated jacobins in my garden seem to feed more often from the red flowers than the yellow ones.

My hypothesis: White-throated jacobins prefer red flowers to yellow flowers.

My prediction: If there are equal numbers of red and yellow flowers, white-throated jacobins will feed more often from the red flowers.

Data collection: Cover plants so that there are equal numbers of red-and yellow-flowered plants in an area. Take regular observations over time or use a trail camera to record white-throated jacobins' feeding patterns. **Analyse** (look at and organise) your data to see if your prediction was correct.

Possible SBAs based on movement of substances

→ Calculate the concentration of the cell contents of different vegetables using osmosis.
→ How do different treatments (cooking, freezing) affect the movement of water into plant cells?
→ Does air temperature affect the rate of diffusion of gases in the air?
→ Does air movement affect the rate of diffusion of gases in the air?
→ Does temperature affect the rate of diffusion of substances in solution?
→ How does the concentration gradient affect the rate of diffusion in water?
→ Investigate a factor that affects the diffusion of a substance dissolved in water.
→ Investigate the effect of different solvents, for example water and ethanol, on diffusion rates.
→ Do some people have more sensitive senses of smell than others? Use diffusion to help you investigate this.
→ Investigate different fruits and vegetables as possible models for demonstrating osmosis in the classroom.
→ Using a microscope, investigate turgor and plasmolysis in cells from different types of plants.
→ Investigate the amount of water that can be removed from samples of different fruits and vegetables, for example aubergines, green mangoes and bananas, using salt crystals.

End-of-chapter summary

In this chapter, you have learnt that:

- diffusion is the net (overall) movement of particles from an area of high concentration to an area of lower concentration
- diffusion takes place because of the random movements of the particles of a gas or of a substance in solution in water
- diffusion is important in many processes taking place in animals and plants; examples include gas exchange in the lungs, the absorption of digested food from the gut and the entry of carbon dioxide into the leaves of plants
- osmosis is a special type of diffusion where only water moves across a partially permeable membrane, from an area of high concentration of water to an area of lower concentration of water
- cell membranes are partially permeable so osmosis occurs frequently in plant and animal cells
- osmosis is very useful to plants and animals, but it also causes many difficulties
- both osmosis and diffusion can be demonstrated experimentally in the laboratory
- water moves from cell to cell in a plant by osmosis
- in active transport, substances are moved against a concentration gradient or across a partially permeable membrane
- active transport uses energy produced by cellular respiration
- cells that carry out a lot of active transport often have many mitochondria to provide the energy they need
- active transport is very important in cells and whole organisms, for example, in the movement of mineral ions into plant root cells and in the movement of excess salt out of the body via the salt glands in some marine creatures.

End-of-chapter questions

Use the following phrases to answer questions 1 and 2:

W   The net movement of particles from an area of high concentration to an area of low concentration

X   The movement of water molecules from an area of low concentration of water molecules to an area of high concentration through a partially permeable membrane

Y   The net movement of particles from an area of low concentration to an area of high concentration

Z   The movement of water molecules from an area of high concentration of water molecules to an area of low concentration through a partially permeable membrane

1   What is the phrase that best describes the process of diffusion?
   A   W
   B   X
   C   Y
   D   Z

2   What is the phrase that best describes the process of osmosis?
   A   W
   B   X
   C   Y
   D   Z

3   What is the process called by which water moves into the tissues of freshwater fish?
   A   Diffusion
   B   Osmosis
   C   Active transport
   D   Pouring

4   Active transport moves substances against a concentration gradient. What does this use?
   A   Enzymes
   B   Partially permeable membranes
   C   Energy
   D   Osmosis

5   Explain the following in terms of diffusion and the movement of particles:
   a   A cake baking in the oven can be smelt all over the house. (2)
   b   When trying on a new perfume or aftershave, you are advised to try on no more than two at a time. (2)
   c   If you cut yourself in water, it looks much worse than a cut on land. (2)

6   a   Explain, using a diagram, what would happen if you set up an experiment with a partially permeable bag containing strong sugar solution in a beaker full of pure water. (4)
   b   Explain, using a diagram, what would happen if you set up an experiment using a partially permeable bag containing pure water in a beaker containing strong sugar solution. (4)

7 Animals that live in fresh water have a constant problem with their water balance. The single-celled organism Amoeba has a special vacuole. It fills with water and then moves to the outside of the cell and bursts, releasing the water from the cell. A new vacuole starts forming straight away. Explain, in terms of osmosis, why the Amoeba needs one of these vacuoles. (5)

8 Experiments on osmosis are often carried out using potato cylinders. You have been asked to find out if sweet potato or breadfruit would be good alternatives. Describe in detail the investigation you would carry out to see if either of these alternatives would be better than the traditional potato. (10)

9 a Explain how active transport works in a cell. (4)

   b Describe two examples of a situation in which a substance cannot be moved into a cell by osmosis or diffusion, and how active transport solves the problem. (4)

   c The processes of diffusion and osmosis do not need energy to take place. Why does the organism have to provide energy for active transport and where does this energy come from? (3)

10 a Explain why cyanide is such an effective poison. (2)

   b Why is active transport so important for animals that live in the sea? (2)

11 You have to produce some CSEC revision sheets on diffusion, osmosis and active transport in living organisms. Use the examples given in this chapter to help you make the sheets as lively and interesting as possible. Use any methods that help you to remember things, and save the sheets to help you revise when exams are approaching. (10)

Chapter 3 Photosynthesis

When you have completed this chapter, you will be able to:
- describe how plants make sugars by photosynthesis
- summarise the process of photosynthesis using word and chemical equations
- identify the chloroplasts as the site of photosynthesis
- carry out experiments to demonstrate that light is needed for the production of starch from photosynthesis
- carry out experiments to demonstrate that chlorophyll is needed for the production of starch from photosynthesis
- describe the structure of a leaf in relation to its adaptations for photosynthesis

What the examiners say

→ Candidates are able to discuss concepts in general terms, but lack the details to convincingly convey familiarity with all the relevant components of the topic. For example, the word equation for photosynthesis and the storage form of sugar in plants prove challenging for many candidates.

Revision tip

Make a revision timetable. There is no need to go to the trouble of ruling one up for yourself. Go to our science website, download a revision timetable and insert all your subjects. Remember to give yourself plenty of time – don't overload yourself.

Like all living organisms, plants need food to provide them with energy for respiration, growth and reproduction. Many organisms, including all animals, take in food to get the energy they need to live. They are called **heterotrophs** (feeding on others). In contrast, plants produce their own food in a process called **photosynthesis**. They are called **autotrophs** (feeding themselves). Photosynthesis takes place in the green parts of plants, especially the leaves, in the presence of light.

What is photosynthesis?

Photosynthesis is the process by which plants make their own food. It is the basis of almost all life on the Earth. During photosynthesis, light is absorbed by a green substance called **chlorophyll**, found in the **chloroplasts** of some plant cells. The energy captured is used to convert carbon dioxide from the air and water from the soil into a simple sugar called **glucose**, with oxygen gas as a by-product.

Figure 3.1 Plants make their own food using photosynthesis and they use it to grow in some spectacular ways. Oxygen is a by-product of photosynthesis, and it is used for respiration by both plants and animals.

We can summarise the process of photosynthesis using a **word equation**. This tells us what is going on.

$$\text{carbon dioxide} + \text{water} \xrightarrow[\text{chlorophyll}]{\text{light}} \text{glucose} + \text{oxygen}$$

We can also summarise photosynthesis using a simple **chemical equation**.

$$6CO_2 + 6H_2O \xrightarrow[\text{chlorophyll}]{\text{light}} C_6H_{12}O_6 + 6O_2$$

> ### The biologist's toolkit: Chemical equations
>
> Chemical equations may look complicated – but they give you very useful information. Keep calm and think carefully when you see a chemical equation. They show you the types and numbers of the different atoms and molecules involved in a chemical reaction. The numbers of the different atoms must be the same on both sides of the equation.
>
> In the case of the summary equation for photosynthesis, you need to remember:
>
> - C = carbon atom
> - H = hydrogen atom
> - O = oxygen atom
> - CO_2 = carbon dioxide, made up of 1 carbon atom and 2 oxygen atoms
> - H_2O = water, made up of 2 hydrogen atoms and 1 oxygen atom
> - $C_6H_{12}O_6$ = glucose, made up of 6 carbon atoms, 12 hydrogen atoms and 6 oxygen atoms.

What happens to the products of photosynthesis?

Plants make glucose and oxygen when they photosynthesise.

The oxygen is used by organisms from tiny microbes to plants, and animals including people, in the process of cellular respiration. Without plants producing oxygen, life on the Earth would not have evolved as it has.

Some of the glucose produced during photosynthesis is used immediately by the cells of the plant for respiration to provide energy for cell functions, growth and reproduction.

The energy released in respiration is also used to build smaller molecules into bigger molecules.

Some of the glucose produced by photosynthesis is always converted into **starch** for storage. Glucose is soluble and so it might affect the water balance within the plant. If the concentrations of glucose vary in different parts of the plant, then water would move by osmosis, upsetting the balance of the whole organism. Starch is **insoluble** (it does not dissolve), so it has no effect on the concentration of solutions in the plant. This means that it can be stored in different places without having any effect on the water balance of the plant.

Starch is also a very compact molecule, so it does not take up a lot of space in the plant. It is easily broken down again into glucose molecules when it is needed for energy by the cells of the plant.

Since so much starch is produced during photosynthesis, we often use it to show that photosynthesis has taken place in a plant.

> **DID YOU KNOW?**
> Plants produce approximately 36.8×10^{13} kg of oxygen every year as a waste product of photosynthesis. This is one reason why plants are so important to the survival of life on the Earth.

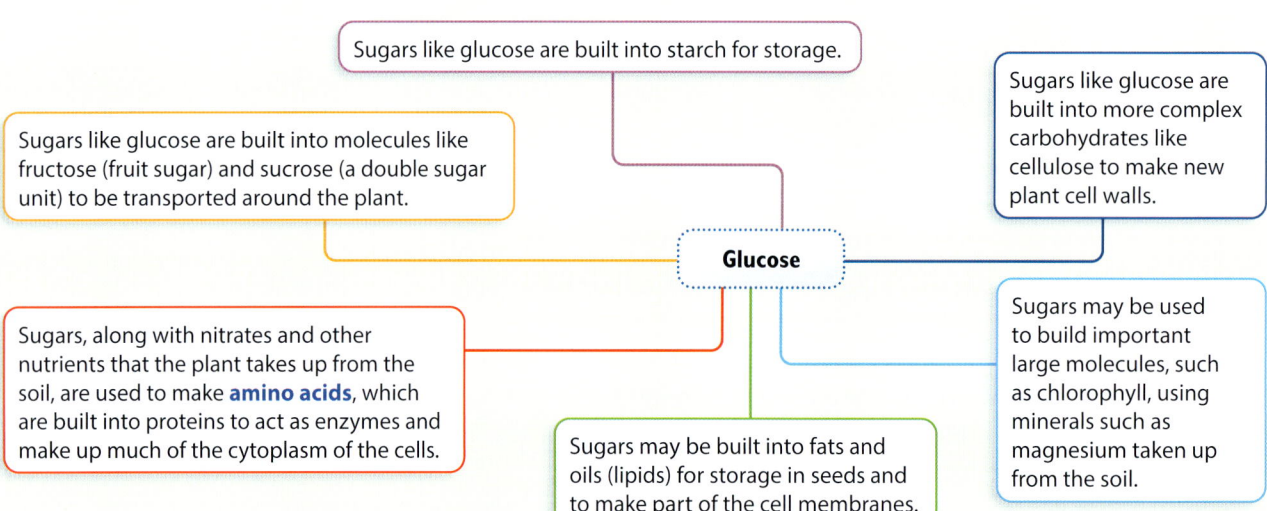

What is needed for photosynthesis?

As you can see from the equation on page 45, for photosynthesis to occur successfully, the plant needs the inorganic molecules carbon dioxide and water, along with a supply of light and the means to capture that energy in the form of the green pigment, chlorophyll. However, to show that certain factors are needed for photosynthesis, we need a way to demonstrate that it has actually taken place.

The simplest way to do this is to look at the end products of the process. We use the presence of starch in the leaf of a plant to show that it has been photosynthesising. Iodine solution is reddish brown, but in the presence of starch it turns blue-black. Unfortunately, you cannot just add a few drops of iodine solution to the leaves of a plant to see if it has made starch. The waxy **cuticle** forms a waterproof layer, which prevents the iodine from penetrating, and the green colour masks any slight colour change. However, there is a simple procedure to test for the presence of starch that you can use in many different experiments to investigate photosynthesis – see Figure 3.2.

Photosynthesis

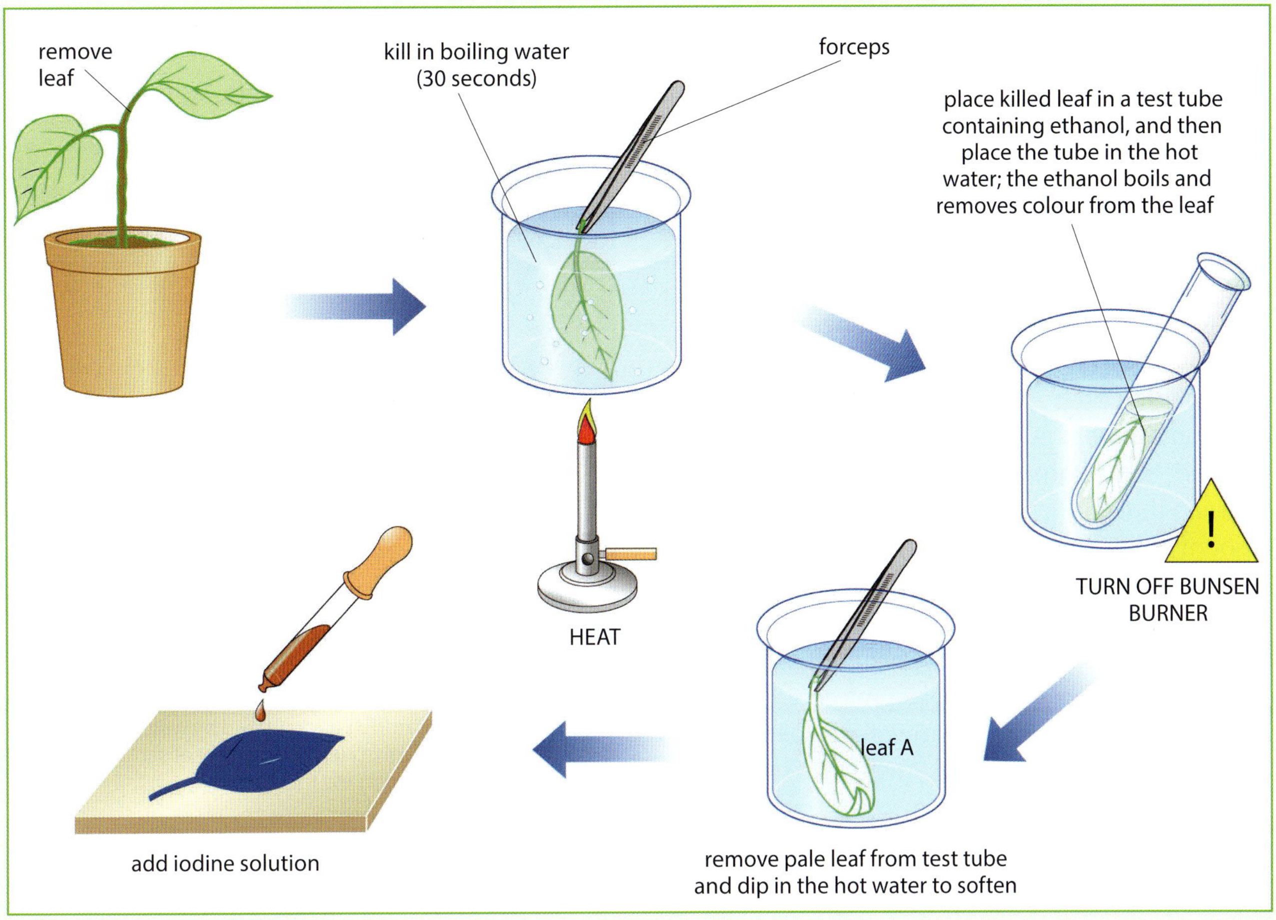

Figure 3.2 We use iodine to test for the presence of starch. A blue-black colour indicates starch is present in the leaf which shows that photosynthesis has taken place.

The need for light

The simple equation we use for photosynthesis is a summary of many different chemical reactions that take place in a plant to convert carbon dioxide and water into glucose and oxygen. For most plants, the source of light energy for photosynthesis is the sun, although in some cases, people supply extra artificial light. If plants are deprived of light for a substantial amount of time, they will die, because once their stores of starch are used up, they are not replaced and so there is no energy available for the metabolic reactions of the cells.

The simplest way of demonstrating that a plant needs light is to deprive it of light and see what happens. Keep a whole plant in the dark for two to three days – this is called destarching the plant – and then compare the leaves with those from a similar plant kept in the light (see Figure 3.3). Alternatively, you can cover either a whole leaf or part of a leaf from a destarched plant with black paper or foil. This prevents light from reaching the covered area. Leave the plant in the light for several hours and then test the covered leaf for the presence of starch and compare it to an uncovered leaf. The difference is plainly visible.

Figure 3.3 The results of the iodine test for starch:
- leaf A – the iodine remains yellow-brown, so there is no starch present. It has been in the dark and is destarched.
- leaf B – the iodine turns blue-black, showing that starch is present. It has been in the light and made starch from photosynthesis.

What is needed for photosynthesis?

However, starch is not the only end product of photosynthesis. Oxygen gas is produced as well. We cannot easily observe the oxygen gas produced during photosynthesis in land plants, but it is a useful way of showing that photosynthesis is taking place in water plants. They give off bubbles of oxygen-rich gas from their leaves and from any cut or broken stems when they are photosynthesising. We can use this to show that light is needed for photosynthesis to take place, simply by collecting the gas given off from a water plant kept in the dark and one kept in the light. The gas given off by the plant in the light will relight a glowing splint, showing that it is rich in oxygen. The much smaller amount of gas collected from the plant in the dark does not do this; in fact it will extinguish the splint as it is rich in carbon dioxide from respiration. You can also use this method to measure the rate at which photosynthesis takes place, by counting the bubbles given off or the volume of gas collected in a specified amount of time.

Figure 3.4 The oxygen-rich gas released by water plants as they photosynthesise gives us another way of showing that light is needed for the process to happen.

The biologist's toolkit: The fair test

Here are some useful terms to learn:
- When you plan a scientific investigation you often carry out an **experiment**. In your experiment, you make changes and observe what happens as a result.
- Anything you change, or that changes during your experiment, is called a **variable**.
- The thing that you change deliberately in your experiment is called the **independent variable**.
- Anything that changes as a result is called the **dependent variable**.
- Scientists always try and make their investigations a **fair test**. This means they change only one variable at a time and keep everything else the same.

Fair tests are particularly important in physics and chemistry. They are harder for biologists to carry out, because all living things vary a bit. Almost every microorganism, cell, animal or plant is slightly different from all the others – so we do our best to make experiments fair tests, but always remember we are using living organisms.

For example:
- If you want to investigate the need for light in photosynthesis, for example Activity 3.1, what will your independent variable be?
- What will your dependent variables be?
- What can you do to make your investigation a fair test?
- What factors will make it difficult to plan a completely fair test?

Activity 3.1

Investigating the need for light in photosynthesis

Method

Follow these steps to plan your own experiment to show that light is needed for photosynthesis to take place.

1. Plan two experiments, one using land plants and one using water plants, which could be used to demonstrate that light is needed for photosynthesis.
2. Make sure that you describe in detail how to show that photosynthesis has taken place in each case, and think carefully about any safety issues.
3. Once you have planned your experiments, ask your teacher to check through them.
4. If you are given permission, set up one of your experiments, and write up your observations and your conclusions on its effectiveness.
5. Make sure you evaluate your method and discuss any ways to improve it.

The biologist's toolkit: Drawing a line graph

In many of the investigations you carry out in biology, you change one variable and measure the effect on another variable. For example, when you are measuring the effect of light on the rate of photosynthesis:

- you change the amount of light falling on your pondweed; this is the independent variable
- you measure the number of bubbles or the volume of gas given off in a specified amount of time; this is your dependent variable.

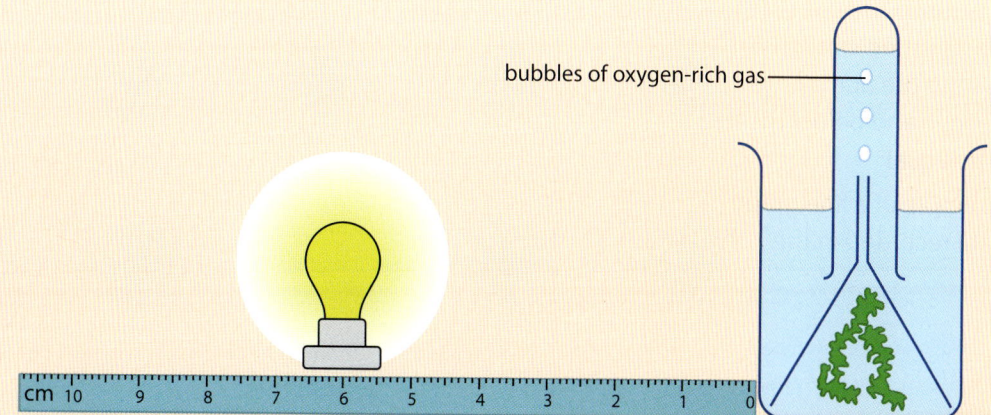

What is needed for photosynthesis?

Record your data in a table as you carry out your investigation. In this example, your table would look like the one below:

Distance of light from plant (cm)	Average number of bubbles per minute
10	100
20	95
30	80
40	62
50	53
60	35

Plotting your data on a graph makes it easier for you to see any links between the two variables.

Remember these points when you draw a line graph:

- Always think carefully about your scales before you draw your graph. Use as much of your graph paper as you can – it makes it easier to see what is happening!
- Make sure the numbers on each axis you draw are spaced evenly.
- The independent variable goes on the horizontal or *x*-axis. Write the name of your variable on the *x*-axis and don't forget to give the correct units, for example: 'distance of light from plant (cm)'
- The dependent variable goes on the vertical or *y*-axis. Write the name of your variable on the *y*-axis and don't forget to give the correct units, for example: 'average number of bubbles per minute'
- Use a pencil cross for each point on your graph.
- Draw a smooth line of best fit. This line might be straight or curved – but the number of points above and below the line should be equal.

Here is an example of a line graph using the data recorded in the table above.

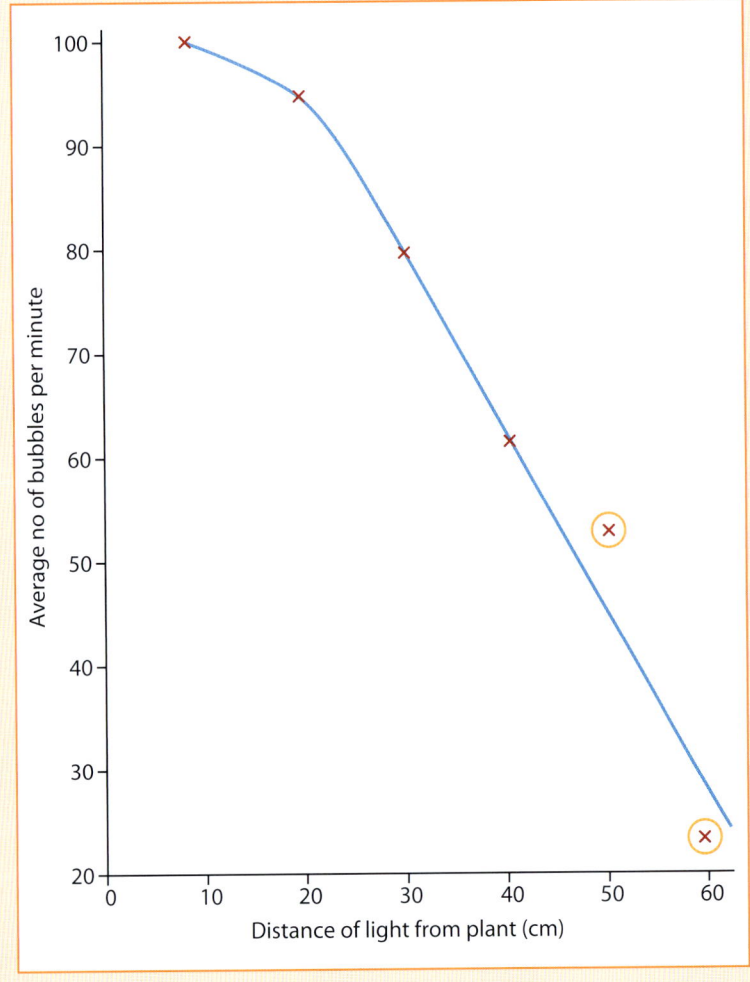

Photosynthesis

The need for carbon dioxide

Plants need carbon to synthesise sugars. There are lots of carbon-containing chemicals in existence, but carbon dioxide from the air or in solution in water is the only form that plants can use in photosynthesis. Carbon dioxide is found more or less everywhere. Plants get the carbon dioxide they need in two ways:

- From the carbon dioxide they make during respiration
- From the air that moves into their leaves through small holes called **stomata**.

The need for water

Carbon dioxide alone is not sufficient to produce carbohydrates. Plants need water too. All the cells of a plant have a constant supply of water, both as a waste product of respiration and travelling up through the plant from the roots to leaves. As you saw in Chapter 2, plant cells need water to stay rigid and work properly, so there is always plenty of water for photosynthesis.

The need for chlorophyll

For photosynthesis to take place, a plant must capture energy, usually sunlight. Light is captured by the green pigment, chlorophyll. The chlorophyll is stored in organelles called chloroplasts in the plant cells. It is the chlorophyll in the chloroplasts of the leaves and stems of a plant that make it look green. The simplest way to demonstrate that chlorophyll is needed for photosynthesis to take place is to look at the leaves of a **variegated plant**. Variegated leaves have areas that contain chlorophyll and areas that do not. The chlorophyll-free areas are usually yellow or creamy-white in colour. Take a destarched variegated plant and leave it in the light for several hours. Then test one of the leaves for the presence of starch. The iodine solution only changes colour in those regions of the leaf that were green. This shows that without chlorophyll, photosynthesis does not take place.

> **DID YOU KNOW?**
>
> There are some plants called resurrection plants that can survive without water. In dry conditions, they can lose up to 95% of their water. They end up as tiny, shrivelled remains. But just add water, and within 24 hours they have recovered and are photosynthesising again as good as new!

Figure 3.5 The leaves on the left and on the right come from plants that have been in bright light for 48 hours. Both plants have had plenty of opportunity to photosynthesise and produce glucose to turn into starch. However, only the green areas of the variegated leaf on the right have been able to make starch because the white areas do not contain chlorophyll. The middle leaf is from a plant that has been in the dark for 48 hours.

Activity 3.2

Showing that chlorophyll is needed for photosynthesis to take place

You can see how important chlorophyll is for photosynthesis by carrying out the following experiment on a variegated leaf. You are going to keep one plant with variegated leaves in the dark and place another variegated plant in bright light for several hours, and then test a leaf from both plants for the presence of starch. What do you expect to see and why?

You will need:

- two potted plants with variegated leaves, for example geraniums (pelargoniums) or ivy, which have been destarched; keep one plant in the dark and bring the other into the light for several hours before the investigation
- a large beaker
- a Bunsen burner, tripod, gauze and heatproof mat
- ethanol (NB: Keep ethanol away from the naked flame.)
- a boiling tube
- forceps
- a white tile

Method

1. Remove one leaf from the destarched plant with variegated leaves that has been kept in the dark and one leaf from the plant that has been exposed to a bright light for several hours.
2. Prepare both leaves to be tested for the presence of starch using the method described on page 47.
3. Spread both leaves on a white tile and add iodine solution.
4. Make careful observations and drawings of your results.
5. Write up your experiment and explain the results you get. Do they fit with your initial hypothesis?

Checkpoint questions

1. Describe how a plant gets the carbon dioxide, water and light it needs for photosynthesis.
2. Relate the path taken by a carbon molecule as it moves from being part of carbon dioxide in the air to being part of a starch molecule in a plant.

A photosynthesising machine

Plants take the inorganic molecules carbon dioxide and water, and use them to produce the organic molecule glucose, along with inorganic oxygen, in the presence of energy from light. This amazing process called photosynthesis is the basis of all life on the Earth – it provides the food we eat and the oxygen we breathe, and it all takes place in the leaves of plants.

Adaptations of a leaf for photosynthesis

Plant leaves are perfectly adapted to allow the maximum possible amount of photosynthesis to take place whenever there is light available.

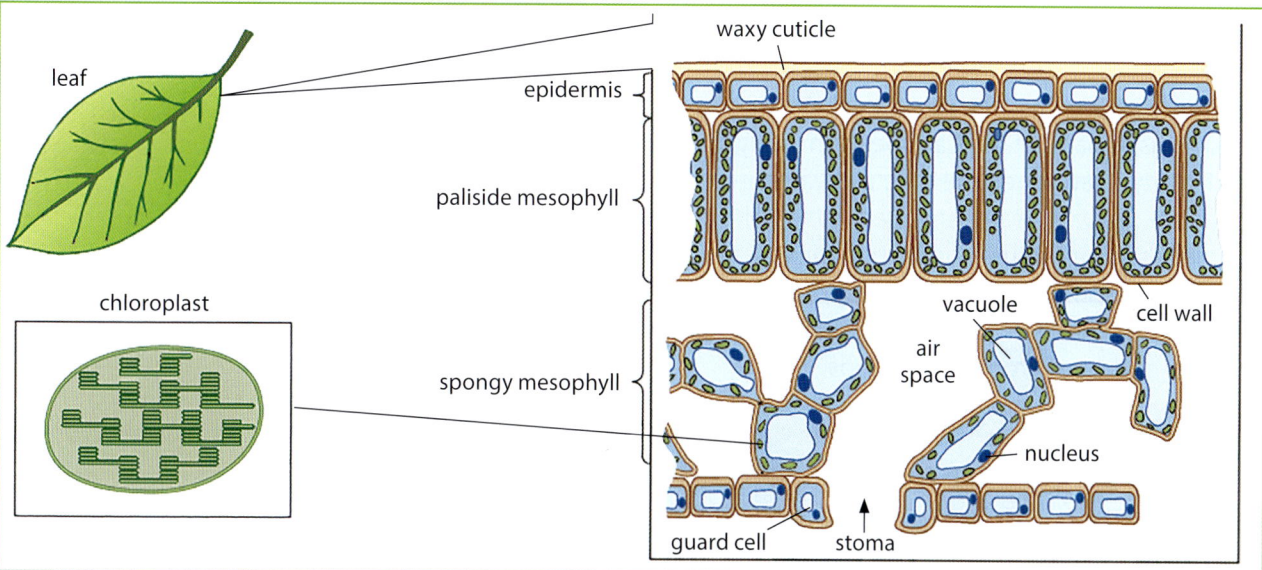

Figure 3.6 The leaves of plants are perfectly adapted to make the best possible use of the light that falls on them.

As shown in Figure 3.6, plant leaves have the following adaptations for photosynthesis:

- The leaf is flat and wide, providing a large surface area to collect light and short distances for gases to diffuse. The veins bring water from the soil to the cells.
- The waxy cuticle is a waterproof layer found on the surface of many leaves, which helps prevent water loss.
- The palisade **mesophyll** is the main photosynthetic tissue of the plant. There are many cells, closely packed together near the surface of the leaf to get as much light as possible. Each cell has many chloroplasts, sometimes hundreds of them, spread out through the cytoplasm of the cell when light levels are high, but which cluster at the top of the cell (near to the light) when light levels are low.
- The spongy mesophyll has fewer cells with fewer chloroplasts. However, there are lots of air spaces and a big surface area for gas exchange. Some photosynthesis takes place here but, more importantly, it is where the carbon dioxide needed for photosynthesis moves into the cells and the oxygen moves out. Water pulled up from the roots evaporates from the cells here as well. The lower epidermis has openings called stomata (singular: stoma) that allow carbon dioxide to diffuse into the leaf, and oxygen and water vapour to diffuse out.

A photosynthesising machine

The **guard cells** surrounding the stomata open and close to control the entry of carbon dioxide into the leaf and also to control the loss of water from the plant.

- The **vascular bundles** contain the **xylem**, which brings water from the soil to the cells of the leaves, and the **phloem**, which carries the products of photosynthesis away from the leaves to all the cells of the plant.
- Each chloroplast contains stacks of membranes and lots of chlorophyll to provide an increased surface area for photosynthesis to take place.

Figure 3.7 In the Caribbean plants like this breadfruit have everything they need to grow fast and well.

Not every country has the same ideal growing conditions as parts of the Caribbean. In many areas of the world, certain crops, particularly salad vegetables, are cultivated in enormous greenhouses. The plants are grown in water enriched with all the minerals they need, and the light levels, temperature and carbon dioxide levels are constantly monitored by electronic sensors that feed information into computers. As a result, the plants have perfect conditions all the time. The plants grow as rapidly as possible from seeds to end up as crops in the shops.

Figure 3.8 A salad vegetable crop being cultivated in a commercial greenhouse.

Checkpoint questions

3. **a** Describe three ways in which a leaf is adapted for photosynthesis.
   b Explain how each adaptation you have chosen makes photosynthesis more efficient.
4. You want to investigate the effect of light on photosynthesis. Suggest two variables you need to control to make this a fair test.

Photosynthesis

Possible SBAs based on photosynthesis

→ Investigate effect of light intensity on the rate of photosynthesis in water plants (*hint:* the further away a plant is from a light source, the lower the light intensity).
→ Investigate the effect of water temperature on the rate of photosynthesis in water plants.
→ Investigate carbon dioxide concentration on the rate of photosynthesis in a water plant.
→ Sometimes plants have plenty of one factor, for example light or carbon dioxide, but not enough of others. The factors that are in short supply are called **limiting factors**. Investigate how one factor limits photosynthesis when the other main factors are in plentiful supply. For example, how different temperatures affect the rate of photosynthesis when there is plenty of light, or how light or carbon dioxide levels limit the rate when the temperature is high.
→ Investigate the structure of different local leaves, for example shape, area and thickness, and how they are adapted for photosynthesis in different conditions.
→ As above, but compare the leaves of Caribbean plants to the leaves of plants in a temperate country, such as the UK, and suggest reasons for the differences.
→ Compare the makeup of the chlorophyll pigments in different types of leaves. (Extract the chlorophyll and run chromatography on them to compare the coloured pigments.)

End-of-chapter summary

In this chapter, you have learnt that:

- photosynthesis is the process by which plants combine carbon dioxide and water to make glucose and oxygen, using energy from light
- a word summary of the process of photosynthesis is:

$$\text{carbon dioxide} + \text{water} \xrightarrow[\text{chlorophyll}]{\text{light}} \text{glucose} + \text{oxygen}$$

- we also summarise photosynthesis using a simplified chemical equation

$$6CO_2 + 6H_2O \xrightarrow[\text{chlorophyll}]{\text{light}} C_6H_{12}O_6 + 6O_2$$

- the light needed for photosynthesis is captured by the green pigment chlorophyll
- photosynthesis takes place in the chloroplasts, which is where the chlorophyll is stored
- the glucose made in photosynthesis is converted to insoluble starch for storage to avoid osmotic problems
- the glucose made in photosynthesis is used in respiration to provide energy for all the cells of the plant
- the energy released in the plant during respiration is used to build up smaller molecules into larger molecules, for example, cellulose, lipids and oils, and chlorophyll
- leaves are well adapted to allow the maximum photosynthesis to take place
- the need for carbon dioxide, water, chlorophyll and light in photosynthesis can all be demonstrated experimentally.

End-of-chapter questions

1. The equation for photosynthesis is:

$$CO_2 + H_2O \xrightarrow[\text{chlorophyll}]{\text{light}} C_6H_{12}O_6 + O_2$$

 This equation is not balanced. Which of the following numbers needs to be put in front of the carbon dioxide, water and oxygen symbols to give a balanced equation for photosynthesis?
 A 2
 B 4
 C 6
 D 8

2. Which substance is a waste product of photosynthesis?
 A glucose
 B oxygen
 C carbon dioxide
 D water

3. In the iodine test, the blue-black colour that appears on the leaves of a plant that has been photosynthesising shows the presence of:
 A starch.
 B chlorophyll.
 C glucose.
 D cellulose.

4. In the middle of a hot sunny day, which of the things needed for photosynthesis is most likely to be missing in a field of sugar cane?
 A water
 B carbon dioxide
 C light
 D temperature

5. The diagram on page 57 shows an unspecialised plant cell from a blade of grass.
 a The diagram shows a section through the leaf of a green plant. Copy the diagram and complete the labels.
 b Name the two areas of the leaf where most photosynthesis takes place.
 c Explain how the leaf is adapted for photosynthesis.

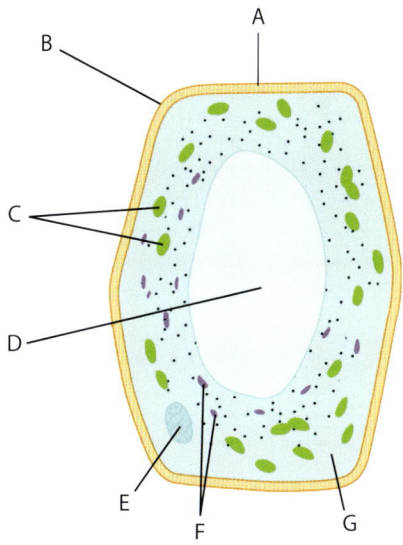

6 Photosynthesis takes place in green plants.
Variegated leaves have areas that are green and areas that are white.
Some students used variegated leaves to investigate photosynthesis.
- They covered part of a variegated leaf with a black paper shape.
- The leaf was left in a sunny place.
- They tested the leaf for starch.

Area of leaf tested	Starch present after test	
	covered	uncovered
Green area	no	yes
White area	no	no

a   Describe how you would test this leaf for the presence of starch.
b   Why is this used as a test to show if photosynthesis has taken place?
c   Explain why starch was not produced:
    i   in the covered green areas of the leaf
    ii   in the uncovered white areas of the leaf.

End-of-chapter questions   57

7 The diagram shows an experiment to investigate the effect of light on the rate of photosynthesis in a water plant.

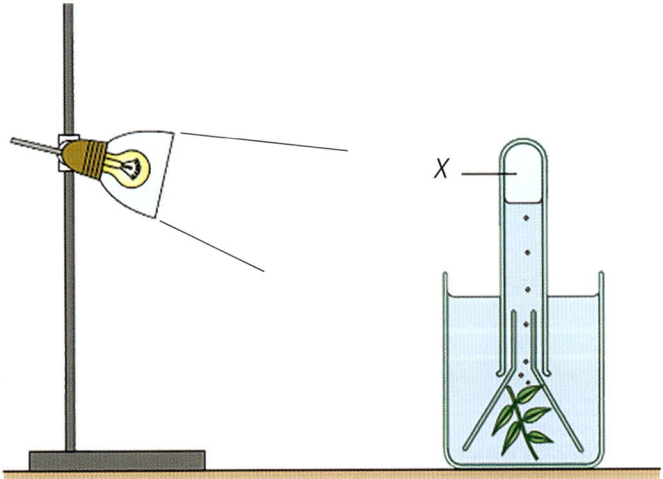

a   What gas would you expect to be collected at *X*?
b   What test would you carry out to show what gas has actually been collected?
c   What would you expect to see if the light was moved closer to the beaker containing the pondweed?
d   Explain why you would expect your answer in question 7c to happen.
e   What factors need to be kept constant in this experiment if it is to be a fair test of the effect of light?
f   Here are the results of this experiment. Plot a graph of the number of bubbles per minute against the distance of the lamp from the plant.

Distance of lamp from plant (cm)	Average number of bubbles per minute
10	130
20	130
30	125
40	105
50	80
60	60
70	42
80	30
90	15
100	10

g   Explain the shape of the graph – suggest what is happening when the light is very close to the plant.

Photosynthesis

Chapter 4
Feeding relationships and the carbon cycle in nature

When you have completed this chapter, you will be able to:

- define a food chain and a food web
- identify the trophic levels of organisms in a food chain and a food web
- define trophic levels
- explain the use and loss of energy at each trophic level within a food chain
- describe the recycling of carbon in nature (the carbon cycle), including the formation and burning of fossil fuels and the impact of global warming on human well-being
- describe and build food chains from different habitats
- explain the ways in which other living organisms – including human beings – depend on plants both directly and indirectly for food

What the examiners say
→ Candidates are often uncertain of the direction of the arrow when drawing a food chain. Additionally, candidates demonstrate a weakness in analysing a food chain.

As you saw in Chapter 3, plants are vitally important because they harness the energy of the sun by photosynthesis and make it available to other organisms in the form of food. They make food from simple inorganic molecules, and without them little else could survive for long. Plants are called producers because of their role in making carbohydrates. What is more, they absorb carbon dioxide from the air and produce oxygen in the process, maintaining the balance of gases in the atmosphere and providing us all with the oxygen that we need to live.

Figure 4.1 Not all plants are quite as spectacular as this wonderful Royal poinciana in Barbados, but even the smallest and dullest green plant makes food by photosynthesis.

The importance of plants

Plants are the main source of food for many thousands of different species of animals, from the aphids that feed on your house plants to the great herds of wildebeest, zebras and elephants of Africa. Animals that eat only plants are known as **herbivores**. Not all animals eat plants. Many of them feed on other animals and they are known as **carnivores**. Some types of animals, ourselves included, eat a diet that contains both plants and animals. These animals are known as **omnivores**.

Table 4.1 Different ways of feeding

Feeding category	Herbivore	Carnivore	Omnivore
Description	Animals that only eat plants	Animals that eat only other animals	Animals that eat both animals and plants
Example	Sphinx moth larvae	American kestrel	Human beings

Around the world, much of the staple diet for human beings comes from plants, for example, cereal crops, nuts, fruits and storage tubers. However, in many cultures, people eat meat as well as plants. Meat comes from animals, but many of those animals eat plants. Cows, sheep, goats, chickens and fish all eat plants to provide them with the energy and materials they need to grow. Sometimes we eat animals that feed on other animals, for example, many of the fish we eat are carnivores. But even when we eat carnivores, we still depend on plants because the animals eaten by the carnivores in turn eat plants.

Plants also provide the raw materials for our drinks, for example, coffee, tea, hot chocolate, lemonade and rum.

It is not only human beings that depend on plants and the process of photosynthesis. Almost all living organisms depend on plants as the producers of food from the raw materials of carbon dioxide and water. The feeding relationships between different organisms are described by looking at models known as **food chains** and **food webs**.

Food chains

The first stage in a food chain involves converting light into stored chemical energy in plants by photosynthesis. This is always done by plants – the **producers**. All the animals that eat plants or other animals are called **consumers**. The different levels in a food chain are called **trophic levels**.

Some of the energy stored in the tissues of a plant is passed on to the animal that eats it. This will usually be a herbivore, although it could also be an omnivore. The herbivore (or omnivore) is called a **primary consumer** because it eats plants. Some of the energy within the herbivore is, in turn, passed on to the animal that eats it. Again, this will usually be a carnivore but it could also be an omnivore. The carnivore (or omnivore) is called a **secondary consumer** because it eats the plant eater. This naming continues along the chain. When you draw a food chain, you use arrows to show the direction in which the biological material moves through the chain – see Table 4.2.

Figure 4.2 Plants not only provide us with the food we eat, but they are often the basis of our drinks, and they are the foundation of our exports and our economies as well.

Table 4.2 Feeding relationships in a Caribbean food chain

Trophic level	Producer	Primary consumer	Secondary consumer	Tertiary consumer
Description	Plants that use light, carbon dioxide and water to produce glucose and oxygen during photosynthesis	Animals that feed on producers (mainly herbivores)	Animals that eat primary consumers (mainly carnivores)	Animals that eat secondary consumers (mainly carnivores)
Example				

Plants ⟶ Insects ⟶ Lizards ⟶ Caracaras

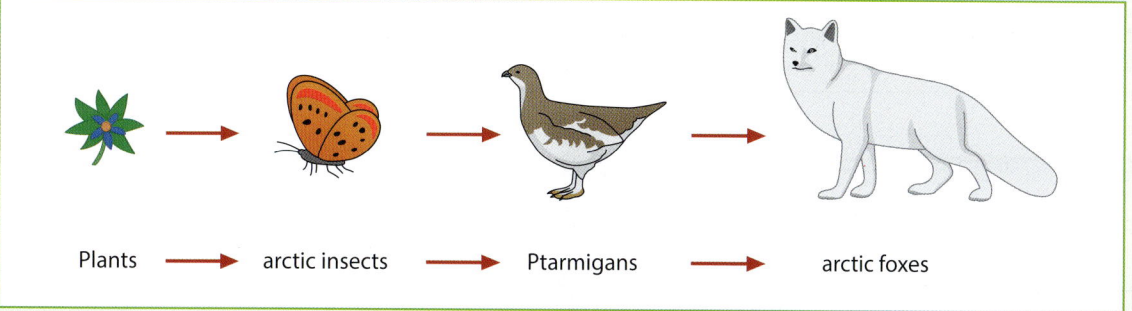

Plants ⟶ arctic insects ⟶ Ptarmigans ⟶ arctic foxes

Figure 4.3 This Arctic food chain was the first one to be observed and recorded by a biologist called Charles Elton. His ideas have formed the basis of much of our study of ecology ever since.

Terrestrial and aquatic habitats

The **habitat** of an animal or a plant is its home – the little bit of the environment where it, and the **population** of which it is a part, actually lives. The habitat must supply everything the organism needs to live, grow and breed successfully. By looking at different habitats, we can see some of the range of food chains that exist as animals and plants interact.

Within any of these different habitats, living organisms depend on plants, the producers, to provide the food on which all the rest of the organisms depend. Some of the simplest food chains have only two trophic levels. They come from terrestrial habitats and they describe the food we eat ourselves. For example, plantain plants photosynthesise and produce fruits that we eat in a wide variety of dishes. This food chain is very simple.

On the other hand, we often eat meat, and this extends the food chain. For example, if you enjoy chicken, the food chain in which you are taking part has three trophic levels.

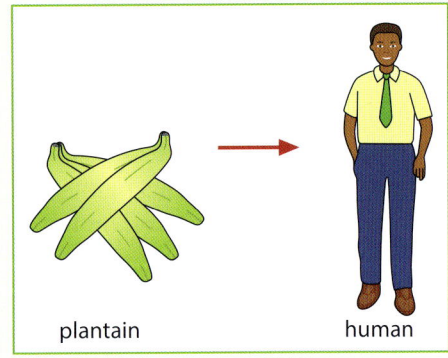

Figure 4.4 Food chain (terrestrial)

Figure 4.5 Food chain (terrestrial)

But if you prefer beef, the chain would look as follows:

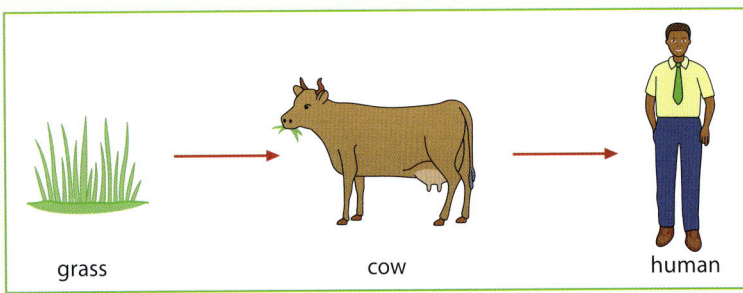

Figure 4.6 Food chain (terrestrial)

Sometimes the chains are even longer. Mountain chicken, which is actually a large frog, has long been a popular dish in Dominica and Montserrat. This terrestrial food chain has five trophic levels.

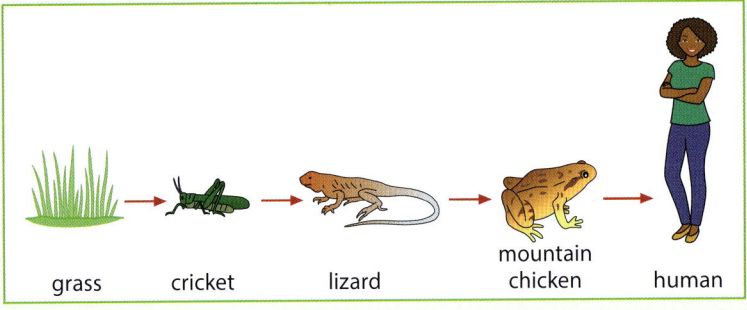

Figure 4.7 Food chain (terrestrial)

Feeding relationships and the carbon cycle in nature

Food chains occur in water as well as on land. For example, the seas around the Caribbean coral reefs are a major habitat. Tiny plant-like organisms called algae grow on the coral, photosynthesising and making food. Parrot fish graze on these algae, before they themselves are eaten by predatory fish such as groupers. But the chain doesn't stop there, as groupers in turn are eaten by larger carnivores such as the barracuda.

> **Revision tip**
>
> Become a peer tutor – sharing what you know is a great way to learn!

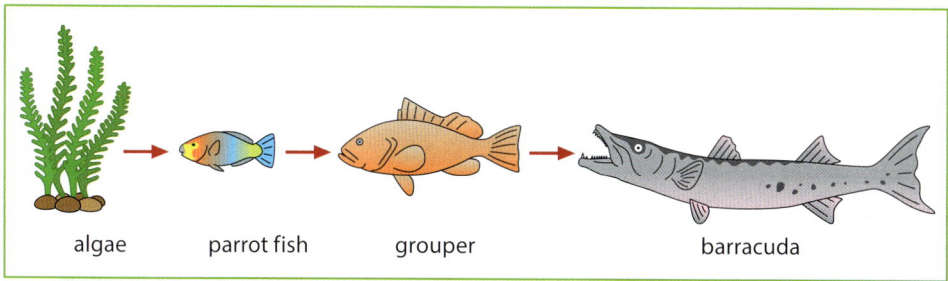

algae → parrot fish → grouper → barracuda

Figure 4.8 Food chain (coral reef)

Many aquatic food chains start with microscopic photosynthetic organisms called **phytoplankton** (plant plankton). These tiny organisms are eaten by the equally microscopic **zooplankton** (animal plankton). These two groups of organisms form the basis of food chains that involve almost every animal in the water, from tiny shrimps to enormous whales. For example, the leatherback turtles, which are such a famous part of Caribbean sea life, feed on jellyfish, and the jellyfish feed on plankton.

Figure 4.9 Coral reefs support a huge variety of life. You can identify many different food chains among the species around a reef.

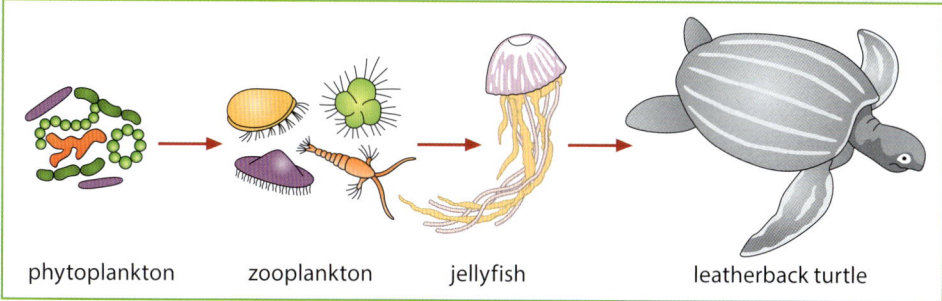

phytoplankton → zooplankton → jellyfish → leatherback turtle

Figure 4.10 Food chain (ocean/saltwater)

The organisms that live in fresh water are also interlinked. For example, a freshwater food chain that you might find in a local pond or river could look like the one below.

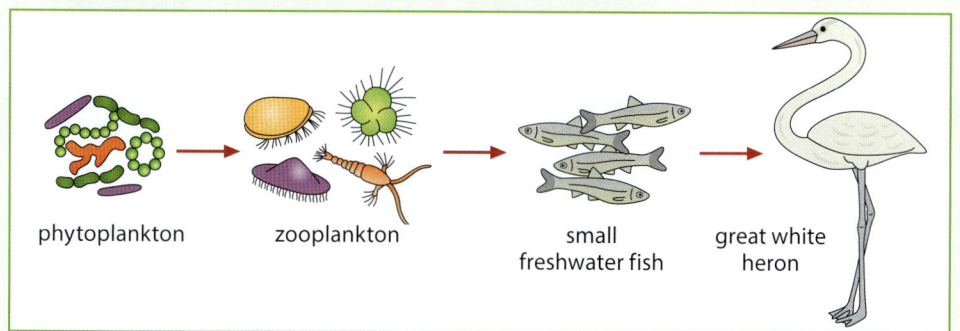

phytoplankton → zooplankton → small freshwater fish → great white heron

Figure 4.12 This great white heron is taking advantage of two different food chains, from the sea and freshwater.

Figure 4.11 Food chain (freshwater)

Food chains 63

The biologist's toolkit: Field studies

Biologists often carry out field studies. A **field study** is an investigation into animals, plants or both in their natural habitat.

Field studies give us a lot of very important information, but we cannot survey everywhere. We must remember to:

- choose carefully the area we are going to sample – do not disturb nesting birds or trample over rare plants
- do a risk assessment – for example, are there stinging plants or insects in the area, are there any poisonous snakes about, or is there a fast-flowing river?
- handle any specimens we collect with care and return them to where we collected them whenever possible.

Field studies cannot be fair tests because they involve many different living organisms, so we need to collect lots of data to make sure our evidence is **repeatable** and **reproducible**.

Activity 4.1

Investigating food chains

Method

Wherever you live or go to school, you will be surrounded by food chains, and animals and plants interacting in their habitats. If you look closely in any small area of habitat – it might be a corner of the school field, a garden or a pond – you will find plants and animals linked together in food chains. In this investigation, you are going to see how many you can find.

You will need:

- trays and containers to store organisms temporarily once you have collected them
- labels for the containers to record where you found the organism and what it is
- a hand lens or viewer
- a net
- forceps
- pooters (if available) to catch small insects

Feeding relationships and the carbon cycle in nature

Method

1. Mark out a small area that you will study (if a land habitat).
2. Carefully collect as many organisms as possible and store them in separate containers (to avoid them eating one another). Remember to collect the plants as well – a single leaf or a sketch will do to help you identify them.
3. Observe each organism carefully. Use the hand lens where it will help. Make a sketch of each organism. Identify each one as well as you can and decide if it is a herbivore or a carnivore.

Hints:
- Herbivores are often slower than carnivores.
- Herbivores are often well camouflaged.
- Herbivores are often found on or close to the plants that they feed on.
- Carnivores often have sharper mouth parts than herbivores.

4. Try and build up as many food chains as you can using the organisms you have found.
5. Then think of several other habitats and try to work out three food chains for each one.

Food webs

Food chains are a great simplification of the situation in the real world. Very few organisms eat only one type of plant or animal, so many organisms appear in many different food chains. Wherever you look, you will find food chains that demonstrate time after time the reliance of animals on plants. The complex interaction of different food chains forms a food web.

For example, let's think about our simple Caribbean food chain (see Figure 4.13).

Plants ⟶ Insects ⟶ Lizards ⟶ Caracaras

Figure 4.13 A simple Caribbean food chain

There are so many different types of plants, and they are not only eaten by butterflies and the caterpillars they produce. Plants, from their leaves and flowers to their fruits and roots, are eaten by a huge range of beetles, bees, crickets, slugs and snails. Hummingbirds feed on the nectar and birds from oropendolas to parrots feast on buds, flowers and fruit. The yellow-headed caracara certainly does not only feed on one type of lizard – it eats lizards, amphibians, snakes, crabs, mammals, large insects, nestlings – and even fruit!

Food chains

So our simple food chain does not really model the feeding relationships of these organisms well. We can try to build up a food web, although even that cannot capture all of the feeding relationships between organisms. However, the food web does give us a better model of what is happening in the real world.

Look at the food web in Figure 4.14 and work out the trophic levels of all the organisms shown.

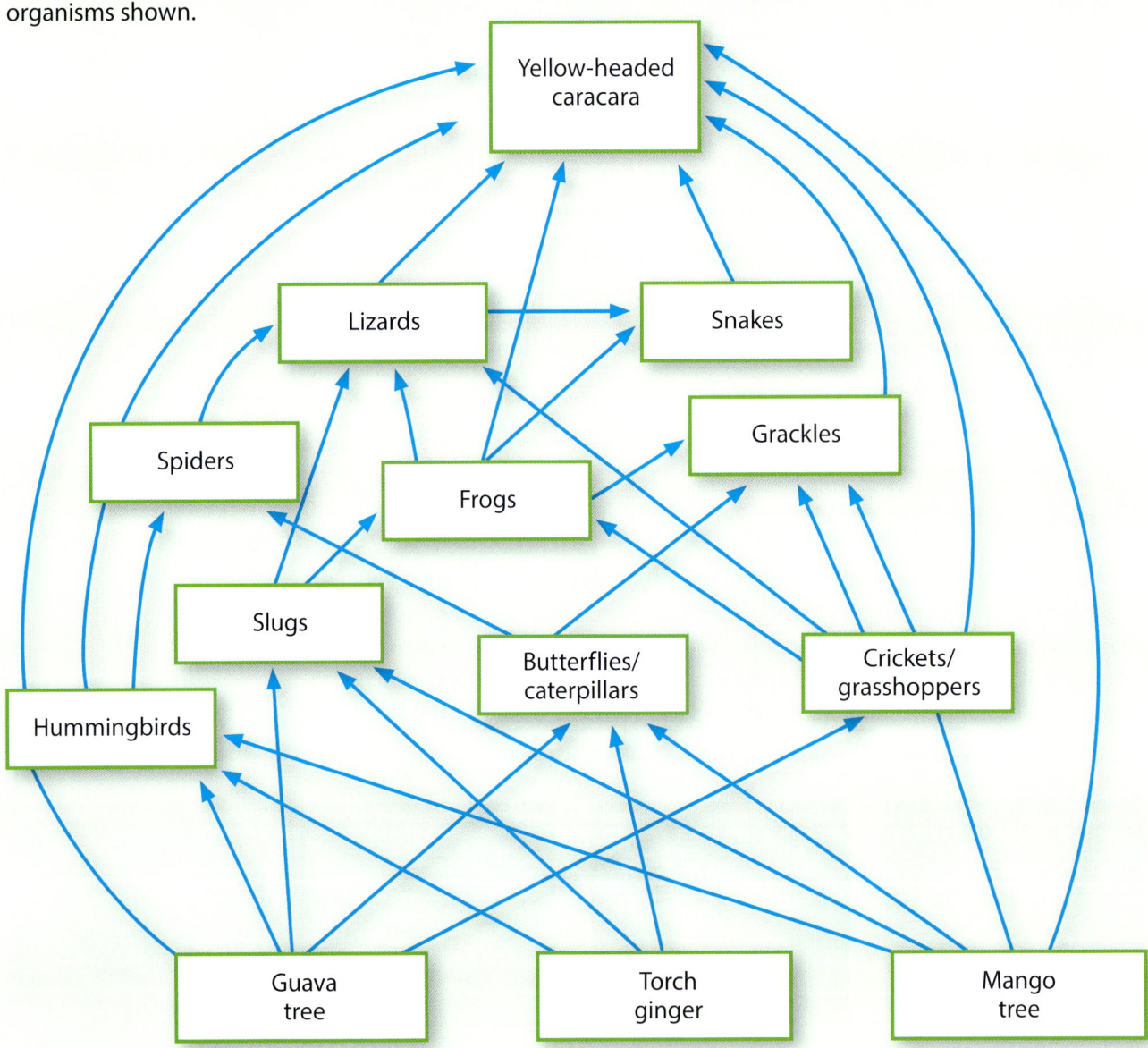

Figure 4.14 This terrestrial food web is made up of many intertwining food chains. It gives a more realistic picture of the feeding relationships between the animals and plants, but it is still very simplified.

Aquatic feeding relationships are also a lot more complex than simple food chains make them look. Figure 4.15 shows you a simple example of a Caribbean marine food web. Can you think of any more organisms to add? Work out the trophic level of all the organisms shown.

Feeding relationships and the carbon cycle in nature

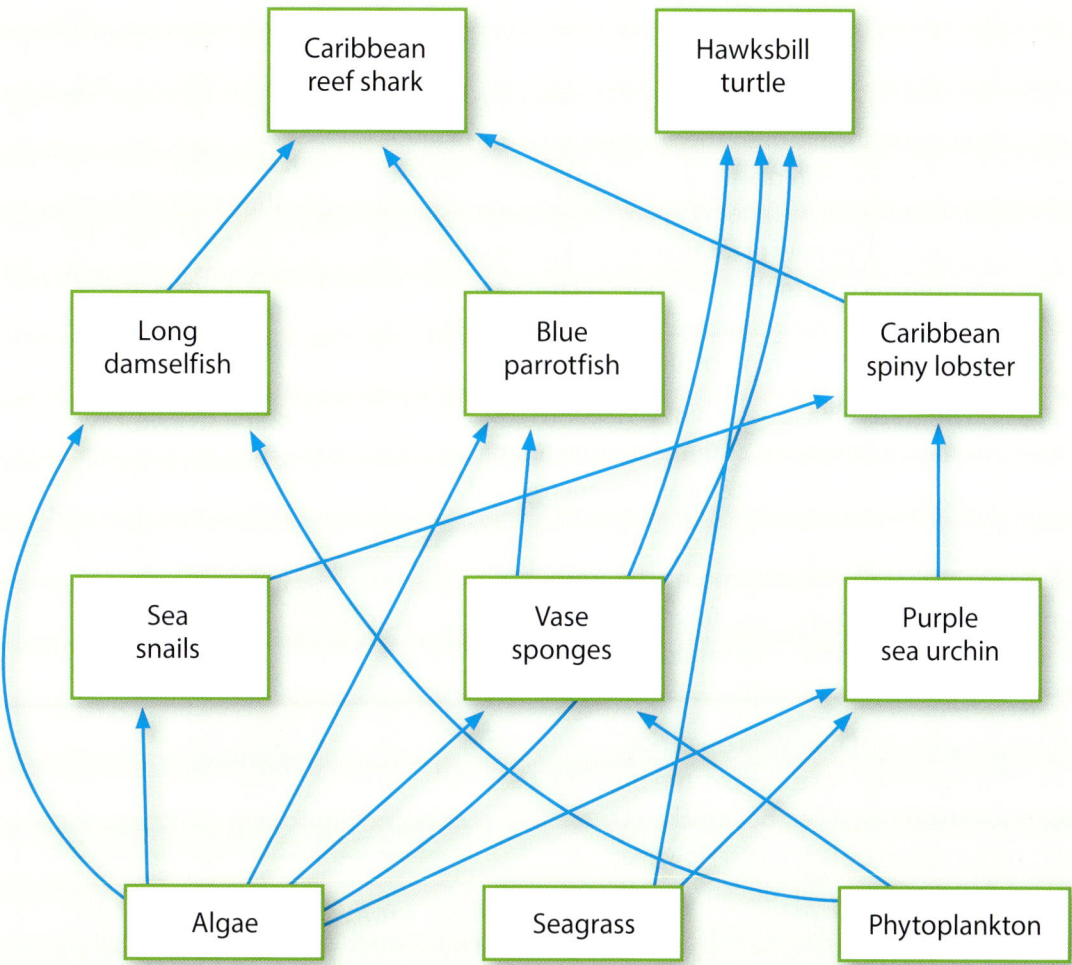

Figure 4.15 A Caribbean marine food web

The biologist's toolkit: Using secondary sources

Sometimes we carry out our own investigations or field studies to collect data. We analyse the data and use it to support our hypothesis, or to show that our hypothesis is not correct.

But often we cannot collect all the data we need. We must use secondary sources.

Secondary sources include data collected by other scientists and presented in different ways. Here are two examples:

- You can observe and investigate feeding relationships in your own country – but you cannot visit every Caribbean country, or every other region of the world. You must rely on data from other sources – from scientific papers or books about different countries to wildlife documentaries – to help you build up models of feeding relationships between organisms in different parts of the world.
- You cannot accurately measure the levels of gases in the atmosphere, or the temperature of the ocean surface around the world. You must rely on large amounts of data collected by scientists to give you the information you need – and you must learn how to identify reliable science from fake news!

Food chains

Energy for life

As you have seen, light is the source of energy for almost all living organisms. Energy from the Sun pours out continually onto the surface of the Earth and a small part is captured by the chlorophyll in plants. It is used in photosynthesis and the energy from the sun is stored in the chemical energy in the substances that make up the cells of the plant. This new plant material adds to the **biomass** of the organism. Biomass is a term that describes all the organic material produced by living organisms. All biomass originally comes from plants as they photosynthesise at the beginning of all food chains.

This biomass is then passed on through a food chain or web into the animals that eat the plants, and then on into the animals that eat other animals. However long the food chain, the original source of all the energy and biomass involved is always the sun.

When you look at a food chain, there are usually more producers than primary consumers, and more primary consumers than secondary consumers. This can be shown as a **pyramid of numbers**.

However, in many cases, a pyramid of numbers does not accurately reflect what is happening. For example, the breadfruit tree can grow to around 20 m tall, yet it can be attacked by mealybugs. They, in turn, are eaten by ladybirds. However, the pyramid of numbers for this food chain doesn't look like a pyramid at all. Cows eat grass, and people eat cows, and this doesn't make a very good pyramid of numbers either.

> **DID YOU KNOW?**
> It has been estimated that plants synthesise around 35×10^{15} kg (35 000 000 000 000 000 kg) of new biological material each year. That's an awful lot of biomass!

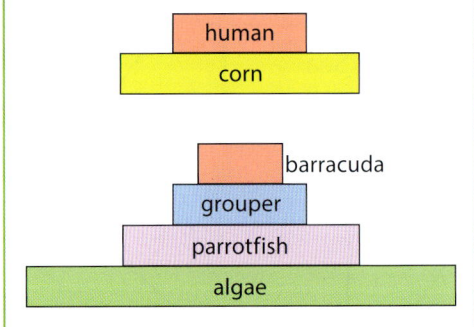

Figure 4.16 A pyramid of numbers like this seems a sensible way to represent a food chain – until we look closer.

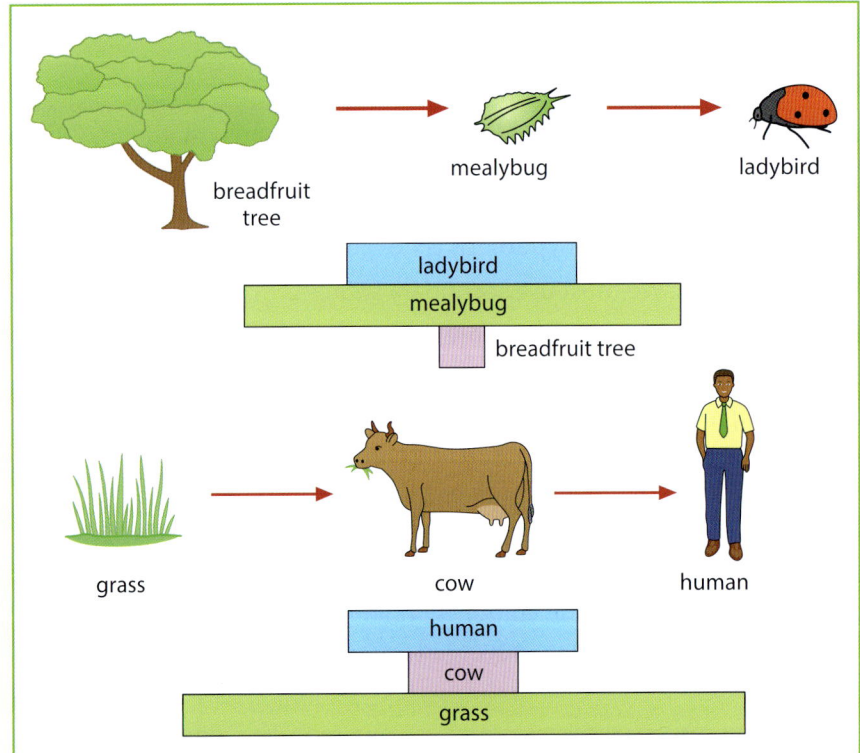

Figure 4.17 These food chains cannot be represented accurately using a pyramid of numbers.

68 Feeding relationships and the carbon cycle in nature

To represent what is happening in food chains more accurately, we use biomass. Biomass is the mass of living material in an animal or plant and, ultimately, all biomass is built up using energy from the sun. The total amount of biomass in the living organisms at each stage of the food chain can be drawn to scale and shown as a pyramid of biomass.

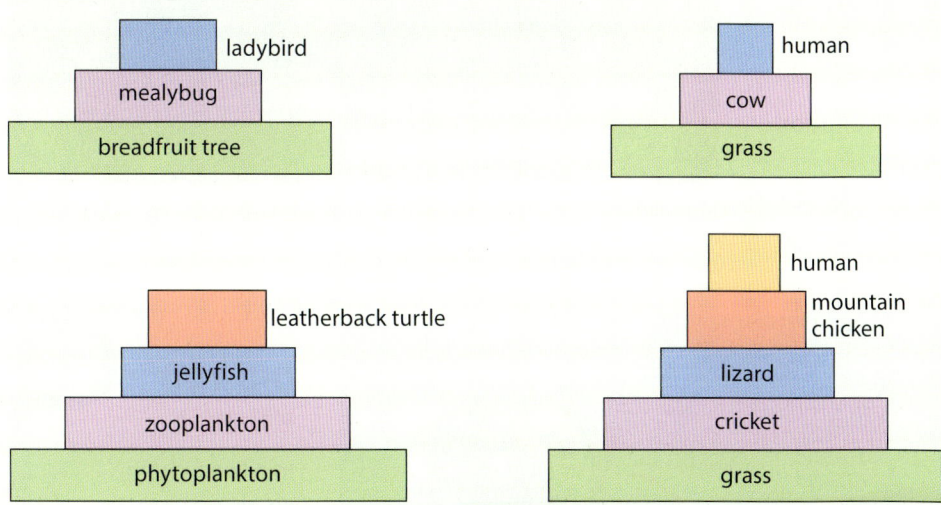

Figure 4.18 No matter what numbers of organisms are involved in a food chain, when the biomass of the different feeding levels is considered, a pyramid of biomass always results.

> ### DID YOU KNOW?
> Counting the number of living organisms is a food chain can be difficult, but measuring biomass is even harder. If the animals and plants are alive, their biomass contains lots of water. Wet biomass is very inaccurate – for example, it is affected by how much water an animal has drunk. Measuring dry biomass is the most accurate measure. Unfortunately to find the dry biomass, the organisms have to be killed and dried, which destroys the food chain you are studying!

The biomass, and so the energy available at each higher trophic level of a food chain, is always less than it was at the previous stage. This is because:

- not all organisms at one stage are eaten by the stage above
- when a herbivore eats a plant, it turns some of the plant material into new herbivore biomass, but much of the biomass from the plant is used by the herbivore to release energy for living and so does not get passed on to the carnivore when the herbivore is eaten.

At each stage of a food chain, the amount of biomass that is passed on decreases – a large amount of plant biomass supports a smaller amount of herbivore biomass, which in turn supports an even smaller amount of carnivore biomass.

Energy reduction between trophic levels

An animal like a zebra eats grass and other small plants. It takes in a large amount of plant biomass and converts it into a smaller amount of zebra biomass. What happens to the rest?

- Not all of the plant material can be digested by the animal, so it is passed out by the body in the faeces.
- Excess protein that is eaten but not needed in the body is broken down and passed out as urea in the urine.

Figure 4.19 The amount of biomass in a lion is a lot less than the biomass in the grass that feed the zebra they prey on. Where does all that biomass go?

Energy for life

- Cellular respiration supplies all the energy needs for the living processes taking place within the body, including movement, which uses a great deal of energy. The muscles use energy to contract, and the more an animal moves about, the more energy (or biomass) it uses from its food.
- Energy produced in cellular respiration is eventually lost as heat to the surroundings. We use energy all the time to keep warm when it is cold or to cool down when it is hot.

Figure 4.20 Animals like horses eat huge amounts of biomass, but they also produce very large quantities of dung containing all of the plant material they cannot digest.

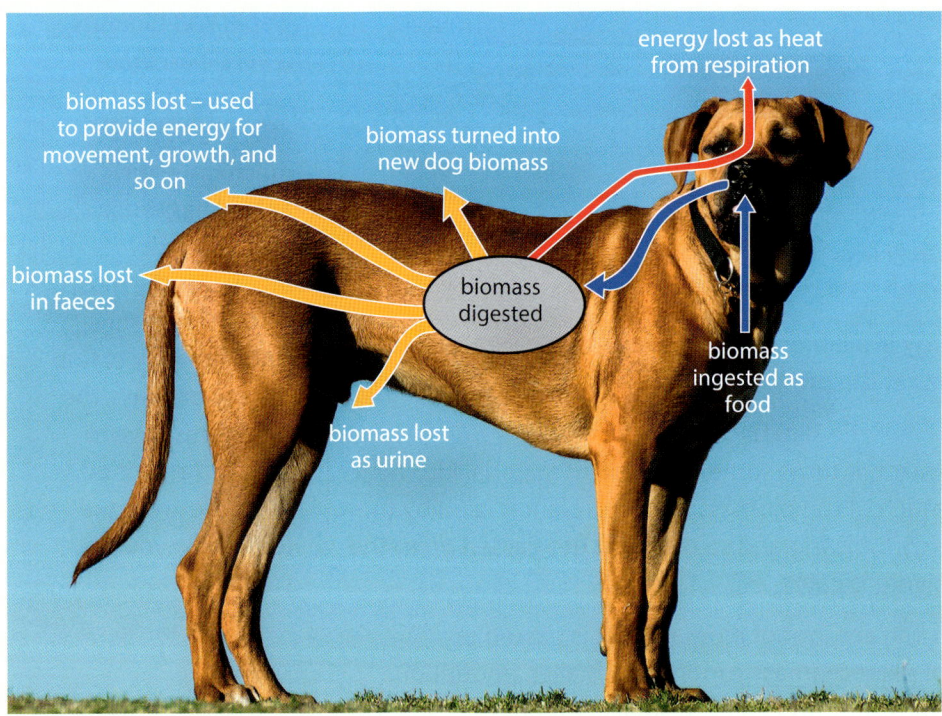

Figure 4.21 Only 2–10% of the biomass eaten by an animal such as this dog will be turned into new dog biomass. The rest will be used or lost in other ways.

What can we learn?

As pyramids of biomass clearly show us, at each trophic level of the food chain less material and therefore less energy is contained in the biomass of the organisms.

There is only a limited amount of the Earth's surface that can be used to grow food. The most efficient way to use this food is to grow plants and eat them directly. If this were the case, then in theory at least, there would be more than enough food for everyone on the Earth to eat their fill.

But at every extra stage we introduce, feeding plants to animals before we eat the food ourselves, the more biomass and energy are lost to humans and the less food there is to go around the human population.

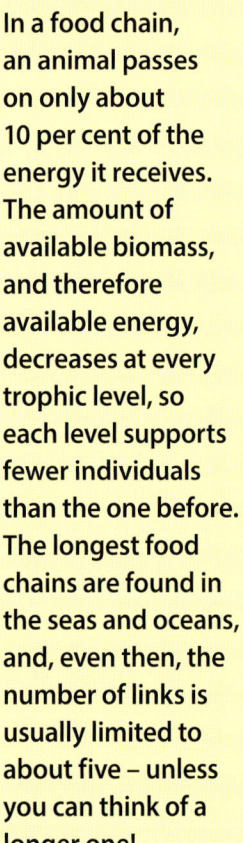

DID YOU KNOW?

In a food chain, an animal passes on only about 10 per cent of the energy it receives. The amount of available biomass, and therefore available energy, decreases at every trophic level, so each level supports fewer individuals than the one before. The longest food chains are found in the seas and oceans, and, even then, the number of links is usually limited to about five – unless you can think of a longer one!

Checkpoint questions

1. Why is a pyramid of numbers not always a useful way to represent a food chain?
2. a. What do pyramids of biomass show about the effect of the number of trophic levels in a food chain on the amount of biomass that is available at the end of the chain?
   b. Explain why the biomass from one stage does not all become biomass for the next stage of the pyramid when it is eaten.

Bioaccumulation

The model of feeding relationships we get through food chains and food webs helps us understand a problem that biologists are observing more and more. Farmers around the world use **pesticides** to kill the animals that affect our crop plants. Pesticides are **toxins** (poisons). Many of these substances break down soon after they are applied, but some remain in the environment and build up in living organisms over time. This process is called **bioaccumulation**.

What happens? Some years ago, scientists observed an increase in the numbers of dead birds of prey such as the American kestrel, and herons. When the scientists tested the dead birds, they found they had high levels of pesticides in their body tissues. They then tested the tissues of the prey animals and found that they also contained pesticide toxins. The older the animal, the more toxins it contained. And the higher up the food chain an animal is, the more toxins it contained.

- The increase in the levels of toxins in an organism over time is called **bioaccumulation**.
- Bioaccumulation happens at every stage of a food chain, so the top predators get the highest levels of toxins – and this is enough to kill them. The passage of toxins along a food chain is called **biomagnification** (see Figure 4.22).

Figure 4.22 Farmers spray crops with pesticides to kill the insects and other animals that eat them.

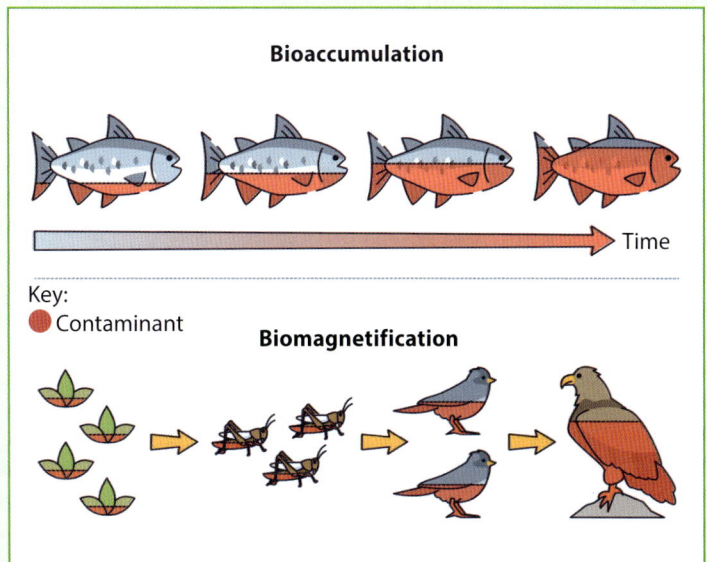

Figure 4.23 Bioaccumulation is when toxins build up in an organism over its lifetime. Biomagnification is how the levels of toxins build up in the organisms along a food chain.

The cycles of nature

Living things are constantly removing materials from the environment. Plants take minerals and water from the soil and carbon dioxide from the air. These materials are then passed on into animals through the food chains and food webs that link all living organisms. If this were a one-way process, then the resources of the Earth would have been exhausted long ago. Fortunately, however, the materials are returned to the environment in the waste products of animals and in the dead bodies of plants and animals.

The nutrients held in the bodies of dead animals and plants, and in animal droppings, are released back into the soil by the action of a group of organisms called **decomposers**. These are microorganisms such as bacteria and fungi. The decomposers feed on waste droppings and dead organisms. They digest them and use some of the nutrients. They also release waste products, which are nutrients that are broken down into a form that plants can use. When we say that things decay, they are actually being broken down and digested by these microorganisms.

As people have developed an understanding of decomposers, they have also developed ways of using them in artificial situations.

As the human population has grown, so has the amount of human waste (**sewage**) produced. Not only is this material unpleasant to live with, it also carries disease. Sewage treatment plants use microorganisms to break down the sewage and make it harmless enough to be released into rivers or the sea for the breakdown to be completed. The treatment plants have been designed to provide the bacteria and other microorganisms with the conditions they need to break down the sewage, particularly a good supply of oxygen.

Another place where decomposers are useful is in the garden. Many gardeners have a **compost** heap. This is where they place grass cuttings, weeds, and sometimes vegetable peelings and trimmings from plants. Then they leave it to let the decomposing microorganisms break down all the plant material into a fine, rich, powdery substance called compost. This process takes at least a year, and the compost produced is full of mineral nutrients released by the decomposers. The compost is then dug into the soil to act as a valuable and completely natural fertiliser.

Figure 4.24 A compost bin – plant material goes in at the top and compost comes out at the bottom.

However, it is in the natural world where the role of the decomposers is most important, and where the cycling of resources plays a vital role in maintaining the fertility of our soil and the health of our atmosphere. In a stable community of plants and animals living in an environment, the processes that remove materials from the soil are balanced by processes that return materials to the soil. In other words, the materials are constantly cycled through the environment. By the time the microbes and detritus feeders have broken down the waste products and the dead bodies of organisms in **ecosystems**, all the energy originally captured by the green plants in photosynthesis has been transferred to other organisms or back into the environment itself as heat or mineral compounds.

You are going to study one of these vital natural cycles – the **carbon cycle**.

Feeding relationships and the carbon cycle in nature

The carbon cycle

All of the main molecules of life – carbohydrates, proteins, fats and DNA – are based on carbon atoms in combination with other elements. There is a vast pool of carbon in the form of carbon dioxide in the air, and dissolved in the water of rivers, lakes and seas. At the same time, carbon is constantly recycled between living things and the environment. This is called the carbon cycle – see Figure 4.24.

- Carbon dioxide is removed from the air by green plants during photosynthesis. It is used to make the carbohydrates, proteins and fats that make up the body of the plant.
- When the plants are eaten by animals, and those animals are eaten by predators, the carbon is passed on and becomes part of the animals' bodies. This is how carbon is taken out of the environment. But how is it returned?
- Respiration is the process by which carbohydrates such as glucose are broken down using oxygen. This process releases the energy needed by the cells of organisms to carry out all the other processes of life. Carbon dioxide and water are the waste products of the reaction. When green plants themselves respire, some carbon dioxide is returned to the atmosphere. Similarly, when animals respire, they release carbon dioxide as a waste product into the air.
- When both plants and animals die, their bodies are broken down by the action of decomposers. When these microbes respire, they also release carbon into the atmosphere as carbon dioxide, ready to be taken up again by plants in photosynthesis.
- When anything that has been living is burnt – whether wood, straw or fossil fuels such as coal, oil or natural gas made from animals and plants that lived millions of years ago – carbon dioxide is also released into the atmosphere in the process of **combustion**.

> **DID YOU KNOW?**
>
> Another important cycle is the **nitrogen cycle**, which plays an essential part in keeping plants and animals alive. Nitrogen is a vital component of amino acids, proteins, chlorophyll, DNA and RNA. As an inert gas, which makes up 80% of the air we breathe, it is added to the soil by nitrogen-fixing bacteria as well as decomposing bacteria, so that it can be absorbed by plant roots. **Legumes** that are rich in proteins actually have nitrogen-fixing bacteria in nodules in their roots. Artificial fertilisers are very rich in nitrogen and even lightning helps make 'fixed' nitrogen available to plants!

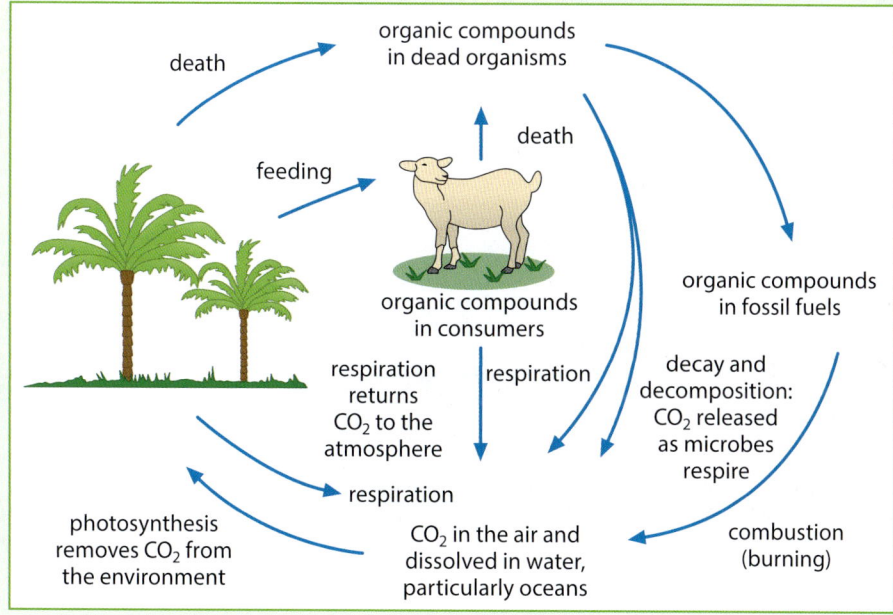

Figure 4.25 The carbon cycle in nature

The cycles of nature

What are fossil fuels?

Fossil fuels are coal, crude oil and natural gas. They are called fossil fuels because they are made from the decomposed remains of organisms that lived millions of years ago. They all contain carbon and hydrogen, and they can all be burned to release usable energy. When we burn fossil fuels, carbon dioxide and water are the major waste products.

Coal is made from fossilized plants. Oil and natural gas are formed from the decomposition of marine organisms trapped under layers of rock.

Figure 4.26 The burning of fossil fuels is one of the major causes of global warming, which could have extreme consequences for Caribbean countries.

Globally, we burn millions of tonnes of fossil fuels every year to generate electricity, cook, power our cars, ships and planes and, when needed, to heat our home. The carbon dioxide produced by this combustion, combined with other human behaviour, is causing huge problems and leading to a level of global warming that may threaten the lives of people everywhere, especially people on island nations like our Caribbean countries.

The carbon cycle and global warming

The carbon cycle regulates itself. However, we humans are having a major effect on the environment. We have cut down millions of acres of trees and burnt them to clear the rainforests for agriculture – for example to grow oil palms and to raise cattle for cheap meat.

Figure 4.27 Clearing the rainforests for agriculture has a negative effect on the carbon cycle, which also contributes significantly to global warming.

Feeding relationships and the carbon cycle in nature

At the same time, we burn huge amounts of fossil fuels, producing enormous quantities of carbon dioxide. As these human activities pour increasing amounts of carbon dioxide into the atmosphere, scientists are observing that the carbon cycle is becoming distorted, and that the climate of the Earth is changing as a result.

The biologist's toolkit: Analysing data

As biologists, you need to analyse data. It is very important to recognise data that is reliable and valid – and to recognise data that you should not trust.
- Who carried out the work – is it a reputable source?
- Who funded the research? If a chocolate company carries out research to tell you that eating chocolate is good for you – be suspicious. If a long-term university study says the same thing – buy that chocolate!
- How big is the sample studied? In general, the bigger the data set, the more reliable the results, so a study involving 20 000 people will be more reliable than a study involving 20 people.
- How long did the study last? Often – although not always – studies that continue over a longer period of time are more reliable than very quick studies.

The data in Figure 4.28 is an example of highly respected, reliable data collected in the same place, in the same way, over many years.

The level of carbon dioxide in the atmosphere is rising steadily. This trend has continued for many years – see Figure 4.28.

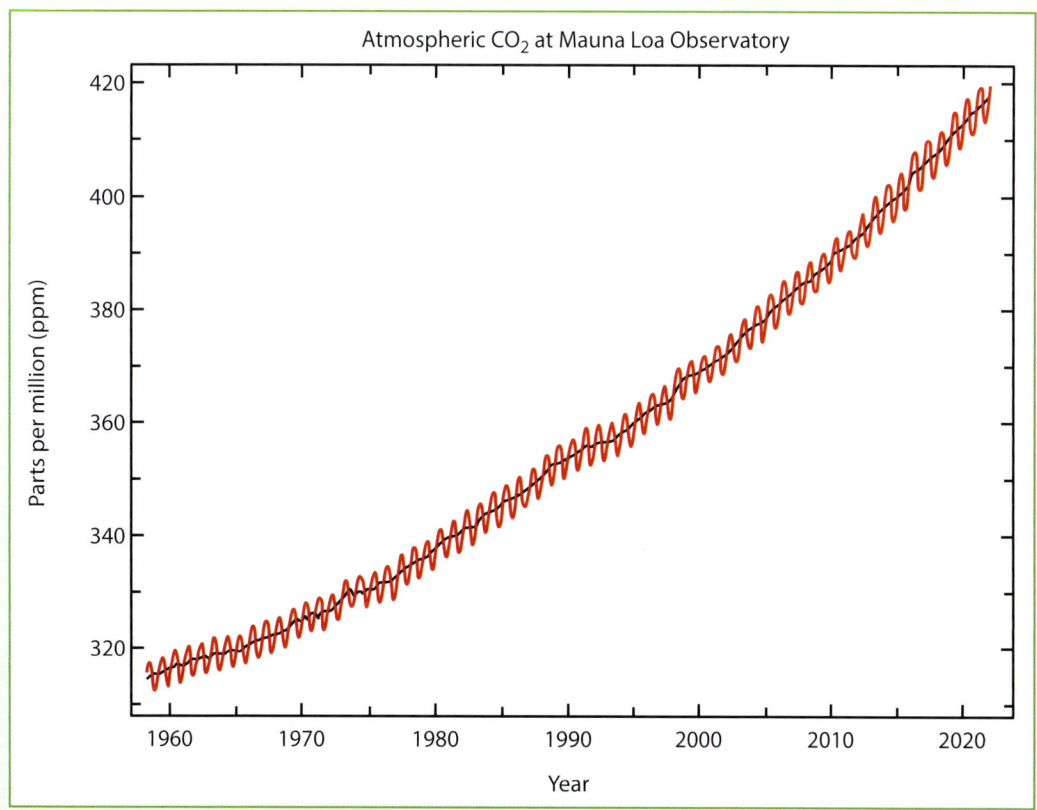

Scripps Institution of Oceanography at UC San Diego.

Figure 4.28 This graph shows the increasing level of carbon dioxide in our atmosphere. Readings have been taken from the same place (Mauna Loa Observatory, Hawaii) for over 60 years. Scientists all over the world respect this data.

An increase in carbon dioxide in the atmosphere acts like a greenhouse, trapping energy from the sun and causing a rise in the temperature at the surface of the Earth. This in turn causes more extreme weather events, such as more frequent and stronger hurricanes. It also causes the ice caps to melt, so sea levels rise. This threatens the existence of some island nations. Global warming also threatens the well-being of humans because some areas of the world are becoming too hot to grow crops. Diseases spread by insects such as malaria and dengue fever are also spreading to more countries around the world. We need to take action to stop global warming now.

Figure 4.29 Global warming causes more extreme weather events like hurricanes, which inflict enormous damage on Caribbean countries.

Checkpoint questions

3. Explain why the carbon cycle is so important for the continuation of life on the Earth.
4. It can be said that no part of our body is truly our own – we have borrowed the materials and at a later date they will be used again elsewhere. Discuss this statement.

Possible SBAs based on feeding relationships and the carbon cycle in nature

- Investigate local food chains and observe how they change between the wet and dry seasons.
- Build up a complex food web of organisms in your local area.
- Use secondary sources to investigate different feeding relationships around the world.
- Investigate an element of a food chain in the laboratory, for example caterpillars feeding on leaves, and measure the changes in biomass of the plant and animals involved over time.
- Investigate the production of carbon dioxide/burning of fossil fuels in your local community/island and plan how the production of carbon dioxide could be reduced.
- Investigate the conditions that affect the rate of decomposition of, for example, particular types of fruit or leaves.

End-of-chapter summary

In this chapter, you have learnt that:

- radiation from the sun is the source of energy for all communities of living organisms. it is captured by green plants in photosynthesis
- people and all other animals are dependent on plants either directly or indirectly for food
- the relationships between plants and the animals that feed on them and one another can be expressed as food chains
- a food chain starts with a producer (a plant) and may have primary, secondary and tertiary consumers, and so on
- the different stages of a food chain are known as trophic levels
- the feeding relationships between living organisms are more complex than simple food chains; we show them using food webs
- the relationships between the organisms in a food chain can be represented by a pyramid of numbers, but this is of limited use
- the mass of living material (the biomass) and the energy at each trophic level of a food chain is less than it was at the previous stage
- the biomass at each trophic level of a food chain can be drawn to scale and shown as a pyramid of biomass
- the efficiency of food production can be improved by reducing the number of stages in food chains
- living organisms remove materials from the environment as they grow and return them when they die through the action of the decomposers
- dead materials decay because they are broken down by microorganisms
- in a stable community, the processes that remove materials from the environment are balanced by processes that return materials to it
- the carbon cycle is an example of the cycling of resources in nature
- human activities, such as burning fossil fuels (coal, oil and gas made from the decomposed remains of organisms that lived millions of years ago), are releasing large amounts of carbon dioxide into the atmosphere
- at the same time people are cutting down trees that naturally absorb carbon dioxide from the atmosphere for cheap farmland
- the increasing level of carbon dioxide in the atmosphere is leading to global warming, which has many negative effects on human well-being.

End-of-chapter questions

1. Which of the following organisms is NOT a producer?
   - A grass
   - B frangipani
   - C breadfruit
   - D mushroom

2. Which of the following animals is a herbivore?
   - A rat
   - B person
   - C sheep
   - D coyote

3. Which of the following animals is NOT an omnivore?
   - A rat
   - B human
   - C chimpanzee
   - D cat

4. There are a number of trophic levels in a food chain. An organism on the third trophic level would always eat:
   - A another animal
   - B seaweed
   - C plants
   - D fungi

5. Which of the following type of organism is NOT a decomposer?
   - A worms
   - B mouse
   - C bacteria
   - D fungi

6.
   - a Describe a food chain.
   - b Draw three food chains that you might find in the Caribbean.
   - c Explain why a food web is a more accurate way of representing the feeding relationships than food chains.

7.
   - a Explain the following terms used in describing feeding relationships:
     - i producer
     - ii primary consumer
     - iii secondary consumer
   - b Name:
     - i **two** producers
     - ii **two** primary consumers
     - iii **two** secondary consumers

8 Pyramids of numbers and pyramids of biomass are both ways of representing a food chain.
   a   Describe a pyramid of numbers.
   b   A pyramid of numbers is not always a useful way to represent what is happening in a food chain. Explain why not and give an example.
   c   A pyramid of biomass is a much more useful way to represent what is happening in a food chain. Describe a pyramid of biomass.
   d   Explain why, for some food chains, a pyramid of biomass is so much more informative than a pyramid of numbers. Give an example.

9 All around the world – including the Caribbean – people eat a huge amount of chicken. In many countries, to supply the millions of chickens eaten each week, the birds are raised in broiler houses, with about 20 000 chickens in each one. There are between 12–18 birds per square metre. The broiler houses are kept warm (around 25 °C), at the same temperature day and night, and there is good ventilation to get rid of the moisture produced by the birds and the smell. There are automatic feeders that deliver the correct amount of food for the size of the birds, and a plentiful supply of water.
   a   Although chickens in broiler houses grow fast – they are ready for the table in around six weeks – the mass of food they take in is far greater than their eventual body weight (biomass). Describe what happens to the food that is not turned into biomass.
   b   Explain how the conditions in the broiler house help to make sure that as much of the food as possible is turned efficiently into chicken biomass.

10 a   Draw and label a diagram of the carbon cycle in nature.
   b   Explain how different elements of the carbon cycle are affected by human activities.
   c   Discuss the impact of burning fossil fuels on the carbon cycle, explaining how human well-being may be affected.

End-of-section questions

1 a List six characteristics of all living organisms. (3)
    b In which three ways do human cells differ from those of plants? (3)
    c Draw a fully labelled diagram of the structure of a typical (unspecialised) animal cell and describe the function of each part. (9)

2 a Draw and label the structure of a typical plant cell. (7)
    b List two structures found only in plant cells and explain their functions. (4)
    c Name two structures that are found in all cells and which can only be seen using an electron microscope. (2)
    d If a plant cell is flaccid, which process has caused this to happen? Explain, using diagrams, what must happen for this cell to become turgid. (4)

3 a Name one part in the human body where osmosis is important. (2)
    b Describe what happens in the part you named in (3a). (4)
    c A membrane is said to be 'partially permeable'. Explain the role that this membrane plays in osmosis. (4)
    d With the help of a labelled diagram, explain what has happened if a plant cell is plasmolysed. (5)

4 a Draw a food chain starting with phytoplankton and ending with humans. Use flora and fauna from the Caribbean as far as possible. (6)
    b List three processes that release carbon dioxide into the atmosphere. (3)
    c Explain why the carbon cycle is so important to organisms on the Earth. (3)
    d Describe fossil fuels and give three ways in which people use them. (5)
    e Explain why it is so important that people reduce the of burning fossil fuels as much as possible and as soon as possible. (5)

5 a Define biomass. (2)
    b An 8-month-old calf eats a large amount of grass each day. Explain what happens to this grass after it has been ingested. (6)
    c How is a food chain different to a food web? (2)
    d Draw a food web using Caribbean organisms. (5)

Data analysis

The diagram on the right shows the rate of photosynthesis against temperature.

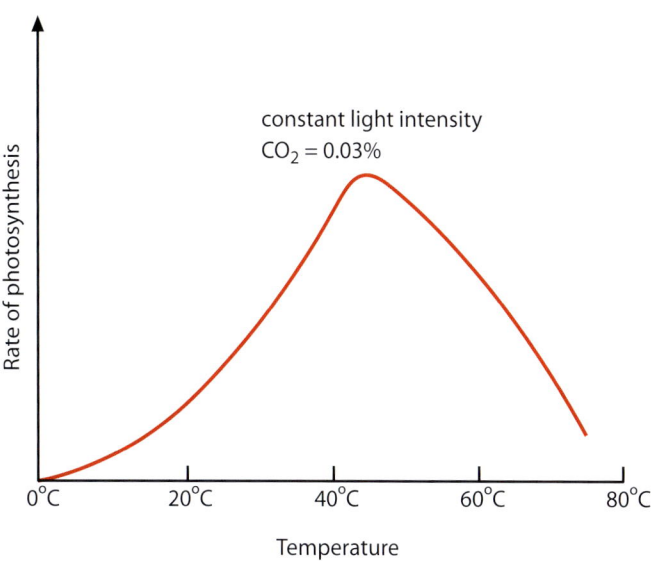

1. Temperature affects photosynthesis as shown in the graph above. At which temperature is the rate of photosynthesis the highest? (2)

2. What is the rate of photosynthesis at 0 °C? Why do you think this is? (3)

3. What happens to the rate of photosynthesis between 60 °C and 80 °C? Explain your answer. (4)

Revision toolkit 1

1. Make a list of the topics in this section of the textbook.

2. Select one topic (you will go through each one later on). On a blank sheet of paper (you may need more), write the topic in a box in the centre.

3. Use this as the centre of a concept map and write every important detail you can remember about the topic. Only write the main ideas. Explain these ideas with examples and illustrations.

4. Look at the information in the textbook and the additional information on the companion website, and match the information there against what you have on the concept map. Is there any missing information?

5. In a different colour, write the main ideas of the information that you missed on the concept map.

6. Read through the information a few times. Remember to include examples and illustrations as you explain these ideas to yourself, especially if you have examples from your own experiences.

7. Now that you feel more comfortable with the topic, it is time to apply it to a question. Select a relevant exam-type question from this book or the companion website, and work through it. Go through the answer with a peer or your teacher, and see how well you did.

Section B: Life processes

Learning outcomes

At the end of Section B, you will:

- understand the role of nutrition in helping humans obtain their energy and satisfy their physical needs
- understand that respiration is the means by which energy is made available for carrying out life processes
- understand the role of transport and defence in humans
- understand the mechanisms of movement and appreciate its role(s) in humans
- understand the process by which humans get rid of metabolic waste and maintain homeostasis
- understand that humans detect and respond to changes in their external and internal environment
- understand the processes by which life is perpetuated.

The human body is an amazingly complex organism. It takes many different systems all adapted to carry out very specific jobs to enable us to live and reproduce successfully over a lifespan that can (occasionally) be over 100 years. What sort of challenges does the human body face – and how do we overcome those challenges?

Our body is constantly working to stay upright against the force of gravity, and at the same time we need to move around to look after ourselves, find food, work and socialise. This is where the skeleton and muscles of our body come in. However, for the muscles to contract and move us around, they need energy. The food we eat provides this energy, but only after our digestive system has broken it down into useable molecules and our circulatory system has carried the digested food to the cells.

Even then, our cells need oxygen to break down the food molecules and make the energy available to use. Our respiratory system brings fresh supplies of oxygen-rich air into our body at regular intervals. Then the blood in our circulatory system carries this oxygen to the cells and removes the poisonous carbon dioxide produced by our cellular respiration.

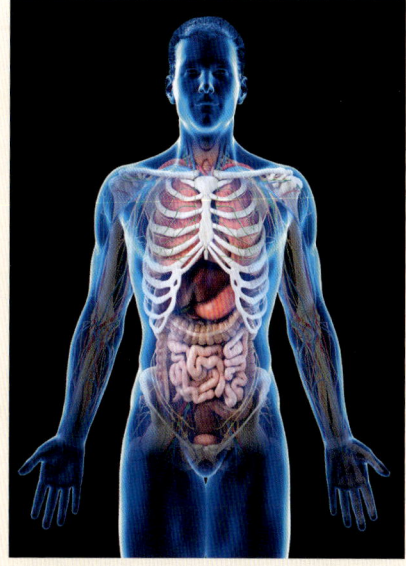

Figure B1 The human body – an amazing combination of complex life processes

For our body to work properly, we need to know what is happening in the world around us and inside us. We need rapid reactions to dangerous situations, and long-term control of body functions such as growth. All of these systems working in our body result in the formation of waste products. Our lungs and kidneys between them manage to get rid of most of the waste that is made.

All this complex body chemistry only works if the internal conditions are kept very stable, within very narrow limits. Our body has a very effective system called **homeostasis**, which makes sure our internal conditions are kept fairly constant almost regardless of what we do.

And finally, to be successful, all living organisms need to be able to reproduce themselves – and human beings are no exception to this rule. Our body is capable of producing an entirely new human being over the course of nine months in an astonishing sequence of events.

Chapter 5: Nutrition

When you have completed this chapter, you will be able to:

- explain what is meant by a balanced diet and what it involves, including macronutrients, micronutrients, fibre and water
- describe the causes, symptoms and treatment of deficiency diseases
- discuss the effects of malnutrition on the body
- determine Body Mass Index and waist circumference, and use both to explain the factors and implications of obesity
- carry out laboratory tests to identify different nutrient groups in a food sample
- describe the major nutrients needed by the human body and their sources
- state the function and the main sources of some of the vitamins and minerals needed by the human body

What the examiners say

→ Candidates seem to have a satisfactory understanding of malnutrition, but seem unable to speak to the effects of the disorders related to nutrition.

→ There seems to be a general belief that water not only provides nutrients, but is also a source of energy.

→ Candidates often confuse the minerals and trace elements, and are generally unsure about the effects of mineral deficiencies.

→ Candidates also display a reluctance to use the appropriate terms to describe scientific processes.

People, like all living organisms, need a source of energy to survive. In our case, this is our food. We are heterotrophs – we cannot make our own energy supply by photosynthesis, so we have to eat other living things.

Throughout human history, almost anything that can be eaten has been eaten, and around the world the variety of food that people consume is still quite amazing. However, it doesn't matter what the food is – from saltfish and plantain to pickled cabbage, from ackee and breadfruit to caviar and cranberries – as long as it contains the right balance of chemicals to provide your body with everything it needs to live, grow and reproduce.

Figure 5.1 The foods we eat need to contain the chemicals our body needs to live, grow and reproduce.

The human diet

Each one of us has to take in all the chemicals we need from the food that we eat. We use our food in three main ways:
- To provide energy for our cells to carry out all the functions of life
- To provide the raw materials for the new biological material needed in our body for growth, and also to repair and replace damaged and worn out cells
- To provide the resources needed to fight disease and maintain a healthy body.

Some types of food are needed in large amounts. These are called macronutrients. The main **macronutrients** are **carbohydrates**, **proteins** and **lipids**. Other substances are equally important in our diet, but only in tiny amounts. They are called the **micronutrients** and include **minerals** and **vitamins**.

In this chapter, you will look at all of the most important components of a healthy human diet, and discover what goes wrong if any of the components are missing.

Macronutrients

The important macronutrients are carbohydrates, proteins and lipids.

Figure 5.2 Food comes in all shapes and sizes, but whatever it looks like, the molecules it contains are surprisingly similar.

> ### The biologist's toolkit: Key chemistry: condensation reactions
>
> To be a good biologist, you need to understand some chemistry. Some types of reaction are very common in biology, for example, **condensation** reactions.
>
> In a condensation reaction, two molecules join together to form one larger molecule and a molecule of water (H_2O).
>
> **molecule 1 + molecule 2 ➜ molecule 3 + water**
>
> Many important biological molecules such as carbohydrates, proteins and lipids are built up in a series of condensation reactions.

Carbohydrates

Carbohydrates provide us with energy. Most of the carbohydrates we take into our body is broken down into glucose, which is used in cellular respiration to produce energy in a form that can be used in all our cells (see Chapter 7). The body stores very little carbohydrate apart from **glycogen**, which is found in the liver and muscles. Any excess carbohydrates that we eat is converted to fat, which is stored all too easily in the body.

Carbohydrates are made up of different combinations of carbon, hydrogen and oxygen. The most commonly known carbohydrates are the sugars and starches.

You will already be familiar with a few types of sugar: the sugar that is such an important product of many Caribbean countries is called **sucrose**; glucose is the sugar you have met as the main product of photosynthesis and it is vital in cells for energy. It is also the source of energy in sports and health drinks.

Another more complex carbohydrate called starch is a storage carbohydrate in plants, and it is commonly found in potatoes and flour. Carbohydrate-rich foods include anything containing sugar or flour, such as bread, biscuits and cakes. Plantains, breadfruit, rice and pasta are also carbohydrate-rich foods.

The basic structure of all carbohydrates is the same. They are made up of carbon, hydrogen and oxygen. They fall into three main types, depending on the complexity of the molecules:

- monosaccharides
- disaccharides
- polysaccharides.

Monosaccharides and disaccharides

The best known **monosaccharide** is glucose, which has the chemical formula $C_6H_{12}O_6$. There are lots of other simple sugars, including fructose, which is the sugar found in fruit. The name 'monosaccharide' means 'single sugar'.

Disaccharides are made up of two monosaccharides joined together, and sucrose (the substance we know as sugar) is one of the most common. The two sugars join together in a **condensation reaction**. The two molecules become joined and a molecule of water is lost.

monosaccharide + monosaccharide → disaccharide + water

When different monosaccharides join together, different disaccharides result. Table 5.1 shows some of the more common ones. Most monosaccharides and disaccharides have two important properties in common: they dissolve in water and they taste sweet.

Figure 5.3 Sucrose is used as a sweetener.

Figure 5.4 All these foods contain a lot of carbohydrates.

The biologist's toolkit: Making a table

Tables are a very useful tool for a biologist. We use them to record the results of experiments and to display information or data clearly so that it is easy for people to read and understand.

A table may be based on words, for example, Tables 5.1, 5.2 and 5.3 in this chapter, OR it may be based on numbers, for example, Table 5.4 in this chapter.

Drawing a table

- Make sure you have the right number of columns and the space to fill in all your data.
- Make sure each column of your table is clearly labelled.
- Remember that, as you are recording measurements or numerical values in a table, you should put the units in the column headings, for example **Mass (kg)**, so that you don't have to fill in the units on every line.

Table 5.1 Three common disaccharides

Disaccharide	Source	Monosaccharide units
Sucrose	Stored in plants such as sugar beet and sugar cane	glucose + fructose
Lactose	Milk sugar – this is the main carbohydrate found in milk	glucose + galactose
Maltose	Malt sugar – found in germinating seeds such as barley	glucose + glucose

Figure 5.5 Sports drinks contain glucose, which gives a tired athlete some energy – fast.

Polysaccharides

The most complex carbohydrates are the **polysaccharides**. Unlike the sugars, they do not taste sweet, and do not dissolve in water. Many sugar **monomers** (single units) join together in condensation reactions to form some complex **polymers** (long-chain molecules made up of lots of smaller repeating units). These polymers or polysaccharides have some very important biological properties. Polysaccharides often form very compact molecules, ideal for storing energy. The sugar units are released when needed to supply energy. Polysaccharides are also physically and chemically very inactive, so storing them does not interfere with the other functions of the cell.

Testing for carbohydrates

There are a number of chemical tests that you can carry out to test for the presence of carbohydrates of different types. You have already met the test for starch in Chapter 3 when you tested plant leaves for it using iodine solution. In the same way, food containing starch will turn the reddish-brown iodine solution blue-black.

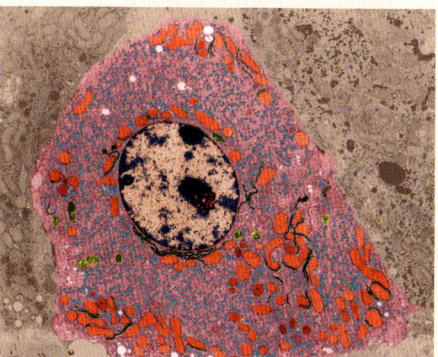
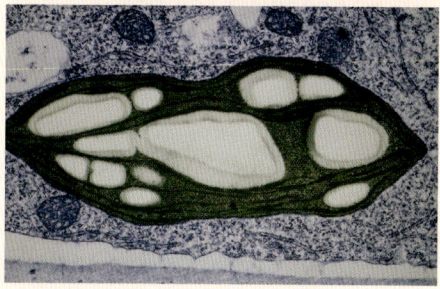

Figure 5.6 Storage carbohydrates are important in both animal and plant cells. The top photograph is of a liver cell with glycogen granules. The bottom photograph shows a chloroplast with starch grains.

The biologist's toolkit: Key chemistry: oxidizing and reducing

Two more important types of reactions for biologists to remember are **oxidation reactions** and **reduction reactions**.

Oxidation is when a molecule **gains oxygen**, **loses hydrogen** or **loses electrons**.

For example: magnesium + oxygen ➜ magnesium oxide
The magnesium has gained oxygen and been oxidised.

Reduction is when a molecule **loses oxygen**, **gains hydrogen** or **gains electrons**.

For example: copper oxide + hydrogen ➜ copper + water
The copper has lost oxygen and been reduced; the hydrogen has gained oxygen and been oxidised.

The human diet

All the simplest sugars, the monosaccharides, and some of the double sugars or disaccharides are called **reducing sugars**, because they reduce a special reagent called Benedict's solution. All the other sugars are called non-reducing sugars.

Activity 5.1

Using Benedict's test for reducing sugars

Some sugars react readily with Benedict's solution. They reduce copper(II) ions to copper(I) ions, and for this reason are called reducing sugars. This provides a straightforward chemical test for the reducing sugars. The reducing sugars include all of the monosaccharides and some disaccharides.

You will need:

- a Bunsen burner
- a tripod, gauze and heatproof mat
- a large beaker half-filled with water
- some glucose powder
- boiling tubes
- Benedict's solution
- different food samples to analyse, for example bread and fruit

Method

1. Bring the water in the beaker to the boil using the Bunsen burner.
2. In one boiling tube, add water to a depth of about 2 cm. This will act as your control.
3. In another tube, add a sample of glucose powder and water to a depth of about 2 cm.
4. Place any food samples to be tested in other boiling tubes in the same way.
5. Add a few drops of Benedict's solution to each boiling tube – enough to colour the mixture blue.
6. Place the tubes in the boiling water and leave for several minutes. TAKE CARE with the boiling water.
7. If a reducing sugar is present, the clear blue solution will change as an orangey red precipitate appears.
8. Write up your method and results, including the different foods you analysed.

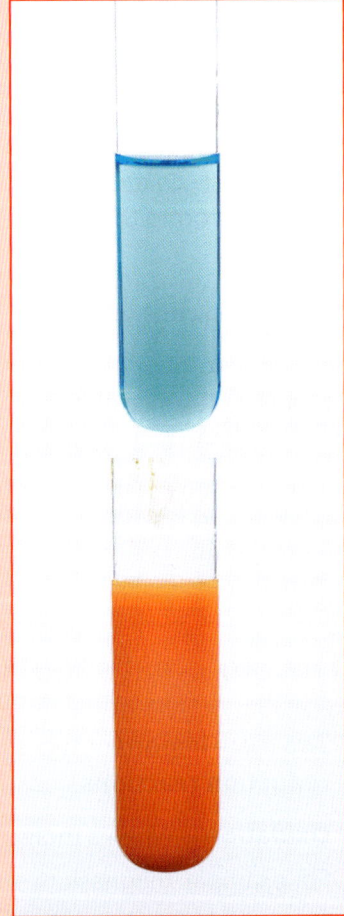

Figure 5.7 Results of the Benedict's test for reducing sugars before and after heating

Nutrition

If you heat pure sucrose with Benedict's solution, it does not react because it is a non-reducing sugar. However, a solution with no carbohydrates in it at all will also give a negative Benedict's test. How can you tell if there is a non-reducing sugar present? Fortunately, hydrochloric acid will break down sucrose into simple, reducing sugar units. It will not have this effect on proteins or lipids. So you can test for the presence of a non-reducing sugar using Activity 5.2.

Activity 5.2

Testing for a non-reducing sugar

You will need:

- a Bunsen burner
- a tripod, gauze and heatproof mat
- a large beaker half-filled with water
- some sucrose powder
- boiling tubes
- Benedict's solution
- dilute hydrochloric acid
- sodium hydrogen carbonate solution
- different food samples from Activity 5.1 that did not show the presence of reducing sugars

Method

1. Bring the water in the beaker to the boil using the Bunsen burner.
2. In one boiling tube, add water to a depth of about 2 cm. This will act as your control.
3. In another tube, add a sample of sucrose and water to a depth of about 2 cm.
4. Place any food samples to be tested in other boiling tubes in the same way.
5. Add a few drops of hydrochloric acid to each tube. Place them in the beaker of boiling water and boil the mixture for about two minutes.
6. Allow the tubes to cool a little and then add a few drops of sodium hydrogen carbonate solution to neutralise the acid in each tube. (When the mixture no longer fizzes when you add the sodium hydrogen carbonate, the contents are neutral.)
7. Now add Benedict's solution to each tube and carry out the test for reducing sugars as normal.
8. If a non-reducing sugar such as sucrose is present, it will have been hydrolysed into its monosaccharide units by the acid, and so these reducing sugars will now react with the Benedict's solution and show up.
9. Write up your method and the results for the different foods you analysed.

Sugar in the diet

People like sweet things, and sweet things are not bad for us. However, if we eat too many carbohydrates, they are converted to fat and stored. So, too many carbohydrates in our diet can lead to health problems like obesity (see later in this chapter).

Sugar is not just the sugar we might add to our tea or coffee. Many processed foods and drinks contain hidden sugars. Sweetened beverages – colas and pops – often contain very large amounts of sugars. These sugary drinks not only make us gain weight, they also damage our teeth as you will see on page 125.

Fruits, like the mangoes, soursop and guava we grow and love, and vegetables, like sweet potatoes, also contain hidden sugars. But these fresh foods also contain lots of micronutrients, fibre and water – so, as you will learn shortly, they have many benefits for our health and we should eat plenty of them every day.

Figure 5.8 These drinks may look fruity and healthy, but they are full of hidden sugar and have very little nutritional value. The soursop and mangoes also contain sugars, but are full of other valuable nutrients as well.

Proteins

Proteins are used for body-building. They are broken down during digestion into small units called amino acids, which are then rebuilt to form the proteins needed by the body. Protein-rich foods include all meat and fish, eggs, dairy products such as cheese and milk, as well as pulses like pigeon peas, black-eyed peas and red beans.

Figure 5.9 Meat is a very good source of protein – but many other foods are high in protein too.

Seventeen to eighteen per cent of our body is made up of protein – a high percentage, second only to water. Our hair, skin, nails, the enzymes that control all the reactions in our cells and digest our food, many of the hormones that control our organs and their functions, our muscles, and many more substances are made of these complex molecules. By understanding the way in which protein molecules are made up and the things that affect their shape and functions, you can begin to develop an insight into the biology, not only of your cells, but also of all living things.

Just like carbohydrates, proteins are made up of the elements carbon, hydrogen and oxygen, but in addition, they also contain nitrogen. Some proteins also contain sulfur and various other elements. Proteins are polymers, made up of many small units called amino acids, joined together. In the same way that monosaccharide units join together to form polysaccharides, amino acids combine in long chains to produce proteins. There are about 20 different naturally occurring amino acids and they can be joined together in any combination.

> **DID YOU KNOW?**
> There are some amino acids that you must eat as part of your diet because your body cannot make them. They are called **essential amino acids** and if they are lacking in your diet for too long, you will show deficiency **symptoms** and can die.

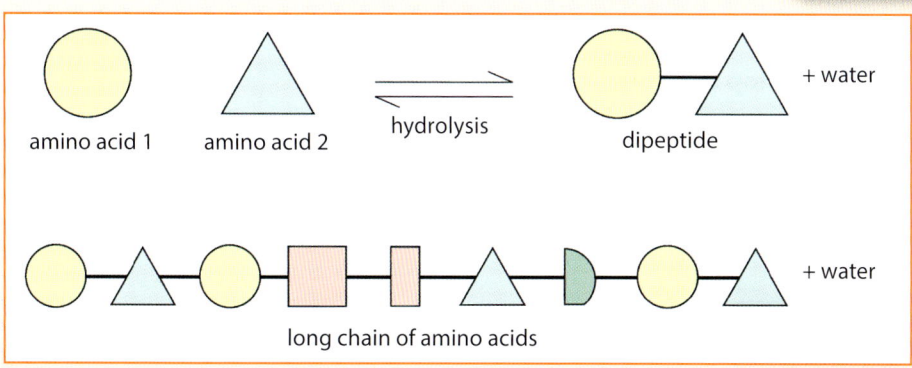

Figure 5.10 Amino acids are the building blocks of proteins and they can be joined in a huge variety of ways to produce an almost infinite variety of proteins.

Amino acids dissolve in water, but the properties of the proteins that they can produce vary greatly. Some proteins are insoluble in water and are very tough, which makes them ideally suited to structural functions within living things. These proteins are found in connective tissue, tendons and the matrix of bones (collagen), the structure of muscles, the silk of spiders' webs and silkworm cocoons, and as the keratin that makes up hair, nails, horns and feathers.

Other proteins are soluble in water. These form antibodies, enzymes and some hormones, and are also important for maintaining the structure of the cytoplasm in the cells.

Activity 5.3

Using the Biuret test for proteins

When we test for proteins, we sometimes add two separate chemicals (5% potassium or sodium hydroxide solution and 1% copper sulphate solution) to our test food. These chemicals are often provided ready mixed, and the mixture is called **Biuret reagent**.

You will need:

- test tubes
- 5% potassium or sodium hydroxide solution and 1% copper sulphate solution or Biuret reagent
- protein powder, for example, albumin
- different food samples

The human diet

Method

1. In one boiling tube, add water to a depth of about 2 cm. This will act as your control.
2. In another tube, add a sample of protein powder and water to a depth of about 2 cm. Shake to mix.
3. Place any other food samples to be tested in other test tubes in the same way.
4. Add an equal volume of dilute potassium or sodium hydroxide solution to each tube and mix.
5. Add a few drops of dilute copper sulphate solution. (NB: If you are using Biuret reagent, steps 4 and 5 are combined into one.)
6. A purple (mauve) colour will develop if protein is present.
7. Write up your method and the results for the different foods that you tested.

Figure 5.11 Results of the Biuret test for proteins

The complicated ways in which the structures of proteins are built up means that they are easily damaged and denatured. The relatively weak forces that hold the different parts of the amino acid chains together are broken easily. The functions of most proteins rely very heavily on their structure. This means that the entire biochemistry of cells and whole organisms is very sensitive to changes that might damage their proteins. A rise in temperature of a few degrees or a change in pH are enough to destroy the 3D structure of proteins and so destroy life itself. This is why our body has so many complex systems that keep the internal conditions as stable as possible (see Chapter 11) and why very high fevers are so dangerous and can lead to death.

Lipids

Another group of organic chemicals that make up body cells are the lipids. Lipid-rich foods include anything containing large amounts of fats and oils. Butter, lard, palm oil, corn oil, soya oil and olive oil are all lipids. Plant seeds like groundnuts and coconuts are also lipid rich, providing an energy-rich store for the embryo plant. Meat, oily fish and eggs are high in lipids too. Fries, crisps, doughnuts, and other foods that are cooked in fat or oil are also rich in lipids and the energy that they supply. The media are constantly reminding us of the importance of a low-fat diet and the dangers of a high **cholesterol** level. But are fats really bad for us, and what is cholesterol anyway?

Figure 5.12 Lipids may be solid fats or liquid oils.

Lipids are an extremely important group of chemicals with major roles to play in the body. They are an important source of energy in our diet and they are the most effective energy store in our body. They contain more energy per gram than carbohydrates or proteins. This is why our body converts spare food into fat for use at a later date. Combined with other molecules, lipids also play vital roles as hormones in our cell membranes and in the nervous system, so they are important molecules for health.

All lipids are insoluble in water but dissolve in organic solvents. This is important because when lipids are present in our cells, they do not interfere with the many reactions that take place in the cytoplasm, because the reacting chemicals are all dissolved in water.

Some of the best-known lipids are the fats and oils. Chemically, they are extremely similar, but fats, for example butter, are solid at room temperature and oils, for example palm oil, are liquid at room temperature. The lipids found in animals are much more likely to be solid at room temperature than plant lipids.

As with the carbohydrates, the chemical elements that make up all lipids are carbon, hydrogen and oxygen. There is, however, a considerably lower proportion of oxygen in lipid molecules than in carbohydrate molecules. Fats and oils are made up of combinations of two types of organic chemicals: fatty **acids** and **glycerol**.

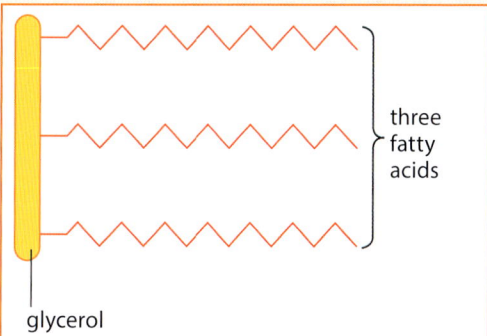

Figure 5.13 It is the combination of fatty acids in a triglyceride that determines what it will be like. Saturated fatty acids give solid fats like butter, whereas unsaturated fatty acids produce a liquid like corn oil.

The biologist's toolkit: Key chemistry: carbon–carbon bonds

Carbon atoms are very important in the chemistry of living organisms.

Each carbon atom can make up to four bonds with other atoms. Carbon atoms join together using **covalent bonds**, which means that they share electrons.

- When carbon atoms bind to each other, they may each share a single electron to form a single carbon–carbon bond: –C–C–
- Sometimes carbon atoms form double bonds with each other, with each carbon atom contributing two electrons to the shared bonds: =C=C=
- Carbon atoms can even share three electrons each to form a triple bond: ≡C≡C≡

The type of bond formed – a single, double or triple bond – affects the properties of the compound that is formed. This is very important when we talk about lipids.

When a molecule of glycerol combines with three fatty acids, a lipid is formed. The molecules combine in a condensation reaction and a molecule of water is produced for each fatty acid that reacts with the glycerol. So to make 1 lipid molecule, we get 3 water molecules:

glycerol + fatty acid 1 + fatty acid 2 + fatty acid 3 ➡ lipid + 3 water molecules

Glycerol is always the same, with the chemical formula $C_3H_8O_3$. On the other hand, there is a wide range of fatty acids. More than 70 different ones have been extracted from living tissues and the nature of the lipid greatly depends on which fatty acids are in it. There are two main ways in which fatty acids vary: the length of the carbon chain can differ, and whether the fatty acid is **saturated** or **unsaturated**.

In a saturated fatty acid, each carbon atom is joined to the one next to it by a single covalent bond. Saturated fatty acids are usually solids. Animal fats are usually made from saturated fatty acids.

In an unsaturated fatty acid, the carbon chains have one or more double bonds in them. Unsaturated fatty acids are more common in plant lipids. Unsaturated fatty acids are often liquids (oils).

Activity 5.4

Testing for lipids

There is a very simple test for lipids: if you rub a food that you think contains lipids on a piece of paper, it will leave a translucent (see-through) spot on the paper that does not dry out and disappear. However, this test, although effective, is not very scientific. Another way to identify lipids in foods is to use the fact that lipids dissolve in ethanol but not in water.

You will need:
- clean, dry test tubes – they MUST be dry
- ethanol
- different food samples

Method
1. Place a sample of ethanol in a dry test tube to a depth of about 2 cm.
2. Place a small sample of food in a dry test tube and add a similar amount of ethanol.
3. Shake the tube to dissolve any lipids in the ethanol.
4. Take two more test tubes and half fill each with water.
5. Carefully pour the contents of the tube containing food into one of the tubes containing water.
6. Pour the pure ethanol into the other tube containing water and compare the two. If lipids are present, a white, cloudy layer forms on top of the layer of water.
7. Write up your method and the results of any foods you tested.

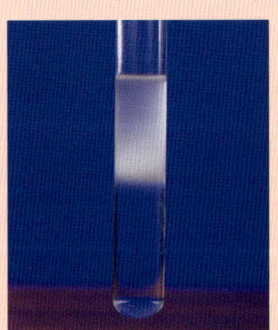

Figure 5.14 Results of the ethanol test for fats

Nutrition

> **Checkpoint questions**
>
> 1 Compare the three main food groups of macronutrients: carbohydrates, proteins and lipids.
> 2 Describe a condensation reaction and explain why they are so important in our nutrition.
> 3 Describe how you would test a food sample to see if it contained:
>   a starch
>   b fat.

Micronutrients

The micronutrients important for a healthy diet are minerals and vitamins.

Minerals

It isn't just carbohydrates, proteins and fats that are important in our food. Mineral salts are only needed in minute amounts, but a lack of them in our diet can lead to a variety of unpleasant conditions, as you can see in Table 5.2. For example, as you will learn in Chapter 9, we need calcium (Ca) in our diet to provide calcium ions to make our bones and teeth hard and strong. Without it, children develop **rickets**. Their bones stay soft and cannot support the weight of the body so the legs become bowed. Milk and other dairy products, such as cheese and yoghurt, are a very good source of calcium. However, calcium alone is not enough to protect you from rickets. You also need vitamin D.

Iron (Fe) is vital to make the **haemoglobin**, found in our red blood cells, needed to carry oxygen around our body (see Chapter 8). Iron is found in foods such as red meat, liver, apricots and eggs. If our diet lacks iron, we suffer from **anaemia**. We do not have enough haemoglobin in our red blood cells and may not even make enough red blood cells, so we don't get enough oxygen to the tissues of our body. This makes us look pale (lack of red blood cells), and feel tired and lethargic (lack of oxygen).

Our mineral needs change throughout our lives – growing children need plenty of calcium for their bones to grow, while girls and women who have menstrual periods need more iron than others to replace the blood lost each month during their period.

Sodium is very important. We get our sodium from the salt we eat with our food. If we do not have enough sodium, we will get severe cramps and our nerves will not work properly. But too much salt increases our blood pressure and may lead to heart disease or strokes. Many people have high blood pressure because they eat too much salt without realising it. Many processed foods, and snacks such as chips and savoury nuts, are salty – which is why they taste so good – but we do not realise how much salt we are taking in. They are hidden sources of salt and can affect our health without us realising. This is another reason why a healthy diet with plenty of fresh fruit and vegetables is so important.

Figure 5.15 The mineral calcium and vitamin D are both needed to prevent the disease rickets.

Iodine is needed for the thyroid gland in your neck to make the hormone thyroxine which helps to control our growth and body metabolism. You will learn more about thyroxine in Chapter 12. However, if your body is deprived of iodine, the gland in your neck gets bigger and bigger as it tries to make enough hormone. This results in a big swelling known as a goitre.

Table 5.2 summarises the main minerals we need in our diet, what they are needed for, the foods we find them in – and the symptoms of the **deficiency diseases** we will suffer from if our diet lacks these vital minerals.

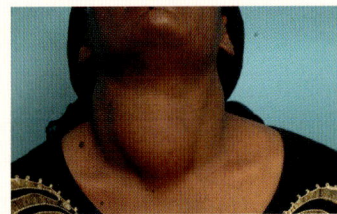

Figure 5.16 A lack of iodine causes a swelling in the neck called a goitre.

> **DID YOU KNOW?**
>
> In Europe in the Middle Ages, a lack of iodine in the diet was so common that almost everyone had a goitre, and the women in many famous paintings have a telltale swelling in their neck!

Table 5.2 Some of the main minerals needed in the diet and the deficiency diseases associated with them

Mineral	Role in the body	Examples of foods rich in this mineral	Symptoms of deficiency diseases
Calcium	Strengthening bones and teeth	Dairy products such as milk, cheese and yoghurt, fish, bread, dark green leafy vegetables, soy beans	Rickets – soft bones and eroded teeth
Magnesium	Making bones; balancing the cell chemistry	Green vegetables	Skeletal problems; problems with the cell chemistry and so defects in metabolism
Fluoride	Strengthening tooth enamel and preventing tooth decay; strengthening bones	Tea, coffee, shellfish, grapes; drinking water often fortified with fluoride	Poorly formed teeth, tooth decay; weak bones
Iodine	Production of thyroid hormone, needed for growth and control of the body metabolism	Fish, shellfish, dairy products and eggs; iodised salt is the simplest way to get iodine in the diet	In children: poor brain development and growth In adults: goitre as the thyroid gland swells trying to make more hormones; slow metabolism, weight gain
Phosphorus	Making bones and teeth; part of many chemicals including the genetic material DNA	Most foods	Bones and teeth do not form properly; the metabolism of the body fails
Sodium	Balance of body fluids; needed for nerves to send electrical messages	Common salt; most foods	Muscle cramps; kidney problems
Iron	Part of the haemoglobin of the red blood cells; carries oxygen around the body	Red meat, liver, eggs, green leafy vegetables	Anaemia – pale, tired, no energy; heart palpitations

Nutrition

Vitamins

Just like minerals, vitamins are micronutrients and only needed in very small amounts. They are usually complex organic substances that can be absorbed directly into the bloodstream from the gut. If any particular vitamin is lacking from our diet in the long term, it will result in a deficiency disease. Different foods are rich in different vitamins and it is important to take in a range of all the important vitamin-rich foods in our diet. For example, vitamin A is needed to make the light-sensitive chemicals in the retina of our eyes (see Chapter 11). If our diet lacks vitamin A – found in fish liver oils, butter and carrots – our eyesight is affected and we find it almost impossible to see in low light levels. This is called night blindness.

Figure 5.17 Foods that contain vitamin A.

> **DID YOU KNOW?**
>
> Mammals store vitamin A in their livers. The liver of a polar bear is so rich in the vitamin that eating only 500 g (half a kilogram) would give you a lethal dose of vitamin A.

Vitamin B1, found in yeast extract and cereals, is needed for the reactions of cellular respiration to take place. If we don't eat enough, we get a condition called **beri-beri**. Our muscles waste away and we become paralysed. It can be fatal.

Lack of vitamin C causes **scurvy**, which used to kill many thousands of sailors as they travelled the world in sailing ships. Vitamin C is needed for the formation of the connective tissues that holds our body together. You find vitamin C in fruits, particularly citrus fruits and green vegetables. Once people started to take limes and lemons on sea voyages, scurvy became a thing of the past.

We need Vitamin D for our bones to take up the calcium they need to grow strong. Vitamin D is found in fish liver oils and it is also made in the skin from sunlight. If children do not have enough vitamin D in their diet, or do not get enough sunlight, they will get rickets even if they have plenty of calcium.

Figure 5.18 Sources of vitamin B1 – lack of this vitamin causes beri-beri.

Figure 5.19 A rough-skinned lemon – just one of the big range of citrus fruits full of vitamin C that we grow here in the Caribbean.

The human diet

Table 5.3 summarises six of the most common vitamins, the best food sources for them and the problems that can arise if they are deficient in the diet. These deficiency diseases can be avoided or remedied by using vitamin supplements if it is not possible to get them all from the food we eat.

The vitamins were given letters to distinguish them in the days before scientists had discovered exactly what each vitamin was. Although we now know all their chemical names, they are still usually referred to as vitamin A, vitamin B1, vitamin C, vitamin D (calciferol), and so on. Some vitamins are soluble in water and these include vitamin B1 (thiamine) and vitamin C. Others are fat soluble, including vitamins A, D, E and K.

> **Revision tip**
>
> *Try to find examples from your own life and surroundings to support the concepts you learn each day. Start a blog with your class and share the information.*

Table 5.3 Some of the main vitamins needed in the diet and the deficiency diseases associated with them

Vitamin	Role in the body	Examples of foods rich in this vitamin	Symptoms of deficiency diseases
Vitamin A (retinol) fat-soluble	Making a chemical in the retina of the eye that allows us to see; protects the surface of the eye	Fish liver oils, liver, butter, margarine, carrots	Night blindness – can't see in dim light; damaged corneas of the eye
Vitamin B1 (thiamine) water-soluble	Needed in cell respiration	Yeast extract, cereals	Beri-beri; muscle wasting, paralysis and death
Vitamin C water-soluble	Sticks the cells together lining surfaces such as the mouth and gut	Fresh fruits, especially citrus fruits, and many vegetables	Scurvy – connective tissue breaks down
Vitamin D fat-soluble	Helps bones absorb calcium and phosphorus	Fish liver oils; made in the skin in sunlight	Rickets (weak bones); poor teeth
Vitamin E fat-soluble	Removes antioxidants that damage cells; helps the immune system; prevents clot formation in the heart	Plant-based oils, seeds, nuts, fruit and vegetables	Damage to the eyesight; damage to the nerves; poor immune system
Vitamin K fat-soluble	Needed for the clotting of the blood	Green leafy vegetables, vegetable oils, fermented foods	Problems with the clotting of the blood

The importance of water

Another vital constituent of a balanced diet is water (see Figure 5.20). An average person can survive with little or no food for days if not weeks. However, a complete lack of water will cause death in 2–4 days, depending on other conditions such as temperature. Your body is actually 60–70% water, depending on your age and how much you have drunk recently, so it is not surprising that water is crucial in your body.

Figure 5.20 Water is a crucial part a balanced diet. Without it, we will die.

Here are some reasons why having water in our body is vital:

- All of the chemical reactions that take place in the body take place in solution in water – it is a vital solvent.
- Water is involved in the transport of substances around the body. Food, hormones (chemical messengers – see Chapter 10), waste products such as urea and many other substances are all carried around the body in solution in water as part of the blood.
- Water is involved in temperature regulation as heat is lost from the body through sweating (see Chapter 11).
- Water is involved in the removal of waste materials from the body in urine and in sweat (see Chapter 11).
- Water is a reactant in many important chemical reactions in the body. For example, as you will discover later in this chapter, many food molecules are broken down in **hydrolysis reactions** where water is added.
- Water is needed for the osmotic stability of the body. The concentration of the chemicals in our cells and in the body fluids surrounding them must be kept constant. If there is not enough water in the blood and **tissue fluid**, the body cells lose water by osmosis and can no longer function, causing death.

Figure 5.21 Water is vital for all our bodily functions.

Fibre in the diet

The final important part of a healthy diet is something that we cannot even digest or absorb. Roughage or fibre cannot be broken down in the human gut, yet it is an essential part of our diet because it provides bulk for the intestinal muscles to work on. It also absorbs lots of water. Wholegrain foods, oats, bran and most fruits and vegetables contain lots of fibre – see Figure 5.22.

If we eat a healthy diet containing the recommended amount of fibre, we will feel full after eating fewer calories. This is because the fibre absorbs water, which gives us that feeling of fullness. Our gut will work well and we will pass soft faeces regularly. This helps us maintain a healthy body mass. People who eat a diet low in roughage generally eat more food overall and are more likely to become obese (see Figure 5.22).

Obesity has many health problems linked to it. You will learn more about this later in this chapter. The movements of the gut, which transport the food through it (**peristalsis**), are sluggish and the food moves through the digestive system relatively slowly. This can result in constipation, which causes the faeces to build up in the gut. The gut absorbs water from the faeces, which become hard and difficult to pass. The opposite of constipation is diarrhoea, when the food passes through your gut very fast, so the gut absorbs little water. The faeces are liquid. Diarrhoea has many causes. It is usually avoided by good **personal hygiene** and the proper preparation of food. You will learn more about diarrhoea and how to prevent it later in this book.

Figure 5.22 Foods containing plenty of fibre.

The human diet

The biologist's toolkit: Bar charts/bar graphs

A bar chart or bar graph is a very clear way of presenting certain types of data. It is usually used when you have a fairly small set of results.

- The horizontal *x*-axis and the vertical *y*-axis will both be labelled so you know what is being shown.
- The heights of the bars show you the frequency of the results – in other words, how often a thing happens.
- The bars should always be drawn separately.
- The title of a bar chart tells you what is being compared.

Figure 5.23 shows two different bar charts. One compares the percentage of the population who eat the recommended amount of fibre in their diet in three different countries. The other shows the percentage of the population who are obese in the same three countries. This data helps you see the effect of fibre in the diet on obesity levels.

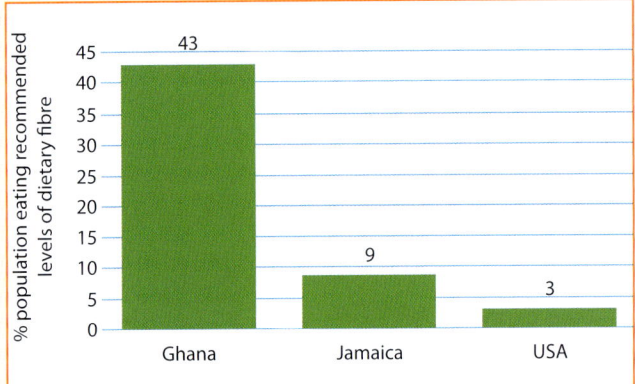
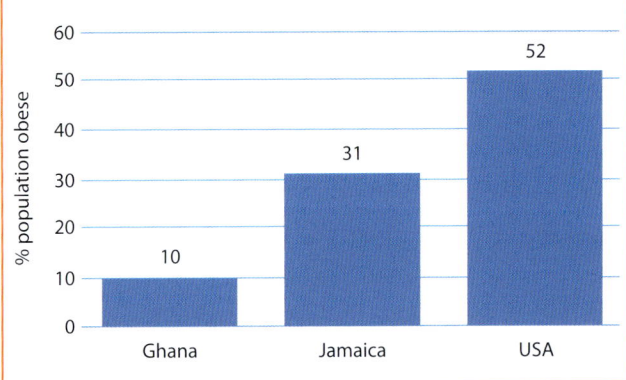

Lie, L.; Brown, L.; Forrester, T.E.; Plange-Rhule, J.; Bovet, P.; Lambert, E.V.; Layden, B.T.; Luke, A.; Dugas, L.R. The Association of Dietary Fiber Intake with Cardiometabolic Risk in Four Countries across the Epidemiologic Transition. Nutrients 2018, 10, 628. https://doi.org/10.3390/nu10050628.

Graph A: Percentage of the population eating recommended levels of dietary fibre in Ghana, Jamaica and the USA

Graph B: Percentage of the population who are obese in Ghana, Jamaica and the USA

Figure 5.23 These bar graphs give you information about the intake of dietary fibre and levels of obesity in the populations of three countries. What is your hypothesis about the link between dietary fibre and obesity from looking at these graphs?

The importance of a balanced diet

Wherever you live and whatever the basis of your diet, the right balance of food is very important to your overall health and well-being. A balanced diet includes enough of all the major food groups (carbohydrates, proteins, lipids, minerals, vitamins, water and fibre) to:

- supply the energy and nutrients needed to maintain the cells, tissues and organs of your body in a healthy state
- support healthy growth and development of your body when it is needed.

If too little food is eaten (**undernutrition**) or too much food is taken in (**overnutrition**), or any one element of the diet is lacking, then you will suffer from **malnutrition**. Malnutrition affects the health of millions of people all over the world.

Figure 5.24 Which meal is more balanced?

Nutrition

One of the most important factors in a balanced diet is eating enough food to supply our energy needs. But how much energy is that? The amount of energy we need to live depends on many different things. Some of these things we can change and some we cannot.

- If you are male, you will need to take in more energy than a female of the same age – unless she is pregnant. During pregnancy, the energy needs of a woman increase steadily as she has to provide the raw materials for a developing baby and supply the energy it needs to live.
- If you are a teenager, you will need more energy than if you are in your 70s.
- The amount of exercise we do affects the amount of energy we use up. If we do very little exercise, then we do not need as much food. The more we exercise, the more food we need to take in to supply energy to our muscles as they work.

People who exercise regularly are usually much fitter than people who do little exercise (see Figure 5.25). They have bigger muscles, and muscle tissue burns up much more energy than fat. However, exercise doesn't always mean time spent training or 'working out' in the gym. Walking to school, running around the house and garden, looking after small children, or having a physically active job all count as exercise too.

Figure 5.25 Athletes who spend a lot of time training and playing a sport will have lots of muscle tissue – up to 40% of their body mass. This means they have to eat a lot of food to supply the energy they need.

The energy we need doesn't only depend on how active we are on a particular day. Two people who are very similar in age, sex and size may still need quite different amounts of energy in their diet. This is partly because the rate at which all the chemical reactions in the cells of the body take place (the **metabolic rate**) varies from person to person.

The proportion of muscle to fat in the body affects the metabolic rate and therefore how much energy is needed – and this also varies from person to person. Men generally have a higher proportion of muscle to fat than women, so they have a higher metabolic rate and need more energy. A person who exercises regularly – for example, an athlete in training or someone doing a very physical job – will have a much higher proportion of muscle than someone who usually moves around very little. So, a fit person will always need more energy than an unfit person, even if they are doing nothing! Exercise increases our metabolic rate for a while even after we stop exercising, and the more muscle we have, the more energy we will use. Finally, scientists think that our basic metabolic rate may be affected by factors we inherit from our parents.

DID YOU KNOW?

About 60–70% of your daily energy needs are used up in the basic reactions to keep you alive. 10% is needed to digest your food – and only the final 15–30% is affected by your physical activity levels!

The importance of a balanced diet

The biologist's toolkit: Pie charts

One way of representing data is to use a pie chart. A pie chart looks like slices of pie!

A pie chart is a circle. Each section of the pie represents the fraction of the whole represented by that data.

There are 360° in a circle, so each section is the fraction of 360 that the data represents.

Pie charts are fun to use and easy to understand. For example, the data in this table represents the way children get to school.

Transport	Walk	Bus	Bike	Taxi
Percentage	50	30	15	5

This can be drawn as a pie chart, as follows:

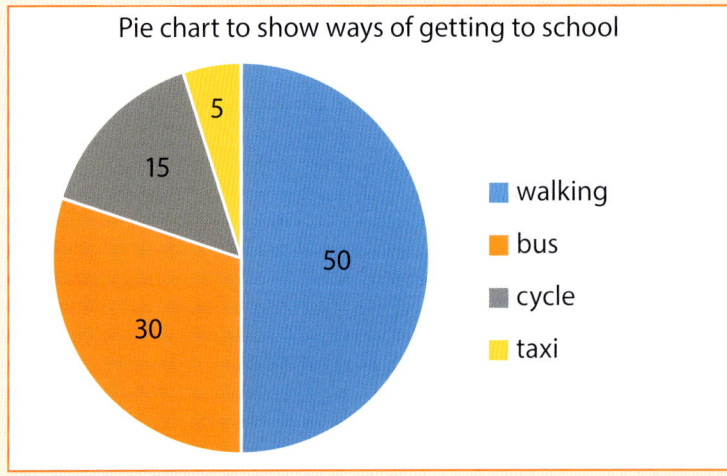

We can represent a balanced diet using a pie chart. This shows the proportion of different foods we need to eat to take in the nutrients and energy we need. It is important to remember that not every meal will be balanced. Some meals have more of one type of food than others. Everyone enjoys treats such as sweets, chips or ice cream from time to time. To be healthy, our diet does not need to be balanced every day – but our intake of different types of food should be balanced over time – and most of our main meals should be balanced like the pie chart in Figure 5.26.

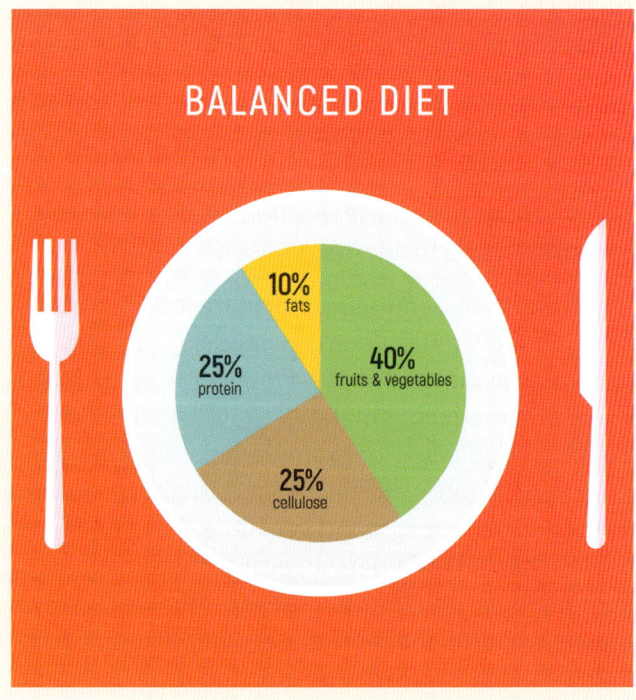

Figure 5.26 This balance of foods will supply everything your body needs.

More health impacts of malnutrition

For vast areas of the world, and in particular the developing world, getting enough food to supply their personal energy needs is a major problem for people. There is simply not enough food available to supply the energy they need. As a result, much of the world's population is seriously underweight, with shortened lifespans and reduced resistance to disease. Hand-in-hand with insufficient energy from food go insufficient essential proteins, minerals and vitamins. Therefore, in areas of undernourishment, malnutrition often results in the deficiency diseases described in Tables 5.2 and 5.3. Lack of protein in the diet may well be linked to an overall lack of energy intake, and results in a number of diseases called **protein-energy malnutrition**. The best known of these are **marasmus** and **kwashiorkor**. Both of these diseases are sadly common in very young children, particularly in the developing nations, and contribute to millions of deaths each year. The elderly are another group in which these diseases are particularly common.

Figure 5.27 The energy needs of individuals vary greatly, but average requirements have been calculated that can be used to help you decide just how much, or little, you need to eat.

- **Marasmus:** This disease appears when both protein and overall energy intake is far below what is needed by the body. An affected child fails to thrive. They will be extremely thin with wrinkled, loose skin and muscle wasting. They will lack energy and have thin, brittle hair. They do not grow properly, and their body weight will be far below the normal level for their height. Usually, if an increase of both protein and calories is added to the diet, the condition can be reversed and the child or adult saved. However, if the disease progresses too far, the body has not got enough protein to make the enzymes needed for digestion and metabolism. Once this stage is reached, even adding protein to the diet cannot save an individual.

- **Kwashiorkor:** This disease is thought to be caused by a lack of protein in the diet even if the overall energy intake is reasonable. It is particularly common around the time a child is weaned, when a diet high in starchy foods and very low in protein is often substituted for mother's milk. The symptoms include weight loss, muscle wasting, weakness, pale, flaky skin and a huge swollen belly due to fluid collecting in the tissues and the growth of a large, fatty liver. In the treatment of kwashiorkor, it is important to introduce protein in limited amounts as the liver is damaged and cannot deal with any excess. Many scientists feel that lack of protein in the diet is only part of the problem in kwashiorkor. And of course, in any condition where an individual is lacking calories and protein, they will almost certainly also be lacking vitamins and minerals as well.

Table 5.4 Daily energy requirements

Age/sex/occupation of person	Energy needed per day (kJ)
Newborn baby	2 000
Child aged 2	5 000
Child aged 6	7 500
Girl aged 12–14	9 000
Boy aged 12–14	11 000
Girl aged 15–17	9 000
Boy aged 15–17	12 000
Female office worker	9 500
Male office worker	10 500
Heavy manual worker	15 000
Pregnant woman	10 000
Breastfeeding woman	11 300
Woman aged 75+	7 610
Man aged 75+	8 770

The importance of a balanced diet

However, malnutrition as a result of a lack of food is not only a problem in the developing world. In countries where there is plenty of cheap and readily available food, many people want to lose weight either for their health or just to look better. We gain fat by taking in more energy than we need, so there are three main ways we can lose it:

- We can reduce the amount of energy we take in by cutting back on the amount of food we eat, particularly energy-rich foods like biscuits, nuts and chips.
- We can increase the amount of energy we use by doing more exercise.
- The best way to lose weight is to do both.

Unfortunately, some people have problems with their mental health that lead them to deliberately starve themselves, often becoming physically ill. People with **anorexia** feel the need to be in control. They become so worried about being fat that they stop eating. Even when their body mass drops dangerously low, anorexics still see themselves as fat. Anorexia leads to all sorts of health problems, just like those seen in people who are starving through lack of food:

- Anorexics become terribly thin and their muscles waste away.
- Their skin becomes thin and papery.
- Their immune systems cannot work properly so they pick up infections.
- In females, their periods become irregular or stop altogether.

Figure 5.28 We can increase the energy we use and lose weight by eating less energy-rich foods and by exercising.

Sadly, anorexia can sometimes lead to organ failure and death as people literally starve themselves.

In another similar condition, people binge eat, consuming huge amounts of food (often junk food). They then make themselves vomit so they have the pleasure of eating without gaining weight. This can easily get out of control and have serious health consequences. It is called **bulimia**.

Obesity

Malnutrition due to too little food is a major problem in many parts of the world. Yet malnutrition also results from eating too much food. As you have seen, the energy requirements of individuals vary depending on their age, sex and levels of activity. If we take in more energy than we need, the excess is stored as fat and **obesity** may result. In the developed world, overeating and the health issues linked to it are becoming more and more of a problem. Up to one-third of the population of the USA is thought to be seriously overweight, mainly due to eating a diet rich in high-energy fat.

We need some body fat to cushion our internal organs and to act as an energy store for when we don't feel like eating. But when this is taken to extremes, and we consistently eat more food than we need, we may end up obese, with a BMI of over 30 (see Figure 5.30).

People who are obese find it uncomfortable and difficult to do exercise, and they are often short of breath. Carrying too much weight is not just inconvenient – it can lead to serious health problems.

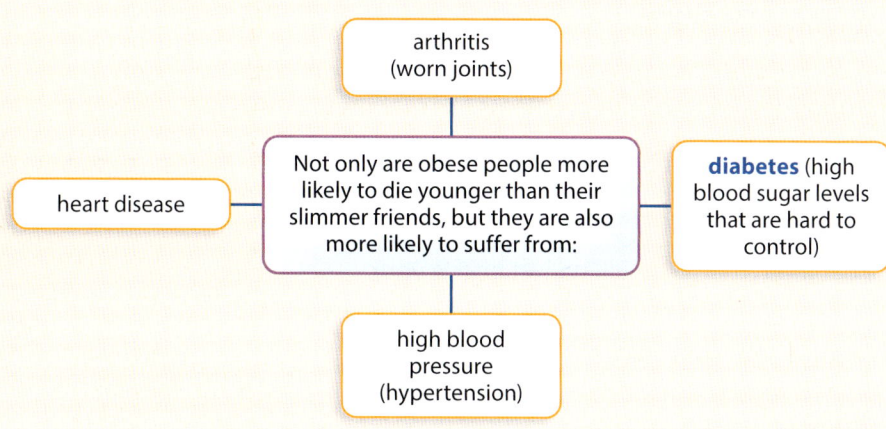

Figure 5.29 Diagram of health issues linked to obesity

DID YOU KNOW?

The heaviest man ever recorded was Jon Brower Minnoch (USA 1941–1983). He was 181 cm (6'1") tall and was overweight all his life. At his heaviest, he weighed 635 kg (1399 lb). The heaviest recorded woman is Pauline Potter who weighed 307 kg or 677 lbs at her heaviest.

Calculating obesity

Most of us look about the right size, but it isn't always easy to tell if we are obese just by looking at us. Some people are very overweight and others appear unnaturally thin. Scientists and doctors don't just measure what you weigh. They look at your **Body Mass Index** or **BMI**. This compares your weight to your height in a simple formula:

$$BMI = \frac{\text{weight (kg)}}{\text{height (m)}^2}$$

Most people have a BMI in the range 20–30. But if you have a BMI of below 18.5, or above 30, then you may have some real health problems. Charts like that shown in Figure 5.30 show the main BMI ranges and they are very useful for working out if your BMI is healthy – but they are only really accurate for adults. Since teenagers are growing and changing rapidly, these charts can give a rough idea, but they do not fully reflect your BMI.

Here are two examples of BMI calculations:

> Willow weighs 80 kg. He is 1.9 m tall. Sal also weighs 80 kg. She is 1.6 m tall. Are they both a healthy body mass?
>
> **Willow** BMI = $80/1.9^2$ = $80/3.61$ = 22.16 Willow has a healthy BMI.
>
> **Sal** BMI = $80/1.6^2$ = $80/2.56$ = 31.25 Sal is obese.

Figure 5.30 Using this BMI chart, you can work out your own BMI – but remember, it is only really accurate for adults, not growing teenagers!

The importance of a balanced diet

Body Mass Index is not the only way to measure obesity – in fact it has some limitations. A simple measurement of the waist circumference is another easy way to identify obesity. The fat most strongly linked to health problems like diabetes and heart disease builds up around your middle. Follow these guidelines to measure the waist accurately:

- Find the bottom of the ribs and the top of the hip bones.
- Place a tape measure halfway between them – it will be just above your tummy button.
- Have the tape tight but NOT digging into your skin.
- Breathe out naturally and take a measurement.
- Repeat this to make sure your reading is accurate.

A healthy waist measurement varies depending on your ethnicity. For adult men of African Caribbean, South Asian, Chinese or Japanese origins, a waist circumference of 35.4 inches (90 cm) or below is healthy. Above this level, you are obese and your health is at risk. For women from these same ethnic groups, below 31.5 inches (80 cm) is healthy. Above this level indicates obesity and a raised health risk.

Another simple way of determining obesity is to measure your height and your waist. Your waist measurement should be around half your height, or less!

All of these methods give you useful tools for maintaining a healthy body mass. This in turn reduces your risks of developing diabetes, high blood pressure or heart disease in the future.

It is not only too much fat in the diet that poses health problems. Too many refined carbohydrates, particularly sugar, can also lead to a gain in weight. What's more, a high proportion of sugar in the diet leads to the formation of dental caries cavities – (tooth decay). Bacteria on the teeth digest the sugar and produce acid, which in turn attacks the enamel of the teeth. Many people have no teeth of their own left as a result of their diet and poor dental hygiene (see Chapter 6 for more about teeth and tooth decay).

Strangely, the diet of the average inhabitant of the developed world, while full of fat and refined carbohydrates, may be low in vitamins, as relatively little fresh fruits and vegetables are consumed. Roughage is frequently missing in any quantity, causing not only constipation but also many diseases of the gut and bowel, including possibly an increased risk of developing some bowel cancers (see Chapter 6). The saying 'You are what you eat' may be truer than many of us would like to think.

Figure 5.31 Fresh fruits and vegetables, as well as high-fibre foods to provide roughage, keep us healthy and can reduce the risk of serious diseases.

> **Checkpoint questions**
>
> 4 a Why does a pregnant woman need more energy than a woman who is not pregnant?
>   b Why do athletes need to eat more food than the average person?
> 5 Plan a menu of meals for one day and show how eating this food would provide a person with a balanced diet.

The biologist's toolkit: Peer-reviewed papers

When scientists carry out investigations, they publish what they find in **peer-reviewed journals**. They write a description of their investigation, called a **scientific paper**. This includes:

- a very short **abstract** – a summary of their investigations and their findings
- their **hypothesis** – the question they are trying to answer
- their **method** – so other scientists can repeat their investigation
- their **results** – the data they collected
- an **analysis** of their data – perhaps graphs or charts that make the findings easier to see.
- a **discussion**, where they share the conclusions they make from the data they have collected.

The scientific paper goes through a process of **peer review** before it is published. This means other scientists who are experts in the same area check the work to make sure it is reliable and can be repeated. Once the paper has been reviewed, it will be published. Since it has been peer reviewed, other scientists trust the findings and can use the data to help in their own work.

Case study

Childhood obesity in the Caribbean

Around the world, obesity is a growing problems, and we in the Caribbean are part of this trend. It is not only adults who are becoming heavier. Globally, children are becoming heavier – the numbers of overweight and obese children are growing all the time. Again, sadly, we are seeing this across the Caribbean too.

In 2020, a group of scientists published a scientific paper on their work on childhood overweight and obesity across the Caribbean, with a special look at trends in Trinidad and Tobago. A study in 2016 on world-wide trends in childhood obesity from 1975–2016 showed that the Caribbean, like most regions of the world, had a growing problem. This was linked to both increased consumption of high-energy foods and sweetened drinks, along with a decrease in physical activity levels.

Here is some of the data from the study, analysed and presented in different ways:

Table 1 Demographic and anthropometric data, and percentage prevalence rates of obesity, overweight, underweight and severe thinness from the 2018 Caribbean Island Urinary Iodine Survey (CRUISE) (Giorgetti and Zimmerman 2019) by country.

Country	n	[a]Age, years	Boys:Girls ratio	[a]Weight, kg	[a]Height, cm	[a]BMI, kg/m (Guthold et al. 2020)	Obesity %	Overweight %	Underweight %
Dominica	192	9.4 (7.9;11.0)	1.09	35.0 (28.0;44.0)	138.5 (129.0;146.7)	18.2 (16.4;20.6)	15.6	44.5	2.9
Jamaica	446	9.1 (7.5;10.9)	1.00	33.6 (26.2;44.4)	137.0 (128.0;148.0)	17.4 (15.4;20.9)	19.8	39.1	2.7
St. Kitts & Nevis	200	10.4 (8.8;11.4)	1.06	34.9 (28.6;44.7)	139.5 (132.0;149.0)	17.6 (15.8;20.8)	15.5	39.5	1.5
Antigua	202	10.2 (8.4;11.3)	1.29	33.1 (27.3;45.4)	139.5 (130.0;148.8)	17.5 (15.5;20.4)	16.2	37.6	2.5
Trinidad & Tobago	417	9.5 (8.2;10.6)	0.90	32.2 (25.7;41.8)	137.0 (128.0;145.0)	16.7 (15.4;20.4)	16.5	35.0	2.1
St. Vincent & the Grenadines	300	9.6 (8.1;11.0)	0.94	32.9 (27.1;42.2)	138.5 (130.0;147.0)	17.1 (15.4;19.9)	16.7	34.0	2.0
Belize	588	9.9 (8.0;11.3)	0.94	31.7 (25.4;41.4)	135.0 (126.0;145.0)	17.5 (15.5;20.0)	15.1	32.9	2.1
Barbados	332	9.3 (7.8;10.4)	1.06	30.1 (25.2;38.0)	135.0 (127.3;143.0)	16.6 (15.0;19.5)	15.7	31.9	4.8
St. Lucia	203	9.6 (7.9;10.8)	0.93	33.3 (27.0;42.7)	139.0 (129.0;147.0)	16.9 (15.5;19.5)	14.3	31.0	3.4
Grenada	200	9.5 (7.6;10.7)	0.85	32.3 (25.1;41.6)	138.0 (128.3;148.0)	16.4 (14.7;19.3)	15.0	28.0	3.0

The importance of a balanced diet

Figure 1 Distribution of combined percentage

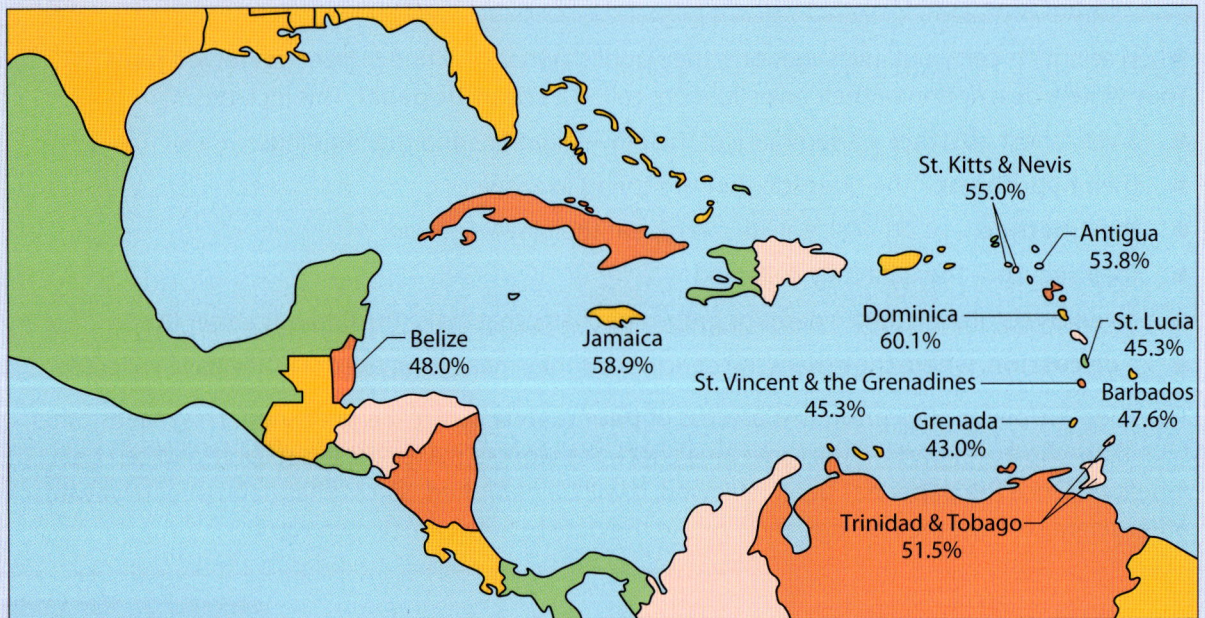

Some of the conclusions the scientists drew from this data are show below. Look at the findings and the data, and make sure the scientists have got the numbers correct!

- The mean level of overweight in children across the Caribbean was 35.1%. The mean level of childhood obesity was 16.3 %

The biologist's toolkit: Calculating the mean (average) of a set of data

When we collect a set of data, whether it is three readings in our school lab experiment or thousands of children across the Caribbean, we often find the **mean** – also known as the **average** – of our data. Using the mean of a set of repeat readings of a very large data set is more accurate than using individual readings.

Follow this process to calculate the mean of a set of data:

- Find the sum of all the values – add them together.
- Divide the total by the number of values in your data set, for example:
  - The BMI of 6 students are: 25   38   22   30   35   25

Follow these steps to calculate the mean BMI of the students:

- Add together the BMI values of the students = 25 + 38 + 22 + 30 + 35 + 25 = 175
- Divide by the number of sets of data (in this case, 6 students: 175/6 = 29.2 (3sf)
- The mean BMI of the students is 29.2 (3sf).

- The highest level of childhood obesity in the Caribbean was seen in Jamaica – 19.8% of 5–19-year-olds are obese, with a BMI of 30+.
- The highest level of childhood overweight was in Dominica – 44.5% of Dominican children were overweight, with BMIs of between 25–29.9.

Nutrition

- Dominica also had the highest combined levels of overweight and obesity, with 60.1 % of all the children being above a healthy BMI.
- Grenada has the lowest combined overweight and obesity score at 43% of children. This is very interesting as Grenada has focused a major 'One Health One Caribbean One Love' campaign on reducing childhood obesity. The evidence suggests that it is working!

You can find and read the whole peer reviewed paper at

www.tandfonline.com/doi/full/10.1080/2574254X.2020.1847632

The title of the paper is:

High prevalence of childhood overweight and obesity in ten Caribbean countries: 2018 cross-sectional data and a narrative review of trends in Trinidad and Tobago

Kirin Rambaran, Surujpal Teelucksingh; Sesh Gowrie Sankar; Michael Boyne; Godfrey Xuereb; Ambra Giorgetti; Michael B Zimmerman

Pages 23-36 | Received 06 Apr 2020, Accepted 02 Nov 2020, Published online: 17 Nov 2020

Kirin Rambaran, Surujpal Teelucksingh, Sesh Gowrie Sankar, Michael Boyne, Godfrey Xuereb, Ambra Giorgetti & Michael B. Zimmermann (2021) High prevalence of childhood overweight and obesity in ten Caribbean countries: 2018 cross-sectional data and a narrative review of trends in Trinidad and Tobago, Child and Adolescent Obesity, 4:1, 23-36, DOI: 10.1080/2574254X.2020.1847632.

Food-borne diseases and food hygiene

It is not only the balance of food in your diet that can affect your health. There are also a number of food-borne diseases. Bacteria growing on food that you eat can make you very ill and even kill you. For example, raw meat and raw eggs can contain bacteria such as salmonella that cause diarrhoea and sickness (vomiting). In most people, this is not too serious, but young children, the elderly and anyone who has other health problems can be very seriously affected by food poisoning. If the food is handled carefully and cooked properly, these bacteria are no problem at all. Problems arise when people eat meat that is not properly cooked or raw egg products. What is more, if raw meat comes into contact with food that is going to be eaten raw, such as salads or fruit, the bacteria can be passed on, even if the meat is then cooked thoroughly. When preparing food, whether in a restaurant or at home, people need to maintain very strict food hygiene.

Figure 5.28 To prepare food properly, at home or in a restaurant kitchen, food handlers must make sure that they and their surroundings are clean.

✔ Store raw meat and eggs separately from salad vegetables and fruit.

✔ Wash the knives used to cut meat and the work surfaces on which it is prepared before using them to prepare salads or to cut cheese.

✔ Disinfect surfaces regularly.

✔ Most important of all, anyone preparing food must wash their hands between handling different types of food *and* when they have been to the toilet. Gut bacteria from faeces can be transferred from the hands to the food very easily and cause stomach upsets to spread around a family or a community.

Listeria, another type of bacteria that turns up in cooked or chilled meals and in cheese made from unpasteurised milk makes adults feel unwell. However, if pregnant women are affected, it can cause serious damage and even kill their unborn child. Again, food needs to be cooked thoroughly, and pregnant women need to think carefully before they eat unpasteurised cheese.

High levels of sanitation in food preparation are vital to avoid food-borne infections. In addition to the measures already described – work surfaces wiped down with disinfectant regularly, regular hand washing and food stored properly – it is also very important to store food at the correct temperatures. Bacteria will grow far more rapidly in food stored at room temperature than they will in food stored in the fridge and, in frozen food, the growth of bacteria is almost completely stopped. So, storing food at the correct temperature – and keeping any meat, eggs and dairy products in the fridge – is another important aspect of food hygiene. In addition, the likelihood of developing a gut infection is often related not just to taking in bacteria, but to the number of bacteria you take in. The fewer bacteria in your food, the better. You will learn more about diseases of the gut, and about food and water-borne diseases later in this book.

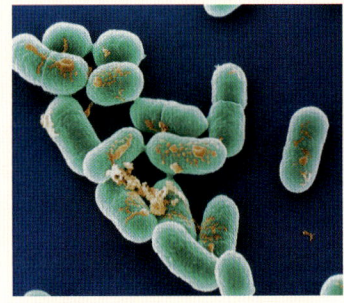

Figure 5.29 Cook food thoroughly to prevent listeria bacteria from contaminating it.

Figure 5.30 Storing meat, eggs and dairy products in the fridge will prevent harmful bacteria growth.

Possible SBAs based on nutrition

- Investigate your own or your family's diet, analysing it over time to see if it is balanced.
- Investigate the balance of school meals and suggest an alternative menu.
- Do an in-depth study of vitamins – how we discovered they were needed in the diet and analysing common foods for vitamins, and so on.
- Do an analysis of the nutritional content or energy values of common processed foods.
- Do a comparison of the energy value of unprocessed and processed foods.
- Use the reagent DCPIP to investigate the vitamin C content of different types of citrus fruits; citrus fruits compared to other common fruits such as pineapples, mangoes, guavas, bananas; and fresh citrus fruits compared to commercial fruit juices or fruit drinks.
- Do an analysis of obesity levels within your class/family/local community.
- Do an analysis of the Grenada campaign to reduce childhood overweight and obesity.
- Plan and develop a campaign to encourage healthy eating/reduced obesity in your family/group of friends/school community and collect data to see if your campaign is effective.
- Investigate the incidence of deficiency diseases within your community/country/the Caribbean.

End-of-chapter summary

In this chapter, you have learnt that:

- a balanced diet contains carbohydrates, proteins, lipids, minerals, vitamins, fibre and water in the right proportions to keep your body healthy and functioning effectively

- carbohydrates are the main energy supply for the body; carbohydrates are found as monosaccharides, disaccharides and polysaccharides; carbohydrate-rich foods include sugar, bread, rice and sweet potatoes

- iodine is used to test for the presence of starch and Benedict's solution for the presence of reducing sugars

- proteins are used as the building blocks of the body; they are made up of chains of amino acids; protein-rich foods include meat, eggs, dairy produce, beans and pulses

- the Biuret test is used to show the presence of protein in the food

- lipids are fats and oils that provide energy for the body; they are made up of fatty acids and glycerol; lipid-rich foods include olive oil, coconut oil, nuts, butter, cream and cheese

- the ethanol test identifies lipids in foods

- minerals such as calcium, magnesium, fluoride, iodine, phosphorus, sodium and iron are micronutrients, needed in small amounts for our body to remain healthy; for example, iron is needed for haemoglobin to carry oxygen in the blood and calcium is needed for strong teeth and bones; a lack of minerals causes deficiency diseases

- vitamins are micronutrients, needed in small amounts for your cells to work properly; vitamins A, B_1, C, D, E and K are all vital for health; for example: vitamin A is needed for good night vision, vitamin D is needed for the body to absorb and use calcium to build strong teeth and bones, and vitamin K is needed for the blood to clot properly; lack of vitamins causes deficiency diseases

- malnutrition is when your diet is unbalanced, which can result from too little food, when you are at risk of deficiency diseases, and also too much food, which can give rise to obesity

- obesity is measured using Body Mass Index or waist measurements; obesity increases the risk of health problems such as heart disease and diabetes, and is a growing problem in the Caribbean and around the world

- good hygiene is important in the preparation of food to prevent food-borne diseases.

End-of-chapter questions

1. Which one of the following is not part of a balanced diet?
   A   Carbohydrates
   B   Proteins
   C   Exercise
   D   Lipids

2. Which one of the following molecules forms the building blocks of proteins?
   A   Monosaccharides
   B   Glycerol
   C   Fatty acids
   D   Amino acids

3. Which one of the following groups are classed as macronutrients?
   A   Proteins, minerals, vitamins
   B   Carbohydrates, proteins, fats
   C   Fats, fibre, folic acid
   D   Carbohydrates, proteins, milk

4. What is vitamin A also called?
   A   Tocipherol
   B   Retinol
   C   Ascorbic acid
   D   Calciferol

5. Fatty acids and glycerol join together to form lipids in what type of reaction?
   A   Hydrolysis
   B   Condensation
   C   Reduction
   D   Oxidation

6. A student carried out Benedict's test on an unknown food sample and the blue liquid turned orange when it was heated. What food substance was present?
   A   Protein
   B   Starch
   C   Reducing sugar
   D   Non-reducing sugar

7. Which one of the following will not cause obesity, even if you eat very large amounts of it in your diet?
   A   Fat
   B   Fibre
   C   Carbohydrate
   D   Protein

8 The three main types of food molecules are carbohydrates, proteins and fats.
  a For each of these substances, give three examples of foods in which you would find them. (6)
  b State what each substance is used for in the body. (3)

9 For many years, it has been known that proteins, carbohydrates and lipids alone are not enough to keep us healthy.
  a What else is needed in a balanced diet? (4)
  b Describe one piece of evidence that shows we need more than proteins, carbohydrates and fats in our diet. (3)
  c Explain how:
   i too little food and (3)
   ii too much food can lead to malnutrition. (3)

10 a Describe two ways of measuring whether an individual is obese. (2)
  b Describe three health problems linked with being obese. (3)
  c Display the data on the percentage of children who are OBESE in different Caribbean countries in the table in the Case study on Childhood obesity in the Caribbean on pages 107–109, as a bar chart. (5)
  d Suggest reasons why levels of obesity are rising in children and adults alike and explain why this trend is such a potential problem for the future. (5)

Chapter 6 — The human digestive system

When you have completed this chapter, you will be able to:

- identify various structures in the human digestive system and relate them to their functions in digestion
- describe the processes of digestion and absorption of food in the digestive system
- describe the structure and functions of the teeth, explain their importance, describe the process of tooth decay and give guidelines for good dental care
- describe the role of enzymes in the digestive process and explain how the activity of specific enzymes are affected by temperature and pH
- explain how the products of digestion are absorbed and assimilated in the body

What the examiners say

→ Candidates are often unable to accurately draw or label the digestive system or discuss the processes involved in digestion.

→ Furthermore candidates are unfamiliar with the various enzymes and their functions

→ One worrying trend is that candidates do not always read and decode the questions accurately, thus providing responses to questions that have not been asked.

Revision tip

Do as many exam-type questions and past paper questions as you can find. Take care to read the questions CAREFULLY – only answer what you have been asked. When in doubt, ask your teacher!

Carbohydrates provide the body with an energy source for respiration to take place. Proteins are needed for building new cells and repairing old ones. Lipids are also an energy source and they provide a way of storing spare energy. However, in the form that they are usually eaten, neither carbohydrates, proteins nor fats are useful to the body. The link between what comes in and what the body needs is the digestive system.

How the human digestive system works

The human body needs small, soluble molecules to use in all the reactions of metabolism such as releasing energy and making new larger molecules. The food we eat usually arrives in our system as large chunks bitten off by the teeth. These chunks contain large insoluble molecules such as starch, protein and lipids, which our body cannot absorb into the blood and use. These large food molecules need to be broken down into smaller, simpler, soluble molecules. This is the main job of the digestive system – food substances are broken down into small, soluble molecules as they pass through the gut.

As you saw in Chapter 5, the large molecules that make up carbohydrates, proteins and fats are built from small molecules that are joined together by condensation reactions, with a molecule of water being lost each time (see *The biologist's toolkit*, page 85). When these large molecules are broken down during digestion, the opposite process takes place – hydrolysis (splitting with water) reactions (see *The biologist's toolkit* below). As water molecules are added to the large food molecules, the monomer units, whether they are monosaccharides, amino acids or fatty acids, are released.

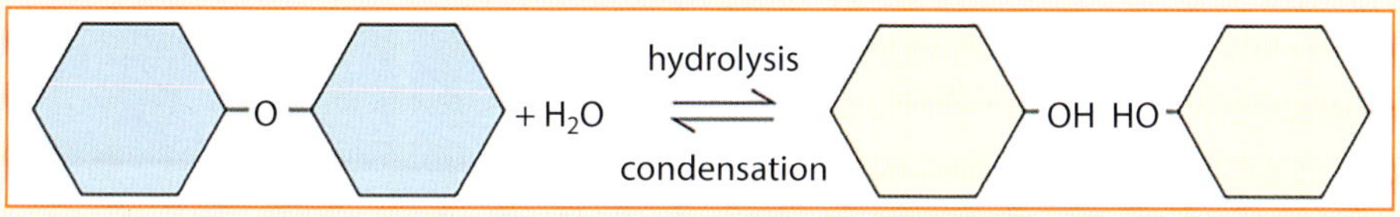

Figure 6.1 Large food molecules are broken down by hydrolysis reactions – the opposite of the condensation reactions by which they are built up in the first place.

The biologist's toolkit: Key chemistry: hydrolysis reactions

Hydrolysis reactions are very common in biology. In Chapter 5, you saw how condensation reactions are used to build up large molecules from smaller ones.

In a **hydrolysis** reaction, a molecule of water (H_2O) is added to a larger molecule, making it split into two smaller molecules. Hydrolysis literally means water splitting.

molecule 3   +   water   →   molecule 1   +   molecule 2

Many important reactions that break down large molecules into smaller ones, in the digestive systems of humans and other animals, and in the cells of all types of organisms, involve hydrolysis.

The working of our digestive system is based on two processes.

- **The physical (or mechanical) breakdown of the food:** The food we eat is physically broken down into smaller pieces in two main ways. Our teeth bite and chew the food up in our mouth. Then our gut, which is a muscular tube, squeezes the food and physically breaks it up, while mixing it with various digestive juices to make it easier to move. Breaking the food up in this way gives the **digestive enzymes** a much larger surface area to work on (remind yourself of the importance of a large surface area by looking at *The biologist's toolkit*, page 28).
- **The chemical breakdown of the food:** The large, insoluble food molecules must be broken down by hydrolysis reactions into small, soluble molecules so they can be absorbed into our body and used by our cells. This chemical breakdown is controlled by enzymes. Enzymes are proteins that speed up (catalyse) other reactions. They do not actually take part in the reaction or change it in any way except to make it happen faster. Enzymes are **biological catalysts** that usually work best under very specific conditions of temperature and pH. To understand digestion we need to understand enzymes – so this is where we will start.

The biologist's toolkit: Key chemistry: catalysts

Many chemical reactions are very slow. A catalyst is used to speed up or control the rate of a reaction. Catalysts:
- increase the rate of a reaction
- do not change the products of the reaction
- are needed in very small amounts
- are not chemically changed themselves during a reaction
- are not used up in a reaction so they can be reused many times.

More about enzymes

Enzymes play a vital role in digestion, but that is not all they do. To stay alive, hundreds of reactions, which involve making new materials and breaking things down, take place in our body all the time in a rapid and controlled way. If you mixed lots of chemicals together, they might explode. Despite all the reactions taking place in them, the cells in our body do not explode because enzymes control these reactions. What do we know about enzymes?

Figure 6.2 A piece of meat is broken down by hydrochloric acid in a few days – but we can't wait that long to get energy and useful chemicals from our food. In the human stomach, a similar piece of meat is broken down in a few hours, thanks to the action of the enzyme pepsin.

- Enzymes are made of protein.
- Enzymes are biological cataylsts, so they control the rate of reactions.
- As with all catalysts, enzymes are not affected by the reaction they speed up, so they can be reused again many times.
- Enzymes are very specific. Each each type of reaction that takes place in our body is controlled by a specific enzyme that does not catalyse any other type of reaction.
- Some enzymes work inside our cells (**intracellular enzymes**) and some of them are secreted into body organs, such as the digestive system, where they catalyse specific reactions (**extracellular enzymes**).

How do enzymes work?

Enzymes are long chain protein molecules coiled and twisted into very complicated shapes. Each enzyme has one area, called the **active site**, which is exactly the right size and shape to fit the chemicals of the reaction it catalyses. Enzymes act by binding the **reactants** of a reaction to their specially shaped active site. Each type of enzyme has a specially shaped active site that only takes the correct molecules for the reaction it catalyses. The enzyme and molecule fit together like a key fitting into a lock, so this model of enzyme action is called the **lock-and-key mechanism** (see Figure 6.3).

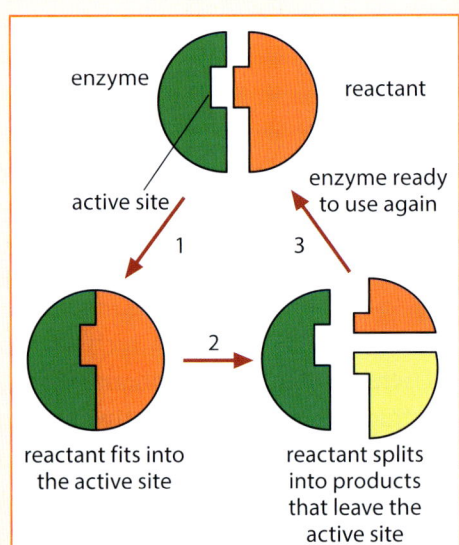

Figure 6.3 Enzymes have their effect as catalysts using the lock and key mechanism shown here. Anything that changes the shape of the protein molecule may change the shape of the active site and stop the enzyme from working.

When the reactants bind to the active site of the enzyme, it lowers the energy needed for the reaction to take place (the **activation energy** of the reaction), so the reaction happens more easily.

Enzyme names usually (but not always) end in –ase, for example:
- **amylase** breaks down starch
- **lipase** breaks down lipids
- **catalase** breaks down hydrogen peroxide
- **pepsin** breaks down proteins (this enzyme still has an old-fashioned name).

What affects enzymes?

Enzymes are vital for the successful functioning of our body. Our normal body temperature is so low that without our enzymes, all the reactions in our cells would be too slow to keep us alive.

Anything that changes the shape of the active site stops the enzyme from working, because it can no longer bind to the reactants. Biological reactions are sensitive to the same sort of things as inorganic reactions, such as concentration, temperature and particle size. But in living organisms, an increase in temperature only speeds up reactions to a certain point. The chemical reactions that take place in the human body happen at their fastest at a relatively low temperature, around 37 °C. The catalytic action of our enzymes makes this possible. However, if the temperature increases too much, then the enzymes cannot work. Since enzymes are made of protein, they unravel and denature if they get too hot. Most enzymes are damaged by temperatures above 45 °C. Once the protein structure is damaged:
- the reactants cannot bind to the active site
- the catalytic effect of the enzyme is lost and the reaction that it controls cannot continue properly.

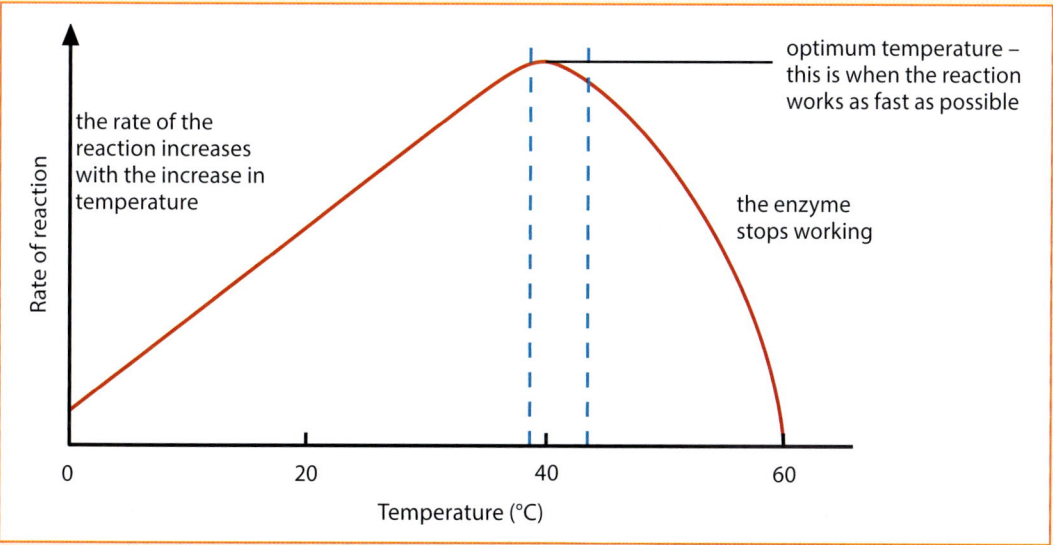

Figure 6.4 Like most chemical reactions, the rate of the enzyme-controlled breakdown of starch by the enzyme amylase increases as the temperature rises – but only to the point where the complex protein structure of the enzyme breaks down. After that, the enzyme no longer works.

More about enzymes

Activity 6.1

Investigating the effect of temperature on the action of amylase

Amylase is an enzyme that is made in the salivary glands in the mouth and in the pancreas. It catalyses the breakdown of starch to the sugar maltose. You can use the reaction of starch with iodine solution to indicate how quickly the enzyme does its job, and use this to investigate the effect of temperature on the way the enzyme works.

You will need:

- amylase solution
- a starch suspension
- iodine solution
- a marker pen
- two spotting tiles
- a beaker of water heated to 30 °C
- boiling tubes
- two 5 cm^3 syringes or pipettes
- a thermometer
- a stopwatch or clock with a clear second hand

Method

1. Place drops of iodine solution in the depressions on both of the spotting tiles.
2. Place 5 cm^3 of starch suspension in each of two boiling tubes, one labelled 'starch' and the other labelled 'starch/amylase'.
3. Place 5 cm^3 of amylase solution in another boiling tube labelled 'amylase'.
4. Place the three tubes in the water bath at 30 °C. Leave for 5 minutes for the temperatures to equilibrate.
5. Measure out 5 cm^3 of amylase solution and add it to the labelled boiling tube of starch.
6. Start the stopwatch, and immediately take a small sample of the starch/amylase mixture and add it to the first drop of iodine on the spotting tile.
7. Take regular samples of the mixture – every 30 seconds – for about 10 minutes, and record the colour of the iodine each time.
8. At the end of the 10 minutes, test a sample of the simple starch suspension in one well of the spotting tile and compare it with the sample that has been mixed with amylase. This will confirm that any change is due to the enzyme rather than the temperature of the solutions.
9. This investigation can be repeated with the starch suspension and the starch/amylase mixture at a range of different temperatures, and the results recorded in a table like the one shown on page 119.

Colour of iodine/starch/amylase mixture at different temperture					
Time (min)	20 °C	30 °C	40 °C	50 °C	60 °C
0.5					
1.0					
1.5					
2.0					
10.0					

10 You can draw a graph of your results, using the time taken to break down all the starch at different temperatures or the rate at which the enzyme breaks down 1 cm³ of starch at each temperature.

11 Write up your investigation, explain your results and suggest ways in which your investigation could be made more reliable.

The biologist's toolkit: Key chemistry: the pH scale

Most solutions are either **acid**, **alkaline** or **neutral**. We measure this using the pH scale. We often use coloured indicators to show us what pH a particular solution is.

- Neutral solutions have a pH of 7.
- Acid solutions have a pH of less than 7.
- Alkali solutions have a pH of more than 7.

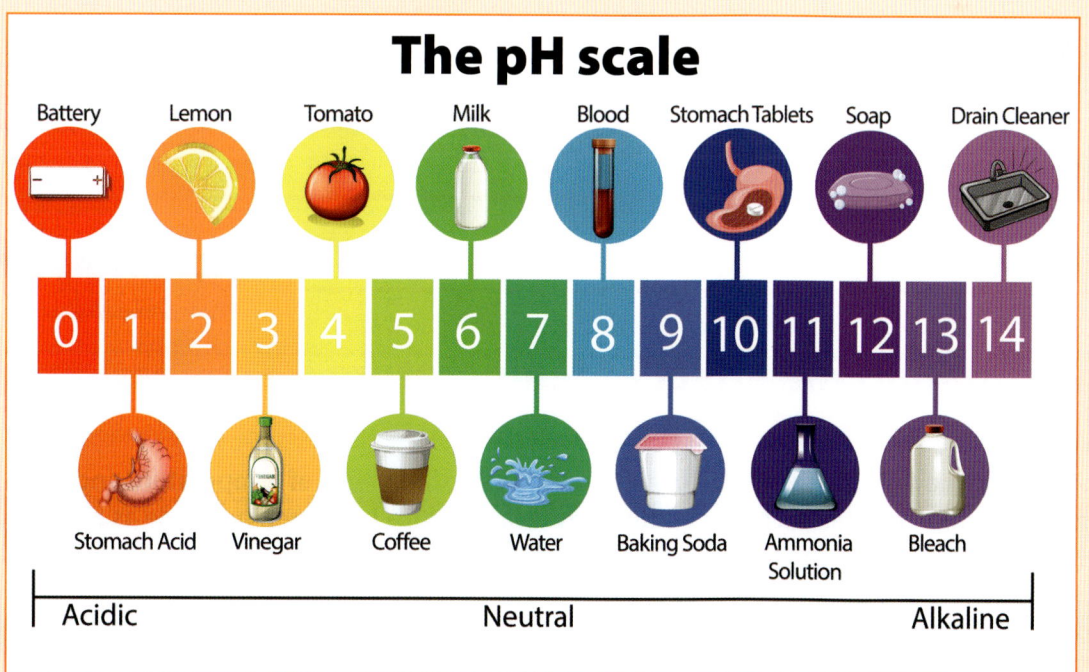

More about enzymes 119

The surrounding pH is another factor that can affect the rate of an enzyme-controlled reaction. A change in the pH affects the bonds inside the protein and changes the shape of the molecule. This in turn alters the shape of the active site so that it can no longer act as a catalyst. Different enzymes have different ideal pHs (see Figure 6.5).

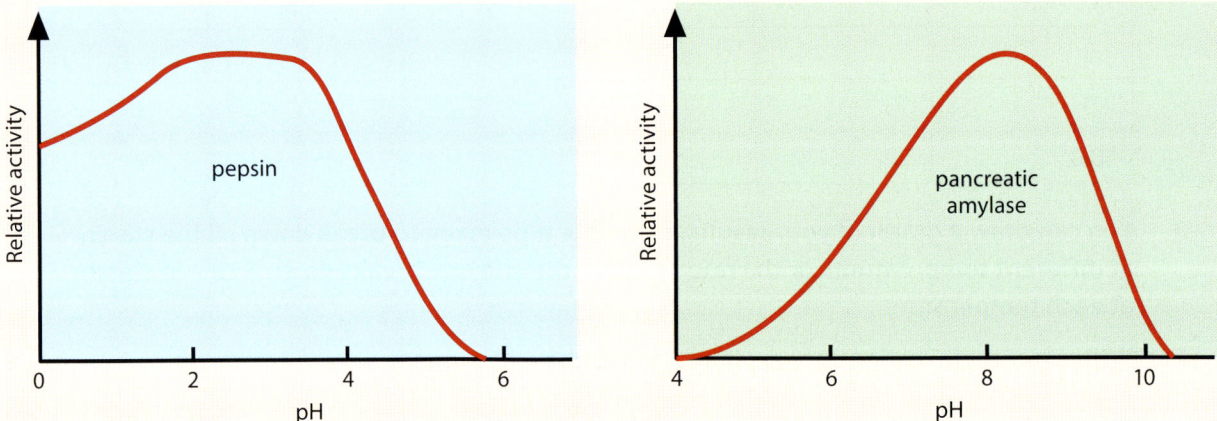

Figure 6.5 These two enzymes are found in quite different parts of the human digestive system, and they need very different pH conditions to work at their maximum rate. Pepsin is found in the stomach, along with hydrochloric acid, while amylase is found in the small intestine along with alkaline bile.

Activity 6.2

Investigating the effect of pH on enzyme activity

You can use the same basic technique as in Activity 6.1 on page 118 to investigate the effect of pH on enzyme activity. You will need to keep the pH constant in each experiment, which involves using special buffer solutions. Investigate the effect of pH in the range 4–9 (from acidic through neutral (pH 7) into basic) on the activity of the enzyme amylase and relate this to the way your digestive system works as you learn about the different areas of activity in the gut. Plan your investigation carefully and then, with the permission of your teacher, carry it out.

Activity 6.3

Investigating the effect of temperature on the action of the enzyme catalase

Another enzyme that can readily be used to investigate the effects of temperature and pH on enzyme action is catalase. Many of our body cells produce hydrogen peroxide as a waste product of some of their processes. Hydrogen peroxide is poisonous and so must be broken down. Fortunately, it breaks down readily into water and oxygen gas, both not only harmless but also useful to our cells.

The human digestive system

Although hydrogen peroxide breaks down very slowly, catalase speeds up the process so the hydrogen peroxide does not have time to poison the cells. Catalase is found in a number of tissues, including liver and potato. If you add a piece of liver or potato to hydrogen peroxide, you will see bubbles of gas and hear a fizzing sound. If you design your apparatus so you can either collect and measure the oxygen gas produced, or simply count the number of bubbles produced, you can use the rate of production of gas as a way to measure the activity of the enzyme. The simplest way to obtain catalase is to liquidise a potato with water or, if a blender is not available, to chop a potato into very small pieces.

You will need:

- potato homogenate (potato blended with an equal volume of water and the mixture left to settle; the liquid layer is removed and used in the experiment – this is the potato homogenate and contains the enzyme catalase from the broken potato cells)
- a boiling tube or flask, bung and delivery tube
- 5% hydrogen peroxide solution
- a beaker of water at 30 °C
- a beaker of cold water
- a stopwatch or clock with a minute hand

Method

1. Place 5 cm^3 of potato homogenate in the boiling tube with the delivery tube attached.
2. Place this tube and the hydrogen peroxide solution in the beaker of water at 30 °C for 5 minutes.
3. Remove the hydrogen peroxide solution from the water bath, measure out 5 cm^3 and add it quickly to the potato homogenate, replacing the bung as quickly as possible. The delivery tube needs to pass into another beaker of water so that any bubbles of gas produced in the experiment can be seen and counted (see Figure 6.6).
4. At the same time, start the stopwatch and count the number of bubbles of oxygen gas produced in the first minute after adding the hydrogen peroxide. You may like to count the number of bubbles in the second minute as well. Record your results. You can modify this basic procedure to investigate the following:
    - The effect of temperature on the activity of catalase (by changing the temperature of the water bath and repeating the method at different temperatures):
        - You can plot a graph to show the number of bubbles produced in the first minute against the temperature of the reacting mixture.
        - You could also plot the number of bubbles formed in the second minute and compare the shape of the graphs.

      Explain the shape of the graphs you have plotted. Is it what you expected? How does temperature affect enzyme activity? How could you improve your experimental technique?

- The effect of pH on the activity of catalase (by changing the pH of the reacting mixture using different buffer solutions):
  - You can plot a graph to show the number of bubbles produced in the first minute against the pH of the reacting mixture.
  - You could also plot the number of bubbles formed in the second minute and compare the shape of the graphs.

  Explain the shape of the graphs you have plotted. Is it what you expected? How does pH affect enzyme activity? How could you improve your experimental technique?

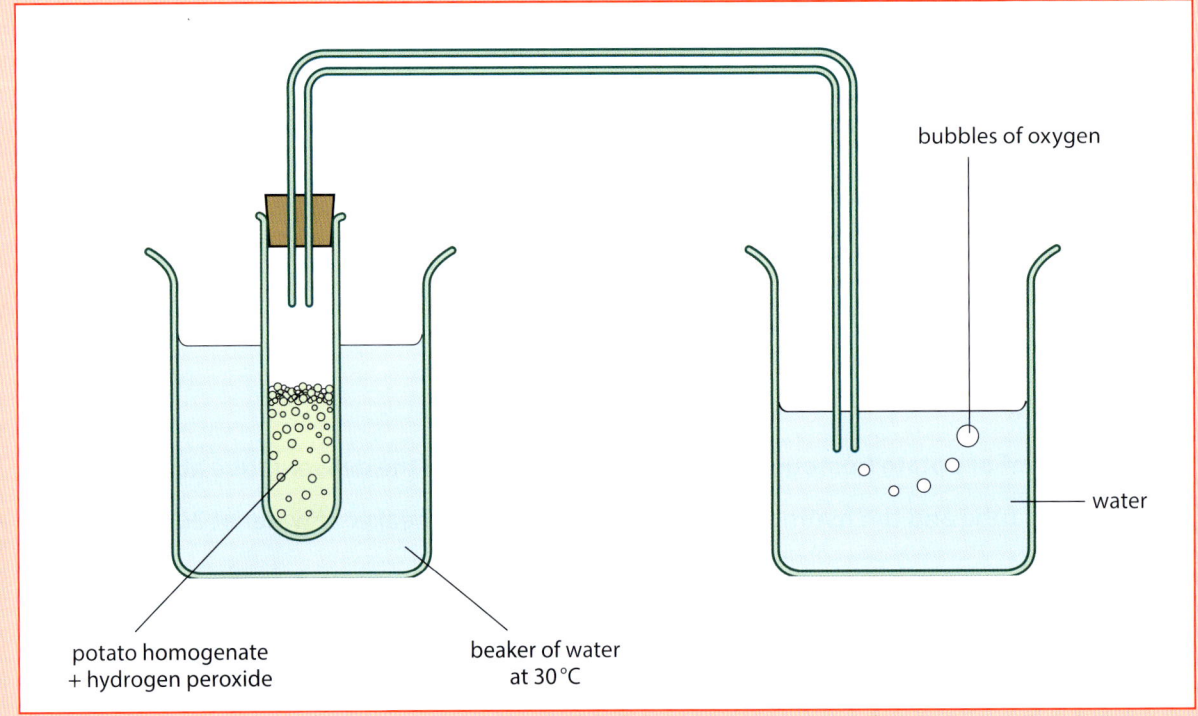

Figure 6.6 Measuring the action of the enzyme catalase.

Checkpoint questions

1. Define an enzyme.
2. a Explain why enzymes are so important in your body.
   b Explain why we have so many different enzymes.
3. Describe with the help of diagrams how enzymes work.
4. a Explain how temperature and pH affect how enzymes work.
   b Using Figure 6.4, state the optimum temperature for amylase.
   c Using Figure 6.5, state the optimum pH for pepsin and for pancreatic amylase.

The structure and function of the human digestive system

The process by which the food on your plate is taken into your body, broken down and used by your cells, with the indigestible material removed, is very complex and it involves the various areas of your digestive system (also known as the **alimentary canal**).

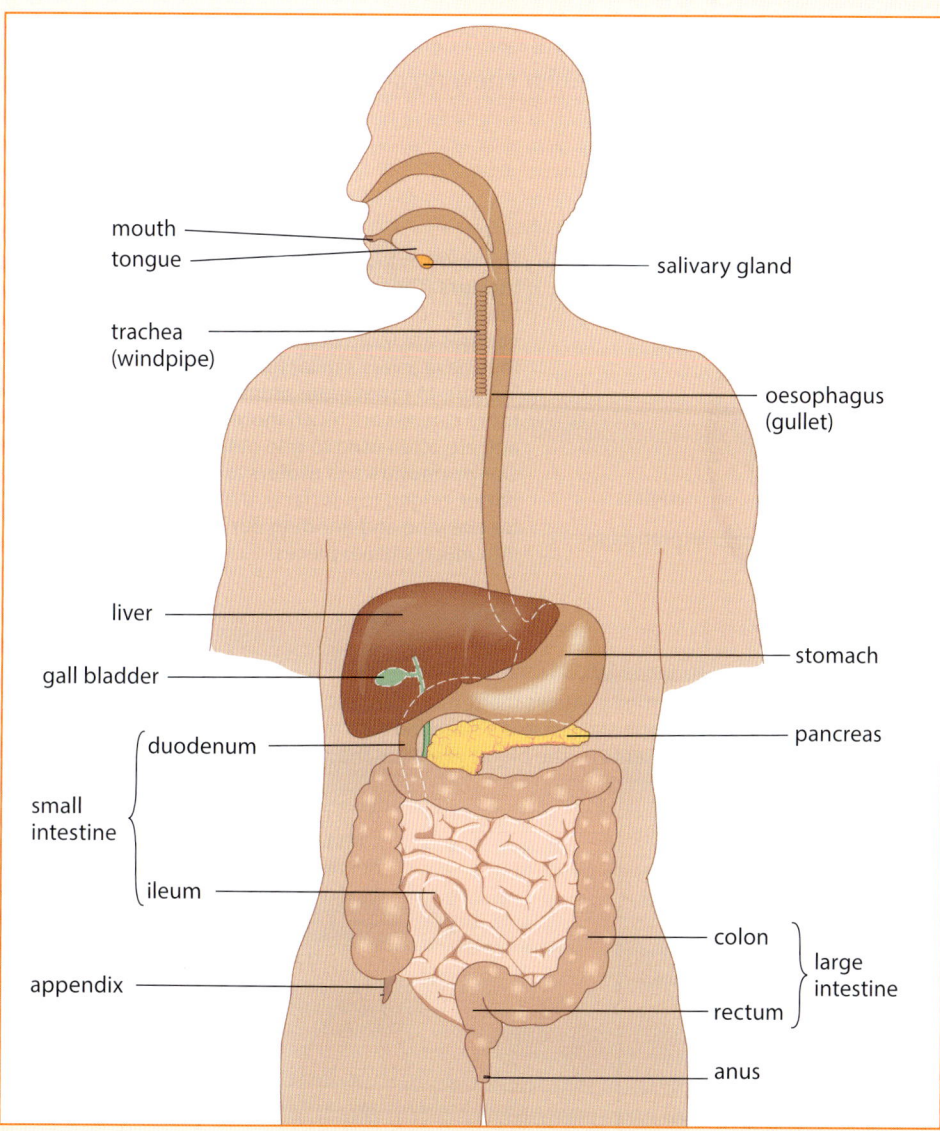

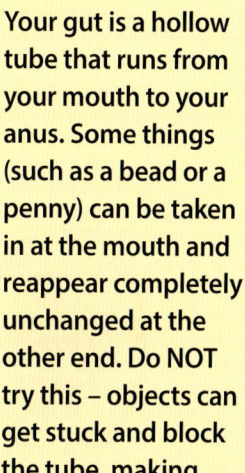

DID YOU KNOW?

Your gut is a hollow tube that runs from your mouth to your anus. Some things (such as a bead or a penny) can be taken in at the mouth and reappear completely unchanged at the other end. Do NOT try this – objects can get stuck and block the tube, making you very ill.

Figure 6.7 The human digestive system

As we eat our food, it sets off on a journey of digestion.

Ingestion

The first stage in the journey of our food is **ingestion**, or taking food into the body through the mouth. We bite off a chunk of food using our teeth, and then physically chop the food up into smaller pieces by chewing it.

Our teeth play a very important role at the beginning of the process of digestion, physically breaking down our food and providing a greater surface area for our digestive enzymes to work on. This process is called **mastication**.

Teeth have evolved to be very strong; in fact, the enamel that covers them is the strongest substance made by the human body. Teeth are needed for a variety of different jobs – gripping, tearing and chewing food, for example. The shape of different teeth means they are ideally suited to their different functions. Since humans have a very varied diet (we are omnivores, so we eat animals and plants), we also have of different types of teeth. The incisors and canines are used for biting, while the premolars and molars are used for chewing and crushing food.

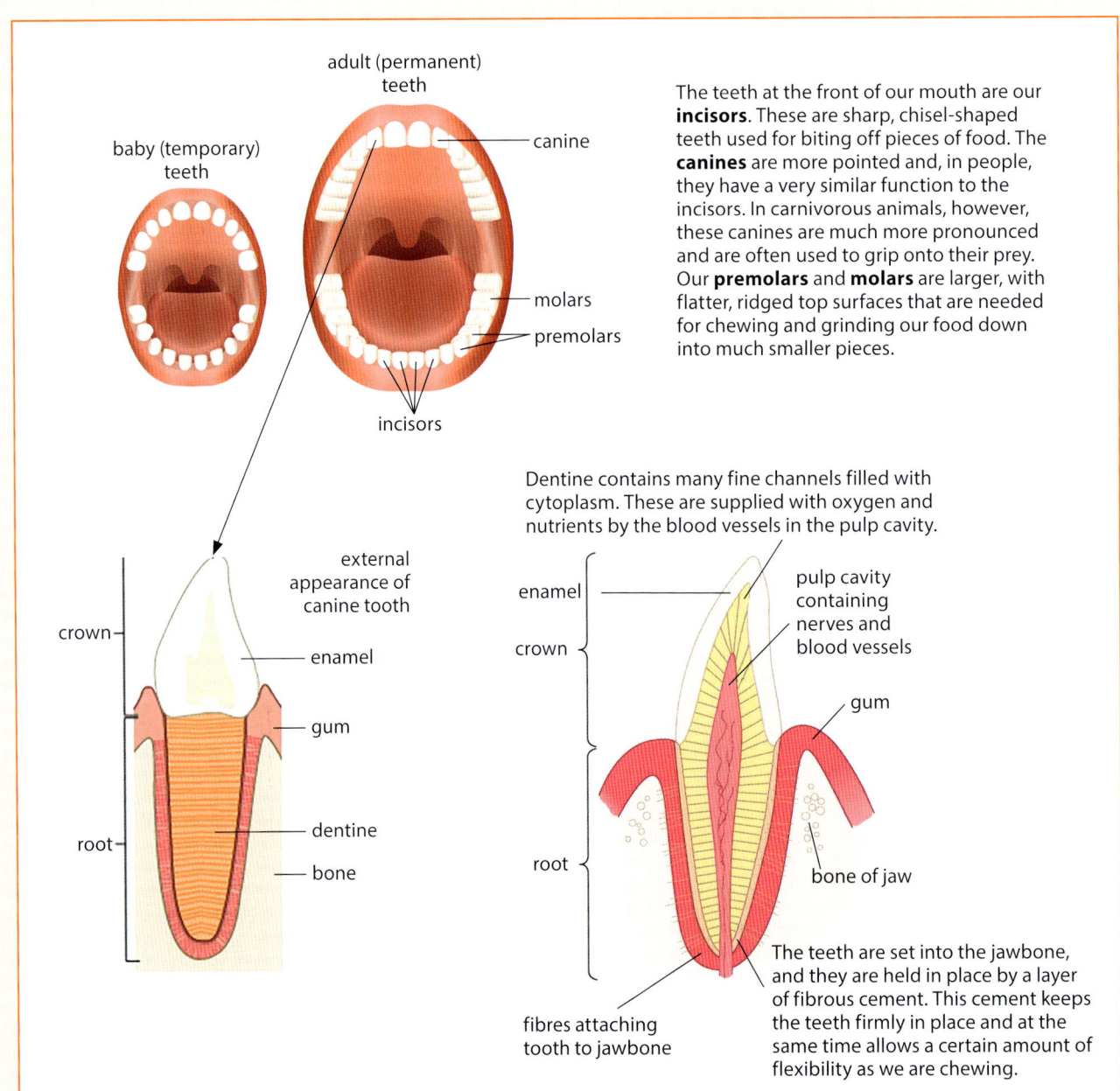

Figure 6.8 The structure of teeth makes them very well adapted to their various functions.

124 The human digestive system

We have two sets of teeth during our lifetime. Our milk, baby or deciduous teeth start to come through during the first year of life, and by the time we are between 2–3 years old, we have a full set of 20 milk teeth. By the time we are 6 or 7, some of these will have started to fall out again, pushed out by our permanent, adult teeth growing through underneath them. We have more adult teeth than milk teeth. There are 32 permanent teeth in a full set. The difference is found at the back of the mouth, where big molars are found in adults that simply wouldn't fit into the mouth of a small child.

The top surface of the teeth is covered by a layer of non-living enamel, and under this is the living dentine. This is not as hard as enamel, but it is still very hard, being similar to bone. In the centre of the tooth is the pulp cavity, which contains nerves and blood vessels.

The process of tooth decay

In theory, our adult teeth should last us all through our lives. Unfortunately this doesn't always happen, because our teeth can be affected by the bacteria that cause dental caries.

There are many different bacteria that are found naturally in the mouth. These bacteria, combined with food and saliva, form a thin film called plaque, on the teeth. If these bacteria are given a sugar-rich diet (in other words, if we eat a lot of sweet, sugary food), they produce a lot of acid waste.

This acid attacks and dissolves the tough enamel coating of the teeth.

Once through the enamel, the acid also dissolves away some of the dentine and then the bacteria can get into the inside of the tooth.

The bacteria then reproduce and feed, eating away at the tooth until they reach the nerves of the pulp cavity, causing toothache. The bacteria and the acid they produce can eat away at the teeth to the extent that they break up completely if we don't get effective dental treatment.

Bacteria do not only attack our teeth. The same bacteria can affect our gums, causing **periodontal disease**. The symptoms include tender gums, bleeding when we clean our teeth and eventually the possible loss of all our teeth from gum disease.

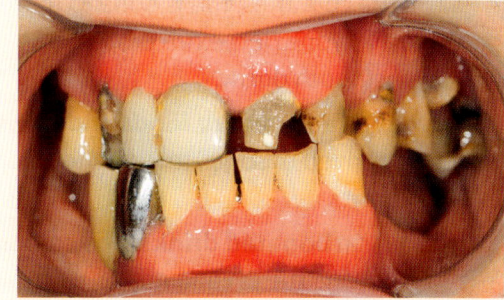

Figure 6.9 Tooth decay not only causes pain and bad breath, it doesn't look very nice either!

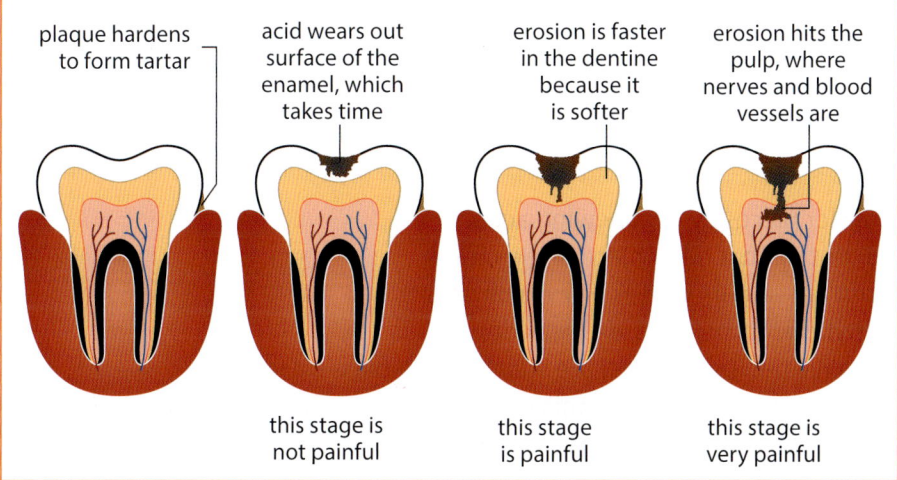

Figure 6.10 The stages of tooth decay

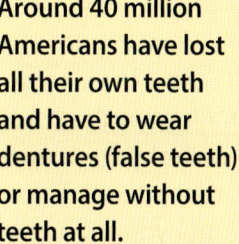

DID YOU KNOW?

Around 40 million Americans have lost all their own teeth and have to wear dentures (false teeth) or manage without teeth at all.

Taking in lots of acidic foods and drink, such as fruits and cola, can also weaken the enamel on our teeth. This is particularly the case if we clean our teeth straight after an acidic drink such as fruit juice or cola, when the softening effect on the enamel is strongest. Brushing the teeth can actually wear the enamel away.

Tooth and gum disease are extremely common all over the world. They cause pain, bad breath, loss of teeth and difficulty eating. The good news is that they can both be avoided, especially if good dental care is available. Ways to avoid tooth decay include the following:

- **Brush the teeth and gums regularly twice a day:** This removes plaque from the teeth, preventing the build up of a sticky, acidic film over the enamel.
- **Use a fluoride toothpaste:** Fluoride helps to make the enamel stronger and more able to resist attack by acids.
- **Floss the teeth regularly:** This helps to remove bits of food and plaque from between the teeth where they can easily build up.
- **Have fluoride in the drinking water:** In some countries, this is done to make sure everyone gets extra protection against tooth decay.
- **Avoid sweet, sugary foods and drinks:** If the bacteria on the teeth are deprived of sugar, they cannot produce acidic wastes and the teeth are safe.
- **Avoid carbonated drinks:** The carbon dioxide used to make them fizzy lowers the pH and makes them acidic, attacking the teeth. If they are sugary as well, then it is even more important to avoid them.
- **Have regular dental checkups:** A dentist can clean our teeth more thoroughly than we can, and any early signs of decay can be treated. The teeth won't heal themselves, but any infected tooth can be removed and cavities can be replaced by a filling.

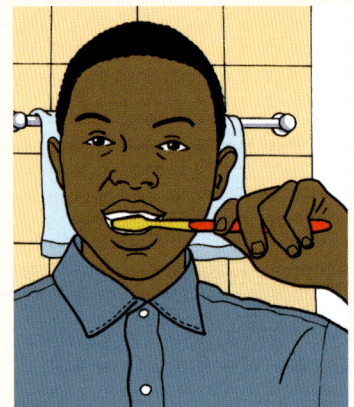

Figure 6.11 Brushing teeth removes plaque.

Figure 6.12 Flossing helps to remove food particles from between the teeth.

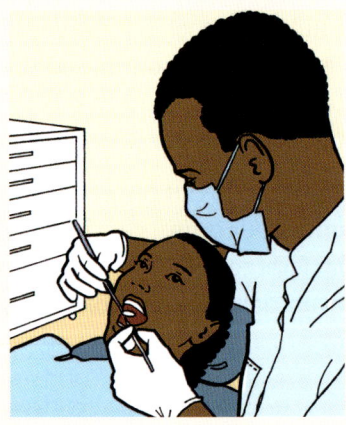

Figure 6.13 Regular dentist visits means any problems are picked up early and dealt with.

Moving the food on

The breaking down of food into smaller pieces by the chewing action of our teeth isn't the only part of digestion that takes place in the mouth. Our food is also coated in saliva from the salivary glands. Saliva contains a **carbohydrase** **enzyme** called **amylase**. Carbohydrases break down carbohydrates. The amylase in our saliva begins to digest the starch in complex carbohydrates such as bread or potatoes, turning it into simpler sugars. The saliva-coated chunk of food (called a bolus) is moved to the back of the throat to be swallowed.

Swallowing is a reflex action that takes place when food reaches the back of the throat. As we swallow, our epiglottis closes over the trachea, preventing food going down into the lungs; we can't swallow and breathe in at the same time. If we try to, we will choke and our body will produce violent coughing and heaving movements to make sure the food doesn't get down into our lungs, where it can cause serious problems (see Figure 6.14).

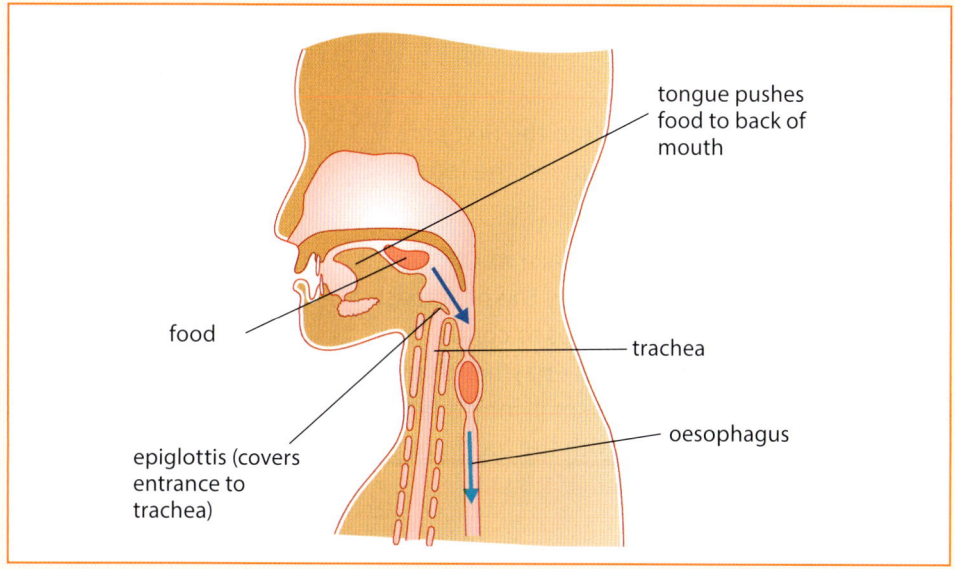

Figure 6.14 The swallowing reflex means you do not get food going down into your lungs – and you can't breathe in while you are swallowing food.

As a result, we can eat at any angle we like – even standing on our head – because food does not rely on gravity to arrive in our stomach. Peristalsis is not confined to our oesophagus – it is important all the way through the gut to move the food along as it is digested, to mix the food with the digestive enzymes produced in the various parts of the gut and to continue the physical break up of the food. When your food is swallowed, it travels down the oesophagus or gullet, in a process called peristalsis.

The walls of our alimentary canal have a layer of circular muscles that form rings around it and a layer of longitudinal muscles that run the length of the gut. Waves of alternate contraction and relaxation of the different muscles (see Figure 6.15) move food from one end of the gut through to the other.

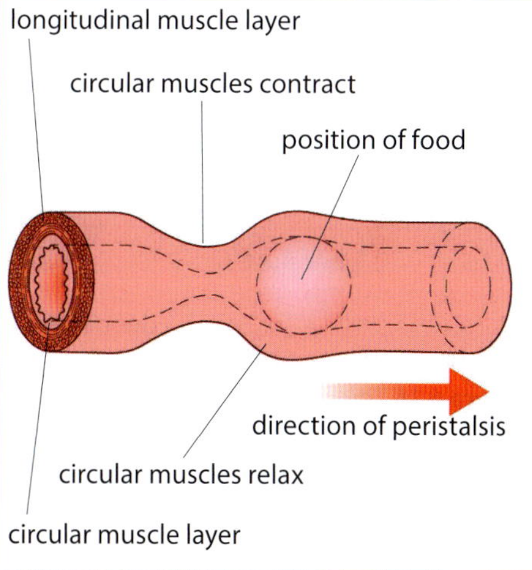

Figure 6.15 Muscular action in your alimentary canal pushes the food along.

Stomach churning activity

At the lower end of the oesophagus, the food passes through a ring of muscle called a **sphincter** into the **stomach**. This sphincter is usually closed except when we are swallowing food or being sick. The stomach is a muscular bag that produces **protease** **enzymes** to digest protein. The main protease made in the stomach is pepsin. The stomach also produces a relatively concentrated solution of **hydrochloric acid**. This acid kills most of the bacteria that are taken in with our food.

The structure and function of the human digestive system 127

The acid also indirectly helps in the breakdown of the protein in our food, because pepsin works best in acidic conditions (see Figure 6.5). Our stomach also makes a thick layer of mucus, which protects the muscle walls from being digested by the protease enzymes and attacked by the acid. The muscles of the stomach squeeze the contents into a thick creamy paste containing partly digested protein along with all the rest of our food.

The enzymes of the small intestine

After spending between one and four hours in the stomach, a paste of partly digested food is squeezed out through another sphincter into the first part of the small intestine called the **duodenum**. As soon as it arrives, the food is mixed with two more liquids – **bile** and enzymes.

The first part of the small intestine (the duodenum) cannot make its own enzymes, but this doesn't matter because they are supplied by the pancreas. Part of the **pancreas** makes the hormone **insulin**, which helps to control our blood sugar level. The rest of the pancreas makes and stores enzymes that digest carbohydrates, proteins and fats. As food enters the small intestine from the stomach, these enzymes are released and are mixed with the food paste by muscle action.

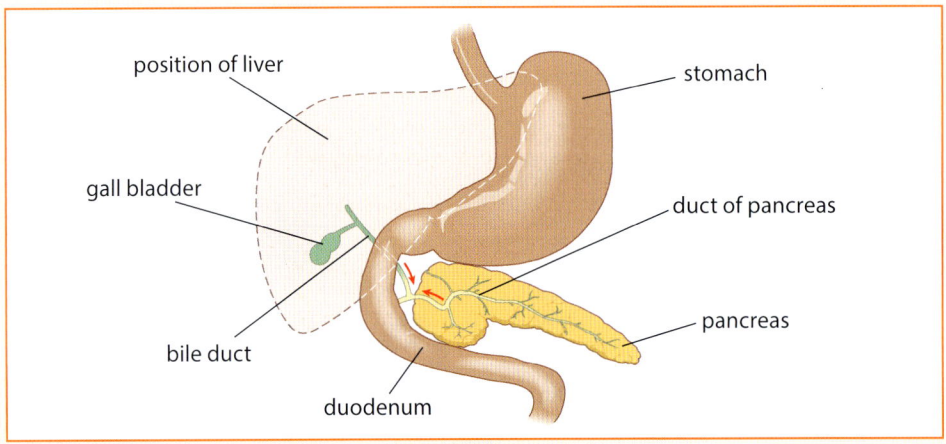

Figure 6.16 The liver and pancreas are important to the successful digestion of food in a number of ways.

Bile is also added to the churned up food in the duodenum. Bile is a greenish-yellow alkaline liquid that is produced in the liver (a large reddish-brown organ that carries out lots of important jobs in the body). It is made by the liver cells and then passes into the intestine or is stored in the gall bladder until it is needed. As fatty food moves into the duodenum from the stomach, bile is squirted into the contents of the duodenum. The bile does two important jobs.

- It neutralises the acid from the stomach and makes the semi-digested food alkaline. This is ideal for the enzymes in the small intestine, which work most effectively in an alkaline environment (see Figure 6.5.).
- Bile also **emulsifies** the lipids in your food. It breaks down large drops of lipids into smaller droplets. This provides a much bigger surface area for the **lipase** enzymes to work on to break down the lipids completely into fatty acids and glycerol.

The human digestive system

The rest of the small intestine (the **ileum**) is a long (6–8 m) coiled tube that produces carbohydrase, protease and lipase enzymes of its own. The tube is coiled up to fit inside the body cavity. The food, which is rapidly becoming completely digested in the alkaline environment, is moved along by peristalsis.

Throughout the small intestine, enzymes speed up the breakdown of large molecules into smaller molecules. The main types of enzymes found in the human digestive system are summarised in Table 6.1.

Table 6.1 Summary of enzyme action in the human digestive system

Types of enzymes	Where it is found in the gut	What it acts on	What the breakdown products are
Carbohydrase, for example amylase, maltase	Salivary glands, pancreas and small intestine	Starch	Glucose
Protease, for example pepsin, trypsin	Stomach, pancreas and small intestine	Protein	Amino acids
Lipase	Pancreas and small intestine	Lipids (fats and oils)	Fatty acids and glycerol

Absorption

To understand **absorption** and the adaptations of the body that make this possible, remind yourself of the processes of diffusion and active transport by looking back in Chapter 2 and of the importance of the surface area : volume ratio by looking into your Biologist's toolkit.

Once the food molecules have been digested, producing glucose, amino acids, fatty acids and glycerol, they are absorbed by the body (**absorption**). They leave the small intestine by diffusion and move into the bloodstream to be carried around the body to the cells that need them. The lining of the small intestine is specially adapted to allow as much diffusion as possible, and as rapidly as possible.

The ileum has many finger-like projections of the lining (called **villi**) to increase the surface area for diffusion, and each individual villus in turn is covered in even smaller projections called **microvilli**. The villi also have a rich blood supply that carries away the digested food molecules and maintains a steep diffusion gradient (see Chapter 2). The diffusion distances are very small, and the whole process takes place in a water-based solution.

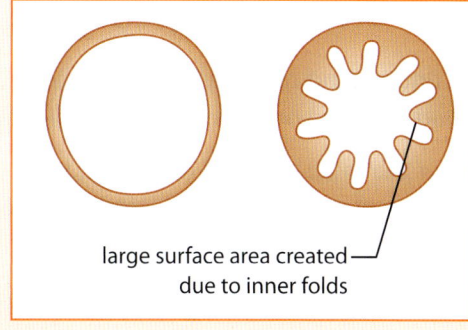

large surface area created due to inner folds

Figure 6.17 By folding the lining of the gut into many tiny villi – far more than are shown here – the surface area over which diffusion takes place is made much bigger, so lots more digested food is absorbed into our blood.

All of these factors make the absorption of the digested food molecules from the small intestine into the blood supply very efficient. Active transport is also used to move glucose and other substances from the intestines into the blood against a concentration gradient if necessary.

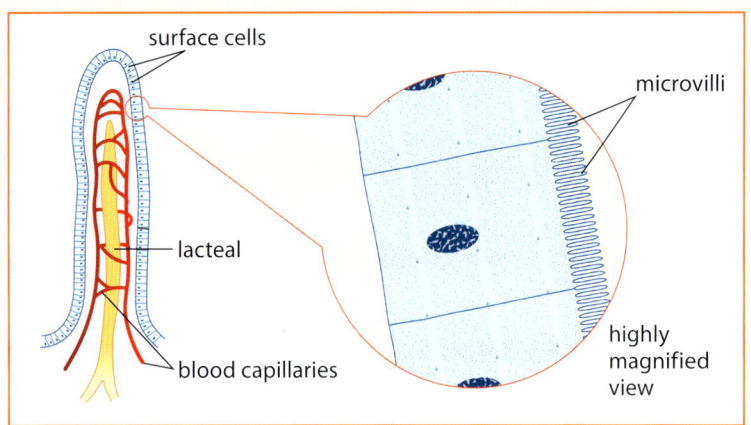

Figure 6.18 A diagram of a single villus magnified greatly. Thousands of villi and microvilli make it possible for all the digested food molecules to move from the small intestine into the blood by diffusion and active transport.

> **DID YOU KNOW?**
> The villi are tiny – each individual villus is only 1–2 mm long. However, as there are millions of villi, and each of them in turn is covered in microvilli, it has been calculated that the surface area of your small intestine is actually around 300 m^2 – that's about half the area of a tennis court.

The glucose molecules and amino acids go directly into the blood in the **hepatic portal vein**, which carries blood directly from the ileum to the liver. The fatty acids and glycerol move initially into lacteals, which are part of the lymph system (see pages 181–2 for more about lymph). The lymphatic fluid with its load of fatty acids and glycerol then eventually drains into the blood as well. Once the digested food molecules have all been taken into the blood, they are taken in the hepatic portal vein to the liver, which processes some of the food, especially the amino acids from the breakdown of protein foods. The remaining products of digestion are carried around the body to the cells where they are needed. They are built up into the molecules required by the cells. This is called **assimilation**.

The end of the story

To understand how water moves into the body from the large intestine, remind yourself of the process of osmosis by looking back at Chapter 2.

After the digested food molecules have been absorbed into the blood, a watery mixture of enzymes, undigested food (mainly cellulose), bile pigments, dead cells and mucus is left in the small intestine, and is moved along by muscle contractions into the large intestine. In this wide, thin-walled tube, water is absorbed back into the bloodstream by osmosis. By the end of the large intestine, the thick paste that remains is called **faeces**. The journey ends as the faeces leave the body through the rectum and the anus due to a final set of muscle contractions.

This removal of the faeces from your body is called **egestion** – the removal of unusable and undigested material. It is *not* **excretion**, because excretion involves the removal of the waste products from the cells, and the final contents of your digestive system have never been inside your cells. The number of times people pass faeces varies from person to person and with the diet that is eaten. Once a day is probably average, but some people go several times a day, while others may go only once or twice a week.

> **DID YOU KNOW?**
> The colour of your faeces comes from the breakdown products of your bile. If you have problems with your liver, gall bladder or bile duct, your faeces may turn silvery-white as they are missing the pigments from your bile.

If the faeces remain in the large intestine for too long, too much water is absorbed from them. They become compacted, hard and difficult to pass out of the body. This is **constipation**, and the most common causes are a lack of fibre in the diet and not drinking enough water (see Chapter 5). Straining to pass faeces can cause haemorrhoids (piles) or a tear in the anus. Constipation can usually be treated relatively easily. This may involve:

- eating more fibre (which gives the muscles of the gut more material to work on)
- drinking more water (so the faeces remain soft)
- sometimes taking laxatives (substances that stimulate the gut to contract and force out the faecal material).

If the faeces become completely compacted (which happens very rarely), they can block the gut. This is a very serious situation that may have to be relieved by surgery.

On the other hand, if an infection causes the gut to contract more strongly or more rapidly than usual, the faeces that are produced may be very loose and watery. This is called **diarrhoea**. Often this condition clears up within 24 hours, but in the very young and the very old – and in anyone if it persists – diarrhoea can be fatal as it causes dehydration of the tissues. It can be treated very simply by giving the sufferer frequent drinks of water with rehydration salts (mainly salt and sugar). These replace the fluids that are being lost and keep the body tissues hydrated until the immune system overcomes the infection.

So, at the beginning of the process of digestion, we eat a meal or snack, taking food into our body. After several hours, the digestive process – including ingestion, physical and chemical digestion, absorption and assimilation – will be complete, although egestion will take longer. The time it takes to digest a meal completely will depend on a variety of things, in particular the size of the meal we eat and the type of food it contains. If the nutritional balance of the food we eat is right, the chances of our body remaining fit and healthy throughout our lives are greatly increased.

Possible SBAs based on the human digestive system

- → Investigate the effect of temperature on different enzymes.
- → Investigate the effect of different pH on different enzymes.
- → Investigate the protease enzymes found in fresh pineapple and what affects their activity.
- → Carry out a survey into the dental health of your classmates or family.
- → Investigate how regularly your classmates or local community visit the dentist.

- → Work with your local dentist and investigate the most common teeth problems in your area or investigate the state of dental welfare in your country.
- → Examine the structure of the different areas of the digestive system using stained, prepared microscope slides.
- → Develop models to demonstrate the benefit of villi (increased surface area) on the rate and amount of diffusion in the digestive system.
- → Plan and carry out investigations into the time it takes food to travel through the gut (using marker foods such as sweet corn or red beets).
- → Carry out a survey into the egestion habits of a group of people over a period of time eg fellow students, family members.

End-of-chapter summary

In this chapter, you have learnt that:

- digestion involves the breakdown of large insoluble food molecules into smaller soluble molecules through hydrolysis reactions is catalysed by enzymes
- enzymes are proteins that catalyse (speed up) specific reactions
- each enzyme has an active site that fits the reactants of the reaction it catalyses
- enzymes are affected by temperature and work best at an optimum temperature; if the temperature goes to high, the structure of the enzyme is denatured and it does not work
- enzymes work best at specific pH levels – pH affects the shape of the active site
- different areas of the gut have different pHs to suit the enzymes involved
- the process of eating your food involves ingestion, digestion, absorption, assimilation and egestion
- the human digestive system is a muscular tube running through the body with specialised areas adapted to carry out different parts of the digestive process
- the teeth are important structures for chewing and biting, and adults have more teeth than small children; good mouth care prevents tooth decay
- peristalsis is a wave of muscular contraction pushing food along the gut
- the liver makes bile, which emulsifies fats, increasing the surface area for enzyme action
- the ileum has a very large surface area due to the presence of villi and microvilli; this enables the digested food products to be absorbed into the blood and lymph systems, and carried to the liver in the hepatic portal vein
- water is absorbed from the remaining undigested food by the large intestine and the remaining material is egested from the body as faeces.

End-of-chapter questions

1. What are enzymes made of?
   - A Carbohydrates
   - B Vitamins
   - C Proteins
   - D Fats

2. Which one of the following does not affect the activity of an enzyme?
   - A pH
   - B Temperature
   - C The surface area of the reactants
   - D Light levels

3. Where do extracellular enzymes work?
   - A Outside of the cells
   - B Inside the cells
   - C Inside the mitochondria
   - D Only in the mouth

4. How many adult teeth do we have?
   - A 20
   - B 30
   - C 32
   - D 34

5. Which part of a tooth contains living nerves?
   - A Enamel
   - B Dentine
   - C Cement
   - D Pulp

6. What are the finger-like projections in the small intestine called?
   - A Bilirubin
   - B Microvilli
   - C Sphincters
   - D Villi

7. The enzyme amylase breaks down starch (a polysaccharide) into maltose (a disaccharide, which is a reducing sugar). A student wants to set up an investigation into the action of amylase on starch. He wants to show that starch is broken down to a reducing sugar by the action of amylase. He also wants to show that the enzyme amylase works at its best at human body temperature (37 °C). Plan and describe how he might carry out both of these investigations successfully.

   (10)

8 a What are enzymes? (2)
   b How do enzymes work? (2)
   c List the types of enzymes made in the salivary glands, the stomach, the pancreas and the small intestine. In each case, say which food substance the enzymes break down. (10)

9

   a Name the parts of the digestive system labelled A, B and C. (3)
   b Explain the role of the muscles in the digestive system. (2)
   c Jenny has a burger and a glass of water for her lunch.
      i Name one macronutrient present in her meal. (1)
      ii Describe what happens to this meal in her digestive system. (6)

10 a Make a table to compare digestion in the stomach and in the small intestine. (6)
   b Explain, using diagrams, why bile produced in the liver is so important in the digestion of fats. (4)

11 a Explain how the gut is adapted to allow digested food to be absorbed readily into the bloodstream. (4)
   b Explain what happens if too much water is absorbed into the bloodstream from the material in the large intestine and the problems this can cause. (3)
   c Explain what happens if too little water is absorbed into the bloodstream from the material in the large intestine and the problems this can cause. (3)

12 a Define the terms 'ingestion', 'digestion', 'absorption', 'assimilation' and 'egestion'. (5)
   b A number of problems can occur with egestion. Explain how these problems can affect the health of the individual concerned. (3)

Chapter 7 The human respiratory system

When you have completed this chapter, you will be able to:

- define gaseous exchange and explain the features of a successful gaseous exchange surface
- explain the importance of breathing in humans
- describe the way air is moved into and out of lungs and explain the concept of vital capacity
- describe the structure and function of the respiratory system
- outline the factors that affect your breathing rate
- explain the effects of smoking
- describe the technique of CPR
- define aerobic and anaerobic respiration, and their importance in the cells, including ATP as the energy currency of the cell

What the examiners say

→ Generally, candidates appear familiar with most of the concepts related to this topic. However, weaker candidates are unable to accurately label diagrams of the respiratory system, or accurately define important terms such as aerobic and **anaerobic respiration**.

Revision tip

Go over topics frequently that you find challenging. Make up study cards of the main points to help you remember them.

The first breath a baby takes when it is born signals the start of a new independent life. Why is breathing so important, and how does it work? In single-celled organisms and other small living things, oxygen diffuses into the cell from the air or water, and carbon dioxide diffuses out. However, human beings are much too large, and have far too many cells for simple diffusion from the air to be enough. Breathing brings oxygen into our body and removes the waste carbon dioxide produced by our cells as they carry out their functions.

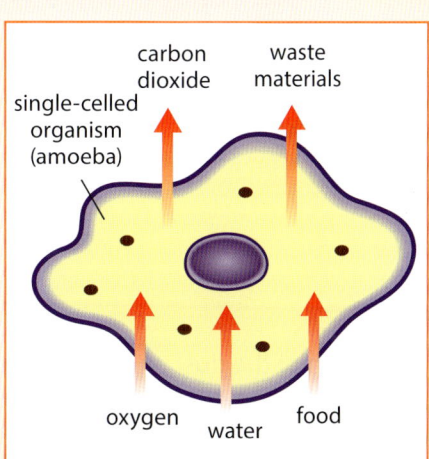

Figure 7.1 Single celled organisms use diffusion to obtain and remove gases.

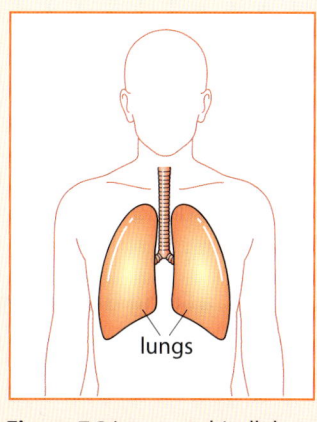

Figure 7.2 Large, multicellular organisms have a respiratory system to move gases into and out of their body.

The human respiratory system

Figure 7.3 shows a diagram of the human respiratory system.

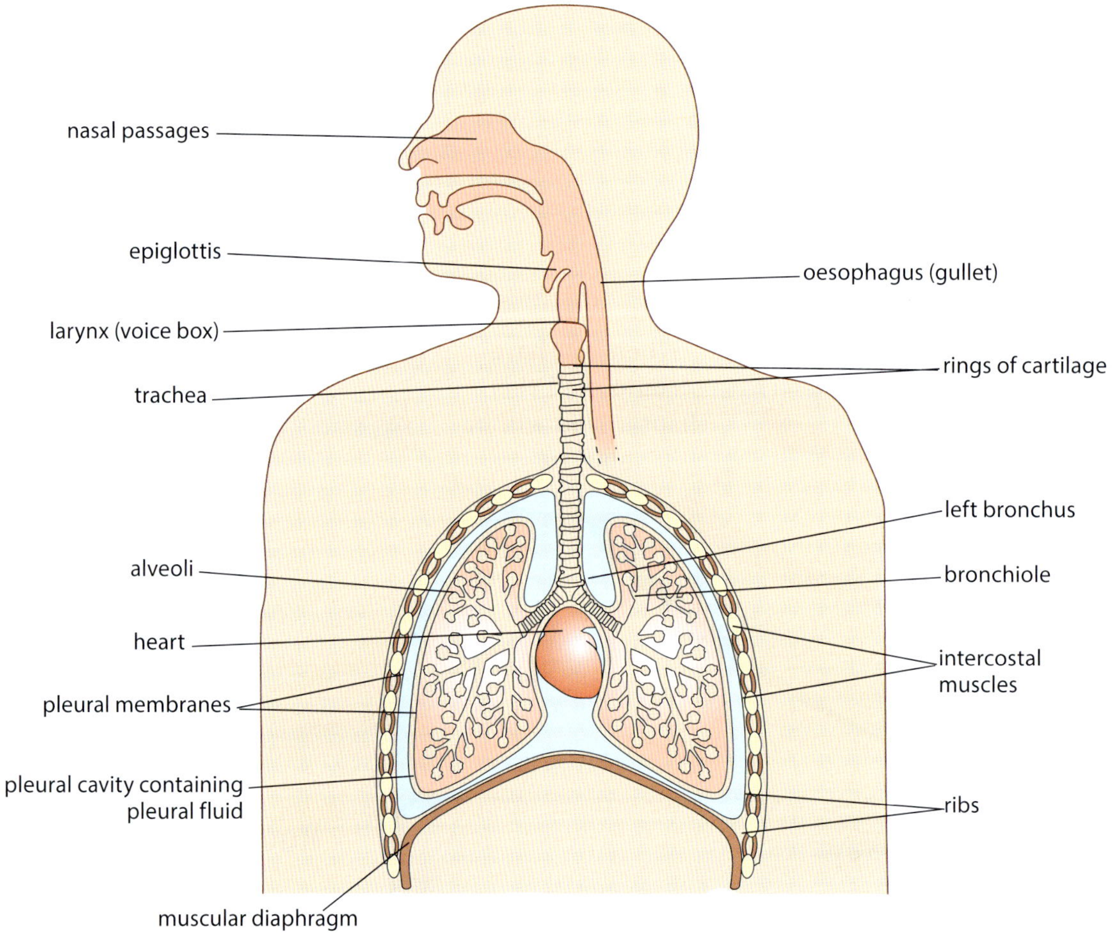

Figure 7.3 The respiratory system supplies your body with vital oxygen and removes poisonous carbon dioxide. The lungs are in the upper part of the body – the thorax. The abdomen contains the digestive system and many other body organs. The diaphragm is a sheet of muscle separating the thorax and the abdomen, keeping the contents of each part of the body quite separate and making breathing movements.

Our respiratory system is beautifully adapted for the job it has to do. The nose contains the nasal passages, which have:

- a large surface area
- a good blood supply
- lots of hairs
- a lining that secretes mucus.

The hairs and mucus filter out much of the dust and small particles such as bacteria and **pollen** that we breathe in, while moist surfaces increase the humidity of the air we breathe into our body and the rich blood supply warms it. All this means that the air we take in is already warm, clean and moist before it gets into the delicate tissues of our lungs.

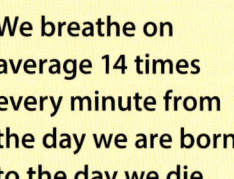

DID YOU KNOW?

We breathe on average 14 times every minute from the day we are born to the day we die.

As air moves down into the trachea, it passes the oesophagus – the entrance to the gut. While air can, and does, make its way down into the gut, this doesn't matter as we can simply bring it back up in the form of a burp. However, it is very important that food does not get into our lungs as it can block the airways or cause a fatal infection, and so the epiglottis closes off the trachea every time we swallow in a reflex action (see Chapter 6). We cannot swallow and breathe at the same time.

At the top of the trachea is the larynx or voice box. By directing air leaving the lungs over the vocal cords (flaps of muscle) in the larynx, we produce the sounds that we use in speech. The trachea itself has a series of incomplete rings of cartilage (shaped like the letter C) that support it and hold it open. They are incomplete so that we can swallow our food. Our oesophagus and trachea run next to each other, so as a lump of food (bolus) moves down our oesophagus, it presses against the trachea. If the trachea had solid cartilage rings, this would be very uncomfortable. However, the open part of the ring faces the oesophagus so food passes by with no problems (see Figure 7.4).

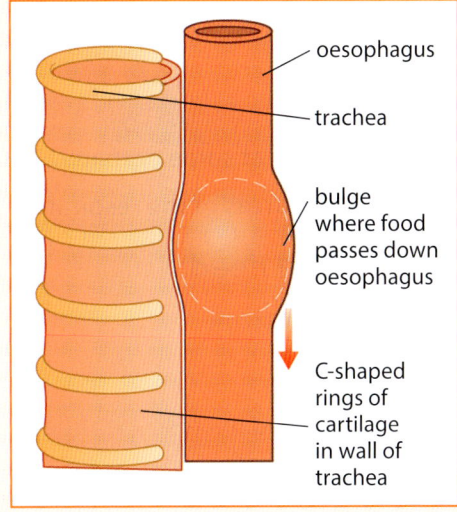

Figure 7.4 The incomplete cartilage rings that support your trachea are a useful adaptation when it comes to swallowing your food successfully.

The lining of the trachea secretes mucus which, like the surface of the nasal passages, collects bacteria and dust particles. The epithelial cells (see Chapter 1) that line the trachea are also covered in hair-like cilia that beat to move the mucus, together with any trapped microorganisms and dirt, away from the lungs and towards the mouth. This mucus is then either swallowed and digested or coughed up.

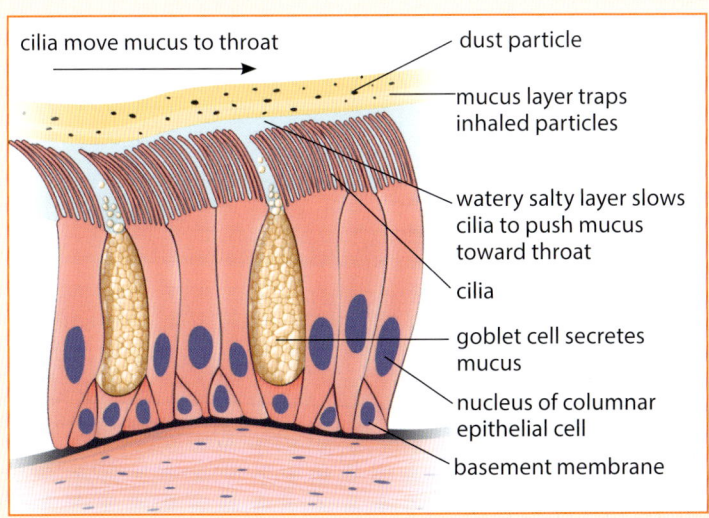

Figure 7.5 The mucus produced in the respiratory system traps dust and bacteria. The cilia beat to remove the mucus and everything trapped in it, stopping it from going into the lungs.

The trachea splits into two tubes: the left and right **bronchi** (singular bronchus), one leading to each lung. The bronchi are also supported by rings of cartilage. Inside your lungs, the bronchi divide into smaller tubes called the **bronchioles**. The bronchioles are much smaller than the bronchi, dividing into ever smaller tubes until they reach the main structures of the lungs – the alveoli (singular alveolus). There are millions of these tiny air sacs, resulting in a massive surface area for the main exchange of gases in the lungs to take place.

DID YOU KNOW?

If you live in or visit a city, the mucus produced when you blow your nose will often be grey or black from the smoke, dust and fumes in the air!

The human respiratory system

Activity 7.1

Investigating the structure of the respiratory system

By looking at some prepared microscope slides, you can see differences in the structure of different areas of the respiratory system.

You will need:

- a microscope
- a lamp
- prepared microscope slides of trachea and lung tissue to show cilia, cartilage rings and alveoli

Method

1. Set up your microscope.
2. Clip the prepared slide into place on the stage and focus carefully.
3. Draw some of the structures that you see and label them as well as you can. Look for the cilia on the epithelium of the trachea and the cartilage rings. In the lungs themselves, look carefully at the structure of the alveoli and try to work out why they are so effective at gaseous exchange.

How is air brought into your lungs?

For your respiratory system to work, you need to move air into your lungs and then move it out again. This is brought about by movements of the ribcage, which you can see and feel, and by movements of the diaphragm, which you cannot.

Breathing movements are controlled by two different sets of muscles that change the pressure in the chest cavity. When we breathe in, the intercostal muscles between our ribs contract, moving our ribs upwards and outwards. At the same time the muscles of the diaphragm contract, flattening the diaphragm from its normal domed shape. These movements increase the volume of the chest (thorax). Since the same amount of gas is now inside a much bigger space, the pressure inside the chest falls. The pressure inside the chest is now lower than the pressure of the air outside. As a result, air moves into the lungs (see Figure 7.6).

When the intercostal and diaphragm muscles relax, the ribs drop down and the diaphragm becomes dome-shaped. The volume of the thorax decreases, so the pressure inside the chest increases, and the air is squeezed and forced out of the lungs – we breathe out. This movement of air into and out of the body is called **ventilation**. These breathing movements can be summarised in a table (see Table 7.1).

Table 7.1 Breathing movements

Inhalation	Exhalation
Intercostal muscles contract.	Intercostal muscles relax.
Ribs move upwards and outwards.	Ribs move downwards and inwards.
Diaphragm muscles contract, allowing it to flatten.	Diaphragm muscles relax, allowing it to return to its normal dome shape.
These movements increase the volume of the chest (thorax).	These movements decrease the volume of the chest (thorax).
Since the same amount of gas is now inside a much bigger space, the pressure inside the chest falls. This means the pressure inside the chest is now lower than the pressure of the air outside it. As a result, air moves into the lungs.	The pressure inside the chest increases, and the air is squeezed and forced out of the lungs; we breathe out. This movement of air into and out of the body is called ventilation.

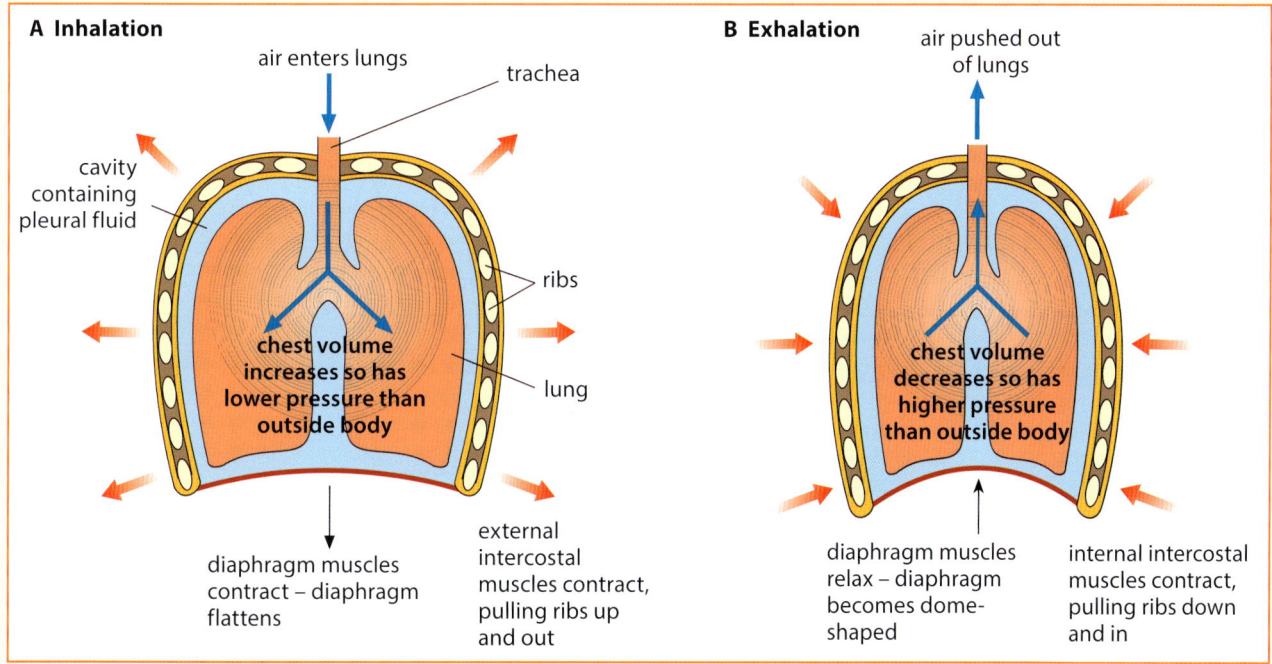

Figure 7.6 Breathing movements bring about changes in the air pressure in the chest. These result in air moving into and out of the lungs. The movements of the diaphragm are hidden but the movements of the ribs can be seen and felt easily.

We have two sets of intercostal muscles. In normal, quiet breathing, we use only our **external intercostal muscles**, which lift our ribs. When these muscles relax, our ribs fall back to their original position due to gravity. However, if we need to breathe out deliberately, forcing the air out of our lungs, or when we are exercising very hard, we also use our **internal intercostal muscles**, which pull the ribs down hard and squeeze more air out of the lungs.

When we are breathing normally, as we go about our daily routine, we do not use the full capacity of our lungs. If we are male, we move an average of 500 cm³ air in and out of our lungs with every breath. If we are female, it is less – we move about 400 cm³. There is always some air in our lungs that we do not move in and out.

DID YOU KNOW?

We can also use the muscles of our abdomen deliberately to increase the amount of air we move into and out of our lungs.

To measure the maximum volume of air we can take in and out of our lungs, we can look at our **vital capacity**. Vital capacity is the maximum amount of air you can take in after you have exhaled (breathed out) as hard as you can (see Figure 7.7). Average vital capacity for men is 4.8 litres and for women, it is 3.1 litres.

The following factors affect your vital capacity:
- Gender – on average, men have bigger vital capacities than women
- Height – on average, taller people have a bigger vital capacity than shorter people
- Age – on average, our vital capacity falls as we get older.

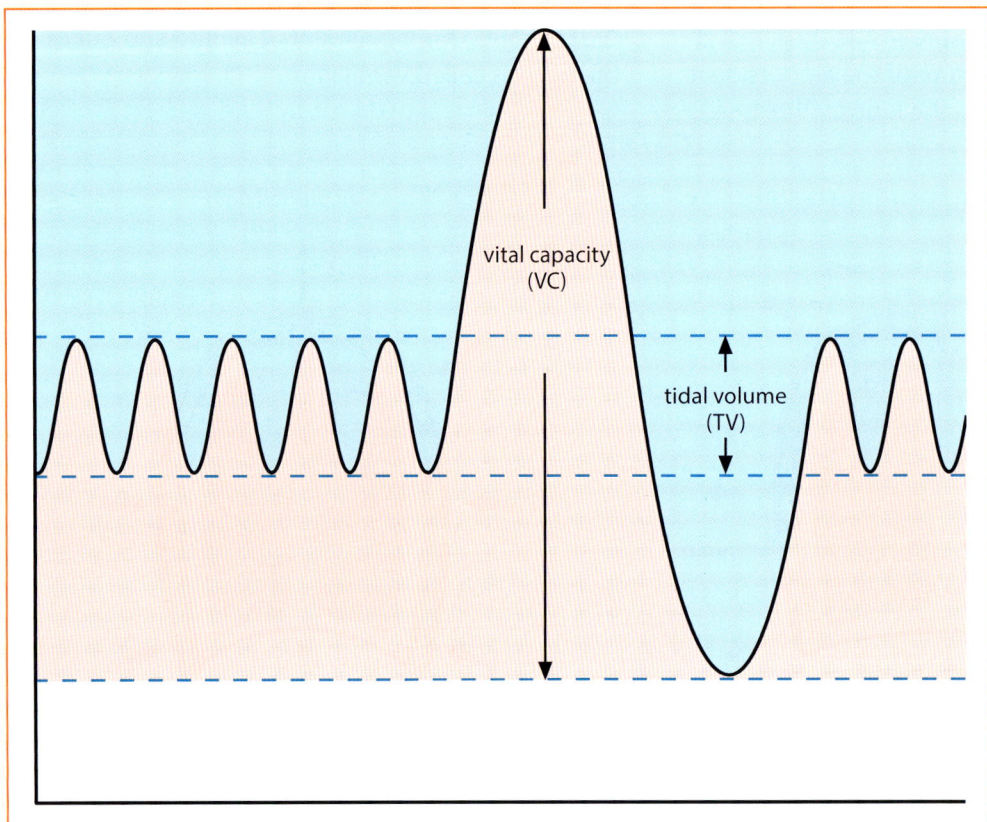

Figure 7.7 Whether you are male or female, for most healthy young people, the vital capacity of your lungs will be several litres of air, compared to your tidal volume of around 0.5 l.

The data in Tables 7.2 and 7.3 show you how different factors affect vital capacity. These data are for males, but similar patterns are seen in females too.

Table 7.2 The effect of height on mean vital capacity in males

Height (cm)	150–155	155–160	160–165	165–170	170–175	175–180
Vital capacity (cm³)	2900	3150	3400	3720	3950	4300

Table 7.3 The effect of age on mean vital capacity in males

Age (years)	15–25	25–35	35–45	45–55	55–65
Vital capacity (cm³)	3425	3500	3225	3050	2850

The human respiratory system

Draw bar charts to represent the data in Tables 7.2 and 7.3 on page 140 and identify the trends seen in the data. This will remind you how bar charts help to make patterns in data easier to identify (see *The biologist's toolkit*, page 100 in Chapter 5 if you have forgotten how to do a bar chart).

Activity 7.2

Investigating breathing movements

It is impossible to see exactly what is happening inside the chest, as we breathe without using special imaging techniques. However, there are two simple investigations you can try in order to build up a useful model of what is going on inside you.

Method

Onion cells do not contain any chlorophyll, so they are not coloured. You can look at them as they are, or stain them using iodine, which reacts with the starch in the cells and turns blue-black.

1. If you stand up and place your hands on either side of your body, on your ribs, you can feel your breathing movements. Experiment with breathing gently and then more deeply, and feel the changes in the size and shape of your ribcage.

2. You can also get an idea of the effect of your diaphragm moving down and up on the pressure in your thorax and the air in your lungs using a model thorax such as the one shown in Figure 7.8. Pull the rubber 'diaphragm' down, then force it up again. Observe the effect this has on the balloon 'lungs'. This gives you an insight into the role of your diaphragm in filling and emptying your lungs.

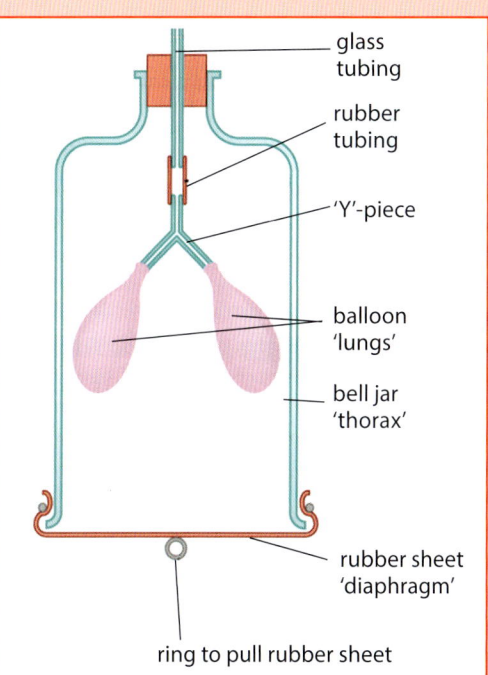

Figure 7.8 Bell-jar model of the thorax

The biologist's toolkit: Recognising the strengths and limitations of models

In Chapter 3, *The biologist's toolkit* introduced the idea of models. We use models in biology to help us visualise things we cannot usually see. Models are very useful, but sometimes they introduce **misconceptions** that make it harder for us to understand what is happening, and not easier.

- When you look at a model, identify its strengths – what does it make it easier for you to understand?
- When you look at a model, try and identify its limitations so that you avoid any misconceptions.

Checkpoint questions

1. a List the main structures of the human breathing system shown in Figure 7.3.
   b Explain the functions of the trachea, ribs, lungs, alveoli and diaphragm.
2. Explain as fully as you can how air is moved into and out of the lungs of the human respiratory system. Draw diagrams if they help your answer.

The process of gaseous exchange

Breathing in supplies us with the oxygen we need for **cellular respiration**, while breathing out removes waste carbon dioxide from the body. But how is this exchange brought about?

When the air is breathed into the lungs, oxygen passes into the blood by diffusion down a concentration gradient. At the same time, carbon dioxide passes out of the blood into the air of the lungs, also by diffusion down a concentration gradient. This exchange of gases takes place in the alveoli, the tiny air sacs with a large surface area that make up much of the structure of the lungs. Remind yourself of the process of diffusion in Chapter 2, and look in *The biologist's toolkit,* page 28 to remind yourself of the importance of surface area.

The movement of oxygen into the blood and carbon dioxide out of the blood takes place at exactly the same time – there is a swap or exchange between the two, so this process is called **gaseous exchange**.

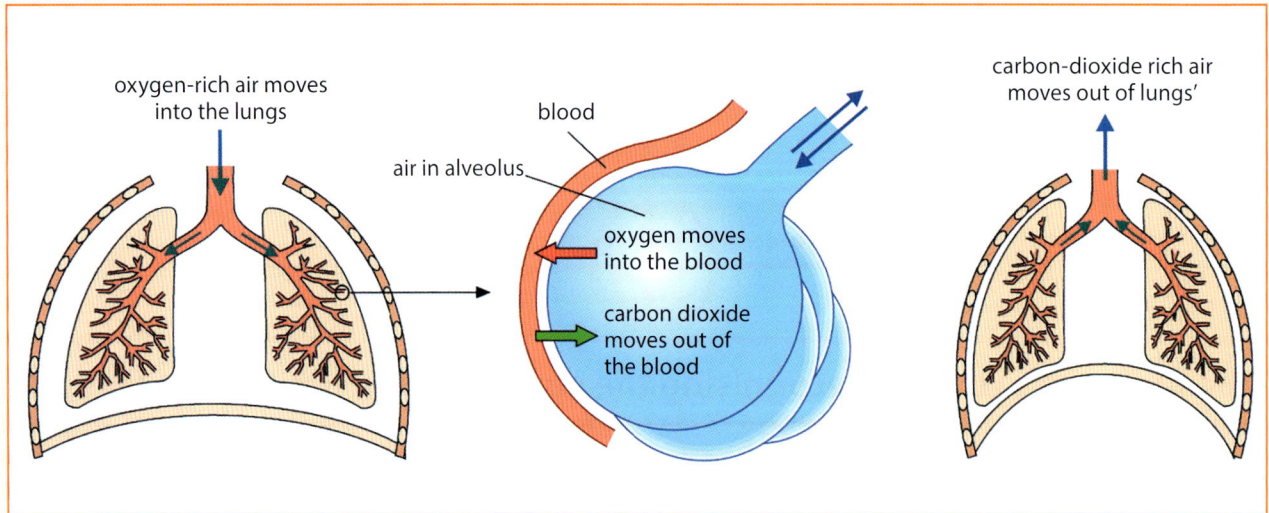

Figure 7.9 An exchange of gases between the blood and the air takes place in the lungs.

Adaptations of the alveoli for gaseous exchange

The tiny air sacs of the alveoli have all the adaptations needed for the most effective possible diffusion of gases into and out of the blood. They have:

- a very large surface area to allow as many gas molecules as possible to diffuse across
- a moist surface, which is important because gases need to be in solution to diffuse into the blood
- thin walls, so that the gases in the air and the gases dissolved in the blood are only separated by two cell layers, a distance of about 0.001 mm, so the diffusion distances are as short as possible
- a rich blood supply, which maintains steep concentration gradients for the diffusion of oxygen and carbon dioxide.

The blood going to the lungs comes from active body tissues, and is low in oxygen and relatively high in carbon dioxide. Oxygen is constantly moved into the blood in the lungs, but more deoxygenated blood immediately replaces it. Similarly, carbon dioxide is constantly delivered to the lungs, where it is diluted in the volume of air, maintaining a concentration gradient between the blood and the air in the lungs.

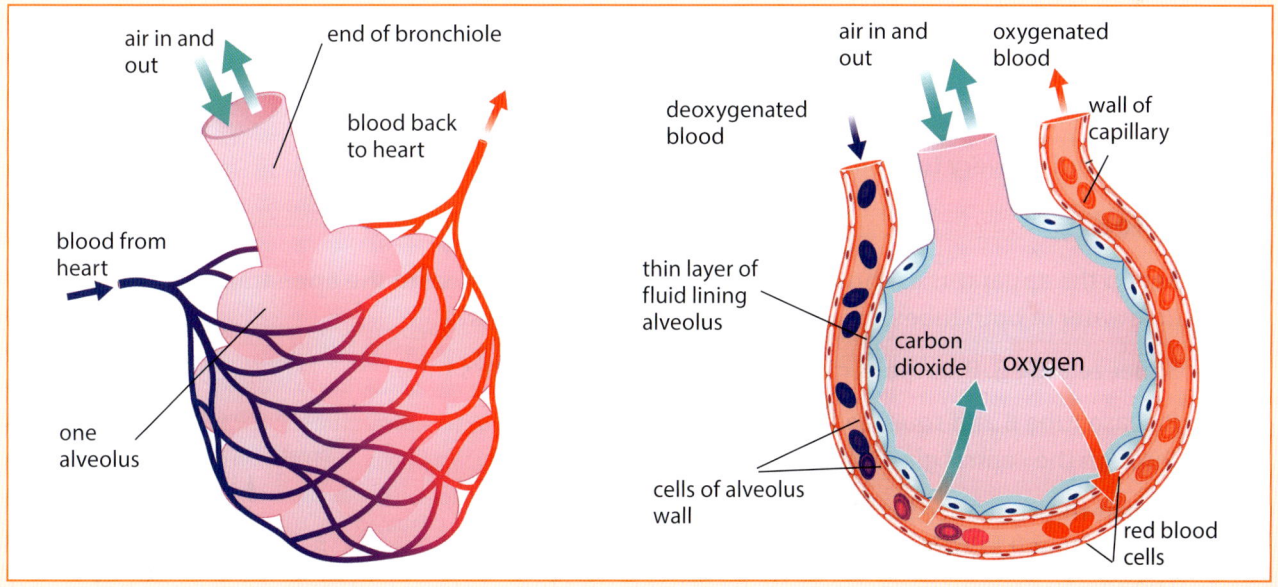

Figure 7.10 The alveoli are the site of very efficient gas exchange in the lungs – they have a large surface area, moist surfaces, short diffusion distances and a rich blood supply maintaining steep concentration gradients.

It is easy to say that there is an exchange of the gases oxygen and carbon dioxide between the air and the blood, but we need evidence to support this idea. The breathing movements only tell us that air is moved into and out of the lungs – they do not tell us what happens to it there. However, an analysis of the gases in the inhaled and exhaled air shows clearly the differences in the quantities of some of the main gases (Table 7.4).

DID YOU KNOW?

It has been calculated that the lungs contain about 500 million alveoli. If all of the alveoli in both of your lungs were spread out flat, they would have an enormous surface area – some people calculate they would cover a tennis court!

Table 7.4 An analysis of the air taken into and breathed out of the lungs shows how the chemical makeup is changed by the diffusion that takes place in the lungs.

Atmospheric gas	Air breathed in	Air breathed out
Nitrogen	About 80%	About 80%
Oxygen	21%	16%
Carbon dioxide	0.04%	4%

Activity 7.3

Comparing the carbon dioxide content of inhaled and exhaled air

A relatively simple experiment can be carried out to demonstrate that air breathed out is different to air breathed in. Limewater is used as an indicator for carbon dioxide. The clear liquid turns cloudy when carbon dioxide is bubbled through it. The faster it turns cloudy, the more carbon dioxide is present.

You will need:

- two boiling tubes
- limewater
- bungs and delivery tubes linked as shown in Figure 7.11

Method

1 Set up the apparatus as shown in Figure 7.11. Observe and record the appearance of the limewater in both tubes.

2 Squeeze tube B. Breathe in gently through the central glass tube.

3 Release tube B and squeeze tube A. Breathe out gently through the central glass tube.

4 Repeat this sequence for several minutes. You will draw air in so that it bubbles through one tube of limewater. The air you breathe out will bubble gently through the other tube of limewater. DO NOT blow too hard. DO NOT SUCK hard on the tube.

5 Observe any changes in the limewater in both tubes.

6 Write up your investigation and record your results. Explain your observations as fully as you can in terms of the air that is inhaled and exhaled from your lungs.

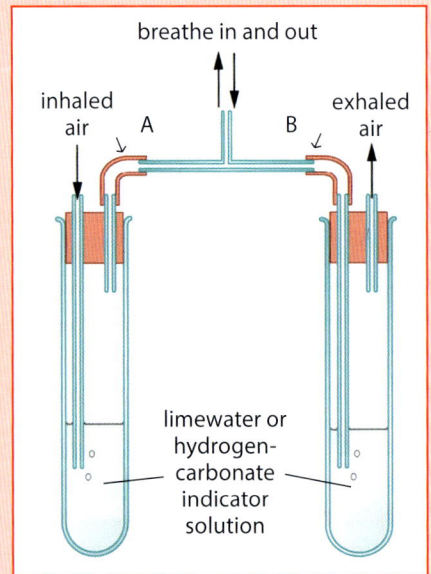

Figure 7.11 Using relatively simple apparatus, it is possible to see differences between the air we breathe in and the air we breathe out.

The human respiratory system

Checkpoint questions

3. Explain the term 'gaseous exchange' and why it is so important.
4. Describe how the lungs are adapted to allow gaseous exchange to take place as effectively as possible.
5. a Draw graphs or bar charts to show the difference in composition between inhaled and exhaled air.
   b On each graph, add an extra bar to show what you would expect the moisture content of inspired and expired air to be.

The biologist's toolkit: Key chemistry: indicators

Indicators are used to tell us that something is present – or absent. An indicator usually changes colour to indicate that something is present. You have already used some indicators in your biology course:

- Iodine is an indicator; if it turns blue black, starch is present.
- Benedict's solution goes from blue to brick orange to indicate the presence of a reducing sugar.
- Biuret solution is used to indicate the presence of protein when it turns purple.
- Universal indicator shows us the pH of a solution; different colours represent different pHs, which in turn are the result of different concentrations of hydrogen ions.
- Limewater turns from colourless to white to indicate the presence of carbon dioxide.

What affects your breathing rate?

The average resting breathing rate for an adult human is around 14 breaths per minute. This supplies the oxygen needed for all of the normal activities of our cells, but as you have seen, it does not use all the capacity of our lungs. When we are breathing normally at rest, we breathe about 500 cm^3 of air in and out with each breath – only about 15% of our possible maximum. This is our tidal volume.

If we need additional oxygen for any reason over and above the normal amount, we have two ways of getting more air into the body:
- Breathe faster.
- Breathe more deeply.

Usually we do a combination of the two. The vital capacity of our lungs is the absolute maximum amount of air we can take into or breathe out of our lungs.

So, what factors affect our breathing rate? Anything that increases the oxygen requirements of the body will tend to increase the breathing rate. The main factors known to have an effect are:

- exercise
- anxiety
- drugs
- environmental factors
- altitude
- weight/obesity
- smoking.

Exercise

Even when we are resting, our muscles use up a certain amount of oxygen and glucose. This is because some of our muscle fibres are constantly contracting to keep us in position against the pull of gravity. Muscles are also involved in life processes such as breathing, moving food through the gut and circulating the blood. However, when we begin to exercise, our muscles start contracting harder and faster. As a result, they need more glucose and oxygen to supply their energy needs. During exercise, the muscles also produce more carbon dioxide, which needs to be removed so they can keep working effectively.

As muscular activity increases during exercise, our breathing rate increases and we breathe more deeply. These changes mean that not only do we breathe more often, but we also bring more air into our lungs with each breath. This increases the amount of oxygen available to be carried to the exercising muscles. It also means more carbon dioxide is removed from the blood in the lungs and breathed out.

Figure 7.12 Our breathing rate increases during exercising to provide the muscles with the extra oxygen they need.

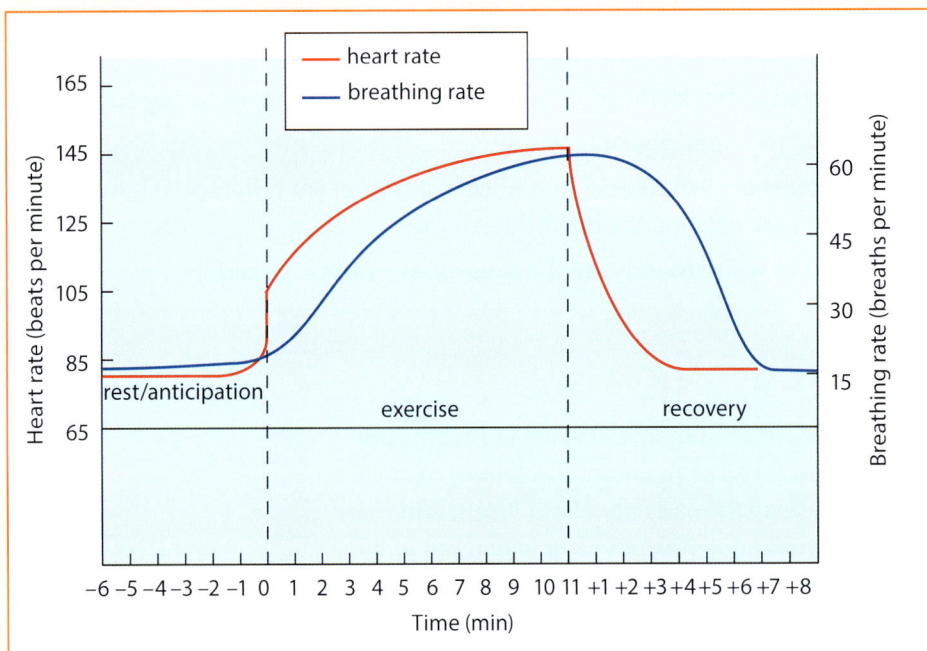

Figure 7.13 During exercise, the breathing rate increases to supply the muscles with the extra oxygen needed and remove the extra carbon dioxide produced.

Exercising and getting fitter means your lungs get bigger. They can supply more oxygen to your muscles so you build up a much lower oxygen debt. As a result, fit people often have a slower breathing rate than unfit people because they take more air in and out with each resting breath.

DID YOU KNOW?

In a 100 m sprint, some athletes do not breathe at all. This means that their muscles use the oxygen taken in at the start of the race and then don't get any more oxygen until the race is over. Although the race only takes a few seconds, lots of energy is used up and a big oxygen debt develops. This is why athletes continue to puff and pant long after the race is over!

> **Checkpoint questions**
>
> 6 Describe the effect of exercise on the breathing rate of a fit person and explain why these changes happen.
>
> 7 Plan an investigation into the fitness levels of your classmates. Describe what you would do, and how you would record and analyse your results. What pattern would you expect to see?

Anxiety

Anxiety affects our breathing rate, because when we are anxious, our body reacts as if we are in danger and need extra oxygen. As a result, our breathing rate increases, ready to supply extra oxygen to our muscles and get rid of carbon dioxide if we have to run away or fight (see Chapter 10 for more about this response).

Drugs

Drugs affect our breathing rate in a number of ways. Some of the drugs we take into our body are medicines designed to make us better. Others are drugs that we take for pleasure, some of which are legal and some of which are not. However, drugs, whether they are medicines, legal or illegal, may affect our breathing rate – sometimes fatally. Amphetamines and cocaine, for example, can cause the breathing rate to increase dramatically, while depressants can cause your breathing rate to drop alarmingly and even stop. Taking any drug that lowers the rate at which we get air into our lungs risks depriving our body and brain tissues of oxygen, with possibly devastating results.

Environmental factors

Certain environmental factors can either change our oxygen needs or change the concentrations of the gases that control breathing. If conditions are particularly hot, the body has to work very hard to keep cool and we may find our breathing rate increases. If the level of carbon dioxide in the air increases, so does our breathing rate, because a build-up of carbon dioxide in the body triggers the breathing response.

Altitude

Height above sea level (altitude) also affects our breathing rate. The higher we go above sea level, the lower both the atmospheric pressure and the oxygen level in the air become. Once we go above 3650 m, there is a noticeable lack of oxygen and our breathing rate will increase to try and keep the oxygen level up. Many people feel ill at high altitudes although with time their body may begin to adapt, taking more air into the lungs with every breath, as well as producing more red blood cells to carry the oxygen. People who are born and live at high altitudes, for example, in the Himalayas and the Andes, don't suffer in this way. They have an increased lung volume with many more alveoli, as well as more blood capillaries and red blood cells to pick up the oxygen from the air.

Figure 7.14 Some hikers have difficulty breathing at high altitudes.

What affects your breathing rate?

Weight/obesity

Excess weight affects the breathing rate. This is a particular problem with people who are obese or very obese. The fat around the abdominal organs, makes it difficult for the diaphragm to contract and lower properly. This in turn makes it hard to breathe in deeply. However, the muscles of an obese person are having to work hard to move the excess weight around. So, people who are very overweight or obese are often breathless, as they cannot get the oxygen they need very easily. One common cause of obesity is lack of exercise, and if an obese person does little or no exercise, they will be very unfit and get breathless easily. However, if obese people begin to do more exercise, they will lose weight, their breathing rate will fall and become more efficient, and they will quite rapidly see the benefits of their improving fitness.

Smoking

Finally, one major factor that affects breathing rate is smoking. Smoking is a habit that directly affects the respiratory system as well as other areas of the body, so we will look at it in more detail.

The effects of smoking on the lungs and the rest of the body

There are 1.1 thousand million tobacco smokers worldwide, smoking around 6 million million cigarettes each year, so smoking is big business. People in the Caribbean tend to smoke less than those in many other countries, including Latin and North America. In spite of this, the scientific evidence suggests that 10% of deaths in the Caribbean are smoking related. Every cigarette smoked contains tobacco leaves that, as they burn, produce a smoke cocktail of around 4 000 chemicals that are inhaled into the lungs. Some of those chemicals are absorbed into the bloodstream to be carried around the body to the brain. Some of the compounds in tobacco smoke cause lasting and often fatal damage to the body cells. These chemicals include: nicotine, carbon monoxide and tar.

Nicotine

Nicotine is the very addictive drug found in tobacco smoke. It is a mild stimulant, although some of its effects are more like a **sedative**. The overall effect of the drug is to produce a sensation of calm well-being and 'being able to cope'. The heart rate and breathing rate are slightly raised by the drug, but blood flow to the skin and areas such as the hands is reduced (see Figure 7.15).

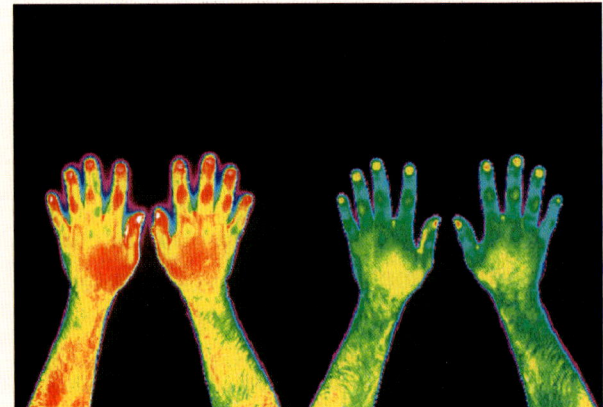

Figure 7.15 Pictures taken with a heat-sensitive camera show the drop in temperature of the hands caused by the reduced blood flow from smoking a single cigarette. This is what causes a smoker's skin to age badly and wrinkle early.

Since nicotine is addictive, we need more and more of the drug to get the same effect, so the number of cigarettes a person smokes daily tends to increase over the years.

Carbon monoxide

Carbon monoxide is a very poisonous gas found in tobacco smoke. It takes up some of the oxygen-carrying capacity of the blood. After smoking a tobacco cigarette, up to 10% of a smoker's blood will be carrying carbon monoxide rather than oxygen. The haemoglobin in the blood has a greater affinity for carbon monoxide than it has for oxygen (see Figure 7.16). This means that carbon monoxide is picked up by the haemoglobin in the lungs in preference to oxygen.

This can lead to a shortage of oxygen for the smoker, and the effect is most marked in pregnant women. During pregnancy, a woman is carrying oxygen not just for her own needs but for her developing **foetus** as well. If the mother's blood does not contain enough oxygen, the foetus is deprived of oxygen and does not grow as well as it should. This can lead to premature births, low birth-weight babies and even stillbirths, which is when the baby is born dead. Educated mothers with high social status are much less likely to smoke, and so are much less likely to have a stillbirth as a result, than mothers who have little education and low social status (see Figure 7.17).

Tar

Tar is a sticky black chemical in tobacco smoke that is not absorbed into the bloodstream. It builds up in the lungs, turning them from pink to grey. In a non-smoker, the ciliated epithelia of the respiratory system are constantly active, moving mucus that has trapped dirt, dust and bacteria from the inhaled air up the respiratory tract away from the lungs. In a smoker, these cilia are anaesthetised (sent to sleep) by each cigarette. They stop working for a time, and this lets dirt and bacteria into the lungs.

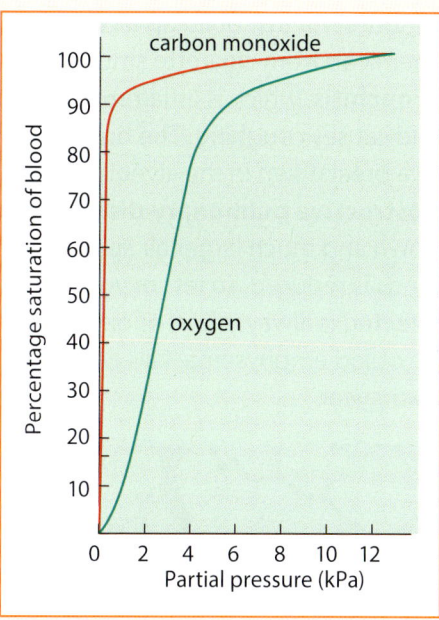

Figure 7.16 This graph shows that carbon monoxide combines with the haemoglobin in the blood more easily than oxygen does, so it prevents the blood from transporting as much oxygen to the tissues.

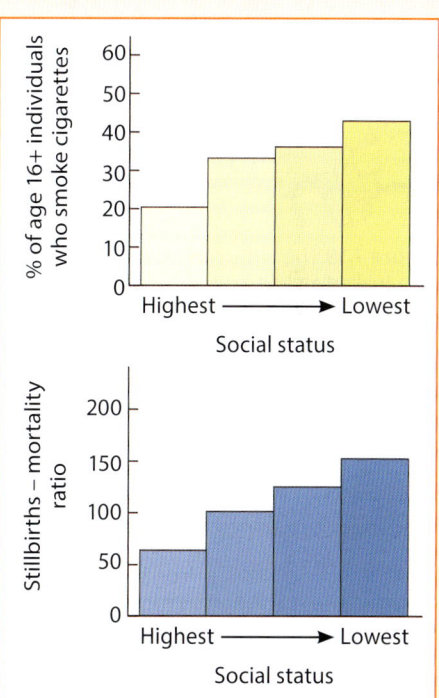

Figure 7.17 Data like these graphs above show that women who smoke are more likely to have stillborn babies than non-smokers. The same pattern applies to having low birth weight babies.

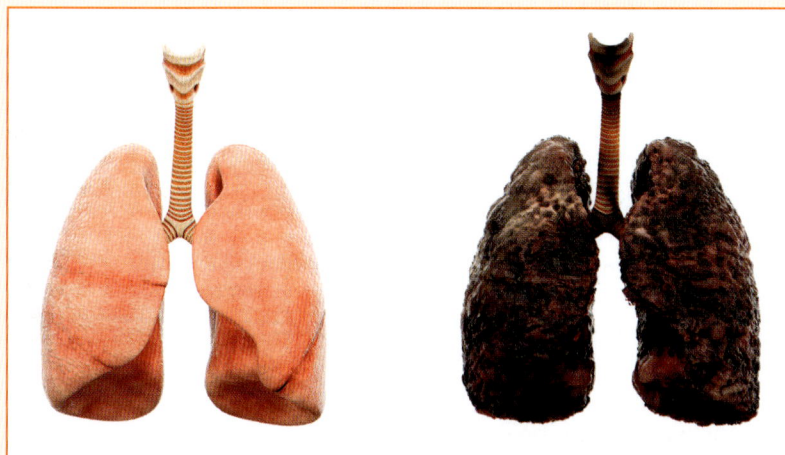

Figure 7.18 You can see instantly the difference between a set of healthy Caribbean lungs, and the set of lungs from a heavy smoker.

What affects your breathing rate?

Tar does not just build up (accumulate) harmlessly in the lungs. Along with other chemicals from cigarette smoke, it makes smokers much more likely to develop **bronchitis**, which is inflammation and infection of the bronchi. Mucus builds up and causes coughing. The build-up of tar in the delicate lung tissue can also lead to a breakdown in the alveolar structure. In people suffering from these **chronic obstructive pulmonary diseases (COPD)**, the structure of the alveoli breaks down and much larger air spaces develop. This means the surface area of the lungs is reduced, so less oxygen diffuses into the blood. As a result, the person affected is always short of oxygen and feels breathless. COPD (which used to be called emphysema) kills and disables millions of people around the world every year.

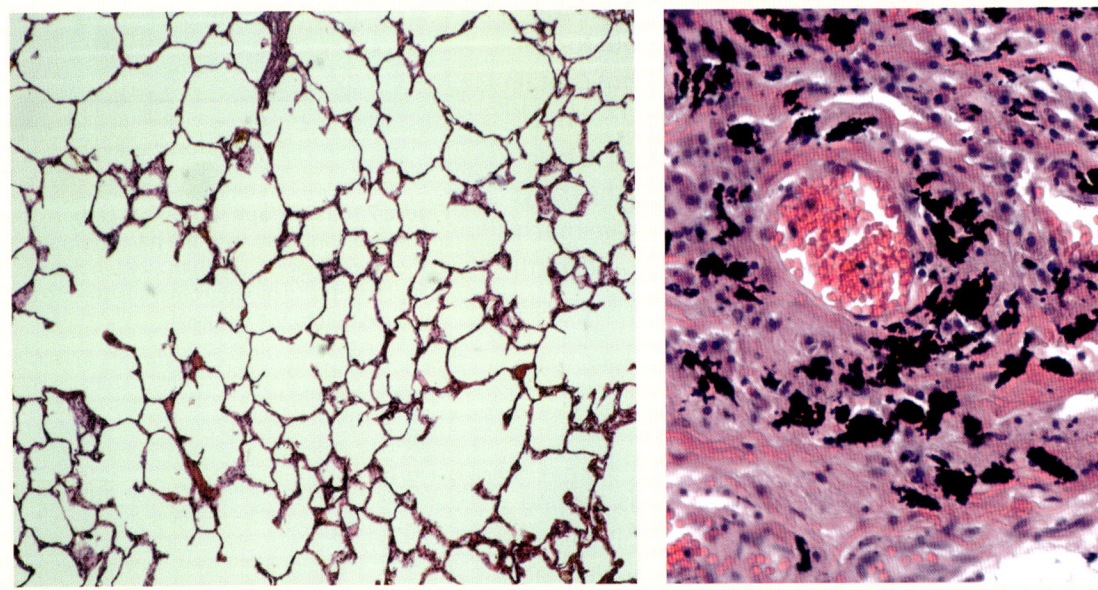

Figure 7.19 The difference in the appearance of the lungs of a non-smoker (left) with plenty of thin-walled alveoli, compared to the lungs of a smoker (right) is clear even to the untrained eye.

Tar is also a major carcinogen, which means it is a cancer-causing substance. Lung cancer, the most well-known disease linked to smoking, is the result of the accumulation of tar. Caught early, lung cancer can be cured, but it often grows silently in the lungs without causing symptoms until it has spread to other parts of the body and becomes fatal. Up to 90% of lung cancers are a direct result of smoking. Cigarette smoking is also linked to cancers of the throat, mouth and larynx – the whole respiratory tract is affected. However, because many of the chemicals in tobacco smoke are absorbed into the blood through the exchange surfaces of the lungs, they are carried around the body and are linked to an increased risk of cancer in areas as diverse as the stomach, the cervix and the pancreas.

The chemicals in tobacco smoke also affect the heart and blood vessels. Smoking raises the blood pressure – known as **hypertension** – and makes it more likely that blood vessels will become blocked, causing heart attacks, strokes and thrombosis. This link between heart disease and smoking will be considered in more detail when you look at the heart and blood vessels in Chapter 8.

Why do people smoke?

It is estimated that up to 3 million deaths a year are a direct result of smoking – in other words smoking kills someone every 10 seconds. Smoking currently causes just under 20% of all deaths in developed countries. So why do people continue to smoke?

Many smokers are strongly **addicted** to nicotine and find it very difficult to give up. The tobacco companies that make cigarettes are immensely rich and powerful, and governments collect large amounts of tax from the sale of tobacco, so selling tobacco products is big business. Cigarettes are marketed as being 'cool', making people believe that smoking gives them a positive image in front of their peers. Advertising drives interest and desire, and once people try smoking, they rapidly become addicted to the nicotine. Most adult tobacco smokers would like to give up the habit. It is expensive, it ages the skin, and it can cost people and their unborn children their lives. In some ways, it seems astonishing that these products are still legal.

Figure 7.20 The evidence for the link between smoking and cancer is so strong that it is now universally accepted. Many governments will not ban smoking but some, like Barbados, insist that warnings like this are printed on every packet of cigarettes. Many people just ignore these warnings and continue to smoke and put their health at risk.

As people in the developed world have become more aware of the health risks of smoking, the number of smokers has begun to fall. However, cigarettes are still being heavily marketed in the developing world, without the health warnings available elsewhere.

While the decision to smoke or not remains a personal one, as evidence mounts that the people around smokers are affected by **passive smoking** (smoke from other people's cigarettes that they breathe in), various countries, including parts of the USA, the UK and the Caribbean, are banning smoking in public places to protect the health of the general population.

Building the evidence

Not everyone who smokes will develop lung cancer, just as not all non-smokers will avoid lung cancer, but as you have seen, the evidence for the link is extremely strong – up to 90% of all lung cancers are thought to be due to smoking (see Figure 7.21). At one stage, smoking was thought to be beneficial and very good for the nerves. Doctors recommended smoking to their patients! But from the 1940s onwards, evidence began to mount that perhaps cigarettes were not so harmless after all.

The link to lung cancer was first noticed in the UK by Sir Richard Doll early in the 20th century, but it has taken many years for the smoking public to accept the link, and even longer for the cigarette manufacturers to accept it. People did not want to believe that something that they did and enjoyed could be damaging their health. Cigarette manufacturers did not want it recognised that their products caused millions of deaths each year. Indeed, despite carrying out extensive research themselves that showed the risks, many cigarette manufacturers denied both the addictive nature of nicotine and the health risks associated with cigarette smoke for years.

For a long time, the evidence was statistical. When figures for the incidence of lung cancer or deaths from lung cancer were considered, smokers could be seen to be far more at risk than non-smokers. Increasingly, specific experiments have identified the chemicals in cigarette smoke that actually cause lung cancer.

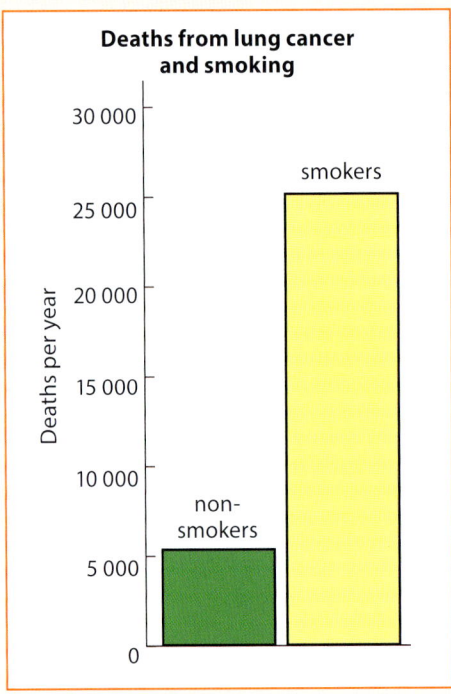

Figure 7.21 Evidence like this shows that smoking is a major risk factor in the development of lung cancer. There is similar evidence linking smoking with cancers of the throat and tongue, and with heart disease.

Hookah smoking and vaping

Not everyone smokes tobacco using cigarettes. Some people smoke tobacco using a hookah – water pipes that pass charcoal-heated air through a tobacco mixture and then through a water-filled chamber. The smoker inhales the smoke through a special tube and mouthpiece. Hookah users often think that this is a safer, less damaging way of smoking, but the scientific evidence shows that tar and at least 81 other toxic chemicals and carcinogens reach the lungs in hookah smoke. People tend to sit with a hookah for a long time, and as a result they end up with higher levels of addictive nicotine and carbon monoxide in their blood than if they had smoked a cigarette.

Basically hookah smoking carries the same risk of damage to the respiratory system, including lung cancer and mouth cancers, as any other form of smoking. People who start using hookah when they are young often also begin smoking cigarettes as a result of their nicotine addiction.

Vaping, or using e-cigarettes, is often seen as a healthier alternative to smoking. It provides a nicotine hit without the other dangerous chemicals in tobacco. People also use vaping to help them give up smoking tobacco – it gives them nicotine and something to do with their hands, and then they can gradually vape less often until they have broken the habit.

Since vaping is relatively new, the long-term health impacts are not well known. In the beginning, vaping appeared a much healthier alternative to smoking. Unfortunately, some of the chemicals in the vape fluid now appear to have health risks of their own. The evidence is still being collected.

Marijuana

Marijuana, also known as cannabis, ganja or weed, comes from the marijuana plant. It is a drug that is illegal in some places and decriminalised in others. In spite of this, people use marijuana in many different ways. Here we are focusing on smoking marijuana, where the smoke from the burning leaves is drawn down into the lungs. Like any smoke, marijuana smoke is harmful to the lungs. Smoke from burning marijuana contains many of the same chemicals as tobacco smoke – and sometimes it is mixed with tobacco as well.

Figure 7.22 Hookahs are a complex way of getting tobacco smoke into the lungs – but when it gets there, it has the same effect as tobacco smoke from cigarettes.

Since smoking marijuana is illegal in many countries, there is much less evidence on the effect that smoking it has on the lungs than there is for tobacco smoking. The fact that, in many places, people smoke a mixture of marijuana and tobacco makes it harder to identify the risk of marijuana alone. The evidence we do have suggests strongly that smoking marijuana damages the lungs, causing both acute and chronic bronchitis. Studies carried out on many thousands of people over a period of 40 years also showed that marijuana use doubles the risk of developing lung cancer.

> **For healthy lungs, it is best to breathe plain air!**

When breathing fails

Sometimes breathing fails. This can be the result of a number of different things, including an accident, drowning or a heart attack. Once breathing stops, death will result in a matter of minutes as the brain, in particular, is starved of oxygen. However, you can take over breathing for a casualty in this situation, and this may be enough to keep them alive until medical support arrives. The way this is done is by expired air resuscitation, which is also more commonly called mouth-to-mouth resuscitation. You keep forcing air into the lungs of the person who has stopped breathing, so that gaseous exchange can continue and their tissues continue to receive oxygen. It is very important that mouth-to-mouth resuscitation should only be given when the casualty has stopped breathing, and not when they are just unconscious. You may chose to carry a disposable device to place over the mouth and nose of the casualty. This closes off their nose, keeping their airway open and also offers some protection against the spread of diseases (see step 4 in Figure 7.23). Sometimes a person not only stops breathing, the heart stops beating as well so you cannot feel a pulse. If this is the case, use cardiopulmonary resuscitation (CPR). This involves mouth-to-mouth resuscitation along with special chest compressions that keep the blood moving around the body of your casualty carrying oxygen to the tissues and removing carbon dioxide. The procedure for these techniques is shown in Figure 7.23.

Step 1: Call the emergency number and shout for help.

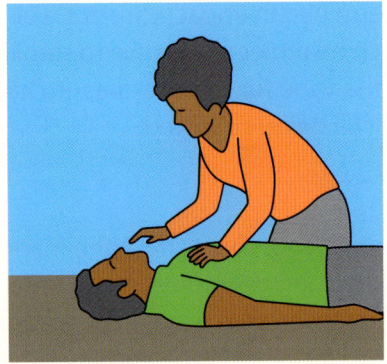

Step 2: Check vital signs – pulse and response to voice.

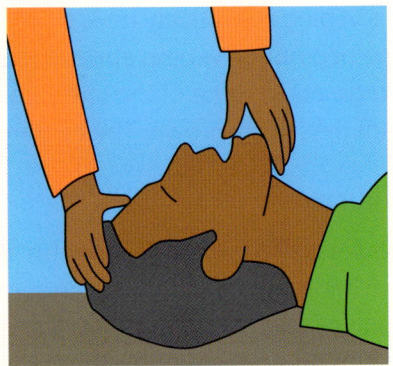

Step 3: Lift the chin and check for breathing.

If not breathing but has a pulse:

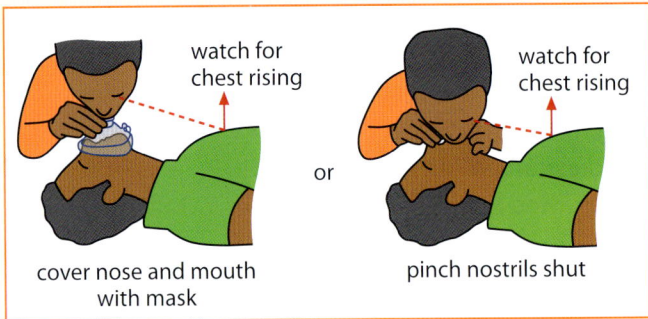

Step 4: Give rescue breaths (mouth-to-mouth resuscitation), with or without a disposable device. Look for the chest rising to show air is going in. Give two rescue breaths and then see if breathing is restored. If not, repeat until the casualty starts breathing for themselves OR help arrives.

Step 5: Give 2 rescue breaths as above in step 4.

If not breathing and no pulse:

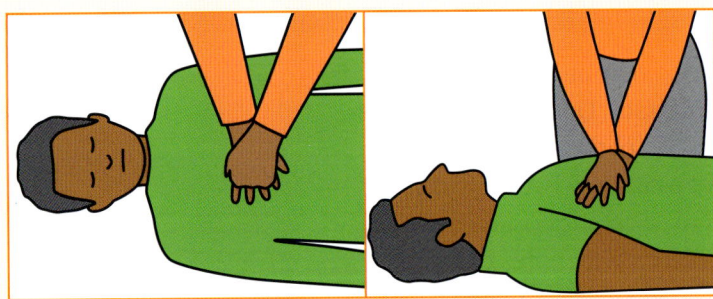

Step 6: Carry out CPR. Place hands one on top of the other in the lower centre of the chest. Push down hard and fast – about 2 times per second – for 30 compressions.

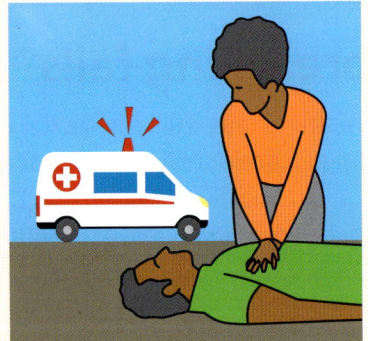

Step 7 Check vital signs again and then repeat rescue breaths and chest compressions until the patient recovers OR help arrives.

Figure 7.23 Following these instructions can, literally, help you to save someone's life.

Cellular respiration

The digestive, breathing and circulation systems all exist to provide the cells of the human body with what they need for respiration. (You will learn more about the circulatory system in the next chapter.) During the process of **cellular respiration**, glucose (a sugar produced as a result of digestion) is broken down to release the energy needed by the cells.

Aerobic respiration

Aerobic respiration uses oxygen from the air. Glucose reacts with this oxygen to release energy to be used by the cells. Carbon dioxide and water are produced as waste products.

The reaction can be summed up as follows:

glucose + oxygen → carbon dioxide + water + energy (ATP)

$C_6H_{12}O_6 + 6O_2 \rightarrow 6CO_2 + 6H_2O +$ energy (ATP)

Aerobic respiration takes place in the mitochondria in cells. These are tiny, rod-shaped bodies (organelles) found in almost all cells (see Chapter 1). They have a folded inner membrane that provides a large surface area for the enzymes involved in aerobic respiration. Cells that use a lot of energy, such as muscle cells, liver cells, and the rods and cones of our eyes, contain many mitochondria because they use a lot of energy.

All of our cells need energy to carry out the reactions of life, and respiration provides this energy. Respiration releases energy from the food we eat so that the cells of the body can use it. The energy used by the cells is stored in the form of a molecule called **ATP**, which stands for **adenosine triphosphate**. This is an adenosine molecule with three phosphate groups attached to it. When energy is needed for any chemical reaction in the cell, the third phosphate bond is broken in a hydrolysis reaction (see Figure 7.25). This results in a new compound, **ADP** or **adenosine diphosphate**, a free inorganic phosphate group – and the release of the all-important energy needed in the cell. This is a reversible reaction – during aerobic respiration, the energy from the reactions of glucose with oxygen is used to produce large quantities of ATP ready for use in the cells, and when the cells need energy, ATP is broken down again to release that energy. This is why cellular respiration is so important – ATP is the single energy-providing and energy-storing molecule for all the processes in living cells.

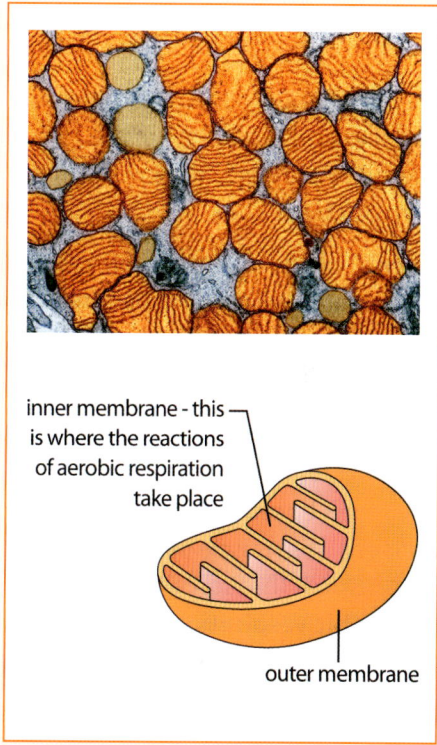

Figure 7.24 Mitochondria are the powerhouses that provide energy for all the functions of a cell.

ADP + P ⇌ ATP (energy produced / energy required)

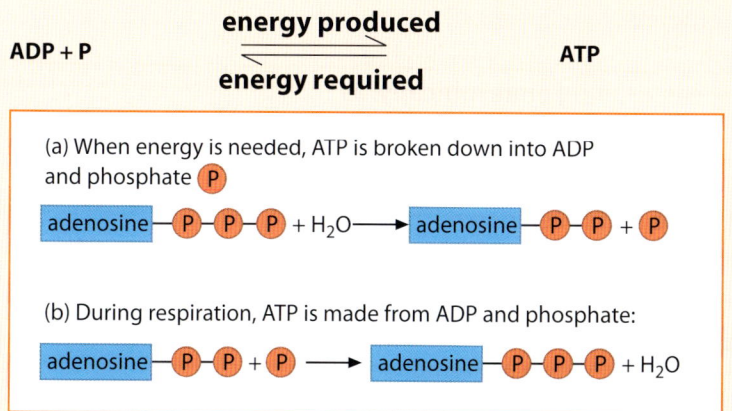

Figure 7.25 ATP is the energy 'currency' of the cell.

Cellular respiration

Your cells need energy from ATP to carry out the basic functions of life, called **metabolism**. One of their main functions is to build large molecules from smaller ones to make new cell material (**anabolism**). Much of the energy released in respiration is used for these 'building' activities. Cells also break large molecules down into smaller molecules. This is called **catabolism** and it also requires energy.

anabolism + catabolism = metabolism

Another important use of the ATP energy from respiration is in making muscles contract. Muscles are working all the time in our body. Even when we are asleep, our heart is beating, our rib muscles and diaphragm contract as we breathe, our gut is churning – and all of these muscular activities use energy.

We are 'warm-blooded'. This means that the temperature inside our body is the same, almost regardless of the temperature around us. On cold days we use energy to keep our body warm, while on hot days we use energy to sweat and keep our body cool.

The ATP produced by aerobic respiration in cells also provides energy for the active transport of some materials across cell boundaries.

Anaerobic respiration

The energy released by aerobic respiration in muscle cells allows them to move. However, during vigorous exercise, the muscle cells may become short of oxygen because the blood simply cannot supply it fast enough. When this happens, the muscle cells can still obtain energy from the glucose, but they have to do it through a type of respiration that does not use oxygen, called **anaerobic respiration**. Anaerobic respiration in animals takes place in the cytoplasm of the cells, and not in the mitochondria.

Anaerobic respiration produces far less ATP than aerobic respiration. It also produces a different waste product called **lactic acid**. When our muscle cells have been used for vigorous exercise for a very long time, they become fatigued, which means they stop contracting efficiently. They switch to anaerobic respiration, and as the level of lactic acid builds up, our muscles really start to ache. Also, anaerobic respiration is not as efficient as aerobic respiration. It does not break down the glucose molecules completely, so far less ATP energy is released than during aerobic respiration. This means our muscles tire more rapidly and cannot work as well when they are respiring **anaerobically**, as there is not enough energy for them.

The body cannot get rid of lactic acid by breathing it out as it does carbon dioxide, so when the exercise is over, lactic acid has to be broken down. This needs oxygen, and the amount of oxygen needed to break down the lactic acid is called the **oxygen debt**. Even though our leg muscles have stopped contracting, our heart rate and breathing rate stay high to supply extra oxygen to oxidise the lactic acid to give carbon dioxide and water, which are removed from the body. In other words, our breathing rate stays high after exercise to pay off our oxygen debt (see Figure 7.26).

The equation for anaerobic respiration is:

glucose ➜ lactic acid + energy (ATP)

The equation for oxygen debt repayment is:

lactic acid + oxygen ➜ carbon dioxide + water

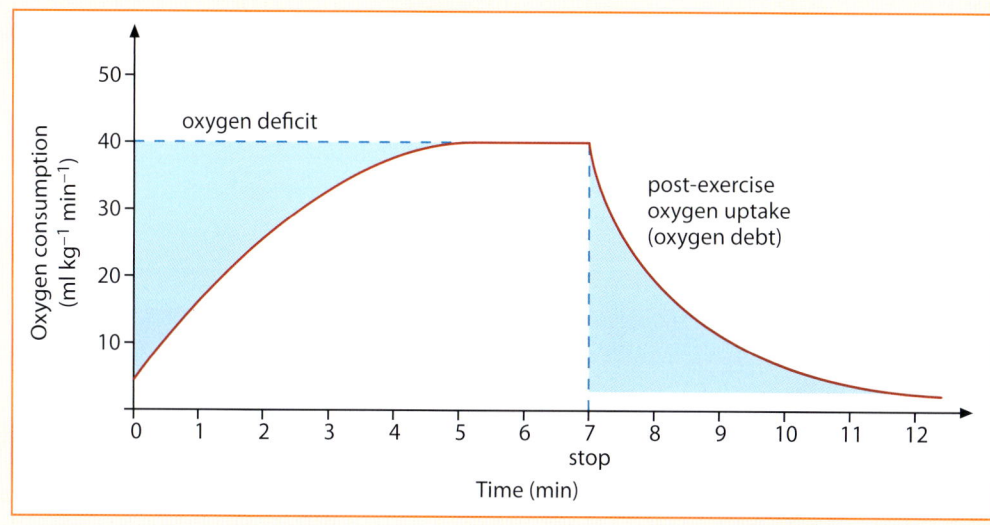

Figure 7.26 When exercise ends, you keep breathing faster and deeper until you have removed all the lactic acid that has built up and paid off your oxygen debt.

Anaerobic respiration in other organisms

Anaerobic respiration isn't simply something that affects people. It takes place in all living organisms, and in a number of cases we have put anaerobic respiration to very good use, both in our industries and in our homes. For example, one of the microorganisms that is most useful to people is yeast, a single-celled fungus.

When yeasts have plenty of oxygen, they respire aerobically in their mitochondria, breaking down sugar to provide energy for the cells, and producing water and carbon dioxide as waste products. However, yeast can also respire anaerobically in the cytoplasm of their cells. When yeast cells break down sugar in the absence of oxygen, they produce ethanol (commonly referred to as **alcohol**) and carbon dioxide. The anaerobic respiration of yeast is sometimes referred to as **fermentation**. The word and chemical equations for anaerobic respiration in yeast are shown below:

glucose ➜ ethanol + carbon dioxide + energy (ATP)

$$C_6H_{12}O_6 \rightarrow 2C_2H_5OH + 2CO_2 + \text{energy (ATP)}$$

Yeast cells need aerobic respiration because it provides more energy than anaerobic respiration, so it allows them to grow and reproduce. However, once there are large numbers of yeast cells, they can survive for a long time in low oxygen conditions and will break down all the available sugar to produce ethanol and carbon dioxide.

We have used yeast for making bread and alcoholic drinks for almost as far back as human records go. We know yeast was used to make bread in Egypt in 4000 BCE, and some ancient wine found in Iran dates back to 5400–5000 BCE.

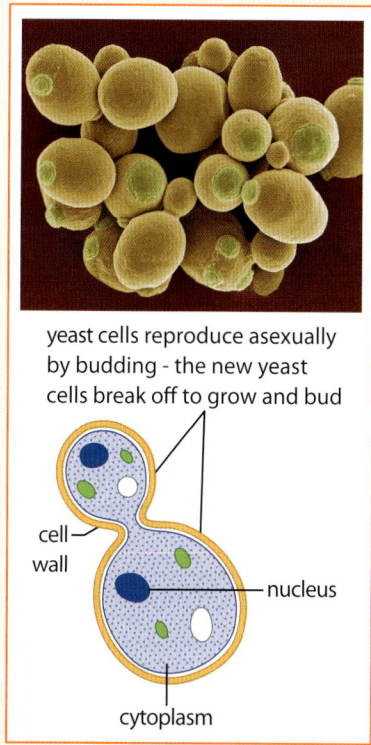

Figure 7.27 Yeast cells – these microscopic organisms have been useful to us for centuries.

The procedure for making simple bread is as follows:

- Mix yeast with sugar to provide it with an energy source for respiration.
- Then mix the yeast and sugar with water and flour, and knead the mixture to improve the texture and to make sure the yeast is evenly spread throughout the dough.
- Leave the mixture somewhere warm. As the yeast grows and respires, it produces carbon dioxide, making the bread rise.
- Place the risen dough in a hot oven. When the bread bakes, the bubbles of gas expand in the high temperature, giving the cooked bread a light texture. The yeast is killed during the cooking process.

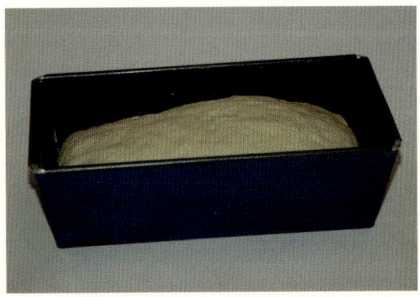  

Figure 7.28 When you make bread, the dough rises as a result of the carbon dioxide made by the yeast as they respire.

In bread making, yeast usually respires aerobically, but we use the anaerobic reactions of yeast too. When fruits fall to the ground and begin to decay, wild yeast present on the skin break down the fruit sugar and form ethanol and carbon dioxide. These fermented fruits can cause animals to become drunk when they eat them, and this is probably how our ancestors discovered alcohol. We now use this same reaction in a controlled way to make beer, wine and spirits. In beer and spirit production, the yeast is supplied with additional carbohydrates to act as an energy source for respiration.

DID YOU KNOW?

There are approximately 500 000 000 yeast cells in one pint of beer!

The biologist's toolkit: Estimation and approximation

In the *Did you know?* box, you are told the approximate number of yeast cells in a pint of beer. An approximate number is roughly correct. Approximation and estimation are similar.

We can't always count things accurately. It is an important skill for scientists to be able to make a rough estimate of numbers. An estimation gives you an answer that is of the right order of magnitude, to the nearest power of 10, but is not precise. For example:

- A pint of beer has approximately 500 000 000 yeast cells in it.
- You are probably between 1–2 m tall.
- A molecule of any substance has a diameter in the order of 1 nanometre (1×10^{-9} m).

Estimating the answer of a calculation – getting a rough idea of the answer you expect – helps you check if your calculated answer is correct.

In contrast, wine-making uses the natural sugar found in fruit, such as grapes, as the energy source for the yeast. The grapes are pressed and the juice mixed with yeast and water. The yeast respires anaerobically until most of the sugar has been used up. At this stage, the wine is filtered to remove the yeast and then bottled, where it will remain for some time to mature before it is sold. Most commercially sold wine is made from grapes, but wine can actually be made from almost any fruit or vegetable.

Interestingly, ethanol in large amounts is poisonous to yeast as well as to people. This is why the alcohol content of wine is rarely more than 14%. Once it gets much higher than this, it kills all the yeast and stops the fermentation process. To make spirits like rum, the basic alcoholic drink is distilled to remove some of the water and increase the alcohol content of the final product.

Figure 7.29 During the production of rum in the Caribbean, billions of yeast cells respire anaerobically, turning sugar into alcohol, and then the liquid is distilled to reduce the water content and make the final spirit more alcoholic.

More recently, the fermentation of maize and sugar cane to produce ethanol for use as a fuel in cars is an important industrial use of anaerobic respiration. It is one way to reduce the amounts of carbon dioxide produced and given off in the atmosphere by burning fossil fuels, because plant-based ethanol is a sustainable resource that removes carbon dioxide from the atmosphere as the plants grow.

It isn't just the anaerobic respiration of yeast that is important to us commercially. The anaerobic fermentation of milk with different types of bacteria can give us cheese and yoghurt – staple parts of the human diet in many areas of the world.

The use of anaerobic respiration to break down human and animal waste products to produce biogas (**methane**) at a domestic and small village level is increasing in many parts of the world. There is increasing interest in using this on a larger scale to produce another, more eco-friendly, fuel.

Checkpoint questions

8 Define aerobic respiration.
9 Write a word and chemical equation for aerobic respiration.
10 Explain how aerobic respiration differs from anaerobic respiration.
11 a Write word equations for the anaerobic respiration that takes place in human cells and in yeast cells.
   b Give the chemical equation for anaerobic respiration in yeast cells.
12 List four different ways in which we use anaerobic respiration.

Possible SBAs based on the respiratory system

→ Using a microscope, investigate and draw the cellular appearance of different regions of the respiratory tract and relate their structures to their functions.

→ Build a model or develop a simple computer animation to show the changes in the chest during inspiration and expiration.

→ Investigate the breathing rates of different groups of people, for example:
  - primary age children and your year group
  - different year groups within your school
  - students in your class and a similar number of adults (teachers, family members, and so on).

→ Investigate the effect of regular exercise on a group of fellow students/family members. Remember that you will need a control group.

→ Survey the numbers of smokers or hookah users in your community and compare respiratory rates.

→ If a peak flow meter is available, measure vital capacity in different cohorts, for example, different ages, different fitness levels, and so on.

→ Use secondary sources to investigate smoking rates and incidence of smoking-related health problems in different regions of your country/ different Caribbean nations/different countries around the world.

→ Investigate factors that affect the rate of respiration in yeast in bread-making.

→ Investigate factors that affect the fermentation of sugar solution by yeast.

→ Develop an emergency first aid training course on CPR.

End-of-chapter summary

In this chapter, you have learnt that:

- the respiratory system takes air into and out of the body to supply oxygen and remove carbon dioxide
- the respiratory system is made up of the mouth and nose, larynx, trachea, bronchi, bronchioles and alveoli; the lungs are surrounded by the pleural membranes and enclosed in the thorax
- the movement of air is brought about by breathing movements caused by the contraction and relaxation of the intercostal muscles that move the ribs and of the diaphragm
- breathing movements cause changes in the volume and pressure of the chest that bring about ventilation of the lungs
- in the lungs, oxygen from the air diffuses into the bloodstream at the same time as carbon dioxide from the blood diffuses out of the bloodstream into the air in the lungs; this is known as gas exchange and takes place in the alveoli
- the alveoli provide a very large, moist surface area, with thin membranes richly supplied with blood capillaries to allow for the most efficient possible gas exchange

- the rate of breathing is affected by a number of factors including exercise, anxiety, drugs, environmental factors, altitude, body weight and smoking
- nicotine is the addictive drug found in tobacco
- tobacco smoke also contains carbon monoxide, which reduces the oxygen carrying capacity of the blood
- in pregnant women, carbon monoxide deprives the foetus of oxygen and can lead to low birth weight babies and stillbirths
- tobacco smoke contains tar and other chemicals, which contribute to lung cancer, bronchitis, COPD (emphysema), and disease of the heart and blood vessels
- evidence for the link between smoking and lung cancer took many years to build up and even longer before it was accepted by everyone
- smoking tobacco using a hookah has the same damaging effects on the lungs as smoking cigarettes; smoking marijuana also damages the lungs; evidence is building that even vaping, which gives nicotine without the other toxins in cigarette smoke, also appears to damage the respiratory system
- aerobic respiration is the breakdown of glucose with oxygen to provide energy for the cells; ATP is produced from ADP, and carbon dioxide and water are the waste products; it takes place in the mitochondria of cells
- ATP is the molecule that releases the energy needed for all of the reactions in the cell
- anaerobic respiration is respiration without oxygen; in humans, glucose is broken down to form lactic acid and a small amount of ATP; in yeast, glucose is broken down to form ethanol and carbon dioxide; anaerobic respiration takes place in the cytoplasm of cells, and not the mitochondria
- aerobic respiration results in the production of much more ATP than anaerobic respiration
- if muscles work hard for a long time, they become fatigued and don't contract properly; if they don't get enough oxygen, they will respire anaerobically
- after exercise, oxygen is still needed to break down the lactic acid that has built up; this oxygen is known as an oxygen debt
- we use anaerobic respiration to make bread, alcoholic drinks, biofuels such as bioethanol, cheese, yoghurts and even to break down human, animal and plant waste to produce biogas (methane).

End-of-chapter questions

1. In which organ of your body does gas exchange take place?
   - A   Liver
   - B   Lungs
   - C   Trachea
   - D   Heart

2. What are the tiny air sacs called where the exchange of gases takes place?
   - A   Ravioli
   - B   Lacunae
   - C   Bronchioles
   - D   Alveoli

3. What is the role of the cilia on the epithelium of the trachea?
   - A   Move dirt and mucus away from the lungs
   - B   Move dirt and mucus down the trachea into the lungs
   - C   Produce mucus
   - D   Prevent food getting into the lungs

4. Which term is needed to complete the word equation for aerobic respiration?
   glucose + oxygen ➜ carbon dioxide + water + _____?
   - A   ADP
   - B   Carbon monoxide
   - C   ATP
   - D   Gas

5. Which one of the following is not yet a major commercial use for anaerobic respiration?
   - A   Production of biogas from human waste
   - B   Beer-making
   - C   Yoghurt production
   - D   Bread-making

6. Which one of the following is not part of the respiratory response to exercise?
   - A   Breathing faster
   - B   Producing more oxygen
   - C   Breathing more deeply
   - D   Producing more carbon dioxide

7. Which one of the following is not a constituent of cigarette smoke?
   - A   Oxygen
   - B   Nicotine
   - C   Carbon monoxide
   - D   Tar

8. Match each definition, i to vi, to the word it is defining, A to F.
   - i   The smallest air passages in the lungs
   - ii   The upper part of the body containing the lungs
   - iii   The main air passage leading in from the mouth and nose
   - iv   The body organs where gas exchange takes place
   - v   Millions of tiny air sacs making up the gas exchange tissue
   - vi   Large sheet of muscle separating the thorax from the abdomen

   - A   Alveoli
   - B   Trachea
   - C   Lungs
   - D   Diaphragm
   - E   Bronchioles
   - F   Thorax

   (6)

The human respiratory system

9 For gas exchange in the lungs to work effectively, we need to move air into and out of the lungs regularly. We do this by breathing. Our breathing movements involve the muscles between the ribs and the diaphragm. Explain fully, using diagrams if you feel they will help, the events that take place:
  a   when you breathe in (4)
  b   when you breathe out. (4)

10 This table shows the effect of exercise on the breathing rate of three people.

Activity	Number of breaths taken per minute		
	Person A	Person B	Person C
Rest	21	15	18
20 step-ups per minute	29	21	25
50 step-ups per minute	40	30	34

  a   Plot a bar chart to show these results and make it easier to compare them. (4)
  b   Which person do you think is the fittest of the three, and which do you think is the least fit? Explain your answers. (6)
  c   What else would happen to the breathing in addition to the rate going up? (1)
  b   Why does our breathing change when we exercise? (3)

11 The air we breathe in contains about 20% oxygen and only 0.04% carbon dioxide. The air we breathe out contains around 16% oxygen and 4% carbon dioxide. What happens in the lungs to bring about this change? (Include details of the alveoli of the lungs in your answer.) (8)

12 Here are descriptions of three different diseases that affect the respiratory system. Use your knowledge of how the respiratory system works to explain why these diseases make breathing so difficult and result in a lack of oxygen.
  a   In **cystic fibrosis**, the lung cells produce a very thick, sticky mucus that fills the alveoli and blocks the bronchioles. It makes lung infections more likely and the lungs have to be cleared of mucus by physiotherapy at least twice a day. People with cystic fibrosis are often short of breath, particularly just before they have physiotherapy. (3)
  b   COPD (emphysema) is a disease often caused by smoking in which the structure of the alveoli breaks down, resulting in lungs with much larger air sacs than normal. These large spaces may fill with fluid. People with COPD are always short of breath and as the disease gets worse, they may need to breathe pure oxygen. (3)
  c   In asthma, the muscles of the air passages spasm, narrowing the airways. The linings of the air passages swell and produce extra mucus. During an asthma attack, people find it very hard to breathe and the air is forced in and out of their chests with a wheezing sound. (3)

13 Some people stop breathing in their sleep. A nasal-intermittent positive pressure ventilation system forces air into the lungs at regular intervals through a small facial mask. The air is under pressure and is forced into the lungs, expanding the chest, then squeezed out again as the chest falls. Explain how this differs from normal breathing. (5)

14 a Define the following words: aerobic respiration, anaerobic respiration, oxygen debt, slow-twitch muscle fibres. (4)
    b Write a word equation for aerobic respiration. (2)
    c How does aerobic respiration differ from anaerobic respiration? (2)

15 a Aerobic respiration provides energy for the cells of the body. Explain why cells need this energy and what they use it for. (3)
    b If you exercise very hard or for a long time, your muscles begin to ache and do not work as effectively. Explain why. (3)
    c If you exercise very hard, you often puff and pant for some time after you stop. Explain what is happening. (4)

16 Draw a table summarising the main components of tobacco smoke and their effects on the human body. (10)

17 Smoking during pregnancy is potentially very damaging, yet many women still smoke when they are pregnant. Prepare an article for a women's magazine explaining why smoking during pregnancy is so harmful and encouraging women to give up the habit. (10)

18 a Smokers are more likely to get infections of their breathing system than non-smokers. Why do you think this might be? (3)
    b In bronchitis, the tubes leading down to the lungs produce a lot of mucus. How does the body of a non-smoker deal with this mucus compared to the body of a smoker? (3)

19 a Draw a table to show the advantages and disadvantages of smoking. (8)
    b Suggest why people smoke. (2)
    c Most anti-smoking campaigns are relatively ineffective. How would you set about targeting young people and reducing the number of smokers? (6)

20 a Look at the data in Figure 7.21 and summarise the evidence it gives for the link between smoking and lung cancer. (6)
    b What additional evidence would you like to see to make the link even more convincing? (2)
    c What do you think is the impact of scientific evidence like this on the public in general? (2)

Chapter 8: The circulatory system

When you have completed this chapter, you will be able to:

- explain the need for a transport system in the body
- relate the components of blood to their functions, including blood clotting and blood groups
- explain the causes and effects of heart diseases
- describe the structure and function of the lymphatic system
- describe the structure and function of the heart, and recognise how the structures of the heart are related to their functions in the cells, including ATP as the energy currency of the cell
- explain the concept of blood pressure, including systole and diastole, as well as modifiable risk factors for hypertension
- describe the structure and function of the circulatory system in humans
- identify the substances that need to be transported around the human body

What the examiners say

→ Generally, candidates seem vague about some of the fundamental concepts associated with this topic. For example, candidates are sometimes unable to differentiate between the left and the right sides of the heart, or adequately discuss the functions of the heart or the differences between arteries and veins.

Revision tip

Link each topic to at least one other topic – for example, the circulatory system to the respiratory system – so you are always reminded of how the different areas of biology are inter-related. This will help you remember both systems!

All cells in an organism need to be supplied with oxygen and food in order to function. As you saw in Chapter 2 and Chapter 7, small, single-celled organisms rely on simple diffusion to exchange materials between the outside world and the inside of their cells. The distances are short, so diffusion works really well. However, as animals get larger and are made up of more and more cells, simple diffusion alone is not enough to supply the body's needs; there is simply not enough surface area available for the exchanges to take place.

This is partly because as animals get bigger, the ratio between the surface area and the volume gets smaller (see *The biologist's toolkit* in Chapter 2 and Figure 8.1). Diffusion takes place through the surface area, but the substances have to reach the innermost volume, so the bigger the organism, the less effective simple diffusion becomes as a means of transport.

Human beings are made up of thousands of millions of cells, most of them a very long way from a direct source of food or oxygen. All of our cells need oxygen and glucose for cellular respiration, the waste products of metabolism must be removed, and the many chemicals needed in the body must be transported to and from the different organ systems. Our surface area to volume ratio means that simple diffusion of substances to and from the outside of our body is not enough. A complex transport system is required to supply the needs of the body's cells and remove the waste materials they produce.

In humans, the transport system is the **blood circulation system**. It has three elements: the pipes (**blood vessels**), the pump (**the heart**) and the transport medium (**the blood**).

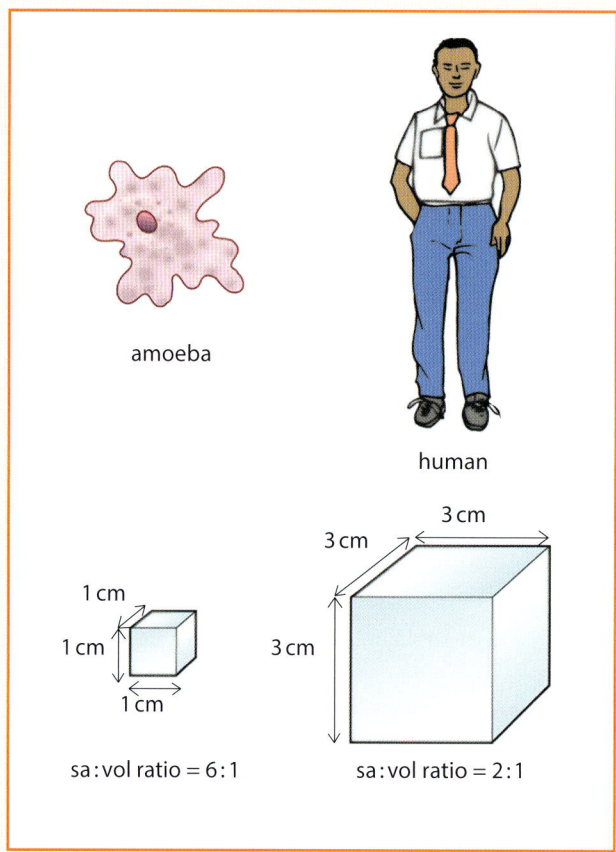

Figure 8.1 The surface area to volume ratio of the small cube is three times bigger than that of the large cube. Imagine the difference between an amoeba and you!

What does our circulatory system transport?

Here are some of the substances that the blood carries around the body:
- Gases – for example, oxygen for cellular respiration and the waste carbon dioxide produced in the process (see Chapter 7 for more details)
- Nutrients – for example, glucose, amino acids, fatty acids, glycerol, vitamins and minerals from the digestive system (see Chapter 6 for more details)
- Hormones – chemical messages that control how our body works (you will learn more about them in Chapter 11)
- Antibodies – molecules made by your immune system to protect you against infectious diseases (you will learn more about them later in this chapter and in Chapter 20)
- Blood proteins – molecules involved in processes such as the clotting or your blood (you will learn more about these later in this chapter)
- Metabolic waste products – for example, carbon dioxide from cellular respiration (see Chapter 7) and urea from the breakdown of excess proteins in the liver (you will learn more about this in Chapter 10).

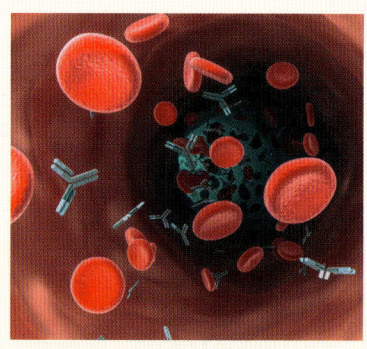

Figure 8.2 Red blood cells and antibodies are two of the substances that the blood transports around the body.

The circulatory system

A double circulation system

We humans actually have not one transport system, but two. We have a double circulation system:

- **The pulmonary circulation** carries blood from the heart to the lungs and back again to exchange oxygen and carbon dioxide with the air
- **The systemic circulation** carries blood all around the rest of the body and back again to the heart.

A double circulation system like this is very important in warm-blooded, active animals like humans, because it is very efficient. The system fully oxygenates our blood in the lungs before it is sent off to the different parts of the body. In animals like fish with a single circulation system, as soon as the blood has picked up oxygen, it starts to lose it again to the tissues, so very few parts of the body receive fully oxygenated blood.

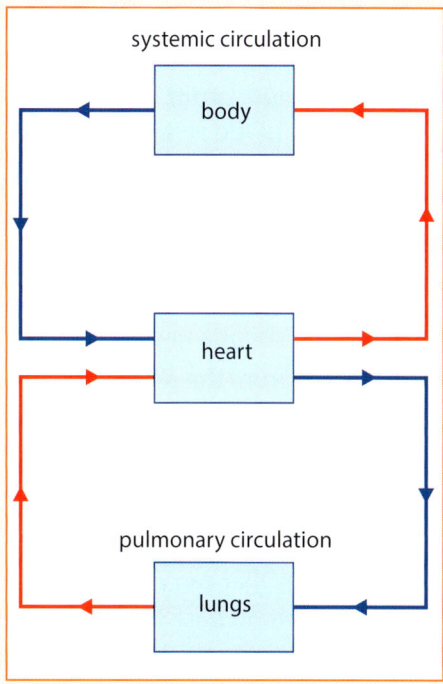

Figure 8.3 Two separate circulation systems supply the lungs and the rest of the body with blood.

The blood vessels

A very important element of any transport system is the pathways along which the transport takes place. In the human body, there are three main types of blood vessels: **arteries**, **veins** and **capillaries**, which are adapted to carry out particular functions within the body, although they all carry the same blood.

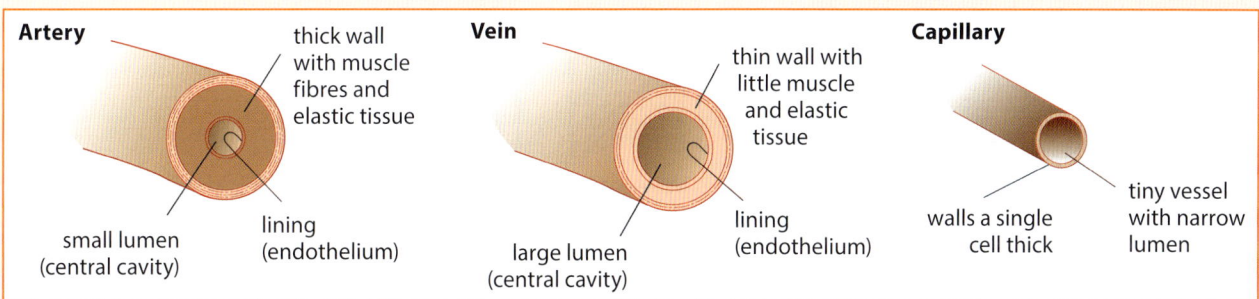

Figure 8.4 Arteries, veins and capillaries – different blood vessels with different functions in the body

Arteries

The arteries carry blood away from the heart, so they have to be able to withstand the pressure caused by the pumping of the heart as it forces the blood out into the circulation. This is usually oxygenated blood, so it is bright red. Arteries have thick walls that contain muscle and elastic fibres so that they can stretch as the blood is forced through them and then return to their normal shape. Arteries have a pulse in them that you can feel at certain places in the body (like the wrist) where they run close to the surface. The pulse is the surge of blood from the heart when it beats. Since the blood in the arteries is under pressure, it is very dangerous if an artery is cut, because the blood spurts out under pressure every time the heart beats.

DID YOU KNOW?

The heart pumps so powerfully that the blood from a severed artery can spurt up to 2 m away from your body!

This means blood is lost very rapidly and the bleeding is difficult to stop. The only arteries that carry deoxygenated blood are the pulmonary arteries, which carry blood away from your heart to your lungs, and the umbilical artery, which carries blood away from a foetus into the placenta (see Chapter 12).

Veins

The veins carry blood towards the heart. Blood in the veins is usually low in oxygen and so is a deep purple-red colour. Veins have much thinner walls than arteries and the blood in them is under much lower pressure because it is a long way from the pumping of the heart. Veins do not have a pulse, but they often have valves to prevent the backflow of blood as it moves from the various parts of the body back to the heart. When the blood flows in the right direction, towards the heart, the valves are open. If the blood starts to flow in the wrong direction, the valves close, preventing a backflow of blood.

The only veins that carry bright red oxygenated blood are the pulmonary veins, which carry blood back from the lungs to the left-hand side of the heart, and the umbilical vein, which carries oxygenated blood from the placenta back to the developing foetus to supply it with the food and oxygen it needs to grow.

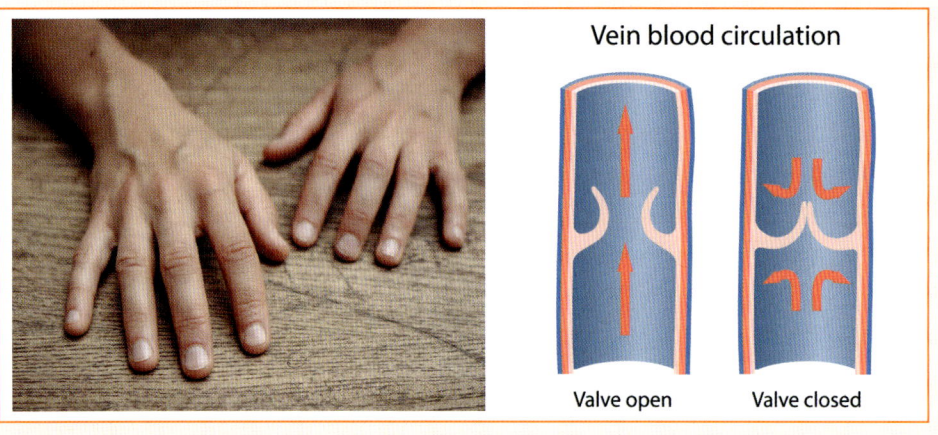

Figure 8.5 The valves in the veins stop your blood from flowing backwards and so move it towards your heart.

Capillaries

Between the arteries, which bring blood from the heart, and the veins, which take it back to the heart, there are very narrow, thin-walled blood vessels called capillaries. The capillaries link the other two types of blood vessels and take the blood into all the organs and tissues of the body. The capillaries are the site of the exchange of substances within the body. Blood from the arteries passes into the capillaries, which have very thin walls and a massive surface area. Substances needed by the cells of the body, such as oxygen and glucose, can easily pass out of the blood by diffusion along a concentration gradient. In the same way, substances produced by the cells, such as carbon dioxide, pass into the blood through the walls of the capillaries. The blood then leaves the capillary network, flowing back into the veins to be returned to the heart and recirculated around the body.

> **Checkpoint questions**
>
> 1. Explain why humans need a transport system.
> 2. Describe the main parts of the human transport system.
> 3. Draw a simple diagram to demonstrate how the arteries, veins and capillaries are linked to each other and to the heart.

The human heart

The human heart is a bag of reddish-brown muscle that beats right from the early days of our development in the uterus until the end of our life, sending blood around the body. The heart is made up of two pumps that beat at the same time, so blood is delivered to the body about 70 times each minute. The heart is made up of a unique type of muscle called **cardiac muscle**, which can contract and relax more or less continuously without fatiguing or producing lactic acid.

The walls of the heart are almost entirely muscle. These muscular walls are supplied with blood by the coronary arteries, so they have a constant supply of glucose and oxygen, and the carbon dioxide produced does not build up in the tissues. The deoxygenated blood is carried away in the coronary veins, which feed back into the right atrium.

The walls of the **atria** are relatively thin so they can stretch to carry a lot of blood. The walls of the **ventricles** are much thicker as they have to pump the blood out through the major blood vessels. The muscular walls of the left ventricle are thicker than those of the right ventricle (see Figure 8.6). This is because the left ventricle has to pump blood around the whole body while the right ventricle only pumps blood to the lungs.

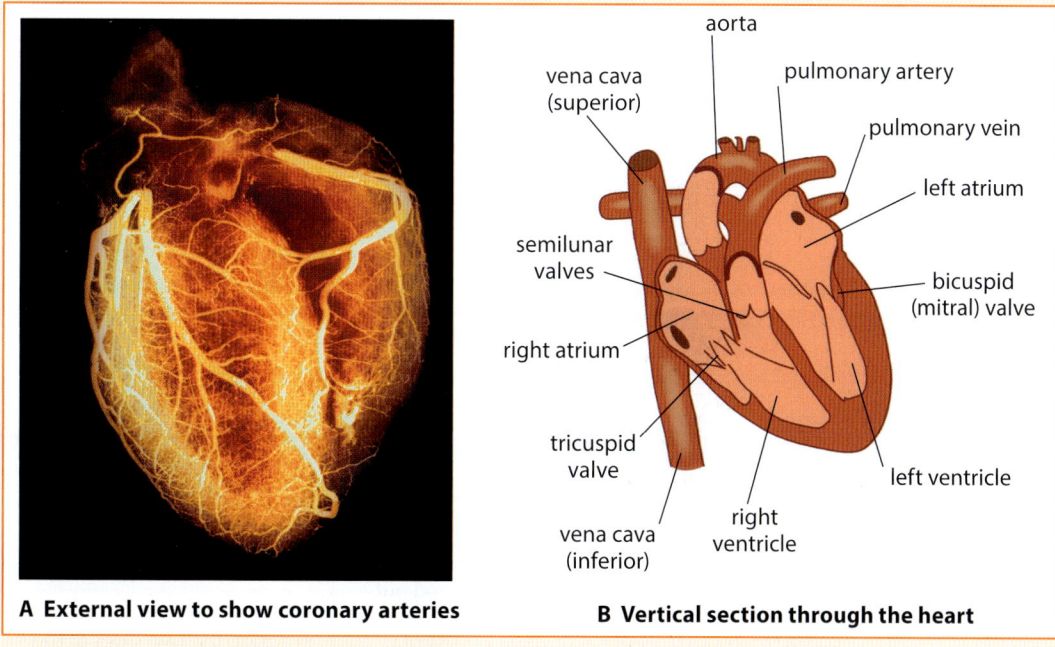

A External view to show coronary arteries

B Vertical section through the heart

Figure 8.6 The structure of the human heart is perfectly adapted to the job it has to do.

The working of the heart

Inside the heart there are many different valves. Their names describe their appearance – bicuspid (two parts), tricuspid (three parts) and semilunar (half moon). Each time the muscular walls of the heart contract and force blood out, some of these valves open to allow the blood to flow in the right direction, and other valves close to make sure that the blood does not flow backwards. The noise of the heartbeat we can hear through a stethoscope is actually the sound of these valves opening and closing. To understand what happens, it is easier to follow a single volume of blood around the heart.

- Deoxygenated blood, which has supplied oxygen to the cells of the body and is loaded with carbon dioxide, enters the right atrium of the heart from the veins of the body, through the vena cava.
- The tricuspid valves open and the right atrium contracts, forcing blood into the right ventricle.
- The tricuspid valve closes and the semilunar valves in the pulmonary artery open as the right ventricle contracts, forcing blood out of the heart and into the lungs where it is oxygenated (picks up oxygen).
- Oxygenated blood returns to the left-hand side of the heart from the lungs in the pulmonary vein and the left atrium fills up.
- The bicuspid valve opens and the left atrium contracts, forcing blood into the left ventricle.
- The bicuspid valve closes and the semilunar valves open as the left ventricle contracts, forcing oxygenated blood out of the heart and around the body, through the aorta.

The two sides of the heart fill and empty at the same time to give a strong, coordinated beat. First both atria fill with blood and contract, filling the ventricles. This is followed by the contraction of both ventricles, emptying the heart. **Diastole** is when the heart muscles relax and it fills with blood. **Systole** is when the heart muscles contract and force the blood out of the heart.

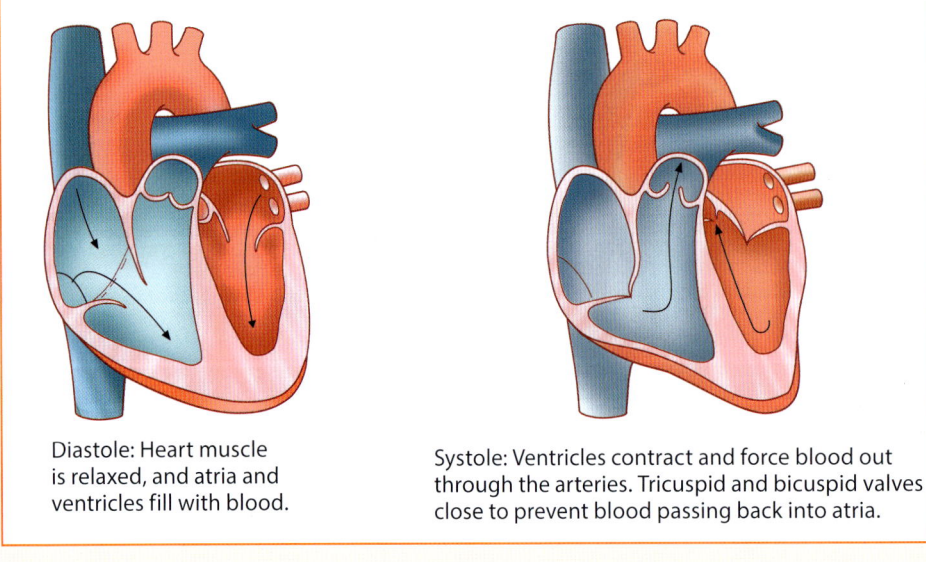

Diastole: Heart muscle is relaxed, and atria and ventricles fill with blood.

Systole: Ventricles contract and force blood out through the arteries. Tricuspid and bicuspid valves close to prevent blood passing back into atria.

Figure 8.7 As the heart fills and empties, the two sides of the heart work together perfectly.

The circulatory system

Activity 8.1

Examining a mammalian heart

If you have the opportunity to dissect the heart of a mammal like a sheep or a pig, you can identify the different features from Figure 8.8 and gain an insight into their adaptations and how the whole heart works. However, the blood vessels and the atria can be damaged by the butcher, so you may not be able to see everything you would like to.

You will need:

- a board for dissection
- dissecting equipment including a scalpel and a mounted needle – TAKE CARE, the scalpel blade is very sharp
- a sheep or pig's heart – you need as many of the tubes intact as possible and any surrounding fat

Method

1. Examine the heart carefully while still intact. Find the blood vessels, the atria, the ventricles, the coronary arteries and any fat. Draw and label what you can see.
2. Make cuts through the wall of the heart as shown in Figure 8.8.
3. Open the heart gently and try to identify as many structures as you can. Compare the thickness of the walls of the atria (if they are present), the right ventricle and the left ventricle, and remind yourself why they are so different. Look for the valves between the atria and the ventricles, and the valves between the ventricles and the great vessels (pulmonary artery and aorta).
4. Draw and annotate your dissection.

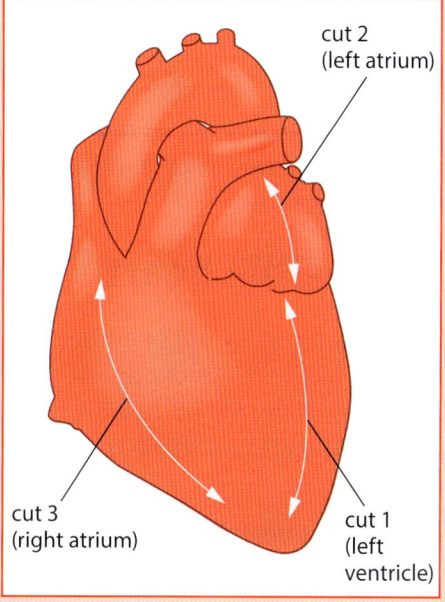

Figure 8.8 Guide to dissecting a mammalian heart

Measuring blood pressure

The pressure at which the blood travels around our arteries varies as the heart beats. When doctors measure blood pressure, they covers the two extremes of the cardiac cycle. At systole, when the heart is contracting and forcing blood out into the arteries, the blood pressure in our blood vessels is at its highest. This is the **systolic blood pressure** and it is the higher of the two readings taken. At diastole, when the heart is relaxed and filling, the pressure is lower. This is the **diastolic blood pressure** and it is the lower reading.

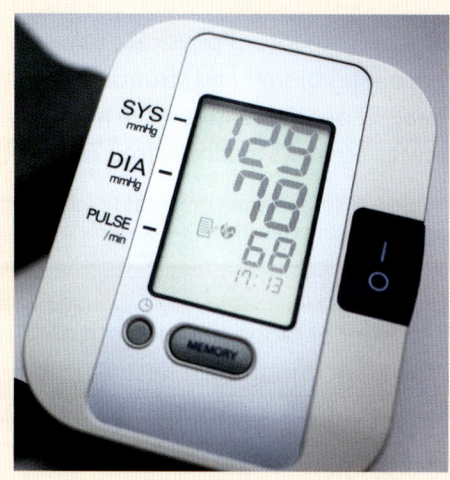

Figure 8.9 Blood pressure readings for systolic, diastolic and pulse.

The human heart

There are a number of ways of measuring blood pressure. It may be done by hand or with a machine. However, they all involve pumping up a cuff around part of a limb and cutting off the blood supply. The point at which blood first forces its way past the cuff is your systolic pressure, while the point at which all of the blood can get through is the diastolic pressure.

A normal blood pressure is **120 mmHg/80 mmHg** – usually quoted as '120 over 80' or '120/80'. Your blood pressure will vary throughout the day and depending on what you are doing. It will be at its lowest when you are asleep and at its highest when you are exercising hard or are very stressed, but in a healthy person it will always return to its normal resting level. If the blood pressure becomes too high, it causes many different problems – you will find out more about this later in the chapter.

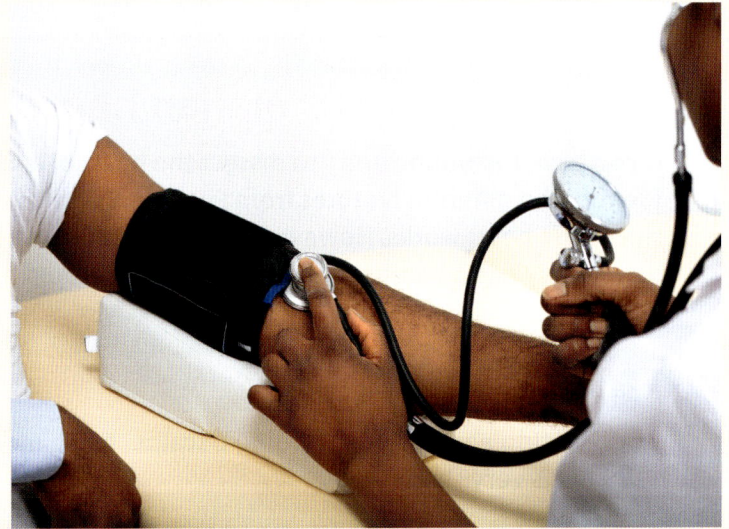

Figure 8.10 Measuring blood pressure

The flexible heart

When we are resting, our heart beats steadily at around 70 beats every minute, supplying all the needs of our cells. When we exercise – whether running to catch a bus, doing the housework or playing sport – our muscles need more food and oxygen, and so the heart needs to supply more blood. It does this in two ways:

- The heart beats faster – the pulse rate can easily go up from resting to over 150 beats a minute, increasing the amount of blood flowing around the body.
- The heart can also increase the amount of blood pumped out with each heartbeat.

If people do lots of physical exercise and are fit, their heart responds by becoming bigger and stronger. Since their heart pumps more blood with each beat, fit people tend to have relatively slow resting heartbeats – some are as low as 50 beats a minute (see Table 8.1).

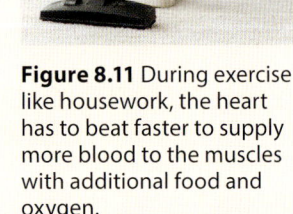

Figure 8.11 During exercise like housework, the heart has to beat faster to supply more blood to the muscles with additional food and oxygen.

Table 8.1 Not everyone can be as fit as a top class athlete, but doing exercise helps keep your heart and lungs healthy at whatever level you take part.

	Before getting fit	After getting fit
Amount of blood pumped out of the heart during each beat (cm^3)	64	80
Heart volume (cm^3)	120	140
Pulse rate	72	63

The circulatory system

Activity 8.2

Investigating the effect of exercise on the heart rate

A good way of telling how fit you are is to measure your resting heart rate. The simplest way to measure your heart rate is to take your pulse. Your pulse simply reflects the surge of blood in the arterial system each time your heart contracts, so it is a good way of recording your heart rate. The fitter you are, the fewer beats per minute you will have. Then, see what happens when you exercise – the increase in your heart rate and how quickly it returns to normal is another way of finding out how fit you are – or aren't!

NB: Anyone who is affected by asthma or has any other illness should take care before taking part in this activity and take any medication they would normally use before a PE session. Anyone who does not usually take part in PE should act as timekeeper and recorder in this investigation and not take part in the physical exercise.

You will need:
- a stopwatch or clock with a clear second hand

Method

1. First practise actually finding and taking your pulse either at your wrist or on the side of your neck.
2. Find your resting pulse rate. Sit quietly without speaking for at least 2 minutes. Then start the stopwatch and record the number of pulse beats in 15 seconds. Repeat this three times to get an average resting pulse rate.
3. Now exercise gently for 2 minutes by gently marching on the spot.
4. As soon as you stop exercising, find your pulse and record the number of beats in 15 seconds. Repeat this every 30 seconds until your pulse returns to your resting rate.
5. Now change the way you exercise. Exercise harder for 2 minutes by gently jogging on the spot. As soon as you stop exercising, start to record your pulse beats. Record the rate for 15 seconds every 30 seconds until it returns to your resting rate.
6. Finally, exercise hard for 2 minutes – run on the spot as hard as you can. As soon as you stop exercising, start to record your pulse. Record it as above until it returns to your resting rate. If you prefer, you can simply extend your period of more gentle exercise, by walking or jogging gently for 4 minutes instead of 2.
7. Write up your investigation, including your results. If you multiply all of your results by four, it will give you your pulse rate per minute. Use a table like the one below to record your results.

		Beats in 15 seconds	Pulse rate (beats per min)
Before exercise	1		
	2		
	3		
	mean		
Time after exercise(s)	0		
	30		
	60		
	90		
	120, and so on		

> 8 Draw a graph of your own personal data and explain what you have observed. In some cases, your pulse rate may drop below your normal resting rate as you recover. Can you explain what is happening?
>
> 9 Collect data from other members of the class and compare the pulse rates and recovery times of the group. Now look for patterns in your data. Are there differences between boys and girls? Do the members of sports teams show different patterns to the rest of the class?

How is the heart rate controlled?

There are lots of different factors involved in controlling your heart rate. The very basic rhythm of the heart comes from the cardiac muscle cells themselves, which beat together from the early days of embryo development. Then, within the heart itself, we have our own electrical **pacemaker** system. An electrical impulse arises in our pacemaker region (found in the right atrium) about once every second. This triggers a wave of electrical excitation that spreads through special tissue in the heart and stimulates the muscle to contract. Our pacemaker keeps the heart beating all the time. Other systems add the 'fine tuning' that allows the body to respond to all the different circumstances we might find ourselves in.

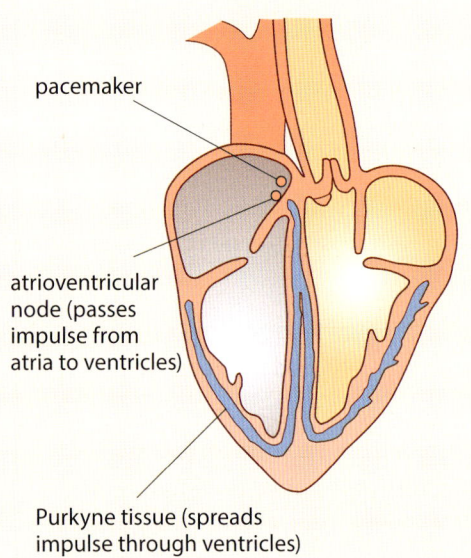

Figure 8.12 As the heart fills and empties, the two sides of the heart work together perfectly.

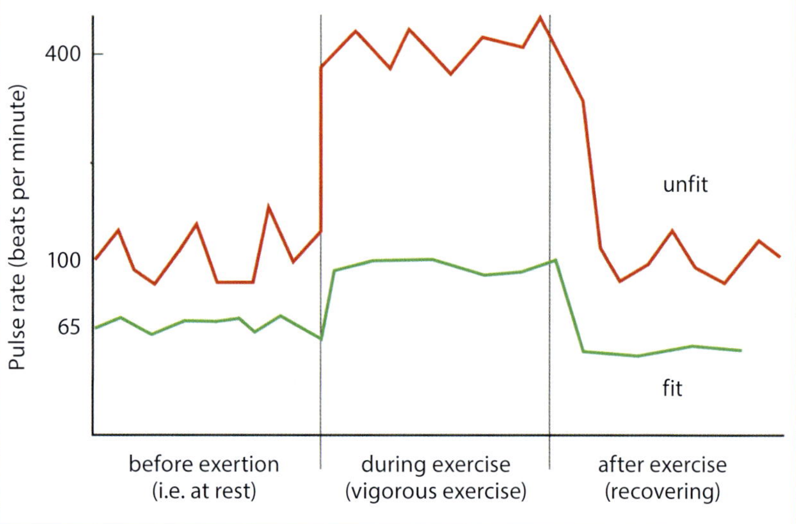

Figure 8.13 A fit heart responds quickly to exercise and returns rapidly to its resting rate when the exercise stops. People with a less fit heart can feel them racing for several minutes after they stop exercising.

The heart doesn't beat with a steady rhythm all the time – it responds to all the needs of the body as they arise.

- As Figure 8.13 shows, when we exercise our heart rate increases.
- If we are worried, stressed or angry, our heart rate will also increase.
- Sitting an exam or having an argument can raise our heart rate as much as if we were running a race!

The change in our heart rate in response to the changing conditions in our body when we exercise is brought about by an area of the brain called the **medulla** (you will be looking at the different areas of the brain in Chapter 11). The medulla responds to the level of carbon dioxide in our blood and makes our heart rate go up or slow down as needed.

The change in our heart rate in response to anger or stress is brought about by the hormone adrenaline. It is called the 'fight or flight' response and you will learn more about this in Chapter 11.

Sometimes, the natural pacemaker region of the heart does not function properly, or the heart stops responding as it should. The heart rate and often the blood pressure falls, and the person affected may suffer from tiredness and even blackouts. In these circumstances, an artificial pacemaker can be implanted. This gives the heart a regular electric shock to stimulate it to contract. An artificial pacemaker can provide a new lease of life for affected patients.

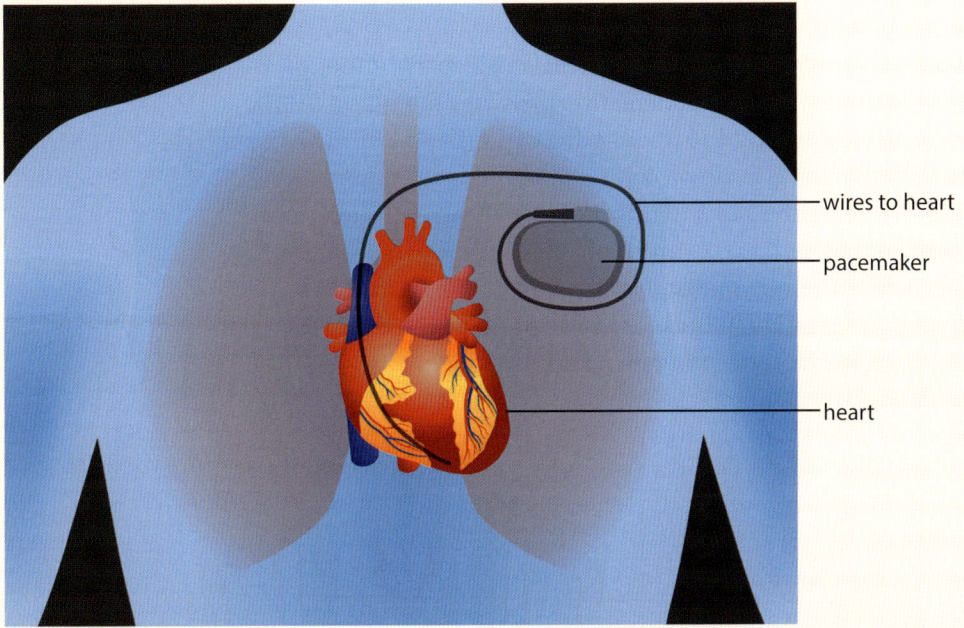

Figure 8.14 An artificial pacemaker can provide a new lease of life for affected patients by keeping the heart beating at a steady rate. The most modern artificial pacemakers change the heart rate when you exercise, and some will even restart your heart if it stops!

Checkpoint questions

4 Draw a diagram of the heart. Then number the areas of the heart to show the route of a single volume of blood returning from the body as it travels around the heart. List the numbers and explain what is happening in the heart at each point you have labelled.

5 An average, not very fit teenager watches a film on TV, and then goes down to the local gym for a workout. Explain what will be happening to his heart:
   a while he watches television.
   b when he is working out in the gym.

6 A member of the school athletics squad goes out for a training run. Discuss how the response of her heart likely to differ from the boy in the gym in question 5.

The blood

The heart and the blood vessels are there to carry the transport medium around the body, and the transport medium is blood. Our blood is a complex mixture of cells and liquid that carries a huge range of substances around the body. It consists of a liquid called **plasma**, which carries red blood cells, white blood cells and **platelets**.

> **DID YOU KNOW?**
> Each adult has approximately 5 litres (10.6 pints) of blood containing about 15 thousand million red blood cells that travel around the body in 80 000 km (50 000 miles) of blood vessels!

Plasma

Plasma is a pale yellow liquid that transports all the blood cells and also a number of other substances. All the small, soluble products of digestion pass into the blood from the gut. They are carried in the plasma around the body to the organs and individual cells that need them. Plasma also contains some large proteins including prothrombin and fibrinogen, which are needed for the clotting of your blood. Hormones, antibodies and any other substances your body needs, and waste products produced by the cells such as carbon dioxide and urea, are carried around your body dissolved in the plasma. Serum is part of the plasma – it is what is left when the blood has clotted and all the cells have been removed. Serum is plasma without the clotting proteins and blood cells.

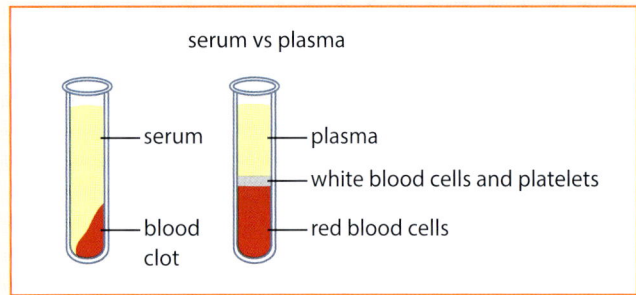

Figure 8.15 Serum is what is left after the blood has clotted – plasma minus the clotting factors and blood cells.

Red blood cells

One of the main components of your blood is the red blood cells. There are more red blood cells than any other type of blood cell. They are superbly adapted to their role in carrying oxygen around your body and supplying it to the cells where it is needed. They can do this because they are packed with a special red pigment called haemoglobin, which picks up oxygen. Haemoglobin is a large protein molecule folded around four iron atoms.

- In a high concentration of oxygen, such as in the lungs, the haemoglobin reacts with oxygen to form oxyhaemoglobin.

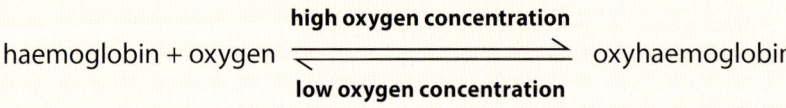

> **Remember**
> Oxyhaemoglobin is bright red, which is why most arterial blood is bright red. Haemoglobin alone is purple-red, which is the colour of the blood in our veins.

The red blood cells are made in the bone marrow:

- Once they are mature, they lose their nucleus. This means that there is more room to carry extra haemoglobin – another adaptation to their all-important function.
- Since they have no nucleus, the red cells only live for 100–120 days in the body, so they are constantly being replaced.

The red blood cells have a unique shape.

- They form biconcave discs. This is another adaptation to their function. The shape gives them a large surface area to volume ratio for the diffusion of oxygen into and out of the cell. It also means they are relatively thin, resulting in short diffusion distances, which again makes the exchange of gases more efficient.
- Their shape also allows them to squeeze easily through the very narrow capillaries.
- Red blood cells also have a thin surface membrane for ease of diffusion.

> **Remember**
> The haemoglobin in our red blood cells is based on iron. This is why iron is such an important mineral in our diet – see Chapter 5. Without iron, we cannot make the red blood cells we need and we suffer from anaemia – we are pale and lack energy as our cells do not get the oxygen they need for the reactions of life.

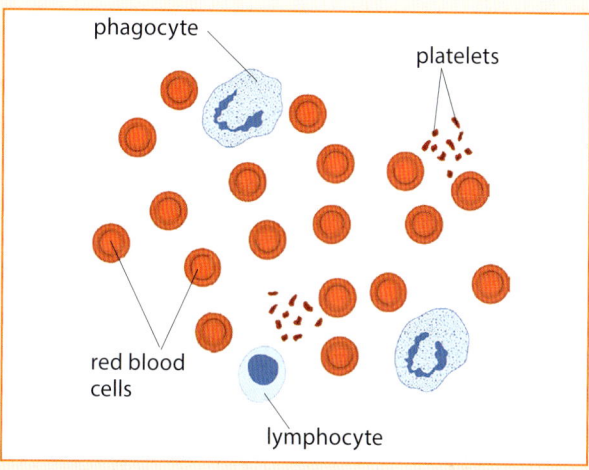

Figure 8.16 The main types of cells and cell fragments making up your blood, along with the rest of the plasma.

White blood cells

Another important component of our blood is the white blood cells. They are much bigger than the red cells and there are fewer of them. White blood cells all have a nucleus and form part of the body's defence system against microbes. Some white blood cells – the **lymphocytes** – protect us by forming antibodies which damage and inactivate the microorganisms that cause many diseases. Other white blood cells, called **phagocytes**, protect us by engulfing invading bacteria and dead cells.

Platelets

Platelets are another component of blood. They are small fragments of cells and are very important in helping the blood to clot at the site of a wound.

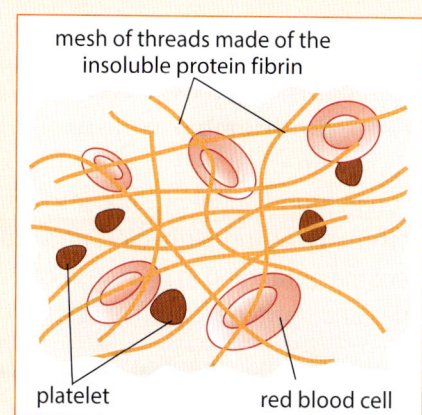

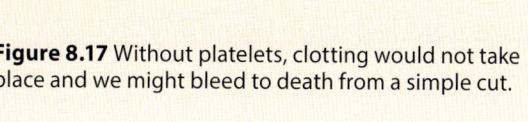

Figure 8.17 Without platelets, clotting would not take place and we might bleed to death from a simple cut.

The blood

Blood clotting

The biologist's toolkit: : Key chemistry: positive and negative ions

In chemistry you will have learned that an atom is made up of a positive nucleus with negative electrons moving around it.

If an atom gains electrons or loses electrons it becomes an ion. Ions dissolve in water. Ions are very important to biologists. Many of the minerals you discovered in Chapter 5 are carried around our body as ions. Calcium ions are important in clotting the blood.

- Metal atoms lose electrons from their outer shell to form **positive ions**, for example: Ca^{2+} (calcium ion), Fe^{2+} (iron ion), Na^+ (sodium ion)
- Non-metals gain electrons to their outer shell to form **negative ions**, for example: Cl^- (chloride ion), H^- (hydrogen ion)

The clotting of our blood is a very important process.

- Clotting prevents us from bleeding to death from a simple cut.
- Clotting protects our body from the entry of bacteria and other **pathogens** (disease-causing microorganisms) through open wounds.
- Clotting protects the new skin from damage as it grows.

It is also important that our blood does not clot when we do not need it to – a blood clot in the wrong place is very serious and can kill us. This why the process by which our blood clots has many different steps (see Figure 8.19). It relies on interactions between the platelets and some of the proteins in the blood. Your body turns soluble proteins dissolved in the plasma into insoluble proteins that form the basis of a clot.

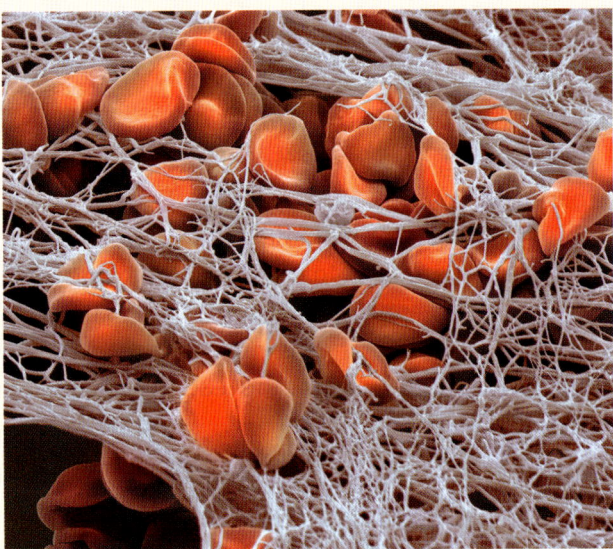

Figure 8.18 Blood clot. Coloured scanning electron micrograph of a blood clot.

When the platelets break open, they release a compound called **thromboplastin**. If there are plenty of **calcium ions** and **vitamin K** present, thromboplastin acts as an enzyme and catalyses the breakdown of the protein **prothrombin** in the plasma to form **thrombin**. Thrombin in turn is an enzyme and it catalyses the breakdown of soluble **fibrinogen**, another big protein found in the plasma, into insoluble **fibrin**. This fibrin forms a network of threads. As more platelets and red blood cells pour out of the wound, they become entangled in the mesh of threads, forming a jelly-like clot. This soon dries and hardens to form a scab, protecting the wound until it has healed.

This may sound complicated – look at the flow diagram in Figure 8.19 on the next page to help you make sense of the process.

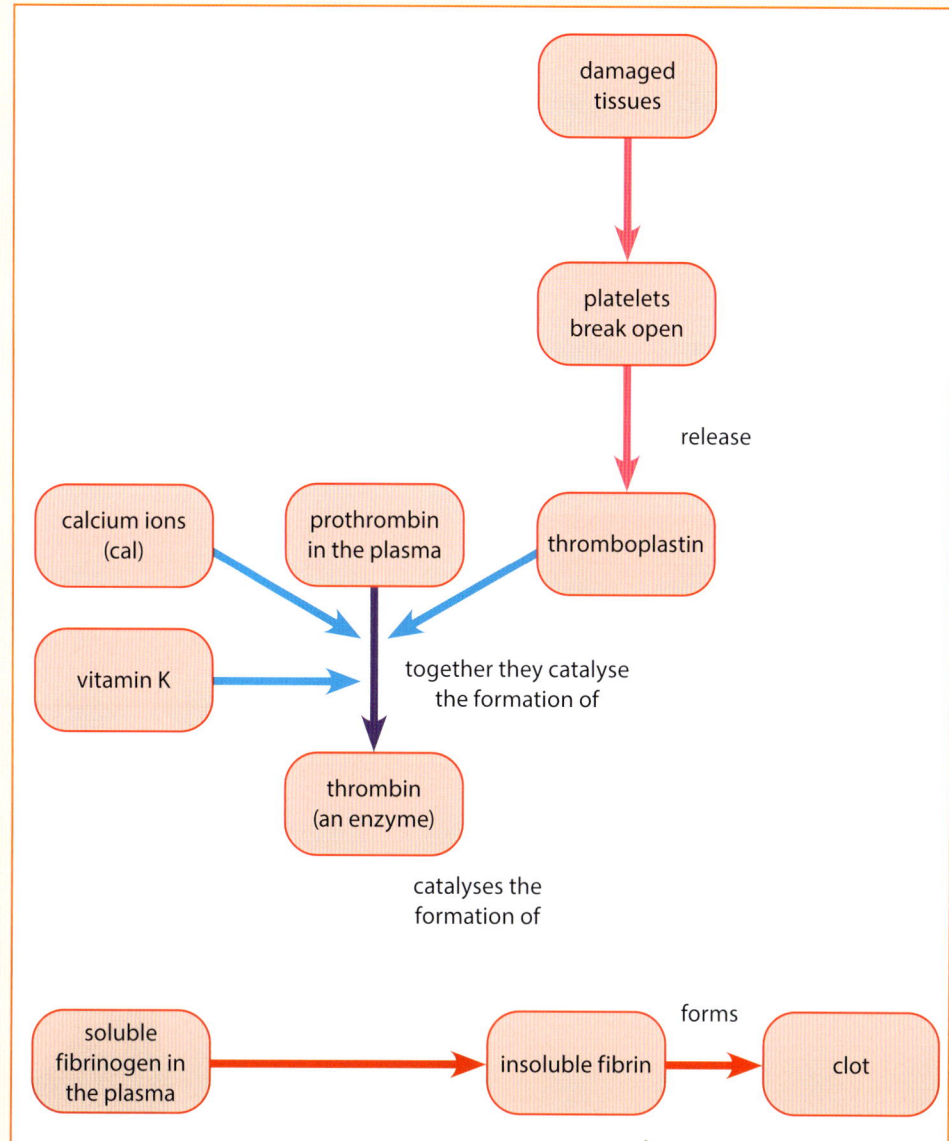

Figure 8.19 Platelets, calcium ions and vitamin K work with the plasma proteins prothrombin and fibrinogen to form a successful blood clot.

Blood groups

All the cells of your body have proteins on their surface membranes (see Chapter 1) called **antigens**. These antigens identify your cells and everyone, except identical twins, has different antigens on their cells. Your body recognises these antigens using your **immune system**. You'll be learning more about that later in this book. For now, you need to know that your lymphocytes recognize the antigens on the surface of your cells. If they meet an antigen that is different, they make **antibodies** that attach to the 'foreign' antigen. These antibodies are carried in the plasma and they either destroy the cell the foreign antigen is attached to OR make it easy for phagocytes to come and digest it.

What has this got to do with our blood? We have a number of different antigens that are found only on the surface of our red blood cells. The antigens on our red blood cells determine our **blood groups**. There are a number of different blood group systems but the best known is the **ABO** one.

There are four ABO blood groups and you will belong to one of them. They are **blood group A**, **blood group B**, **blood group AB** and **blood group O**. Blood group O is the most common in the Caribbean – see Figure 8.20. How do blood groups work?

- There are two possible antigens on your red blood cells – **antigen A** and **antigen B**.
- There are two possible antibodies in your plasma – **antibody a** and **antibody b**. These antibodies are there all the time.

It is the combination of antigens and antibodies that you inherit from your parents (see Chapter 13) that determines which blood group you will have – see Table 8.2.

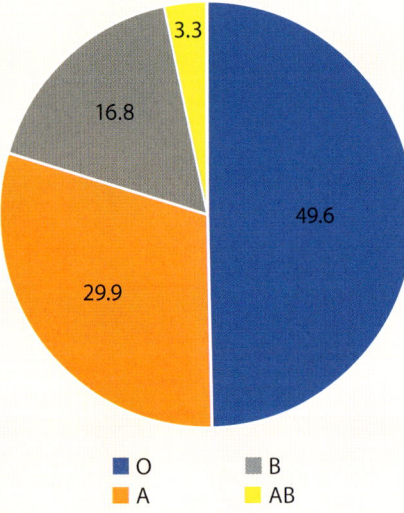

Figure 8.20 As on data.

Table 8.2 Blood groups

Blood group	Antigen on red blood cells	Antibody in plasma
A	A	b
B	B	a
AB	A & B	none
O	none	a & b

Usually, we each have our own blood and it does not mix with the blood of anyone else, so any differences between the blood groups do not matter. But sometimes, after an accident or during surgery, one person will be given blood donated by someone else in a process known as a blood transfusion. Blood transfusions save lives – see Figure 8.22.

However, blood transfusions must be carried out very carefully. If blood from different blood groups is mixed, there may be a reaction between the antigen on the red blood cells and the antibodies in the plasma. If there is a reaction, the red blood cells stick together and no longer work. They will block up capillaries or even larger blood vessels. Usually this isn't a problem, as we all have our own blood in our veins. But during a blood transfusion, one person is given blood from several other people. It is vitally important that the blood groups are matched correctly. Blood containing a particular antigen must NOT be given to someone with the matching antibody. The blood group of the recipient (the person receiving the blood transfusion) and the donor (the person who gave the blood) are always matched very carefully before a life-saving blood transfusion is given. Blood group O is particularly useful for transfusions, as its red blood cells have no antigens so it does not react with other types of blood. As Table 8.3 shows, other blood groups have to be matched more carefully. Mixing the wrong blood groups can cause the blood cells to stick together (agglutination), which be fatal.

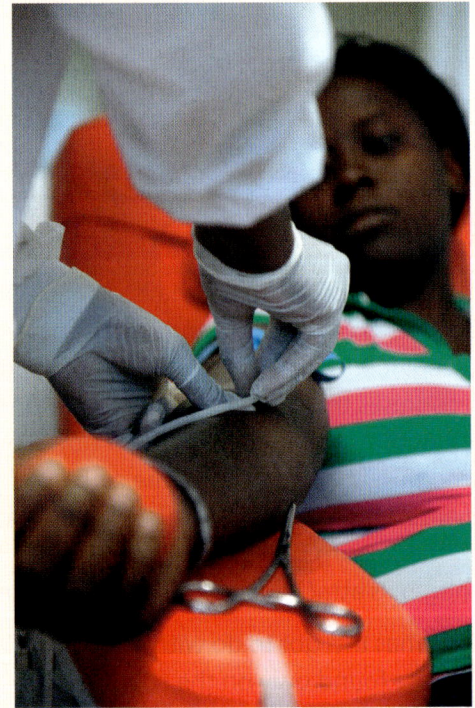

Figure 8.21 Almost anyone can give blood – and blood donors give the gift of life to others.

The circulatory system

Table 8.3 Matching blood groups

Donor	Recipient			
	A	B	AB	O
A	Yes	No	Yes	No
B	No	Yes	Yes	No
AB	No	No	Yes	No
O	Yes	Yes	Yes	Yes

Another important blood group system is the **Rhesus** one. Everyone either has the Rhesus antigen on their red blood cells – they are **Rhesus positive or Rh+** OR they do not have the antigen and they are **Rhesus negative** or **Rh–**. About 8% of the Afro-Caribbean population is Rhesus negative, compared to 15% of the Caucasian population.

- If you are Rhesus positive, you do not have any Rhesus antibodies in your blood.
- If you are Rhesus negative, you have Rhesus antibodies in your blood.

Usually it doesn't matter if you have Rhesus antibodies in your blood or not, unless you need a blood transfusion OR you are pregnant. Blood is matched for Rhesus factors as well as ABO groups before a blood transfusion is given. Problems may arise if a Rhesus negative woman is expecting a Rhesus positive baby (you will learn more about the inheritance of blood groups in Chapter 13). The first pregnancy is usually safe – but because some of the cells from the growing fetus often pass into the blood of the mother, she develops an immune response to the antigen and makes antibodies against it. In the worst case, the mother's antibodies will destroy the blood of her next fetus.

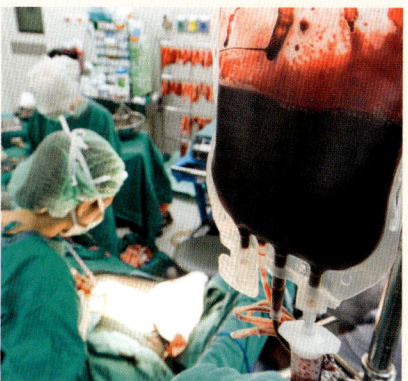

Figure 8.22 Patients who need blood transfusions during an operation must be matched to receive the correct blood group.

Doctors now monitor the pregnancies of Rhesus negative women very carefully. They can test the unborn baby to see if it is Rh+ or Rh–. They can even give the developing baby a blood transfusion BEFORE it is born!

Best of all, complications from Rhesus factors can be completely prevented by giving the Rhesus negative mother an injection immediately after the birth of each baby she has, to neutralise and remove any antigens she has made. This makes the next pregnancy safe.

Tissue fluid and lymph

The capillary walls are permeable to everything in the blood apart from the red blood cells and the large plasma proteins. As blood flows through the capillaries from the arterial system, it is under pressure and fluid is squeezed out of the vessels. This fluid fills the spaces between the cells of the body and it is called tissue fluid. Tissue fluid is plasma *without* the blood cells but *with* the clotting proteins. It is through this fluid that all the diffusion between the blood and our cells takes place.

Most of the tissue fluid is eventually returned to the blood. Some of it moves back into the blood capillaries as they near the veins, but a lot of it drains into a series of blind-ending tubes that are part of the **lymph system**. Once in these tubes, the fluid is called **lymph**. The lymph vessels join together and get larger. The lymph is moved along the vessels by the squeezing effect of skeletal muscle movements, and the system has a series of valves, like those found in our veins, to prevent the backflow of liquid. The lymph is finally returned to the blood in the neck area, where the lymph vessels join the large veins.

Along the lymph vessels are the **lymph glands**. These are very important in helping our body to fight infectious diseases. When you were younger, if you felt ill, your mum may have felt in your neck to see if your glands were swollen as a sign that you really were poorly!

- White blood cells called lymphocytes gather in the lymph glands and produce antibodies against invading pathogens. These antibodies are carried in the lymph to the blood.
- The lymph glands filter out bacteria and other microbes from the lymph, to be taken in and digested by a different type of white blood cell called a phagocyte.
- Enlarged lymph glands are a sign that the body is fighting off an invading pathogen. You can see the main sites of the lymph glands in Figure 8.23. Now you will understand why doctors often examine the neck, armpits, stomach and groin of their patients. This is where the lymph glands are near the surface of the body – and enlarged lymph glands suggest an infection.

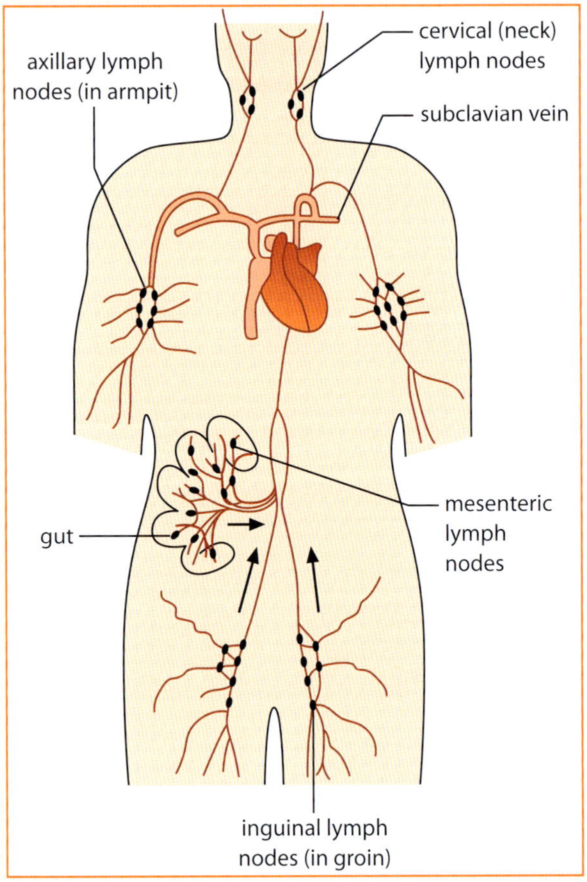

Figure 8.23 The lymph system is important both because it returns tissue fluid to your blood and because it is involved in the defence of your body against infection.

Diseases of the circulatory system

The circulatory system is very important, and a healthy circulatory system is vital for life. Unfortunately, sometimes parts of the circulatory system do go wrong. There are lots of ways we can treat the problems – and many ways in which we can help ourselves to avoid these diseases. You will learn more about the causes and effects of some of these diseases, how often they occur in the Caribbean and what we can do to avoid them in Chapters 19 and 20, but here is a brief summary of some of the more common problems.

Hypertension (high blood pressure)

Blood pressure is used as a measure of the health of both the heart and the blood vessels. A weakened heart may produce a low blood pressure, whereas damaged blood vessels that are closing up or becoming less elastic will give a raised blood pressure. Once the diastolic pressure is regularly over 90, or the systolic over 140 when you are at rest, you have high blood pressure, or **hypertension**. Hypertension is often a silent problem – it causes few obvious symptoms itself, but it greatly increases your risks of developing other potentially fatal circulatory problems, such as heart attacks and strokes.

There are many factors that influence the development of hypertension, but obesity is one of the main ones. Most doctors will suggest treatment to reduce hypertension. This may involve simple steps to begin with, such as losing weight and taking more exercise, or it may involve taking medication to lower the blood pressure (see Chapter 19).

As you saw in Chapter 5, obesity is a growing problem across the Caribbean – Figure 8.24 shows you how hypertension is linked to obesity. Look back to *The biologist's toolkit* in Chapter 5 if you need help in interpreting bar graphs.

Atherosclerosis and heart disease

There is an old saying 'The way to a man's heart is through his stomach.' This is a relic from the days when girls were educated with a view to catching a husband, and it was thought that it was of particular importance to be a good cook. However, there is an element of truth in this old saying, although it certainly does not apply only to men. One of the biggest killers in the developed world is heart disease, and one of the major contributory risk factors appears to be our level of obesity (see Chapter 5).

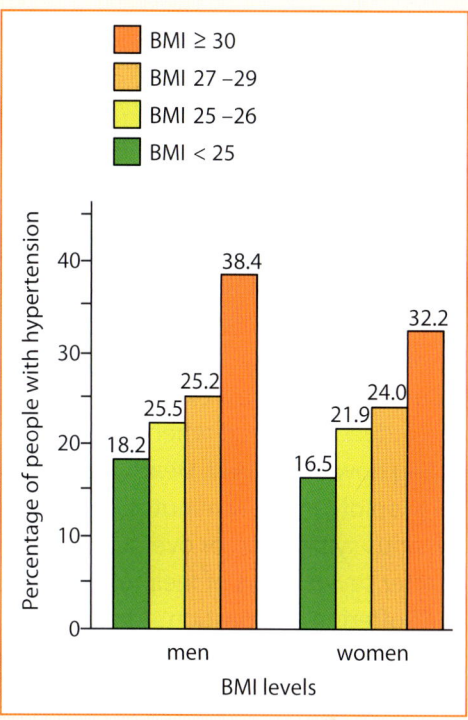

Figure 8.24 Raised blood pressure or hypertension is a complication of obesity as the figures on this graph clearly indicate – the higher the BMI, the higher the percentage of people with hypertension, which means a higher percentage of people at higher risk of heart disease and strokes.

Atherosclerosis (hardening of the arteries) is a build-up of yellowish fatty deposits (plaque) on the epithelial layer of the arteries. It can begin in late childhood and continues throughout life. The plaque builds up until it restricts the flow of blood through the artery or even blocks it completely. Plaque is particularly likely to form in the arteries of the heart (coronary arteries) and the arteries of the neck (carotid arteries). The plaque often causes a blockage of the blood vessels, frequently resulting in heart attacks when the coronary artery is affected, or strokes when the carotid arteries are blocked. In the USA alone, atherosclerosis accounts for half of all adult deaths each year.

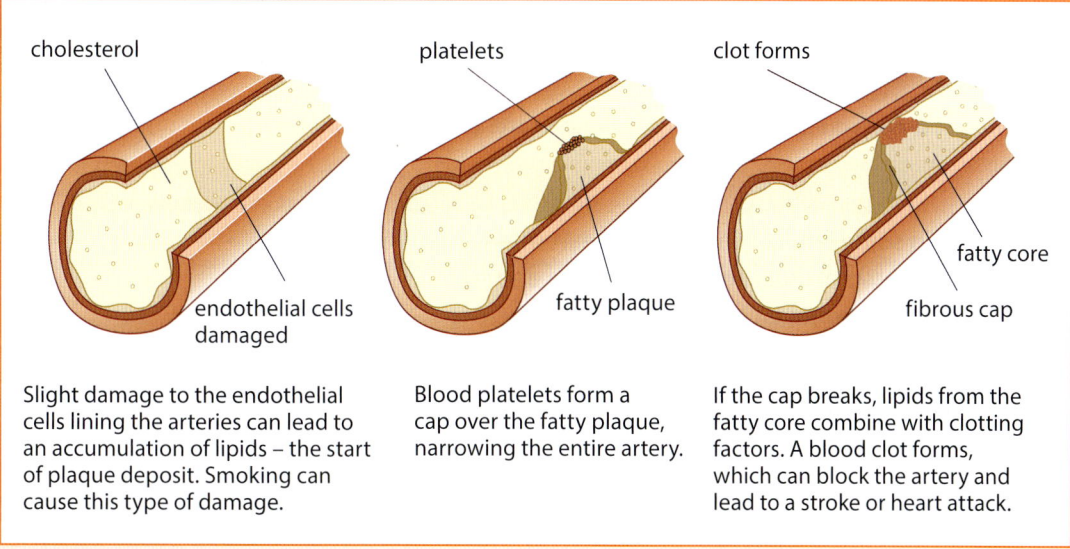

Figure 8.25 Atherosclerosis builds up in your arteries without you knowing, but the long-term effect can be fatal.

Diseases of the circulatory system

If the blockage in one of the branches of the coronary artery becomes complete, part of the heart muscle will be completely starved of oxygen and the patient will suffer a **myocardial infarction** or **heart attack**. Often the blockage is due to a blood clot that forms as a result of the plaque (also called a **coronary thrombosis**). During a heart attack, there is severe chest pain. It does not improve if exercise stops, and indeed may start when you are sitting down or in bed. Death may occur very rapidly, in a few minutes, with no previous symptoms, or over a few hours, or after several days of feeling 'tired' and suffering 'indigestion'. Anyone with a suspected heart attack needs to be admitted to hospital as quickly as possible for intensive care and treatment with anti-clotting drugs. An electrocardiogram (ECG) will confirm a diagnosis of heart attack. You need to react quickly when someone has a heart attack, as up to 45% of heart-attack victims will die within 5 years of their initial attack. Coronary heart disease causes about 27% of all deaths in the countries of the developed world, mainly from heart attacks.

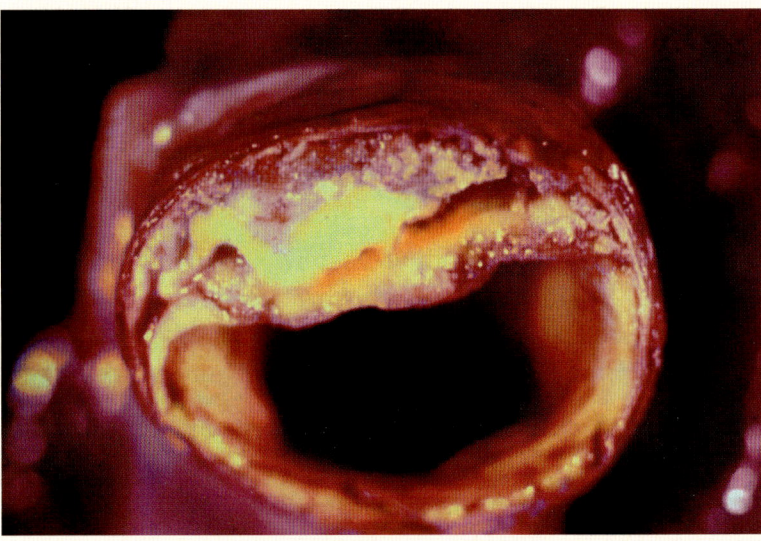

Figure 8.26 Magnified view of a slice through the aorta showing a thick blockage of plaque (top) caused by the disease atherosclerosis.

The risk factors involved in heart disease break down into two main groups – those we cannot help and those we can do something about.

Figure 8.27 A heart attack is usually extremely painful and may be fatal – but some people mistake one for indigestion!

Table 8.4 The risk factors involved in heart disease – some are under our control, and others are not.

Factors that cannot be controlled	Factors we can control
There is a genetic tendency in some families to develop heart disease or hypertension (high blood pressure) that can cause damage and so precipitate heart disease. This we cannot (at present) control.	Smoking
Age is also a factor that we cannot control. As we get older, our blood vessels begin to lose their elasticity and narrow slightly, making us more likely to suffer from high blood pressure and heart disease.	Exercise
The final factor that we cannot control is our gender – males are more likely to suffer from heart disease than females. This is particularly obvious before the female menopause, as the female hormone oestrogen appears to offer some protection against the build-up of plaque.	Stress and diet

The circulatory system

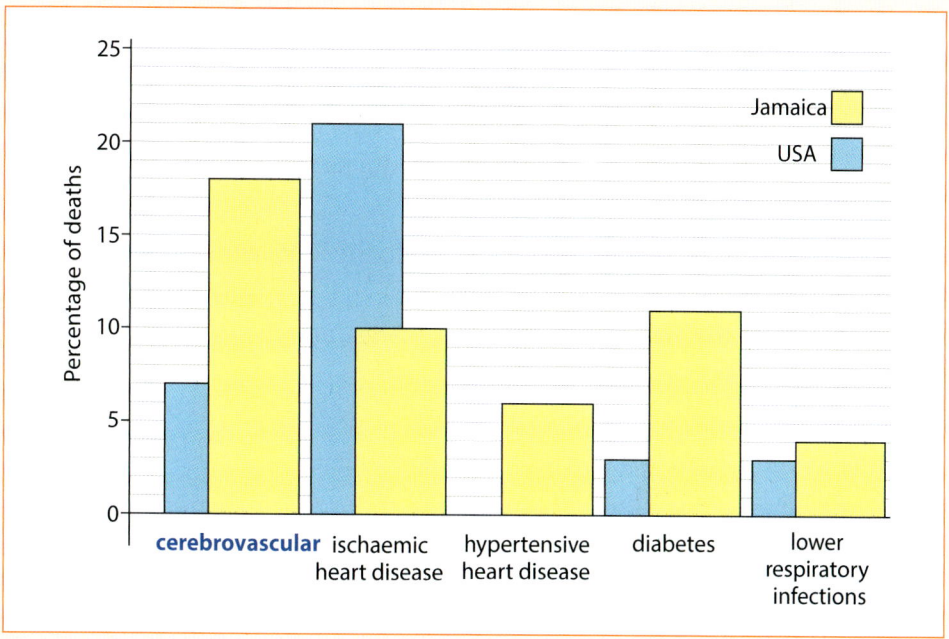

Figure 8.28 Figures for percentage of deaths from cardiac disease in Jamaica compared to the USA

For more information on heart disease and how it can be treated, see Chapter 15.

Possible SBAs based on the circulatory system

→ Investigate the size and anatomy of different animal hearts from a butchery, including dissections, photos and drawings.

→ Investigate the microscope anatomy of the main types of blood vessels.

→ Investigate the microscopic appearance of normal blood and the blood of people affected by diseases such as anaemia, sickle cell anaemia, and so on.

→ Investigate the level of oxygenation of blood of different groups of people, for example different ages, before and after exercise, and so on using a finger oximeter.

→ Investigate the average pulse rates of different cohorts of students, for example sports team and non-team members, and use as a measure of fitness.

→ Investigate the average pulse rates of different age ranges of individuals.

→ Design and carry out an investigation to see if taking up regular exercise has an impact on the average pulse rate of yourself/a group of individuals who agree to take part.

→ Using secondary sources, investigate the incidence of heart disease across the Caribbean and what measures are being put in place to reduce it.

→ Investigate the numbers of blood donors in different countries in the Caribbean/survey to discover people's attitudes to giving blood/plan an evidence-based campaign to increase the number of blood donors.

Diseases of the circulatory system

End-of-chapter summary

In this chapter, you have learnt that:

- the human body needs a transport system because the surface area to volume ratio of the organism is too small for diffusion between the cells and the outside world to be effective
- the human transport system consists of the blood vessels (the pipes), the heart (the pump) and the blood (the transport medium)
- human beings have a double circulation – the pulmonary circulation to the lungs and the systemic circulation to the body
- the three main types of blood vessels are the arteries, veins and capillaries, and they are each adapted for different functions in the bod.
- the heart is mainly made of muscle
- the heart pumps blood around the body in response to the needs of the tissues
- blood enters the atria of the heart, which contract to force blood into the ventricles; when the ventricles contract, blood leaves the heart to go to the lungs (from the right) and around the body (from the left)
- valves control the flow of heart in the blood
- the blood pressure measured in the arteries has a systolic reading, when the heart pumps, and a diastolic reading, when the heart is relaxed and filling
- blood has four main components:
  - – plasma, which transports blood cells, nutrients such as dissolved food molecules, gases such as oxygen and carbon dioxide, antibodies, blood proteins and metabolic waste products such as urea
  - – red blood cells, which transport oxygen
  - – white blood cells, which defend against attacks by microbes
  - – platelets, which help clot the blood
- oxygen is carried by haemoglobin, which becomes oxyhaemoglobin in a reversible reaction
- the blood clotting mechanism stops us bleeding to death from a cut; it involves platelets, calcium ions, vitamin K, thromboplastin, prothrombin, thrombin, fibrinogen and fibrin
- blood group systems, such as the ones with the ABO groups and the Rhesus positive and negative groups, are the result of antigens on the surface of the red blood cells and antibodies in the plasma
- great care must be taken to match the blood groups in blood transfusions
- tissue fluid is forced out of the blood in the capillaries and bathes the cells of the body; exchange of substances by diffusion between the blood and the cells takes place through the tissue fluid
- when the tissue fluid passes into the lymph system, it becomes lymph; lymph eventually returns to the blood enriched with antibodies.
- lymph glands contain lymphocytes and phagocytes
- diseases of the circulatory system kill many people; in hypertension, the blood pressure is higher than it should be, which increases the risk of heart attacks and strokes; in atherosclerosis, a fatty deposit (plaque) builds up on the lining of the arteries, particularly the coronary artery and the aorta, also increasing the risk of heart attacks and strokes
- heart attacks occur when one or more of the coronary arteries are blocked completely by plaque or by a clot formed as a result of a plaque build up (coronary thrombosis); an area of the heart muscle is then deprived of oxygen and may die.

End-of-chapter questions

1. Why can humans not get the oxygen they need by diffusion alone?
   A   The surface area to volume ratio of the body is too small.
   B   The surface area equals the volume of the body.
   C   The surface area to volume ratio of the body is 1.
   D   The surface area of the body is much larger than the volume.

2. What are the main parts of the human transport system?
   A   The heart, blood vessels and blood
   B   The heart, blood and lymph
   C   The heart, arteries and veins
   D   The arteries, veins and capillaries

3. What is the main job of the arteries?
   A   To carry deoxygenated blood away from the heart
   B   To carry oxygenated blood away from the heart
   C   To carry deoxygenated blood to the heart
   D   To carry oxygenated blood to the heart

4. Which type of vessel has a pulse?
   A   Capillaries   B   Lymph vessels
   C   Veins   D   Arteries

5. Which chamber of the heart has the thickest walls?
   A   Right atrium   B   Left atrium
   C   Right ventricle D   Left ventricle

6. What is the main role of the platelets in your blood?
   A   The clotting mechanism
   B   The carriage of oxygen
   C   The carriage of carbon dioxide
   D   The production of antibodies against invading organisms

7. Copy and complete the following sentences. Use the words in the box to fill in the gaps.

   | blood   blood vessels   capillaries   circulatory system   glucose |
   | heart   oxygen   transported   waste products |

   Substances are _____ around the body in the _____ . Food molecules such as _____ and substances such as _____ are carried to the cells where they are needed. The blood also collects and carries away _____ from the cells. Blood flows around your body through a network of _____ , which make up the _____ .

   Your blood is kept flowing round this system by the action of the _____ , which is a muscular bag. The blood vessels are the arteries, veins and _____ .

8   Copy and complete this table to show the main components of blood, their appearance and what they do in your body. (12)

Part of the blood	Description

9   Plasma is very important for transporting substances around the body. Three of the main substances transported are carbon dioxide, urea and digested food. For each substance:
   a   say where in the body it enters the plasma (3)
   b   say where it is transported to, and what happens to it when it gets there. (6)

10 The red blood cells carry oxygen around the body.
   a   Draw and label a typical red blood cell. (2)
   b   Explain how red blood cells carry oxygen around the body and release it in the tissues where it is needed. (5)
   c   Explain how red blood cells are adapted for their role in the body. (3)

11 The number of red blood cells in 1 mm^3 of normal human blood is around 5000 million. However, there are certain situations where the numbers of red blood cells in the blood may be particularly high or low. For each of the examples below, explain the difference in the red blood cell count and the effect it will have on the person.
   a   A person living at high altitude will have a higher red blood cell count than someone living at low altitude. (3)
   b   A person who does not have sufficient iron in their diet will be anaemic, feel very tired and lack energy. Their red blood cell count will be lower than normal. (3)

12 a   Explain why blood clotting is so important.
   b   Describe how the blood clots, using words or a flow diagram.

13 a   Compare and contrast the ABO and the Rhesus blood group systems in people.
   b   Describe why a Rhesus negative woman who is expecting a Rhesus positive baby needs careful medical attention.

14 Here are descriptions of two heart problems and how they may be overcome. In each case, use what you know about the heart and the circulatory system to explain both the problems caused by the condition and how the treatment helps.
   a   Sometimes babies are born with a 'hole in the heart'. This means there is a gap in the central dividing wall of the heart. They may look blue in colour and have very little energy. Surgeons can close up the hole. (4)
   b   The blood vessels supplying blood to the heart muscle itself may become clogged with fatty material. The person affected may get chest pains when they exercise or even have a heart attack. Doctors may be able to replace the clogged up blood vessels with bits of healthy blood vessels taken from other parts of the patient's body. (4)

Chapter 9: The skeletal system

When you have completed this chapter, you will be able to:

- identify the main bones of the human skeleton
- relate the structure of the skeleton to its functions
- identify, draw and label hinge, fixed and ball-and-socket joints
- explain how the skeletal muscles are involved in moving the limbs
- distinguish between the main tissues of the skeletal system – bone, cartilage, tendon and ligaments
- evaluate some of the factors that adversely affect the skeletal system
- recognise that bone is living tissue and relate its structure to its functions

What the examiners say

→ Some candidates are comfortable with questions that relate to this topic, but weaker candidates demonstrate an inability to define key terms such as locomotion or state the effects of certain occurrences such as the effects of thinning cartilage.

→ Additionally, candidates confuse terms such as 'tendons' and 'ligaments'.

→ In some instances, candidates are not prepared enough to answer questions about how various muscles enable movement.

Revision tip

Create a catchy tune or rap to help you remember the bones of the skeleton, or the different tissues of the skeletal system. Share it with your friends to see what they think.

Figure 9.1 Arthropods like this common Caribbean spider have rigid skeletons on the outside of their body.

Skeletons are very important to most animals for support and movement. Animals need support against the force of gravity and they need a way of moving their body around. In soft-bodied animals such as worms, the skeleton usually depends on fluid trapped within the body of the animal (called a **hydrostatic skeleton**). Sometimes the soft body is combined with a protective shell (like a snail). The arthropods (which include insects and crustaceans) have developed a rigid external skeleton (**exoskeleton**) to do the same job.

In vertebrates like us, the skeleton is a strong, rigid structure found inside the body.

The functions of the skeleton

The human skeleton has a number of important roles in our body. These include:

- support
- protection
- movement including breathing movements and **locomotion**
- the production of blood cells
- the storage of minerals.

Support

One major function of the skeleton is obviously support. It provides a framework for our body and helps to determine our shape. What's more, many organs are attached to our skeleton for support and stability. The vertebrae that make up our spine and the bones of our legs are particularly important in supporting our body against the pull of gravity.

Protection

Since it is strong and rigid, the skeleton protects our delicate internal organs from damage. The bones of the skull protect the brain, the vertebrae (backbones) protect the spinal cord, the ribs protect the heart and lungs, and the pelvis protects some of the digestive organs and the reproductive organs in the female.

Movement including locomotion

We need to move to feed, mate, scratch, escape from harm, and so on. The human skeleton is well adapted to enable movement to take place, giving our muscles something to act against and often to join on to, with different types of joints allowing differing degrees of movement.

- One of the important functions of the skeleton linked to movement is breathing. As you saw in Chapter 7, the up-and-down movement of the ribs changes the volume of the chest and results in air being drawn into or forced out of the lungs.
- Another important function of the skeleton linked to movement is locomotion. Locomotion is the movement of the whole body from one place to another.

As you look in detail at the human skeleton, its adaptations for all these functions will become clear.

DID YOU KNOW?

Even the biggest land invertebrates, like the giant Caribbean millipede, cannot match the size and mobility of land vertebrates like ourselves, because their skeletons simply do not offer enough support.

Production of blood cells

The skeleton is very important for the production of blood cells. Inside the bones of the skeleton is tissue called bone marrow, and this is where both red and white blood cells are formed.

Storing minerals

The bone itself contains a lot of minerals, particularly calcium. So the bone acts as a store or reservoir of calcium and phosphorus, and other minerals. If these minerals are needed in other parts of the body, they will be taken from the bone.

> **The biologist's toolkit: Core physics: forces**
>
> - A force is a push or a pull that acts on one object as it interacts with another.
> - You can't see forces, but you can see their effects.
> - A force has a **magnitude** that can be measured and a **direction**. A force is a **vector quantity** and we show forces using arrows that give both the size and direction of the force.
> - Force is measured in **Newtons**.
> - A contact force, for example thrust, acts between two objects that are touching.
> - A non-contact force, for example gravity, acts between two objects that are not physically touching.
> - A basic understanding of forces is useful when you are learning about the skeletal system, because locomotion, lifting things and staying upright against gravity all involve forces generated by our body.

The human skeleton

The human skeleton is an internal structure made mainly of **bone** and **cartilage**, which provide a framework for the body. **Ligaments** and **tendons** are tissues that are also closely associated with the skeleton.

The living bone

When we look at skeletons, we always see a dead, white structure. However, living bone is very different. It has a rich blood supply and is constantly being broken down and built up where it is needed in response to the different stresses the body imposes on it. Any new activity will cause the bones to be altered slightly. This is brought about by the bone cells. Your skeleton is a surprisingly fluid structure, constantly changing even as you sit reading this book.

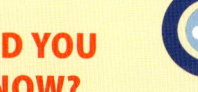

DID YOU KNOW?

An adult human skeleton contains 206 bones. Over half of these bones are found in your hands and feet. You have 26 bones in each foot and 27 bones in each hand!

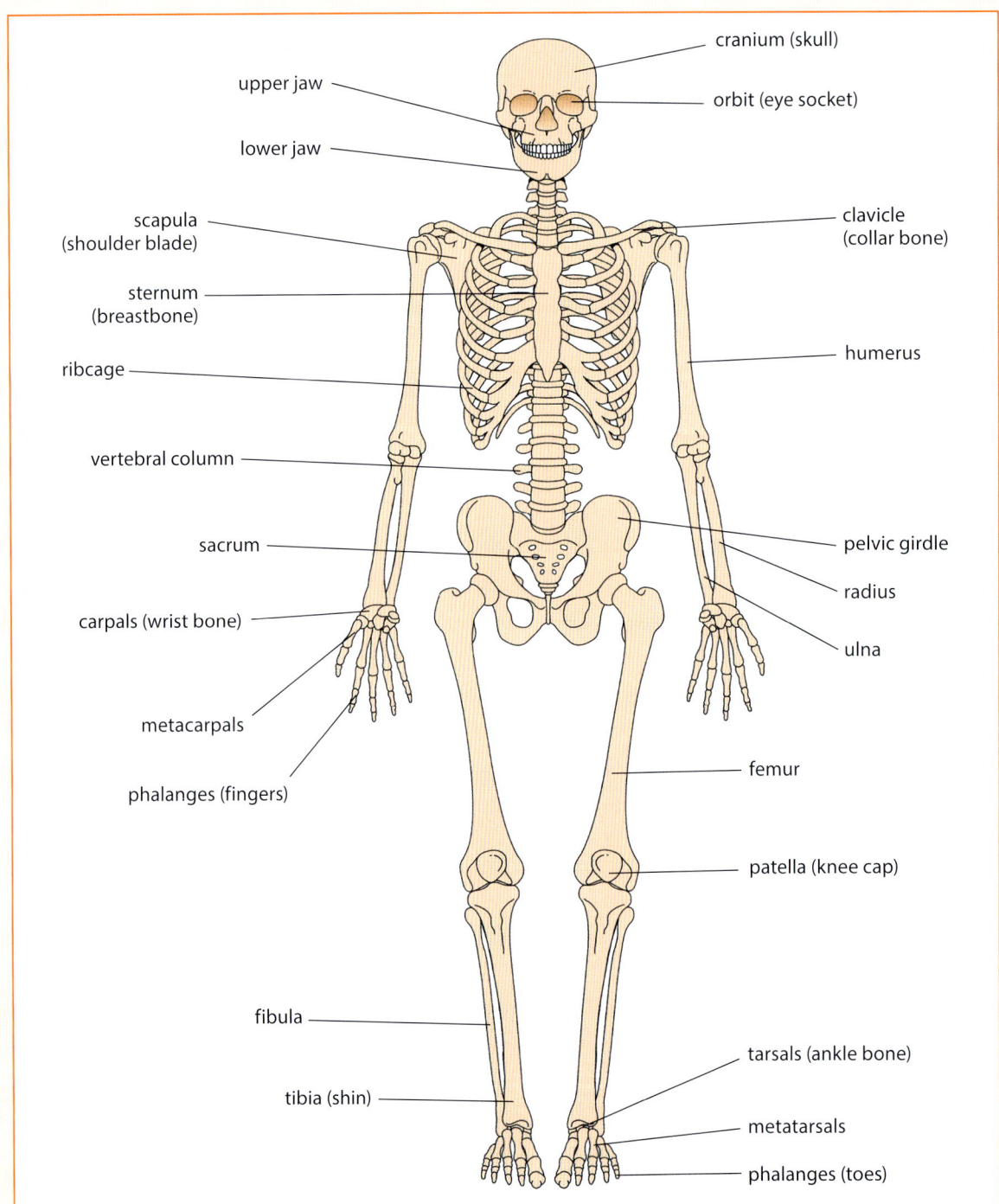

Figure 9.2 The human skeleton

Bones are also very strong and rigid, which allows them to carry out their function of supporting the weight of the body and providing a rigid framework for muscle. Living bone is an active, dynamic tissue. It is a **composite** material (containing more than one substance) made up of bone cells embedded in a matrix of **collagen** (protein) fibres hardened by calcium phosphate, which is deposited within the structure. These hard calcium salts make up 70% of the bone. This is how your body manages to mend any breaks in your bones. The bone-building cells quickly produce more cells and more matrix, and then calcium salts are deposited to harden things up until the bone is as good as new.

The skeletal system

Bone is extremely strong; it is particularly strong under compressive (squashing) forces, which means it stands up well to the stresses of supporting and moving our body. The cells and the collagen help bone to be slightly flexible, while the calcium salts make it strong, so that almost no matter what we do, our bones remain intact.

Bone needs to be hard and strong to support the weight of our body as we move around, but it also needs to be as light as possible. If bones were solid structures, they would be so heavy that they would be impossible to move around. However, our bones are not solid – as you see in Figure 9.3, bone has a mesh-like structure and the bones themselves are hollow tubes made up of two different types of bone (see Figure 9.4).

The long bones of the body – like the femur and the humerus – have bone marrow in their hollow regions. This is where blood cells are made.

How do bones grow?

In small children, there are areas of cartilage in many of the bones, particularly at the top and the bottom of the long bones of the arms and legs. These are the sites where growth takes place, and once the final growth spurt is over in our teenage years, they become **calcified** (calcium salts are deposited) and the whole structure is from then on made of bone. This is one way of aging a skeleton – if there are any cartilage areas, the person was younger than about 18–25 years old because by that stage, the bones are always fully formed.

If children lack either calcium or vitamin D when they are growing, they will suffer from the deficiency disease called rickets and their bones may not form properly. If older people lack calcium, they may suffer from **osteoporosis**, which is when their bones become weak and brittle (see Chapter 5).

> ### DID YOU KNOW?
> The smallest bones in your body are found in your middle ear. Three tiny bones called the malleus, incus and stapes (hammer, anvil and stirrup) transfer sound waves across from your outer ear to your inner ear and play a vital role in your hearing.

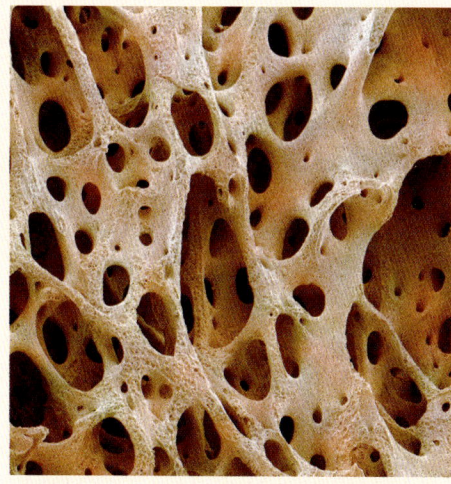

Figure 9.3 The normal appearance of bone under the electron microscope shows a structure that is strong and hard but light. If the calcium salts are removed from bone, it becomes a relatively soft, floppy tissue.

Much of the rest of the skeleton contains large amounts of spongy bone that has a much lighter, open structure. It is found in growing regions and in large masses of bone such as the head of the femur (thigh bone).

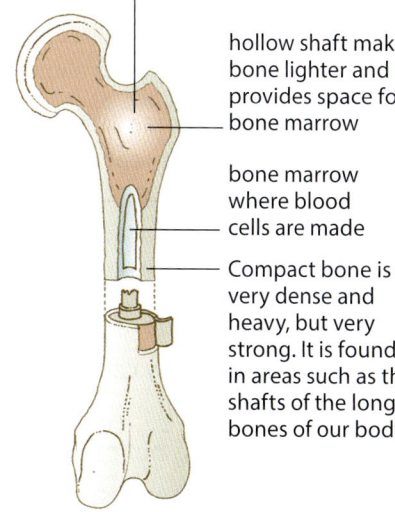

- hollow shaft makes bone lighter and provides space for bone marrow
- bone marrow where blood cells are made
- Compact bone is very dense and heavy, but very strong. It is found in areas such as the shafts of the long bones of our body.

Figure 9.4 The arrangement of different types of tissue within a long bone like the femur gives an ideal combination of lightness and strength, so it is strong and supportive but it can also be moved easily and make blood cells.

The human skeleton

Other skeletal tissues

The skeletal system is made up of the skeleton and a number of other skeletal tissues. These tissues hold the bones together, stop them wearing each other away and move individual bones or the whole body about. They are cartilage, ligaments and tendons.

Cartilage

Cartilage is a very strong tissue found in the skeleton, but it is flexible, not rigid. Its structure is rather like bone without the calcium salts, with cartilage cells embedded in a collagen matrix. Much of the skeleton is first formed as cartilage, which is then hardened when calcium salts are deposited in it. Since cartilage is not rigid, it can be slightly compressed, and this makes it very important to the body as a shock absorber. Cartilage is found covering the bones in most joints, and pads of cartilage that act as shock absorbers are found between the vertebrae. Your ears and nose are made of cartilage too.

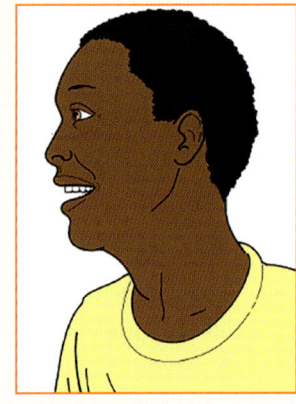

Figure 9.5 Our nose and ears are made of cartilage covered by skin.

Ligaments and tendons

Ligaments attach bones to other bones in joints – they form the **joint capsule** holding the whole arrangement of bones in place. Ligaments have a lot of **tensile strength**. This means they can withstand a lot of stretching or pulling. They also have a certain amount of **elasticity** so that as the bones move within a joint, the ligament capsule stretches to allow that to happen without damage.

Tendons, on the other hand, attach muscles to bones. Like ligaments, they have a great deal of tensile strength – they can withstand being stretched – but they have very little elasticity. This is very important, because if tendons stretched when muscles contracted, the bones wouldn't move.

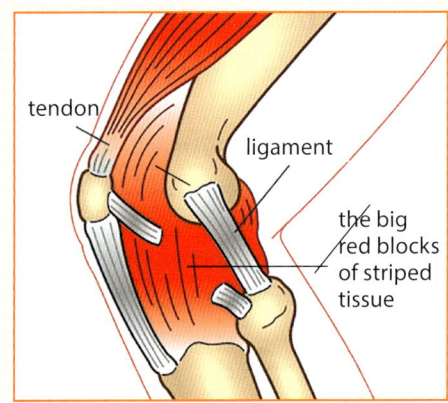

Figure 9.6 Tendons and ligaments in a joint

Here is a summary of the main properties of the skeletal tissues.

- Bone is hardened by calcium salts and resists compression, bending and stretching.
- Cartilage is strong but not rigid, so it can be compressed and absorb shock.
- Ligaments have tensile strength and some elasticity, so joints can bend without the bones dislocating.
- Tendons have tensile strength and little elasticity, and they attach muscles to the bone.

Checkpoint questions

1. Describe the main functions of your skeleton.
2. Describe the structure of bone and explain how this makes it well adapted for its roles in the body.
3. Describe the main functions of:
   a cartilage   b ligaments   c tendons

Joints

Joints are of vital importance in our skeleton. If the bony structure was all in one piece, our body would be supported and internal organs would be protected, but movement would be completely impossible. Fortunately, the body contains a large number of joints of different types, which allow us (and all other vertebrates) to move freely.

The structure of a joint

At a joint, two bones meet and in most cases they move against each other.

Table 9.1 shows the different types of joints – fixed joints, hinge joints and ball-and-socket joints. The only joint that doesn't move is the fixed joint.

The fixed joints in our skull and pelvis are fused together and allow little or no movement. The joints in our skull are not fused when we are born – the bones are separate. This allows the bones to be squeezed together as a baby is born to make sure the head can pass along the birth canal. It also allows our brain to grow when we are children. As we get older, the joints grow together, so adults have fused bones in their skulls.

Table 9.1 Summary of the different types of joints

Type of joint	Type of movement	Diagram	Examples
Fixed joint	None	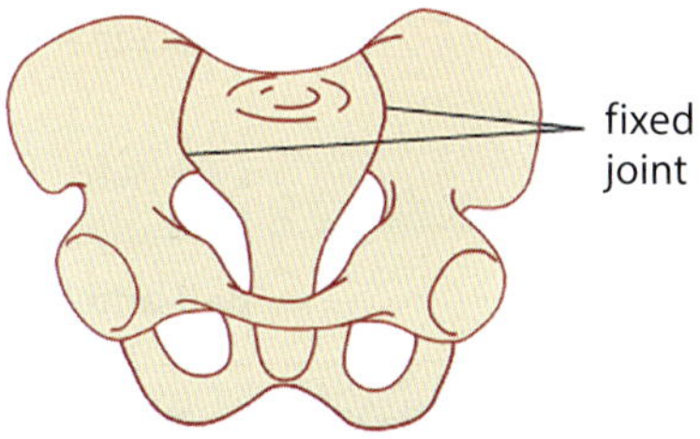	Pelvis (seen here) Skull
Hinge joint	• Works like a hinge, allowing an opening and closing movement • Type of synovial joint	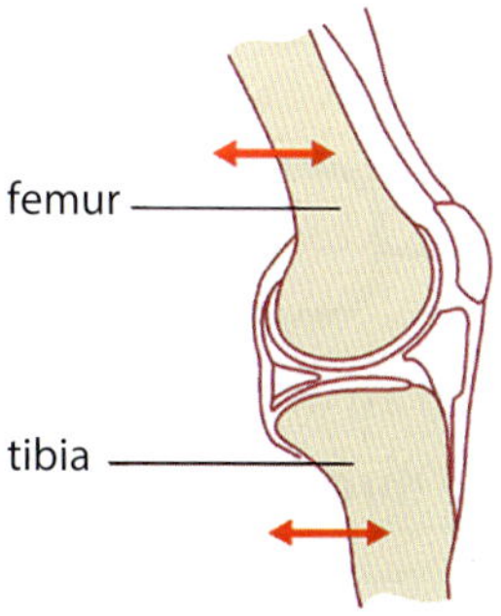	Knee (seen here) Elbow
Ball-and-socket joint	• Allows very free rotational movement due to a ball at one end of one bone fitting into the smooth socket in the other bone • Type of synovial joint	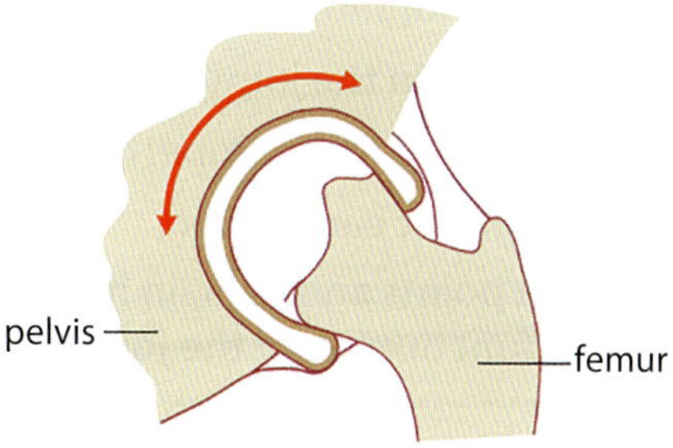	Hip (seen here) Shoulder

When materials rub against each other constantly, as they do in most joints, **friction** tends to wear them away. This would happen if bones rubbed against each other. However, the ends of the bones in most types of joint are covered with a layer of smooth, rubbery cartilage. This cushions the joint and makes sure that the bone ends do not come into contact with each other. The cartilage is continually being replaced.

In very active joints, like the hip, knee, shoulder and elbow, the movements are almost constant. These joints are called **synovial joints**. Even the cartilage cover is not sufficient to keep these joints moving smoothly.

A special membrane (the **synovial membrane**) in the joints secretes an oily fluid called **synovial fluid**. This bathes the joint and acts as a **lubricant**, making the surface of the cartilage slippery and making sure that the joint moves easily. It works in a similar way to the oil that bathes the moving parts of a car, preventing friction and sticking, and keeping the whole system working smoothly. Unlike a car, we don't need regular 'oil changes' because the body continually makes new synovial fluid.

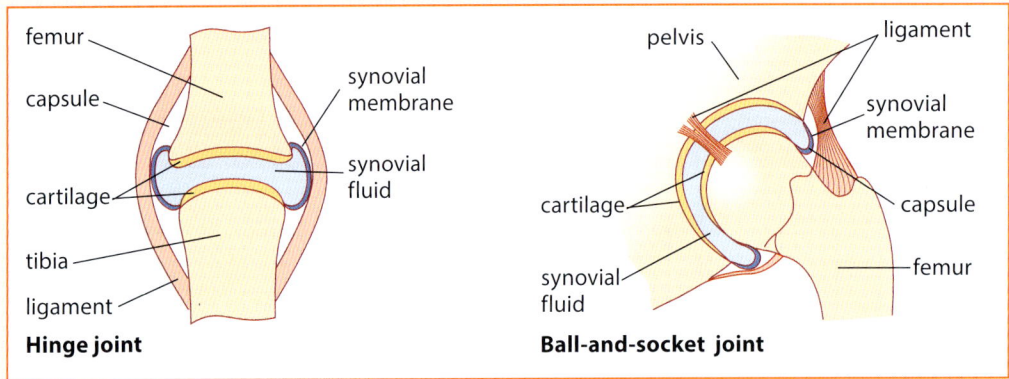

Figure 9.7 Synovial joints like these found in the knee and the hip are perfectly adapted to allow smooth, free movement where we really need it. In these diagrams, the size of the fluid-filled space has been exaggerated to make it easier for you to see.

Checkpoint questions

4. Explain why joints are such an important part of the skeleton.
5. Compare and contrast the three main types of joints in the human skeleton.
6. If a person suffers from osteoarthritis of the hip or knee, it is very painful and the joint does not move properly:
   - the cartilage wears away in the joint
   - the synovial fluid does not form properly or leaks away
   - the bones may rub against each other.

   Take each of these bullet points and explain how it osteoporosis has its effect on the normal functioning of the joint.

Muscles and movement

The skeleton is well adapted for support and protection and has a system of joints for movement – but on its own it cannot move at all. Muscle tissue is needed for movement and support. **Muscles** are joined to bones by tendons, and they move the bones at a joint when they contract.

The structure of muscle

Muscle tissue is made up of protein fibres that contract when they are supplied with energy from respiration. They contract in response to messages from the nervous system and the hormones in our body. In other words, they are **effector organs** that enable us to respond to the world around us. **Skeletal muscles** are the muscles involved in moving our skeleton. Skeletal muscle fibres usually occur in big blocks or groups called muscles.

The contraction of muscles holds us upright against the pull of gravity and allows us to make the millions of tiny adjustments to our posture that keep us upright throughout the day. The contraction of muscle fibres is also responsible for the much larger movements our body makes as we move around carrying out all the normal functions of life. Not all of our muscle is skeletal muscle. You have already met **cardiac muscle**, which makes up the heart, and **smooth muscle**, which lines the arteries and the veins. Table 9.2 summarises the three types of muscle tissue found in humans. In this chapter, we are focusing on skeletal muscles.

Table 9.2 Summary of enzyme action in the human digestive system

Type of muscle	Where it is found	What it does
Skeletal muscle	Attached to the bones of the skeleton	• Contracts to move the bones of the skeleton • Involved in moving parts of the body and in locomotion • Controlled by the voluntary and involuntary nervous systems (see Chapter 11) • Gets fatigued
Cardiac muscle	Makes up the structure of the heart	• Contracts in a regular rhythm to pump blood around the body • Controlled mainly by the involuntary nervous system • Does not get fatigued
Smooth muscle	Important part of the structure of organs such as the digestive system and the blood vessels	• Controlled by the involuntary nervous system • Contracts to carry out internal functions of the body, for example, moves food through the gut, widens and narrows blood vessels and changes the opening of the pupil of the eye

How do skeletal muscles work?

Movement, whether the tiniest twitch of a finger or the explosive burst of speed of a sprinter, is brought about by the action of skeletal muscles on bones. Each of our skeletal muscles is attached by tendons to two different bones, spanning at least one joint. The attachment nearest to the heart is called the **origin** of the muscle and it does not move when the muscle contracts. The attachment furthest away is called the **insertion** and it is on the bone that will be moved (see Figure 9.8).

Muscles are attached to bones in pairs. Muscle fibres are always either **contracted** or **relaxed**. When a muscle contracts, it shortens and pulls on the bones to which it is attached, moving one of them. When it relaxes, it can be stretched out to its original length, but it cannot push the bone back into place. This can only be done by another muscle, contracting and pulling in the opposite direction to the first one. This is why muscles always come in pairs – one pulls the bone in one direction, the other pulls it back to its original position. Because they work in direct opposition to each other, these muscles are called **antagonistic pairs**. You can see a clearer picture of how movement is brought about in Figure 9.8.

Locomotion is very important for human beings – we need to move around to find food, find a partner, look after our babies, and run away from predators. This is why the role of the skeleton, as well as the muscles that move the bones, and the cartilage, ligaments and tendons that form the joints and attach the muscles, are so important.

Although all of the skeletal muscles in our body do the same job, contracting to move bones and then relaxing again, they are not all exactly the same. When you use a set of muscles repeatedly, more muscle is built up. This is why people who do a lot of sport and exercise, or who do hard physical work, have large, well-defined muscles. However, if you don't exercise, the amount of muscle you have is reduced. Your body reabsorbs the protein and so it becomes much harder to do things and exercise. This may be the result of laziness, but it also happens when people are ill and have to rest a lot. It is surprising how quickly your muscles start to waste away if you have to stay in bed for a few days.

Muscles throughout the body are constantly partially contracted to support the body against gravity. This constant tension in our muscles against gravity at rest is called **muscle tone**. If muscle tone is lost – for example when you are unconscious – you collapse as you cannot support yourself. The level of muscle tone increases if you exercise regularly, and this enables you to respond faster when you need to.

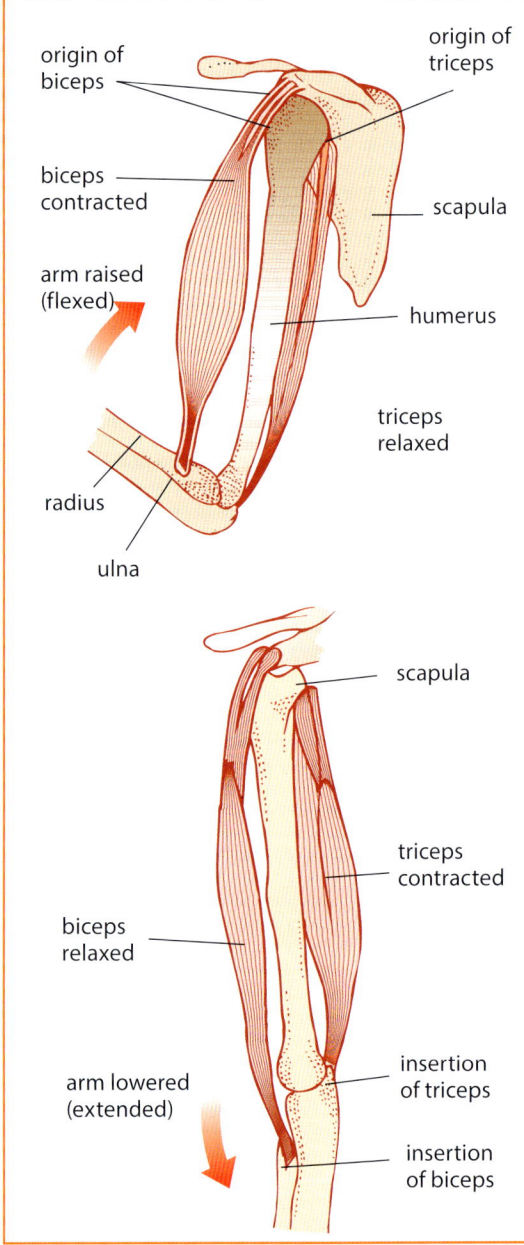

Figure 9.8 The biceps and triceps form the antagonistic pair of muscles that we use to raise and lower our arms.

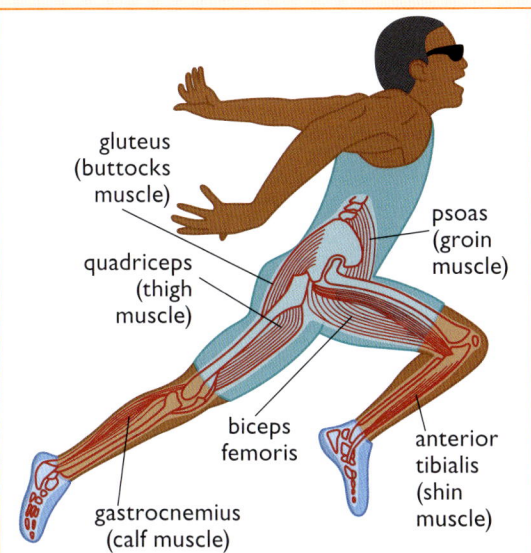

Figure 9.9 It isn't just your arms that have antagonistic pairs of muscles. You can find them everywhere and they are very clear in your legs. Look carefully at the pairs of muscles drawn over the legs of this running athlete and try to work out which pairs of muscles move which bones.

Activity 9.1

Examining bones

By examining individual bones, you can learn a great deal about the way they work in a skeleton. Look out for the surfaces where bones articulate with each other. Find the rough surfaces where tendons attach muscles to the bone. Look for examples of bones that protect parts of the body.

You can use your own body to help you in your observations. Carry out controlled movements of your own limbs and body. See if you can recognise how the movement is produced by the bones of your skeleton

You will need:

- a sheet with an unlabelled drawing of a human skeleton
- individual bones, for example, vertebrae, leg bones, arm bones and skull, and so on

Method

1 Examine as many individual bones as you can.
2 Draw and label several of the bones.
3 Annotate these diagrams to show where the bones are involved in joints with other bones, any areas for muscle attachment and any protective or support functions you can identify. (It is a good idea to have a look at an example of a vertebra as they are involved in protection, support and movement.)

Damaging the skeletal system

Problems with the skeletal system, from broken bones to arthritis and from back pain to bunions, are some of the most common human ailments. Many of the problems of the skeletal system come from accidents, sporting injuries or wear and tear, and are hard to avoid. Some, like rickets or muscle-wasting conditions such as kwashiorkor or marasmus, result from malnutrition. However, there are some factors that have an adverse effect on the skeletal system every day. By taking some simple precautions, you can avoid many common skeletal system aches and pains.

You have 206 bones and hundreds of different muscles. If they are all aligned properly, they work smoothly with the minimum of strain. However poor posture – hunched shoulders, slouching in seats, and so on – is very common. If a person has poor posture, then the bones, muscles and joints are not aligned properly. Not only are sprains and strains more likely, but aching and tension in the muscles as they work in the wrong direction can lead to chronic pain and even to the development of osteoarthritis.

Figure 9.10 Poor posture may cause problems in your spine, hips and shoulders.

Our feet are particularly vulnerable. Each foot has 26 small bones and around 30 joints, and they carry the weight of our body all day every day. Crushing the bones of your feet into poor footwear – shoes that are too small, too tight or have very high heels, for example – can lead to many difficulties. The foot itself may develop calluses, bunions, hammer toes and other problems that are both painful and can lead to a lack of mobility. But poor footwear can also affect the way you walk, which in turn affects the joints of the knees and hips, and may lead to the development of osteoarthritis. In osteoarthritis, the cartilage cushion in your joint wears away and the ends of the bones rub together, wearing the bone away. It is very painful and limits movement. If this happens, you may eventually need an artificial hip or knee. These joint replacements are very successful but they involve major surgery.

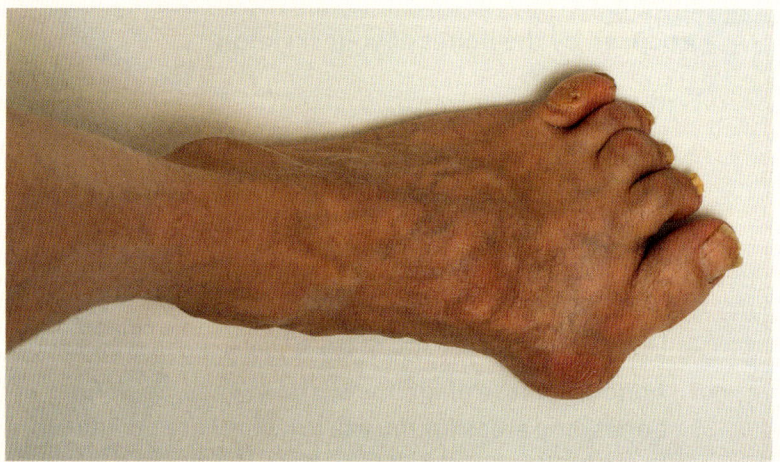

Figure 9.11 The effect of poor footwear can be both painful and unsightly.

By maintaining good posture, keeping your weight down so the load on the feet is minimised and by buying good, well-fitting shoes, many of these problems can be avoided.

The skeletal system

Another way in which many people damage their skeletal systems is by lifting heavy objects incorrectly. The forces we exert on our body are very different if we lift heavy objects correctly than if we lift them incorrectly. Regular, careful heavy lifting will not cause any harm. If you bend your knees, keep the curve of your spine natural and use the power of your thigh muscles to help you lift, all will be well (see Figure 9.12). But if you lift badly, with a curved back and using your arms, you risk straining the muscles of your back or even crushing or displacing one of the cartilage discs between your vertebrae – what is commonly called a slipped disc. This causes bad back pain and difficulty moving about, and may take a very long time to heal.

Figure 9.12 It is important to lift heavy weights correctly to avoid damaging your spine and back muscles.

In general – if you look after your skeletal system and treat it well it will work well for you throughout your life.

Possible SBAs based on the skeletal system

→ Study a collection of different bones – either different bones from the same type of animal or the same bone from different animals. Measure dimensions, make drawings and compare features.

→ Study the different vertebrae of a human skeleton or an animal skeleton. Research the different vertebrae and their roles in the body, draw the different shapes and dimensions, and so on.

→ Make a model skeleton using different materials.

→ Survey the incidence of skeletal injuries in your class/school/community. See if you can find patterns in – for example – the group most likely to have suffered a broken leg, or the most common broken bone in boys vs girls, or the incidence of arthritis in different age groups in the community.

→ Investigate the appearance of different skeletal tissues under the microscope and in dissection.

→ Investigate the strength of muscles in different groups by what they can lift or the number of repetitions for the lifting a book from the table or similar. Can practice increase the time before a muscle fatigues?

→ Investigate the effect of a specific exercise on the size of a muscle, for example, biceps measurement.

→ Develop a web resource encouraging people to look after their skeletal system with advice on good practice.

> **End-of-chapter summary**

In this chapter, you have learnt that:

- the human skeleton has a number of major roles in the body – support, protection, movement and locomotion, making blood cells and storing minerals
- the main material in the human skeleton is bone, and all the bones of the skeleton have specific names
- bone is a composite material made up of bone cells in a matrix made of collagen fibres and calcium salts
- bone is strong and hard yet has some flexibility
- cartilage is another tissue found in the skeleton; it consists of cartilage cells in a collagen matrix and is very flexible; it is a very good shock absorber
- bones articulate against each other at the joints; there are several different types of joints including fixed joints, hinge joints, and ball-and-socket joints
- synovial joints have a synovial membrane that secretes synovial fluid; this acts as a lubricant in the joint capsule; hinge joints, and ball-and-socket joints are often synovial joints
- joints are held together by ligaments; ligaments have tensile strength and some elasticity so joints can bend without the bones dislocating
- tendons have tensile strength and little elasticity; they join muscles to bones
- bones are moved by muscles; muscles are made of protein fibres that can contract and relax
- there are three main types of muscle tissue: skeletal muscle that moves the bones, cardiac muscle that makes up the heart, and smooth muscle found in organs in the body, for example, the digestive system, blood vessels and respiratory system
- muscles work in antagonistic pairs; one contracts to move a bone in a particular direction; it then relaxes and the other contracts to return the bone to its original position
- muscles work to maintain the body in an upright position against the force of gravity (muscle tone) and to move the limbs and body (locomotion)
- each muscle is attached to two bones; the attachment nearest to the heart is known as the origin of the muscle and it does not move when the muscle contracts; the attachment furthest away is called the insertion and this is on the bone that will be moved
- it is easy to damage the skeletal system, for example, by poor posture, wearing poor footwear or lifting heavy objects without taking care.

End-of-chapter questions

1. Which one of the following is not a function of the human skeleton?
   A   Support
   B   Making blood cells
   C   Protection
   D   Reproduction

2. Which one of the following materials is not found in the bones of the human skeleton?
   A   Collagen fibres
   B   Calcium salts
   C   Elastic fibres
   D   Bone cells

3. Which one of the following types of joints allows movement in all directions?
   A   Ball-and-socket
   B   Hinge
   C   Fused
   D   Semi-moveable

4. What is the main role of synovial fluid in a joint?
   A   Producing blood cells
   B   Lubrication
   C   Water for osmosis
   D   Noise reduction

5. In which way do muscles usually work?
   A   Agonistic pairs
   B   Antagonistic pairs
   C   Angled pairs
   D   Unconnected pairs

6. Which of the following are functions of cartilage?
   i   Form a barrier between bones
   ii   Lengthen the space between bones
   iii Reduce friction between bones
   iv   Prevent shock between bones

   A   i and ii only
   B   i and iv only
   C   ii and iii only
   D   iii and iv only

7. Make a table to compare and contrast the properties and the functions of the four main skeletal tissues – bone, cartilage, ligaments and tendons.
   (8)

8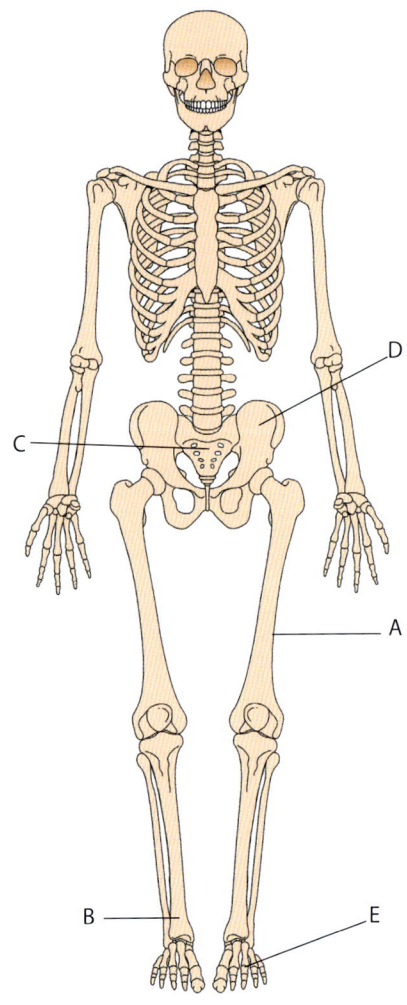

- **a** Name the bones A to E in this skeleton. (6)
- **b** Explain how your skeleton grows when you are young, and how it can heal itself if you break a bone.

9
- **a** List as many advantages of a bony internal vertebrate skeleton as you can. (5)
- **b** Bone is particularly good at withstanding compression forces. Why is this so important to its role in the body? (2)
- **c** Look at the diagram of the skeleton. Make a table to show which bones are involved in movement and the type of movement involved (for example, ribs – breathing movements), which bones have a protective function and which bones are important in support. Some bones may appear in more than one column of your table. (10)

10 The long bones of the body like the femur in the leg are very strong.
- **a** Describe how the bones are made both as strong and as light as possible. (4)
- **b** The top and bottom of the long bones in children and young people has an area of cartilage. By adulthood, these areas of cartilage have gone. What might be the function of these cartilage areas in the bone? (1)

11 a Draw and annotate a diagram to explain how a synovial joint works. (10)

b Arthritis is a disease that affects the joints. Often associated with old age, it can affect even very young children. It is associated with very painful joints that may be swollen and with a reduction in the mobility of the joint. In osteoarthritis, the cartilage in the joints is worn away, while in rheumatoid arthritis, the synovial membranes are destroyed as well as the cartilage. In both cases, bony growths may develop within the joint capsule between the two surfaces of the bones. Explain why these two conditions cause problems in relation to the normal way in which the joint works. (8)

12

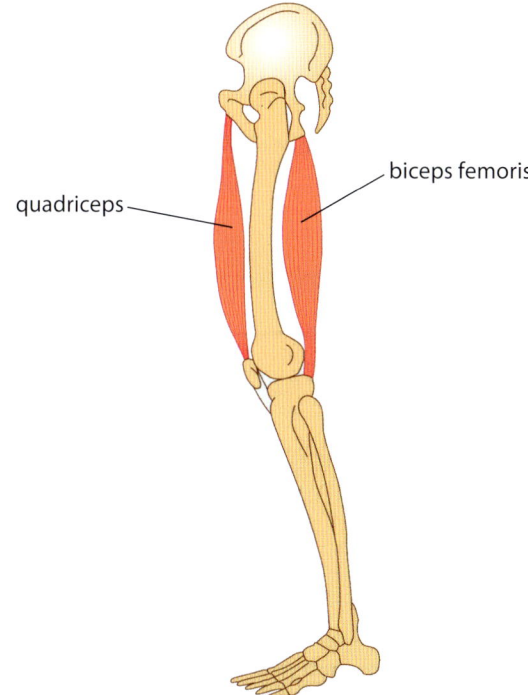

The lower leg is bent and straightened through the knee joint by a pair of muscles attached to the pelvis and the top of the tibia. Draw diagrams to show how these muscles work to:

a straighten (4)

b bend the leg at the knee joint. (4)

Chapter 10 Excretion and homeostasis

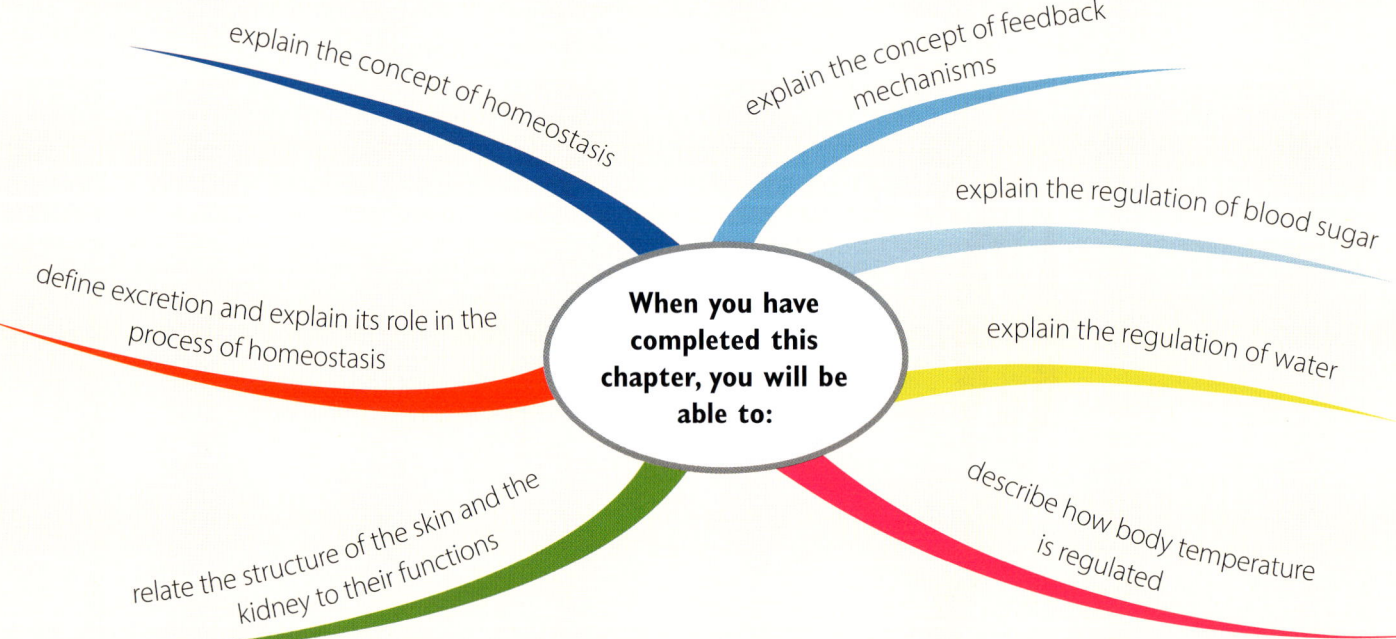

When you have completed this chapter, you will be able to:
- explain the concept of homeostasis
- explain the concept of feedback mechanisms
- explain the regulation of blood sugar
- explain the regulation of water
- describe how body temperature is regulated
- relate the structure of the skin and the kidney to their functions
- define excretion and explain its role in the process of homeostasis

What the examiners say
→ Candidates tend to provide superficial responses to questions, which indicate a lack of familiarity with the details required by the syllabus. Candidates also seem to experience difficulties in appropriately or accurately labelling diagrams.
→ The popular inaccuracies and misconceptions include the functions of sweat and what causes the surface of the skin to appear flushed; explanations of terms such as 'ultrafiltration' and 'selective reabsorption', which candidates use interchangeably; definition of the term 'homeostasis'; explanations relating to why the glucose concentration of the dialysis fluid needs to be the same as that of the blood; and the belief that the hypothalamus controls the water content in the blood.

Think about the different conditions your own body finds itself in during a single day. You may eat lots of food, or you may eat very little; you may spend time training for sport in the sun, swim in the sea, or sit in the shade. You may be ill and have a fever, or you may spend time moving stock in the chill rooms of a supermarket. You may even take a flight to a country with a very different climate to your own country. But, however much your external environment may change, things inside your body need to stay the same.

Revision tip

Practise drawing and labelling diagrams so you know them really well.

Homeostasis

For our body to work properly, the conditions surrounding the millions of cells that make up the body must stay as constant as possible. A stable internal environment, independent of the external conditions, is vital to life as we know it. How do we achieve this? Many of the functions that go on in our body are actually involved in keeping our internal environment as constant as possible. This maintenance of a stable internal environment is called homeostasis.

Figure 10.1 Our beautiful Caribbean beaches bring pleasure to local people and tourists alike – and everyone maintains a stable internal environment, regardless of the external conditions they are used to.

Here are some of the main threats to a stable state inside the body:

- Our blood sugar levels goes up and down depending on when we eat (see Chapter 5), but our cells need a constant supply of glucose for respiration.
- Respiration releases carbon dioxide, which may change the pH of tissues if its levels rise (see Chapter 7).
- As we break down the products of digestion, poisonous waste products such as urea are produced. If levels get too high, these poisonous waste products could kill us.
- Exercising hard, being in the sun too long or having a fever all raise our core body temperature. If it gets too high, our enzymes denature and all our cell chemistry comes to a halt and we may die.
- If our body temperature gets too low, the cellular reactions slow down and we may die.
- The water balance inside our body varies as the amount of water and salt we take in and lose varies from day to day. If it varies too much and the concentration of our tissue fluid changes, our cells will lose or gain water by osmosis. This can destroy our cells and so kill us.

As you can see, keeping our internal conditions in a stable state is very important, but it isn't easy, because we are constantly changing what we take into our body and what we are doing. But our body manages to keep our internal environment constant 24 hours a day, every day of our life. How does our body do it?

Maintaining homeostasis

Inside the body, the **nervous system** and the **hormone system (endocrine system)** work together to maintain homeostasis.

- The nervous system is a complex system of nerve cells that carry electrical messages around your body at high speed. The nervous system gives us rapid, coordinated responses. You will learn the details of how it works in Chapter 11.

- The hormone system is a chemical control system, with chemicals called **hormones** produced by special **endocrine glands** released directly into your blood. Some parts of the hormone system produce rapid responses, and these are the ones you will meet in this chapter. Most hormonal control is slower and more long term, as you will see in Chapter 11.

Our kidneys are vitally important in two aspects of homeostasis: **excretion** and **osmoregulation**. For both processes, the kidneys rely on feedback mechanisms to maintain the balance of the body.

Feedback mechanisms

Many control systems in the body involve **feedback mechanisms**, and most of these systems are examples of **negative feedback**. This means that when levels of a substance in your body rise, changes are made that *lower* the levels again. Similarly, when levels of a substance *fall*, changes are made so that it rises again to the original levels. The control of your breathing rate (see Chapter 7 to remind yourself about breathing) is an example of a negative feedback mechanism. For example, when you start to exercise, your carbon dioxide level goes up. In response, your breathing rate goes up, which makes the carbon dioxide level fall. As the carbon dioxide level falls, your breathing rate also falls and returns to normal. Look at Figure 10.3 showing a general negative feedback loop and use it to model the control of your breathing rate. You will see negative feedback mechanisms in this chapter when you study the control of the water balance, the control of the blood sugar level and temperature control in the body.

> **DID YOU KNOW?**
> The word 'homeostasis' comes from the Greek words '**homoios**', which means 'like' or 'the same', and '**stasis**', which means 'state'. So the word tells you exactly what it means – keeping the conditions inside of your body (the internal environment) in the same state all the time.

Figure 10.2 Wherever we go and whatever we decide to do, conditions inside our body must be kept stable if we are to survive, even if we want to celebrate carnival for hours inside a very hot and heavy costume!

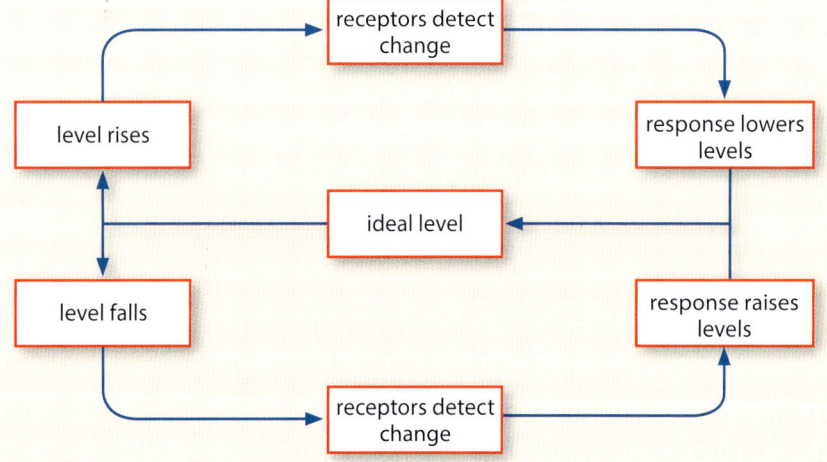

Figure 10.3 A model of a negative feedback mechanism – you will apply this model to many of the control systems you meet while studying homeostasis.

There are a few examples of **positive feedback** in the human body. In a positive feedback mechanism, a small change is magnified or made bigger – this does not restore homeostasis! Two important examples of positive feedback in the human body are the following:

- The blood clotting mechanism (see Chapter 8) – when the clotting factors start to be released, this process triggers the release of even more clotting factors to speed up the process until the bleeding stops.
- During the birth of a baby, once the signals that make the uterus (womb) contract begin, they keep increasing in speed and strength until the baby is born. You will learn more about this in Chapter 12.

Look out for more information on feedback mechanisms in this chapter, as you learn about homeostasis, and throughout the book.

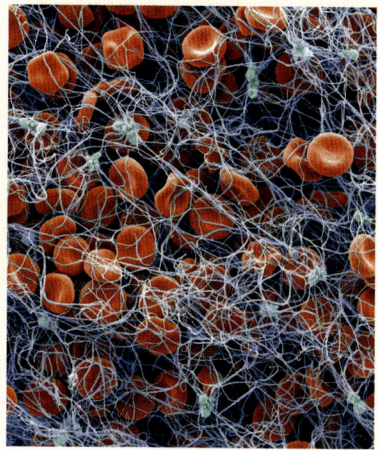

Figure 10.4 Coloured scanning electron micrograph showing the formation of a blood clot: the red blood cells (red) and platelets (green) in a fibrin mesh (blue) are part of the clotting mechanism.

Excretion: getting rid of metabolic waste

Excretion – getting rid of the waste products that can build up in our body and damage the cells – is one of the most important aspects of homeostasis. Metabolic wastes are materials that are produced by the metabolic processes of life. There are two main metabolic waste products: **carbon dioxide** and **urea**. They are toxic so, if they build up in our body, they cause cell damage and death. Our body deals with them both very differently. The organs that are involved in getting rid of these metabolic wastes are called excretory organs. The main excretory organs in the body are the lungs, kidneys and skin.

The carbon dioxide produced during cellular respiration is almost all removed from the body via the lungs when we breathe out (see Chapter 7). So, the lungs are not only the site of gas exchange for respiration, they are also an excretory organ, removing carbon dioxide waste very effectively from the body.

Another metabolic waste that can cause serious problems is urea. Urea is produced in the liver when excess amino acids are broken down. These excess amino acids come from protein in the food we have eaten and from the breakdown of worn-out body tissue. Our body cannot store excess protein or amino acids, so any excess is always broken down. The amino acids are converted into carbohydrates (which can be stored or used) and urea. The urea is filtered out of the blood by the **kidneys** and removed in the urine. The kidneys important in homeostasis because they remove toxic urea. They are also important because they maintain the water balance of our body, so the concentration of our tissue fluid is stable at all times.

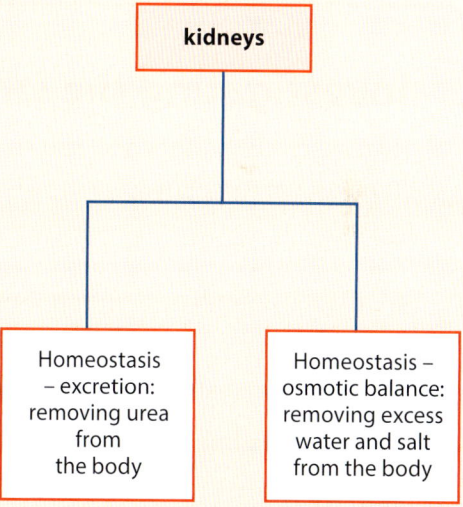

Figure 10.5 Functions of the kidneys

> ## Checkpoint questions
>
> 1. State what is meant by the 'internal environment of the body'.
> 2. Define homeostasis.
> 3. Define excretion and explain why it is so important.

The kidneys

Your kidneys are a pair of red-brown organs with a very rich blood supply (see Figure 10.6). Put your hands on your waist with your thumbs towards your back. Your thumbs are about where your kidneys are found. Your kidneys are surrounded by fat to protect them from damage. When the fat is cleaned off, you can see the blood vessels and tube to the bladder coming out of the organ (see Figure 10.6).

How do the kidneys remove urea and control the levels of water and ions in the body? Blood flows into the kidney along the renal artery (see Figure 10.7). The blood is filtered, so fluid containing water, salt, urea, glucose and many other substances is forced out into the kidney tubules. Then, everything the body needs is taken back (reabsorbed), including some water and all the sugar and mineral ions needed by the body. The amount of water reabsorbed depends on the needs of the body. The waste product urea, as well as excess ions and water not needed by the body, are released as urine. Each kidney has a very rich blood supply and is made up of millions of microscopic tubules (**nephrons**). This is where all the filtering and reabsorption takes place – look at Figure 10.7 and read the following information to discover the roles of the different areas of a single nephron (kidney tubule) in the production of urine inside each kidney tubule.

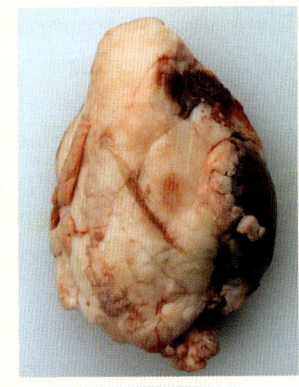
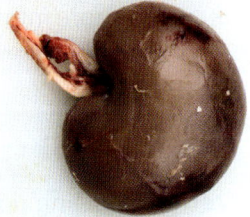

Figure 10.6 Our kidneys are very important organs of homeostasis, involved in controlling the loss of water and mineral ions from the body as well as getting rid of urea.

Activity 10.1

Investigating the structure of the skin

By dissecting a kidney you can see the way the different tissues are arranged. Remember to indicate the magnification of your drawing each time – if it is life-size, it is ×1.

You will need:
- a kidney (lamb or pig) from the butcher, preferably with the fat surrounding the kidney in place
- dissecting instruments – a scalpel, forceps and a seeker
- dissecting board

Method

1. Observe the outer appearance of the kidney with the fat on if possible. Draw and label what you see.
2. Carefully remove the fat, clearing the tubes leading into and away from the kidney. Again, draw and label what you see.
3. Slice the kidney in half **longitudinally** (along its length) and open it out to see the internal structure. Again, draw and label the regions carefully – use Figure 10.7 to help you identify them.
4. You may have the opportunity to look at prepared slides of kidney tissue under the microscope – if so, keep the drawings you make with these drawings from a fresh kidney to build up a record of the whole organ from your own observations.

Excretion and homeostasis

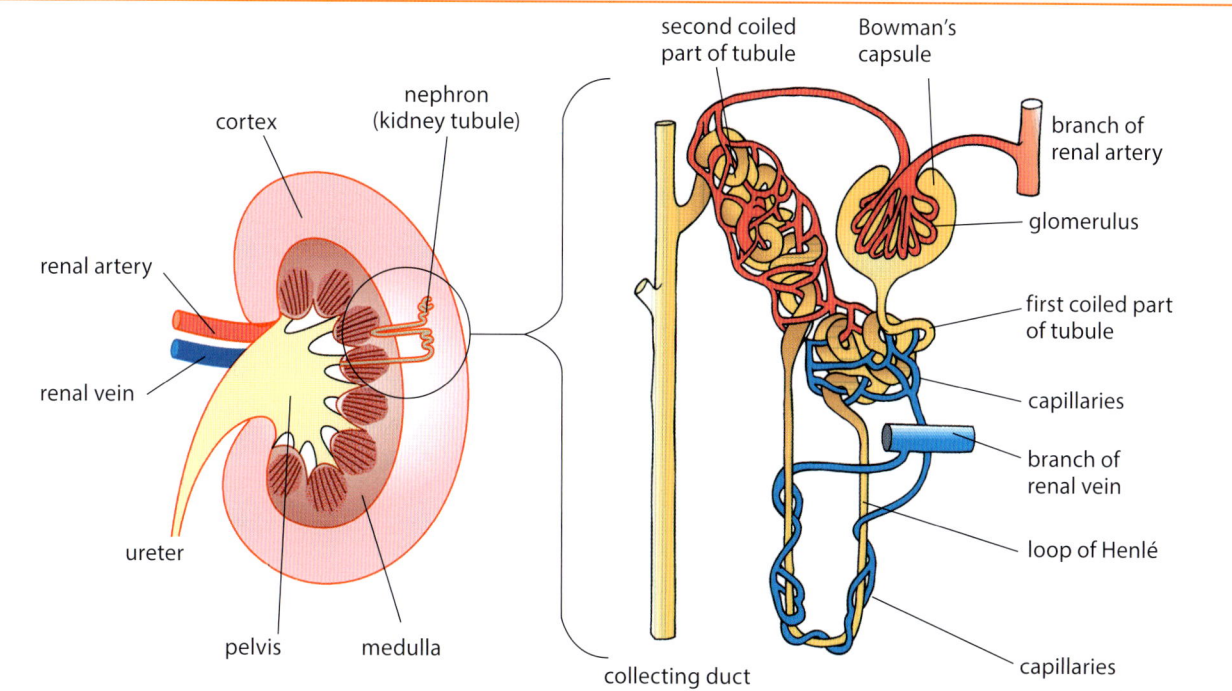

Bowman's capsule: This is the site of **ultrafiltration** of the blood. In ultrafiltration, smaller molecules pass through the filter down a pressure or concentration gradient but larger molecules remain in the original solution. In the Bowman's capsule, the blood vessel feeding into the capsule is wider than the vessel leaving the capsule, so the blood in the capillaries is under a lot of pressure. Some of the water, salt, glucose, urea and other substances are forced out into the start of the tubule. The large proteins, platelets and blood cells remain in the capillary with much of the water.

Glomerulus: This is knot of blood vessels in the Bowman's capsule where the pressure builds up so that ultrafiltration occurs.

First coiled (convoluted) tubule: The liquid that enters this first tubule is called the **glomerular filtrate** and is where much of the **selective reabsorption** takes place. In selective reabsorption, some molecules are selected to be taken back into the blood, but other substances are not. All the glucose is actively taken back into the blood along with around 67% of sodium ions and about 80% of the water. The first tubule has lots of microvilli to increase the surface area for absorption.

Loop of Henlé: This is where the urine is concentrated and more water is conserved.

Second coiled (convoluted) tubule: This is where the main water balancing is done. If the body is short of water, more is reabsorbed into the blood in this tubule under the influence of the **antidiuretic hormone** or **ADH** (see below for more about this).

Collecting duct: This is where the liquid (essentially urine) is collected. It contains about 1% of the original water, with no glucose.

DID YOU KNOW?

Each nephron is 12–14 mm long, but only microns wide, and there are about 1.5 million of them in each kidney!

Figure 10.7 The kidney filters the blood and removes materials that are not needed to form urine. The nephrons, or kidney tubules, are the units that carry out all the work.

The processes of ultrafiltration and selective reabsorption mean that the composition of our blood is kept relatively constant, but the concentration of our urine varies a lot depending on how much water and salt we have taken into our body.

Table 10.1 Contents of plasma, filtrate and urine

Substance	Plasma glom. (mg/100 ml)	Filtrate (mg/100 ml)	Urine (mg/100 ml)
Glucose	100	100	0
Urea	26	26	1 820
Uric acid	4	4	53

Excretion: getting rid of metabolic waste

Urine is formed constantly in our kidneys, and it drips down to collect in our bladder. The bladder is a muscular sac. We can control the opening of the bladder thanks to a strong ring of muscle called a **sphincter** at the entrance to the **urethra**, the tube that leads from the bladder to the outside world. We can open and close this sphincter voluntarily, although it also opens as a reflex action if the bladder is too full or if we are very frightened! When we are young, we have to learn to control our bladder sphincter voluntarily.

The mechanism of water regulation

If the concentrations of our body fluids change, water moves into or out of our cells by osmosis. If too much water moves in or out, our cells will be damaged or destroyed. Yet, some days we may drink several litres of liquid and other days much less. How is the balance maintained?

We gain water when we drink and eat. When we exhale, water evaporates into the air in the lungs and is breathed out. This water loss from the lungs is constant. Whenever we exercise or get hot, we sweat and lose more water (see Figure 10.8).

The ion concentration of the body, particularly ordinary salt (sodium chloride), is also important. We take in mineral ions with our food. Some are lost via our skin when we sweat. Again, the kidney is most important in keeping an ion balance. Excess mineral ions are removed by the kidneys and lost in the urine. The balance of water and salts in the body is very important because of the osmotic impact on the cells if the balance is wrong (see Chapter 2) Controlling this balance is called **osmoregulation**.

> **DID YOU KNOW?**
>
> Your blood passes through your kidneys at the rate of 1200 cm^3 per minute, which means all the blood in your body passes through your kidneys and is filtered and balanced approximately once every five minutes.

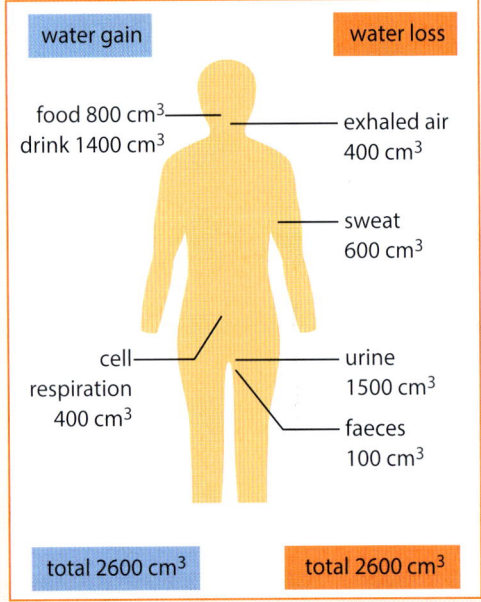

Figure 10.8 The daily water balance of an adult human

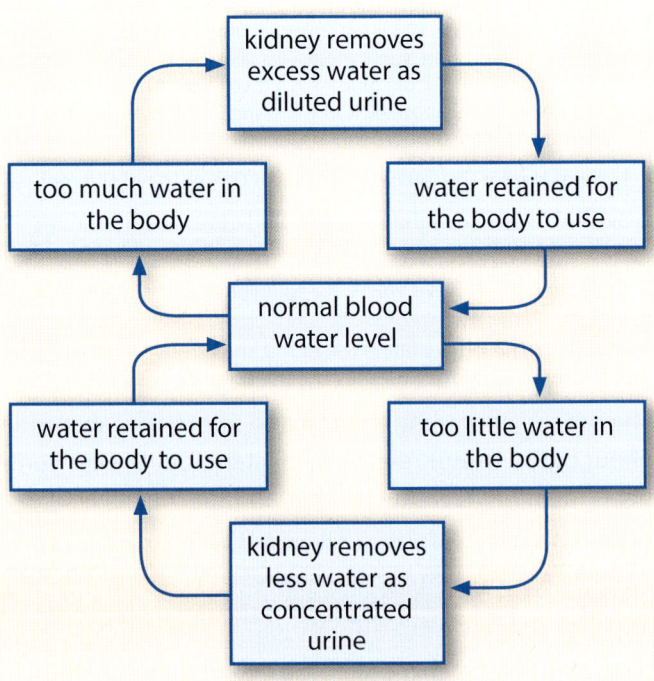

Figure 10.9 Our water balance is maintained by the kidneys.

212 Excretion and homeostasis

Negative feedback and osmoregulation in the kidneys

The amount of water lost from the kidneys in the urine is controlled by a sensitive negative feedback mechanism involving antidiuretic hormone or ADH. Diuresis means passing urine, so anti-diuresis means preventing or reducing urine flow.

If the water content of the blood is too low (so the salt concentration of the blood increases), special sense organs called **osmoreceptors** in our brain detect this. They stimulate the pituitary gland in the brain (see Chapter 11 for details) to release ADH into the blood. This hormone affects the second coiled tubules of the kidneys, making them more permeable so more water is reabsorbed back into the blood. This means less water is left in the kidney tubules and the urine is more concentrated. At the same time, the amount of water in the blood increases and so the concentration of salts in the blood returns to normal.

If the water content of the blood is too high, the concentration of the salts in the blood may become too low. Again, the osmoreceptors in the brain detect this and the pituitary gland releases much less ADH into the blood. The kidney tubules are much less permeable, so less water is reabsorbed back into the blood, producing a large volume of dilute urine. Water is effectively lost from the blood and the concentration of salts returns to normal (see Figures 10.10. and 10.11).

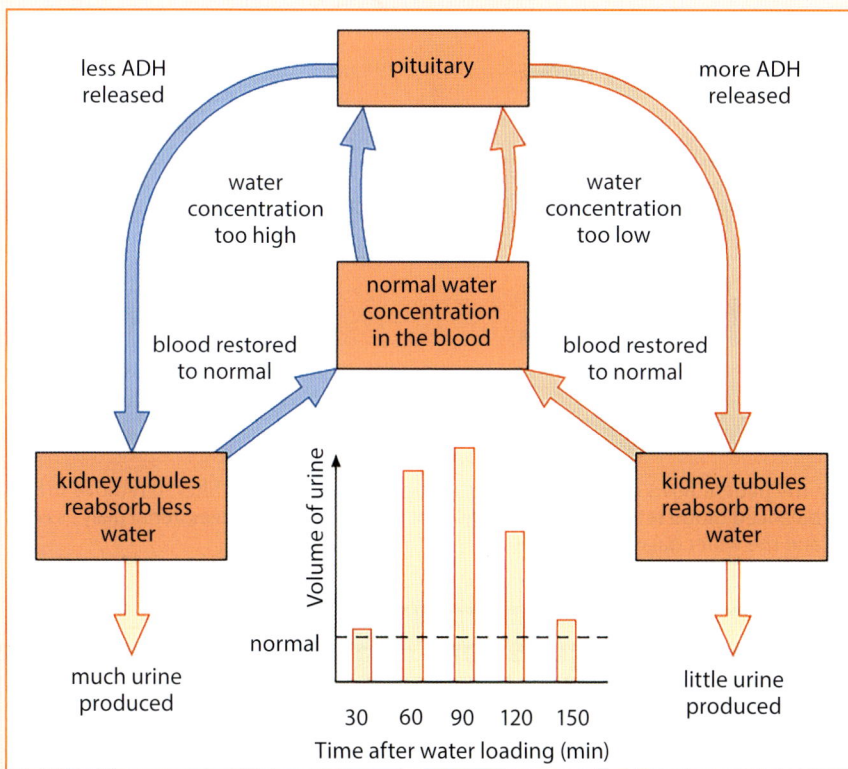

Figure 10.10 The negative feedback system that controls the amount of water that the kidney removes from the blood means that we can cope with temporary shortages or excesses of water surprisingly well.

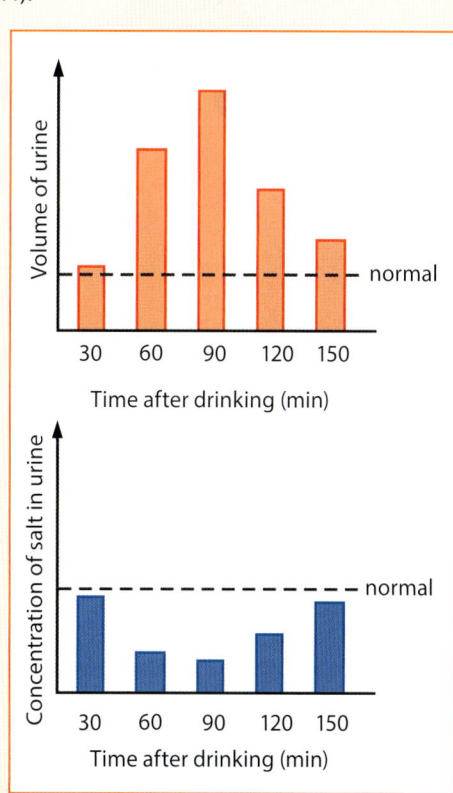

Figure 10.11 The graphs show the effect of drinking a given volume of water on both the volume and the salt concentration of the urine produced. Urine was collected at 30-minute intervals after the drink was given and it clearly shows how sensitive the response of the body is. The water load is removed, but without losing much-needed salt.

Renal dialysis

Our kidneys are vital organs. If they fail, we will die because we will be unable to:

- remove the toxic urea we produce
- regulate the concentration of our blood.

People with failing kidneys need to use a special 'artificial kidney' called a **dialysis machine** (see Figure 10.12). Patients have to be connected to the machine several times a week, for several hours each time, to clean and balance their blood. Blood leaves the body and goes into the machine where the metabolic wastes and excess salt are removed by diffusion across a special membrane. The cleaned blood is then returned to the body (see Figure 10.13). People who need renal dialysis have to be very careful what they eat as well. The only other alternative is a kidney transplant.

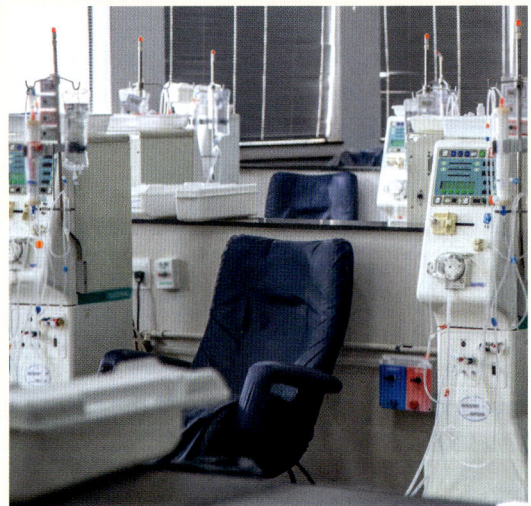

Figure 10.12 A dialysis machine

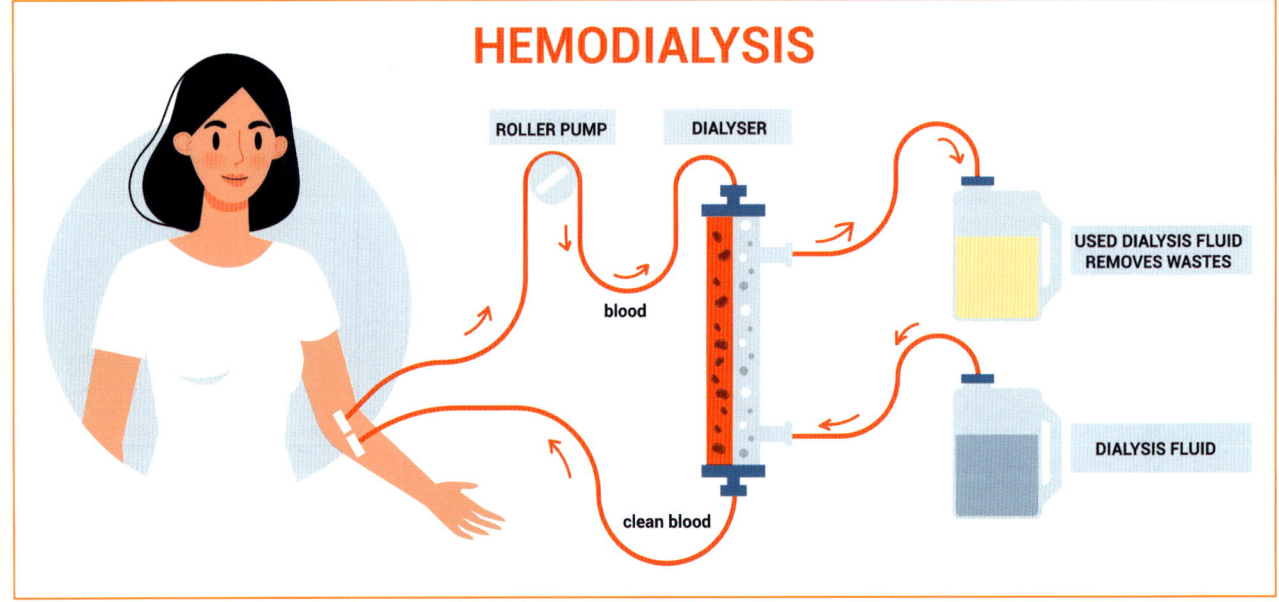

Figure 10.13 During kidney dialysis, blood from the patient flows out of their body, through a special dialysis filter to remove the excess salt, urea and water. Then the clean and balanced blood returns to the patient. Like the kidneys it replaces, renal dialysis is involved with both the excretion of waste urea and the water balance of the blood.

Checkpoint questions

4. Describe what happens in the Bowman's capsule of a kidney tubule.
5. Describe what happens to the ultrafiltrate as it passes through a nephron.
6. Compare and contrast the ultrafiltrate that moves into the kidney tubule in the Bowman's capsule and the urine that leaves the kidney.

Excretion and homeostasis

The importance of temperature regulation

It isn't only our blood concentration and urea levels that need to be controlled. Other aspects of our internal environment also need to be carefully maintained within small limits, and one of the most important is our internal or core body temperature.

It is vitally important that, wherever we go and whatever we do, our body temperature is maintained at the temperature (around 37 °C) at which our enzymes work best. It is not the temperature at the surface of the body that matters – our skin temperature can vary enormously without causing harm. It is the temperature deep inside the body, the core body temperature, which must be kept stable. We can get a good measure of our core body temperature by taking the temperature in the mouth or on the surface of the eardrum.

Human beings have an internal control mechanism, a system of homeostasis, which enables us to lose excess heat if our core temperature starts to rise, yet generate and conserve heat when our core temperature starts to fall.

Figure 10.14 People live in conditions of extreme heat and extreme cold and still maintain a constant internal body temperature.

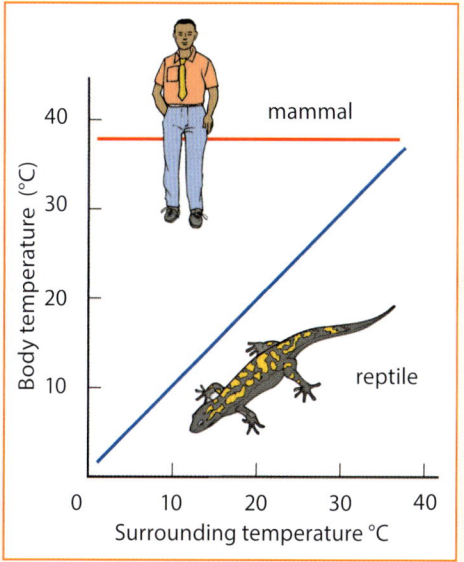

Figure 10.15 The difference between endotherms like us and ectotherms like this lizard is clear to see when you look at the comparison between the internal and external temperatures of both organisms.

> **DID YOU KNOW?**
>
> Human temperature receptors are so sensitive they detect a difference of as little as 0.5 °C in the temperature in our surroundings.

The biologist's toolkit: Core physics: heat, temperature and energy

One problem for us as scientists is that some of the words we use scientifically are also used in everyday language – but not always in the same way! Be very careful when you talk about **temperature**, **heat** and **energy** in biology to use them correctly.

- **Temperature** is a measure of how hot something is. We measure temperature using a thermometer, usually calibrated in degrees Centigrade (°C).
- In biology, we **heat** something to make it hotter and increase its temperature.
- **Energy** is the capacity to do work. It is the currency we use in science to measure how things change or move. Energy is always transferred – it is never made or produced.
- **Heating** is the process by which things get hotter/the temperature increases.
- **Cooling** is the process by which things get colder/the temperature decreases.

How does our temperature control mechanism work? Control of the temperature relies on the **thermoregulatory centre** in the brain (you will learn more about the areas of the brain in Chapter 11). This centre contains receptors that are sensitive to the temperature of the blood flowing through the brain itself. Extra information comes from the temperature receptors in the skin, which send impulses to the thermoregulatory centre containing information about the skin temperature.

The skin and thermoregulation

Our homeostatic temperature control mechanism depends heavily on the skin, a remarkably complex organ, which carries out a number of important functions in the body:

- It allows sensitivity – there are nerve endings that form special receptors sensitive to touch, temperature, pressure and pain.
- It forms a waterproof layer around the body tissues, which protects us against the loss of water by **evaporation** and prevents us gaining water by osmosis every time we have a bath.
- It protects us from the entry of bacteria and other pathogens.
- It protects us from damage by UV light from the sun.
- It is an excretory organ (nitrogenous wastes are lost in sweat).
- It is vital in controlling our body temperature.

The layer of the skin particularly involved in temperature control is the **dermis**.

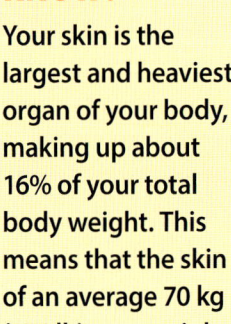

DID YOU KNOW?

Your skin is the largest and heaviest organ of your body, making up about 16% of your total body weight. This means that the skin of an average 70 kg (154 lb) man weighs almost 13 kg (29 lbs or 2 stone 1 lb).

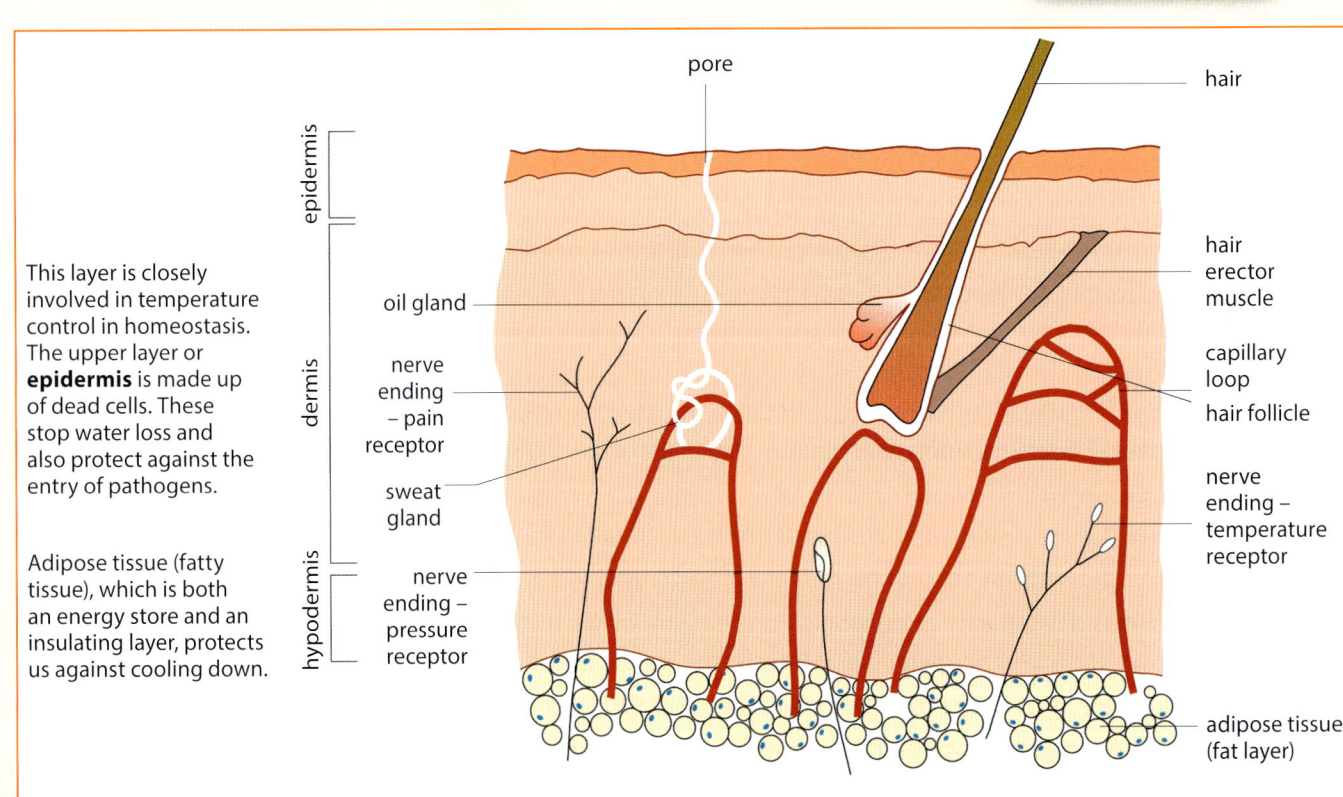

Figure 10.16 This cross-section through the human skin helps you understand how the structure of the skin is related to its functions.

Excretion and homeostasis

How the skin controls core body temperature

The feedback control of body temperature involves the thermoregulatory centre in the brain, and the skin.

Cooling mechanisms

If our core body temperature begins to rise and gets too high, impulses are sent from the thermoregulatory centre, which cause a number of changes to take place (see Figure 10.19).

- Sweat (made up mainly of water and salt, but also containing a small amount of nitrogenous waste) oozes out of the sweat glands and spreads over the surface of the skin. It evaporates from the surface, cooling the skin as it does so (see *The biologist's toolkit: Core physics: evaporation and cooling* below).
- The blood vessels supplying the capillaries in the skin dilate, so that more blood flows through the capillaries. The skin flushes and energy is lost through radiation from the surface, cooling us down. This is called **vasodilation** and it is particularly obvious in pale-skinned people. Less blood flows through the slightly deeper vessels in the skin as a result.
- The hair erector muscles that move our body hair all relax and our hair lies very flat against our skin. In humans, this has very little effect, but in hairy animals this reflex action is important because it reduces the layer of insulating air trapped in the fur and so makes it easier to lose energy by convection.
- The metabolic rate drops so less energy is released.

Figure 10.17 Flushed skin to radiate energy and sweating are two ways in which skin helps the body to cool down.

The biologist's toolkit: Core physics: evaporation and cooling

Evaporation is when a liquid changes state to become a gas. A liquid evaporating to a gas is a physical state change that needs energy to take place.

This is the principle behind the cooling effect of sweating. The water in the sweat evaporates. The energy needed for the state change comes from your skin, cooling it down.

Example:

During the COVID pandemic, we all used lots of alcohol-based hand gel. When you rubbed it on your hands, you may have noticed your skin felt cooler.

- Energy from your skin is transferred to the gel.
- The alcohol in the gel evaporates.
- The transfer of energy from your skin cools it down.

The importance of temperature regulation

Warming mechanisms

If, on the other hand, our core body temperature starts to drop and gets too low, the thermoregulatory centre sends messages to various parts of the body to reduce cooling and keep the system as warm as possible (see Figure 10.19).

- The blood vessels that supply the skin capillaries constrict (close up) to reduce the flow of blood through the capillaries. This reduces cooling through the surface of the skin by radiation and makes us look paler. This is called **vasoconstriction**. More blood flows through the deeper blood vessels of our skin as a result.
- Our metabolic rate speeds up, releasing energy so the body temperature starts to increase.
- We may start to shiver. When we shiver, our muscles contract rapidly, which involves lots of cellular respiration. This transfers some energy from your muscles, raising the body temperature. As we warm up, shivering stops.
- Sweat production is reduced to a very low level to prevent cooling by evaporation.
- The hair erector muscles contract. In furry animals, this pulls the hairs upright, trapping an insulating layer of air, which is very effective preventing cooling (see Figure 10.18). Our hairs are also pulled upright, but we have so little body hair that it has little or no effect on keeping us warm. The most obvious effect is that we get goosebumps on our skin – each bump is the contracted erector muscle pulling on a hair.

Figure 10.18 The Arctic fox stays warm in freezing temperatures by trapping an layer of air between the hairs in its fur.

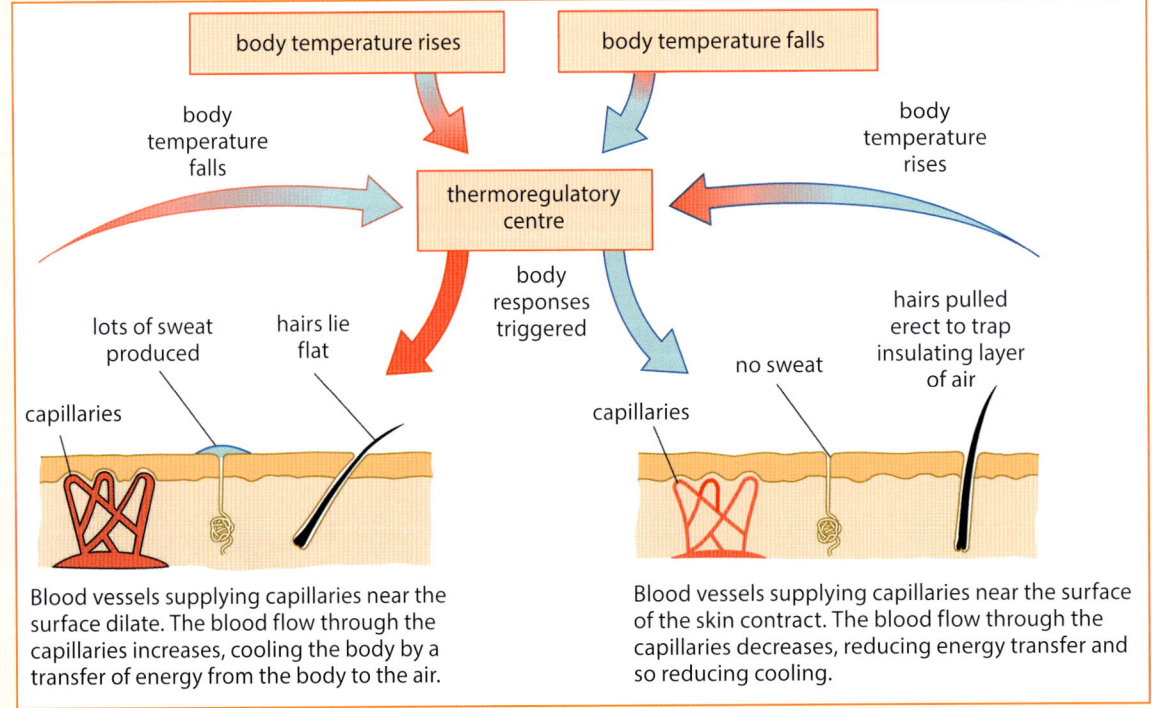

Figure 10.19 The thermoregulatory centre in the hypothalamus of the brain acts as our body's thermostat. As a result of all these sensitive control mechanisms, the core temperature of the body is usually kept constant (the same) to within 1 °C.

Excretion and homeostasis

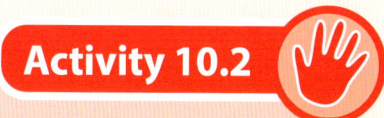

Activity 10.2

Investigating factors that affect cooling

Different factors affect how quickly a person cools down – surface area to volume ratio, insulation, whether they are wet, and so on. You can mimic these situations in a number of ways using beakers or conical flasks as model organisms, and use this to investigate factors that affect cooling.

You will need:

- 250 cm³ beaker or conical flask (your control)
- 100 cm³ beaker or conical flask
- 250 cm³ beaker or conical flask wrapped in cotton wool or fabric
- 250 cm³ beaker or conical flask wrapped in wet cotton wool or fabric
- For each container you use, a cardboard lid with a thermometer poking though it; if you have cotton wool or wet cotton wool around your container, cover the lid with it as well
- Hot water – 60–70 °C is hot enough – handled carefully
- Stopwatch or clock with a minute hand

Method

1. Arrange your containers – work with a control and at least one other container each time.
2. Add the same volume of hot water – for example, 75 cm³ – to each container and gently place the lids on. Make sure the bulb of the thermometer is in the water.
3. Take and record the temperature in each container.
4. Repeat the temperature readings at 1 minute intervals for 20 minutes.
5. If you have time, repeat the experiment using the same control but a different combination of containers – or vary the conditions, for example, by placing all the containers in a draught or outside in the sun.
6. Record your results on a graph, plotting temperature against time. Use the same axes for all your results – use different colours or clear labels to identify the different containers. This allows you to compare your results very easily.
7. Write up your experiment and explain the results you have obtained. Which container loses heat the fastest? Which retains the most heat? How does this relate to temperature control in humans? Do you think these containers are good models? How could you improve or extend this investigation?

Checkpoint questions

7. Describe the feedback system that is used to control body temperature to within narrow limits.
8. Explain why it is so important to control our body temperature within narrow limits.

The importance of temperature regulation

Controlling blood sugar levels

It is very important that our cells have a constant supply of the glucose they need for cellular respiration. Glucose is transported around the body to all the cells by our blood. The level of sugar in our blood is controlled by hormones produced in our pancreas.

When we digest a meal, large amounts of glucose pass into our blood. Without a control mechanism, our blood glucose level would fluctuate wildly. After a meal, it would get so high that glucose would be removed from the body in the urine A few hours later, the level would fall and our cells would not have enough glucose to respire.

This type of chaos is prevented by the **pancreas**. The pancreas is a small pink organ found below the stomach (see Chapter 6). It constantly monitors the blood glucose concentration and controls it using two hormones called **insulin** and **glucagon**. A normal blood glucose level ranges between 4.0–7.8 mmol/litre of blood.

Main stages of blood sugar control

The two main stages of blood sugar control involve rising and falling blood sugar levels.

Rising blood sugar level: After a meal, the blood glucose level starts to rise from about 4.0 mmol/l.

- The rising blood glucose level is detected by the **insulin-secreting cells of the pancreas**.
- The pancreas releases insulin into the blood.
- The insulin makes the body cells take up more glucose, and the liver takes up glucose and stores it as **glycogen**.
- The blood glucose level falls as a result. This is detected by the insulin-secreting cells and the release of insulin stops.

Falling blood sugar level: Some time after a meal, a low blood glucose level is detected by the **glucagon-releasing cells of the pancreas**.

- The pancreas releases glucagon into the blood – the liver is the target organ.
- The liver breaks down its glycogen stores and releases glucose into the blood.
- The blood glucose level rises as a result. This is detected by the glucagon-secreting cells and the release of glucagon stops.

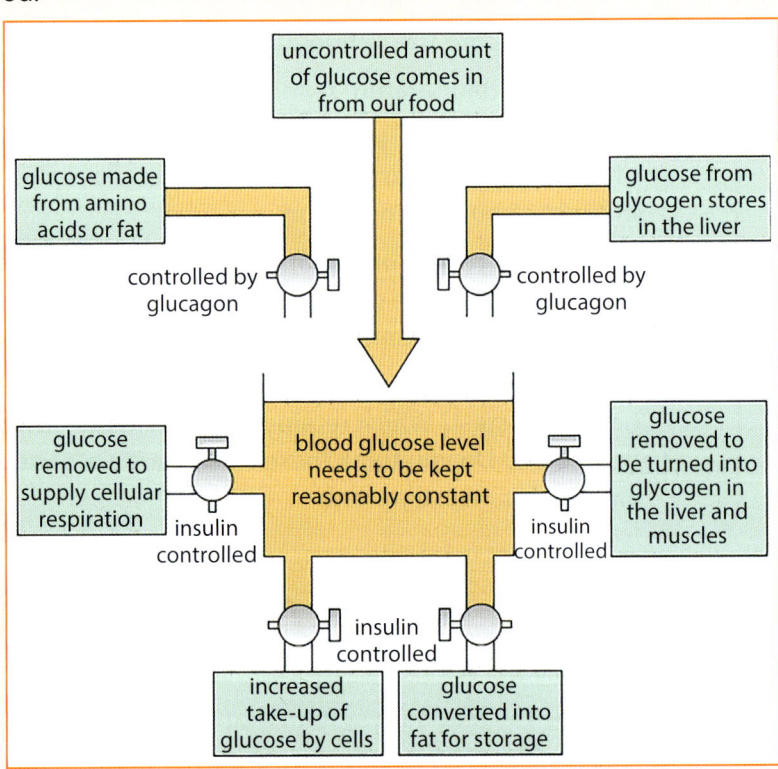

Figure 10.20 This model of our blood glucose control system shows the blood glucose as a tank. It has both controlled and uncontrolled inlets and outlets. In each case, the control is provided by the hormones insulin and glucagon.

Excretion and homeostasis

> **Checkpoint questions**
>
> 9 State why the level of glucose in your blood is so important.
> 10 Name, compare and contrast the two hormones involved in the control of your blood sugar level.

The causes and treatment of diabetes

Most of us never think about our blood sugar level because it is perfectly controlled by our pancreas. But for some people, life isn't quite this simple, because their pancreas does not make enough, or any, insulin or their body cells do not respond to the insulin they make. Without insulin, our blood sugar level gets higher and higher after we eat food. Eventually our kidneys produce glucose in the urine. We produce lots of urine and feel thirsty all the time. Without insulin, glucose cannot get into the cells of the body, so we lack energy and feel tired. We break down fat and protein to use as fuel instead, so we lose weight. This condition is called diabetes. Before there was any treatment for diabetes, people would waste away, fall into a **coma** and die. Fortunately, there are now some very effective ways of treating diabetes.

For people with Type II diabetes, managing their diet is enough to keep them healthy. Losing weight, doing plenty of exercise and avoiding carbohydrate-rich foods keeps their blood sugar levels relatively low so the reduced amount of insulin can cope with them.

However, everyone with Type I diabetes and some people with Type II diabetes needs replacement insulin before meals. Insulin is a protein, so it is digested in the stomach. This is why it is given as an injection to get it directly into the blood. The injected insulin allows glucose to be taken into the body cells and converted into glycogen in the liver. This stops the concentration of glucose in the blood from getting too high. Then as the blood glucose level falls, natural glucagon makes sure glycogen is converted back to glucose. As a result, the blood glucose level is kept as stable as possible (see Figure 10.20 on page 220).

Insulin injections treat diabetes successfully but they do not cure it. Diabetes can still cause problems such as blindness (see Chapter 11 for more about this effect) or neuropathy, when the nerve cells die leaving parts of the body – often the feet and hands - without sensation. Until a cure is developed, diabetics have to inject insulin several times every day throughout their lives, and take care of their health to stay well. But managed carefully, people with diabetes can do anything, including competing in sport at the highest levels. Ways of monitoring blood glucose levels and of delivering insulin to the blood are improving all the time, with modern technology in the form of portable devices and smartphone apps, and so on, helping people manage their condition.

> **DID YOU KNOW?**
>
> Scientists have transplanted the pancreatic cells that produce insulin from dead donors into people who have diabetes. 75% of the people who receive a transplant no longer need insulin a year later, and may not need it for many years. The technique is so new that we do not know yet how long the transplanted cells will last.

Figure 10.21 A glucometer is a portable device that diabetics can use to check their blood glucose level quickly and easily.

Controlling blood sugar levels

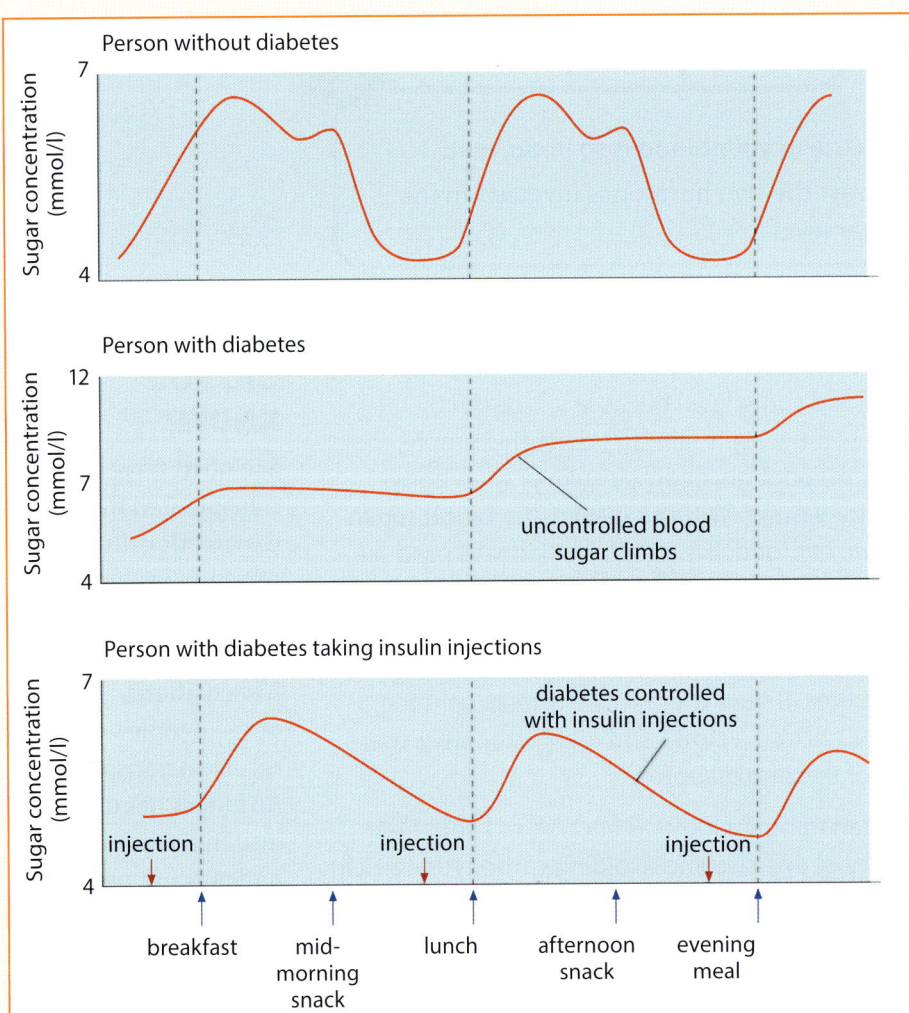

Figure 10.22 Insulin injections keep the blood sugar level within safe limits. They cannot mimic the total control given by the natural insulin production of the pancreas, but they work well enough to let people lead a full and active life.

Possible SBAs based on excretion and homeostasis

→ Study kidneys from different species of animals – investigate the similarities and differences.
→ Investigate the impact that drinking different amounts of liquid/doing different amounts of exercise/eating food like red beet has on the amount and colour of urine produced over time.
→ Make a model to demonstrate ultrafiltration and the working of the kidney tubule.
→ Use secondary sources to carry out a research project into the adaptations of the kidneys of animals living in extreme environments such as in deserts, in fresh or saltwater, and so on.
→ Investigate the appearance of skin under the microscope.
→ Investigate factors that affect how quickly model organisms cool down or warm up and relate these findings to how different species of animals are adapted to warm up or cool down in different conditions, including different behaviours.
→ Survey people in your community who are affected by diabetes and discover what support is available.
→ Use secondary sources to investigate the most modern methods of monitoring blood glucose levels.
→ Investigate the levels of diabetes in different Caribbean countries/the diabetes-related illnesses of neuropathy or limb amputation in different Caribbean countries.
→ Develop a diabetes awareness campaign for young people across the Caribbean to try and reduce the incidence of Type II diabetes in the future.

Excretion and homeostasis

End-of-chapter summary

In this chapter, you have learnt that:

- human beings need to maintain a constant internal environment
- negative feedback mechanisms and a few positive feedback mechanisms are important in the body
- waste products are removed from the body by the lungs and the kidneys, and in tiny amounts through the skin
- the main metabolic wastes are carbon dioxide and urea
- water is lost through the lungs, the skin and the kidneys
- the ion and water content of the body is controlled by the kidneys
- the kidneys filter the blood and control the amounts of ions and water that are reabsorbed
- the amount of water reabsorbed by the kidneys is controlled by a negative feedback system involving the hormone ADH produced by the pituitary gland in the brain
- core body temperature is controlled by the thermoregulatory centre in the brain; this affects the dilation of the blood vessels supplying skin capillaries, the position of the body hairs and the amount of sweat produced
- if the core body temperature goes up, sweating and vasodilation are used to increase the transfer of energy through the skin and reduce the body temperature; in humans, flattening the hairs has little effect; the metabolic rate is also reduced to keep the temperature down
- if the core body temperature goes down, sweating stops and vasoconstriction occurs to reduce cooling; the hairs are raised to trap a layer of insulating air – again this is ineffective in humans; the metabolic rate is increased to warm up the body – the liver becomes very active and shivering starts
- the blood glucose concentration is monitored and controlled by the pancreas
- insulin and glucagon are the hormones involved in controlling blood sugar concentration. Insulin converts glucose → glycogen, glucagon converts glycogen → glucose; between them, they maintain the blood glucose concentration at around 4–7.8 mmol/l
- in diabetes, the blood glucose may rise to fatally high levels because the pancreas does not secrete enough insulin; it can be treated by injections of insulin before meals.

End-of-chapter questions

1. Which one of the following is not an example of homeostasis?
    - A   Control of the blood sugar level
    - B   Control of the body temperature
    - C   Control of the water content of the blood
    - D   Control of the swallowing mechanism

2. Which one of the following areas is not part of the nephron (kidney tubule)?
    - A   Bowman's capsule
    - B   Urinary bladder
    - C   Loop of Henlé
    - D   First coiled tubule

3. Which one of the following statements about ADH is true?
    - A   ADH is a hormone produced in the brain that affects the second coiled tubules of the kidneys, making them more permeable so more water is reabsorbed back into the blood and a small amount of concentrated urine is formed.
    - B   ADH is a hormone produced in the brain that affects the first coiled tubules of the kidneys, making them more permeable so more water is reabsorbed back into the blood.
    - C   ADH is a hormone produced in the kidneys that affects the coiled tubules of the kidneys, making them more permeable so more water is reabsorbed back into the blood.
    - D   ADH is a hormone produced in the brain that affects the second coiled tubules of the kidneys, making them less permeable so less water is reabsorbed back into the blood and more dilute urine is formed.

4. Here is a jumbled list of the events through which your body temperature is controlled when it starts to increase. Which sequence of events is correct?
    - i   Winston exercises hard and his body temperature starts to rise.
    - ii   Winston takes a long, cool drink to replace the liquid he lost through sweating.
    - iii His temperature returns to normal.
    - iv   Winston's skin reddens and he sweats heavily so the amount of heat lost through his skin goes up.

    - A   i, ii, iii, iv
    - B   i, iii, iv, ii
    - C   i, iv, iii, ii
    - D   i, iv, ii, iii

5. Two hormones are particularly important for the control of the blood sugar level in the body. Which are they?
    - A   Insulin and glucagon
    - B   Insulin and adrenalin
    - C   Insulin and glucose
    - D   Glucose and glucagon

6 The two main metabolic waste products that have to be removed from the human body are carbon dioxide and urea.
- a What is meant by the term 'metabolic waste'? (1)
- b For each waste product named above, describe:
  - i how it is formed. (4)
  - ii why it has to be removed. (2)
  - iii how it is removed from the body. (8)

7 Around 500 people take part in the Trinidad and Tobago Marathon each year. Although it is run at the coolest time of year, and early in the day, running 26 miles in Caribbean conditions is a test for any athlete.
- a List the ways in which running this marathon will affect their:
  - i water balance. (3)
  - ii ion balance. (3)
  - iii internal temperature. (3)
- b How can runners in this marathon maintain a constant internal environment? Include as many ways as possible in your answer. (5)

8
- a Summarise the way in which the kidney functions, including the processes of ultrafiltration and reabsorption in the nephron. Draw and label diagrams to help your explanation (10)
- b Explain carefully how the body responds to water loading, for example, if you drink several pints of liquid in one go. (5)

9
- a Why is it so serious if the kidneys fail? (4)
- b What must an 'artificial kidney' or dialysis machine be able to do? (4)
- c Most patients only receive dialysis three times a week. The food they eat and the amount they drink must be carefully controlled except for the first hour or two they are on dialysis. Explain why this is the case. (4)

10 Write the correct letter with its corresponding number to match each term to its definition.
- A Hormone
- B Insulin
- C Diabetes
- D Glycogen

- i A condition in which the pancreas cannot make enough insulin to control the blood sugar level.
- ii A chemical message carried in the blood that causes a change in the body.
- iii An insoluble carbohydrate stored in the liver.
- iv A hormone made in the pancreas that causes glucose to pass from the blood into the cells where it is needed for energy. (4)

11 Explain the roles of the following in maintaining a constant core body temperature:
- a The thermoregulatory centre in the brain (5)
- b The temperature sensors in the skin (3)

12 a Explain how the pancreas keeps the blood glucose level of the body constant. (8)
b Why is it so important to control the level of glucose in the blood? (2)
c What is diabetes and how can it be treated? (4)

13

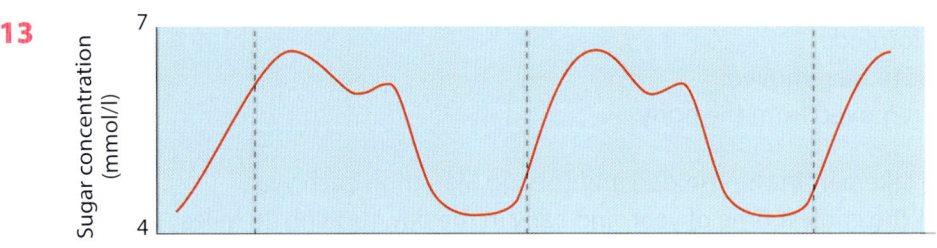

A Person without diabetes

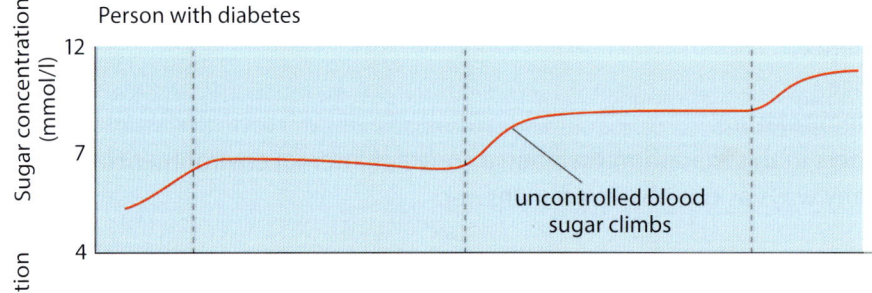

B Person with diabetes

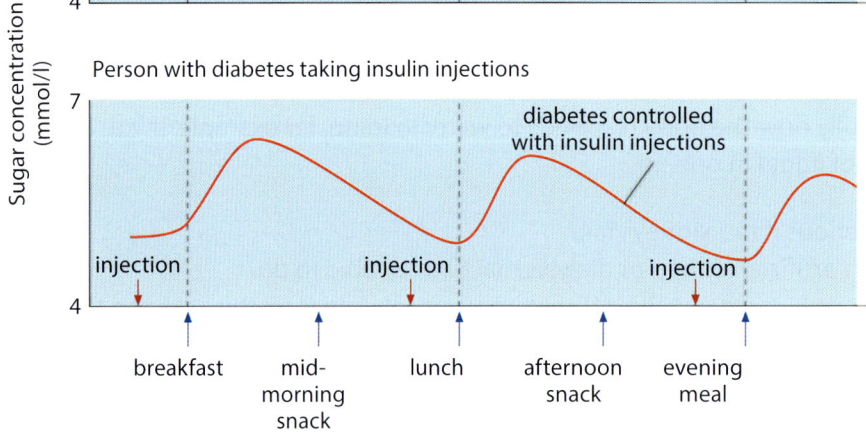

C Person with diabetes taking insulin injections

a Look at graph A. Why does the level of insulin increase after a meal? (2)
b Explain how the blood sugar levels change between meals. (3)
c Graph B shows the blood sugar pattern of someone who has just developed diabetes and is not yet using injected insulin. What differences are there between this pattern and the one shown in A? Explain why these differences occur. (3)
d People with untreated diabetes like that shown in graph B feel increasingly tired and lethargic. Why do you think this is?
e Graph C shows the effect of regular insulin injections on the blood sugar level of someone with diabetes. Why are the insulin injections so important to their health? (3)

Excretion and homeostasis

Chapter 11
Coordination and control

When you have completed this chapter, you will be able to:
- describe the main divisions of the nervous system
- describe a synapse
- define a reflex system and understand the importance of reflexes in the body
- explain the process by which voluntary actions occur
- distinguish between nervous and hormonal control systems
- explain the roles of specific hormones in the body
- have a basic understanding of the human brain and what happens in mental illness
- identify the sites of hormone production
- distinguish between a neuron and a nerve, and explain the difference between sensory and motor neurons
- describe the human eye in detail and relate the structures to their functions

What the examiners say
→ Candidates demonstrate several areas of weakness and are often unable to relate the information covered in this topic to other areas of the syllabus.
→ Specifically, candidates seem unable to differentiate between exocrine and endocrine glands; compare and contrast the nervous system and the endocrine system; accurately label diagrams related to the nervous system; make the connection between reflex and involuntary actions; name the glands, instead naming the hormones associated with the glands; draw and label diagrams that show where images are formed for short-sightedness and long sightedness; differentiate between concave and convex lenses; and discuss the functions of the rods and cones in the eye.

Revision tip
Create a quiz about each topic and challenge your friends to complete it in 15 minutes. Make sure it can be done in that time, and that you know all the answers. Ask your friends to do the same for you!

All living organisms need some level of awareness of their surroundings to avoid danger, find food and, in some cases, find a mate. Whatever the level of awareness, an organism requires coordination and control to respond to changes in the surroundings. In large and complex organisms like ourselves, it is also very important that the different systems within our body are coordinated and work together. As you saw in Chapter 10, the:
- nervous system is involved in all rapid responses with the passage of electrical impulses around the body.
- hormonal system involves the movement of chemical messages around the body. It is slower than the nervous system but is responsible for coordinating most of the functions of the body.

The nervous system

Our nervous system is a highly complex system that provides us with rapid, coordinated responses to the situations we meet in our life. It allows us to react to our surroundings and coordinate our behaviour.

Figure 11.1 Human beings not only control their own bodies, but they expect their coordination systems to be able to control fast and complicated machines as well.

A nervous system needs the following requirements to work:

- A **stimulus** – a change in the internal or external environment
- A **receptor** to pick up the stimulus
- **Neurons (nerve cells)** to carry information about the stimulus to …
- A **coordinator**, where information from a number of receptors is analysed to give the best response
- More neurones to carry messages to appropriate parts of the body where specialised **effectors** (the muscles and organs) respond

Figure 11.2 Requirements for a nervous system to work

Your sensory receptors pick up stimuli – changes in the environment. For example, our eyes respond to light, our nose responds to smells and the insulin-producing cells in the pancreas (which you learnt about in Chapter 10) respond to the level of glucose in our blood. The information from these receptors feeds into the nervous system. The human nervous system has two main divisions:

- The **central nervous system or CNS**, made up of the **brain** and **spinal cord**. This is where the information brought in from the body is coordinated, and instructions are sent out.
- The peripheral nervous system, made up of the **spinal nerves**, the **cranial nerves** and the **autonomic nervous system**. This huge network of nerves brings information from your sensory receptors to the brain and spinal cord, and carries instructions back to the different parts of the body.

How nervous coordination works

Neurons (nerve cells) are the basic unit of the nervous system. They are cells specialised for the transmission of electrical impulses.

The properties of neurons include:

- **irritability** – they react to stimuli from their surroundings
- **conductivity** – they carry electrical impulses.

The structure of neurons include:

- a cell body that contains the nucleus, mitochondria and other organelles
- slender, finger-like processes called dendrites that connect to neighbouring nerve cells
- an axon or nerve fibre, which is extremely long and thin and carries the nerve impulse.

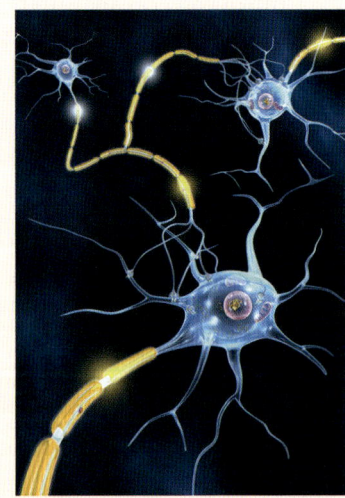

Figure 11.3 Neurons carrying electrical impulses

Coordination and control

Some neurons have a **myelin sheath**, a layer of fatty material that provides a layer of insulating material so the nerve impulse travels as fast as possible.

Sensory neurons (see Figure 10.4) carry impulses from the receptors to the **central nervous system (CNS)**.

Motor neurons (see Figure 10.4) carry impulses from the CNS to the **effector organs**, which are the muscles or organs of the body that bring about the response to the original stimulus.

Our nervous system relies on nerve impulses travelling along the neurons. Each nerve impulse is a minute electrical event that is the result of charge differences across the membrane of the axon. The wave of positive charge inside the axon when the neuron is stimulated is called the **action potential**.

> **DID YOU KNOW?**
> The average axon is only 10 μm in diameter, but in a large animal like an elephant or a giraffe, they can be up to 4 m in length. It has been calculated that a piece of spaghetti on the same scale would be 400 m long!

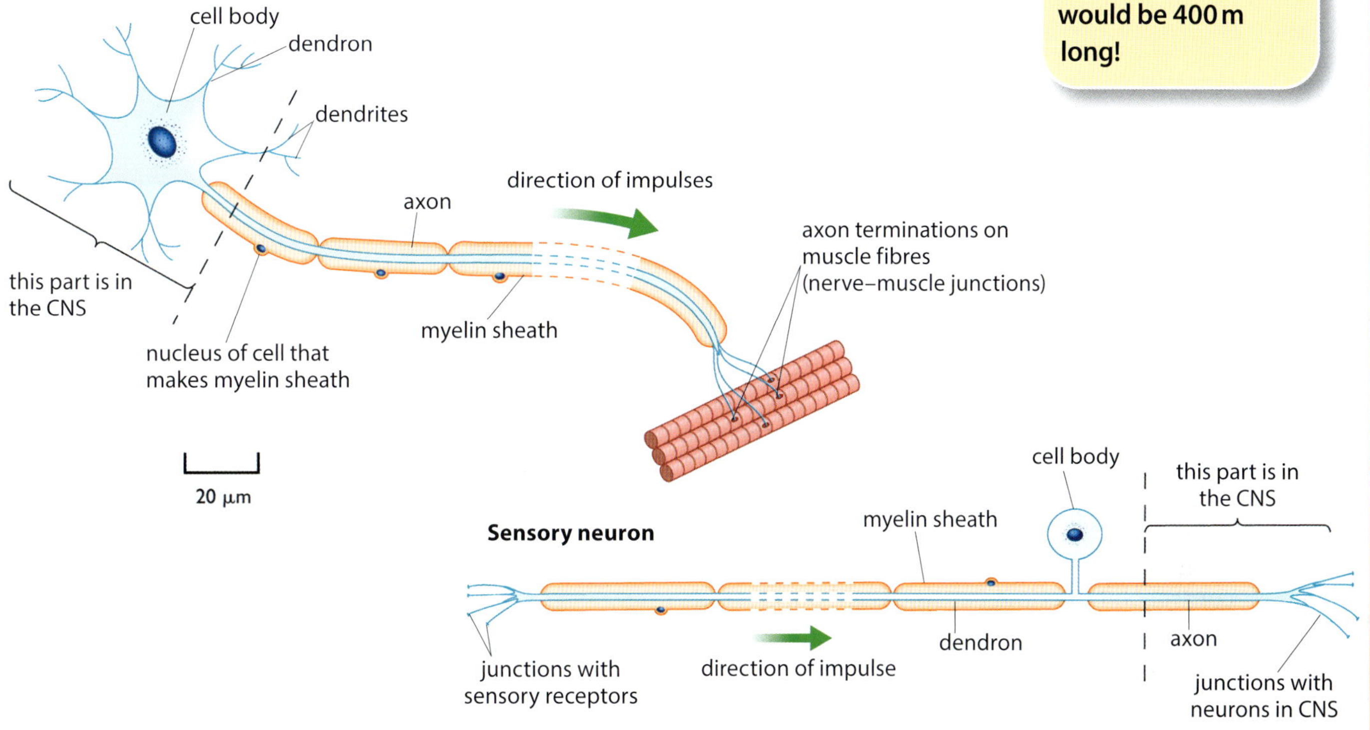

Figure 11.4 Motor neurons and sensory neurons carry electrical impulses to and from your central nervous system and are vital for coordination in your body.

Synapses

Whenever one neuron ends and another begins, there is a minute gap called a **synapse**. The electrical impulses have to cross these synapses, but an electrical impulse cannot leap the gap. So when an impulse arrives at the end of a neuron, chemicals are released. (See Figure 11.5.)

These **chemical transmitters (neurotransmitters)** cross the synapse and are picked up by special receptors in the end of the next neuron. In turn, this starts up an electrical impulse, which then travels along the next neuron. There are many different chemical transmitters in the synapses including **acetyl choline** (see Figure 11.6) and **adrenaline** in the peripheral nervous system and **serotonin**, which is found only in our brain.

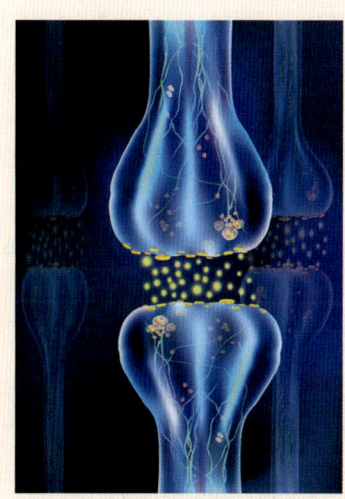

Figure 11.5 A synapse

The nervous system

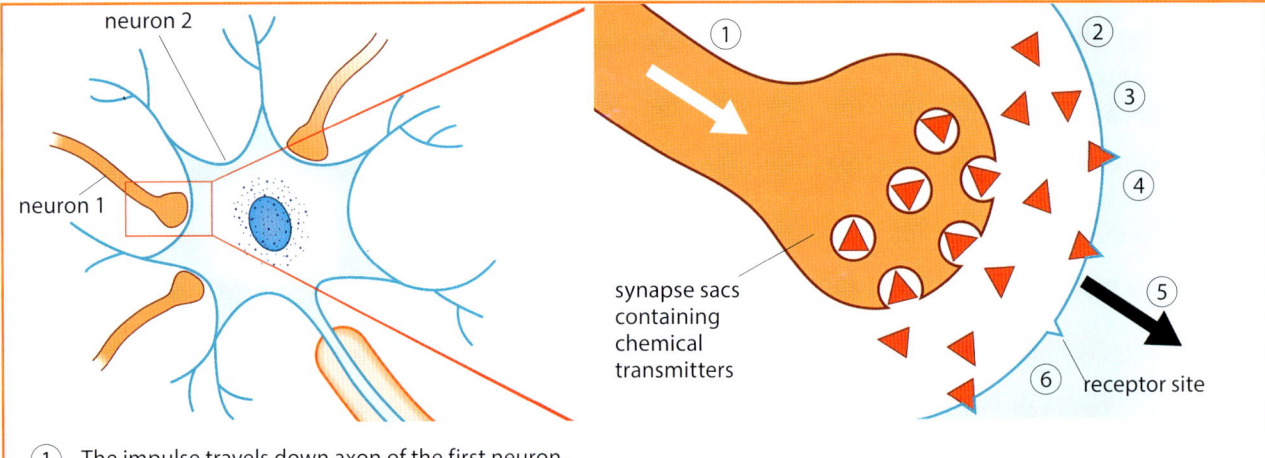

① The impulse travels down axon of the first neuron.
② The arrival of the impulse at the end of the neuron causes the release of the chemical transmitters into the synapse.
③ Chemical transmitters diffuse across the synapse.
④ Chemical transmitters attach to the receptors on the membrane of the second neuron.
⑤ Impulse is set up in the second neuron.
⑥ The transmitter substance is broken down by enzymes from the second neuron.

Figure 11.6 A synapse is a junction between one neuron and another. The electrical impulse cannot cross the gap. It relies on chemical transmission across the synapse.

DID YOU KNOW?

A number of well-known poisons work by affecting the synapses in our nervous system.

- Botulinus toxin stops the release of one type of neurotransmitter, blocking the nervous system and causing death.
- Strychnine inactivates the enzymes that break down one neurotransmitter so the muscles contract all the time causing rigid paralysis.
- Curare blocks the second part of the synapse so muscles can no longer respond to the nerve impulses, causing paralysis and death.

The difference between neurons and nerves

As you have seen, a neuron is a type of cell specialised for carrying electrical impulses from one place to another.

A nerve is a bundle of many neurons. Some nerves carry only motor neurons and are called **motor nerves**. Some nerves carry only **sensory neurons** and they are called sensory nerves. Some nerves contain a mixture of motor and sensory neurons and they are called **mixed nerves**.

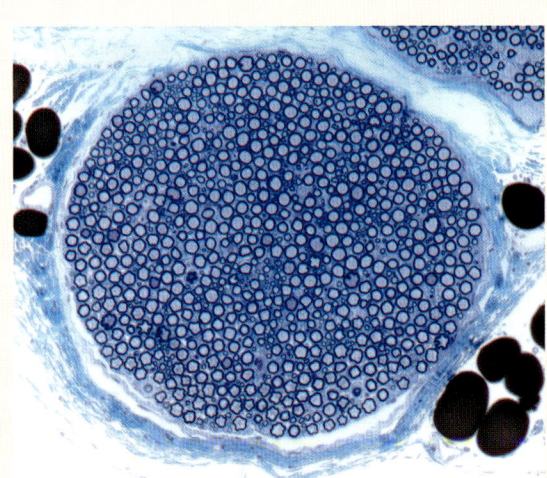

Figure 11.7 A light micrograph of a section through a nerve bundle, showing the myelin sheaths (dark blue circles) surrounding the axons (light blue dots) of the neurons.

Coordination and control

> **Checkpoint questions**
>
> 1 Define the central nervous system and the peripheral nervous system.
> 2 Compare and contrast a neuron and a nerve.
> 3 Draw and label a motor neuron and explain the differences between motor neurons and sensory neurons in both structure and function.
> 4 Compare and contrast a neuron and a nerve.

The central nervous system

Information from the sensory receptors is no use without some sort of coordination, and this is where the central nervous system (CNS) comes in. The central nervous system consists of two main regions: the brain and the spinal cord.

The brain is enclosed in membranes and protected by the bones of the skull in a space called the **cranium**. The spinal cord is a thick, mixed nerve that runs out from the brain down the body. It is encased and protected by the **vertebrae** that makes up the spine.

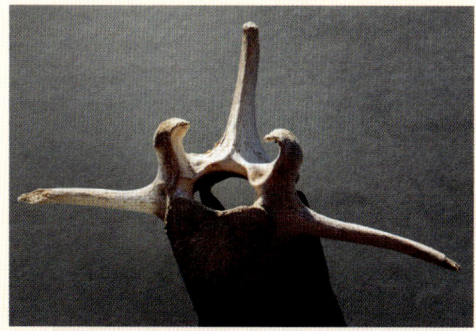

Figure 11.8 This photograph of a vertebra shows the channel that protects the spinal cord.

As we have seen, the nerves that run to and from the CNS make up the peripheral nervous system.

- **Cranial nerves** come out of the brain and go mainly to structures in the head and neck, like our eyes, tongue and jaws.
- **Spinal nerves** come out of the spinal cord and these are the majority of our nerves. They go to the arms, legs and trunk (that is, the rest of the body).

The brain and spinal cord together act as coordinators that:

- process the information coming from sensory receptors and neurons
- instruct motor neurons and effectors to react. They also contain **relay neurons**.

These are short neurons which connect the sensory and motor neurons directly in the CNS, without input from other areas.

The human brain

The brain is a very complex structure. The different areas of the brain carry out very different functions, from the basic reflexes that keep us breathing to the complex ideas needed to create a story, or write and play music.

The bulk of the brain is made up of **grey matter**, which consists of the cell bodies of neurons and the synapses that connect them. Inside this is **white matter**, which is made up of the axons that lead into and out of the brain.

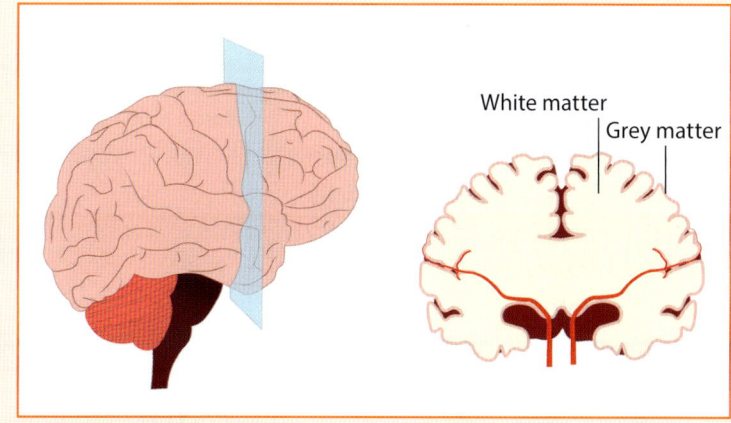

Figure 11.9 The brain is made up of grey matter and white matter.

> **DID YOU KNOW?**
>
> An average human brain weighs 1300–1400 g (about 3 lb), yet the cerebral cortex, which controls most of our conscious thoughts, is only about 3 mm thick.

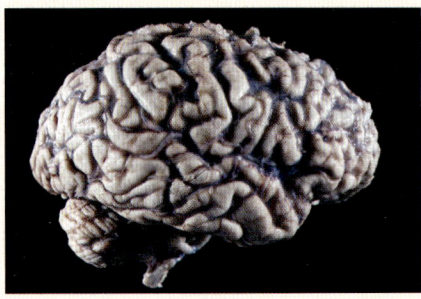

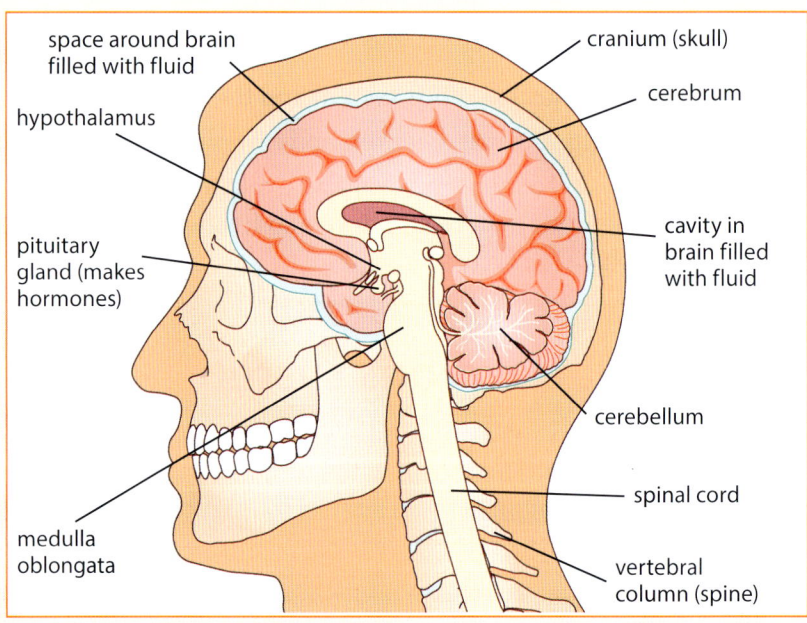

Figure 11.10 The human brain; it may not look very exciting, but it is capable of amazing things!

Table 11.1 summarises the roles of the main areas of the brain shown in Figure 11.10.

Table 11.1 Main functions of each area of the brain

Area of brain	Main functions
Cerebrum (cerebral hemispheres)	Higher levels of thought and intelligence, coordination of sensory information and sending out motor responses
Cerebellum	Balance and orientation
Medulla oblongata	Basic reflexes of life, for example, breathing
Hypothalamus	Controlling homeostasis and many endocrine glands
Pituitary	Produces hormones, for example, controlling growth and metabolism

The spinal cord

The spinal cord has a much simpler structure than the brain. The grey matter (cell bodies and short relay neurons – see page 229) is in the middle, and the white matter (axons) is on the outside. At regular intervals along the spinal cord, there are entry points for sensory nerves bringing information into the CNS and exit points for motor nerves carrying instructions from the CNS.

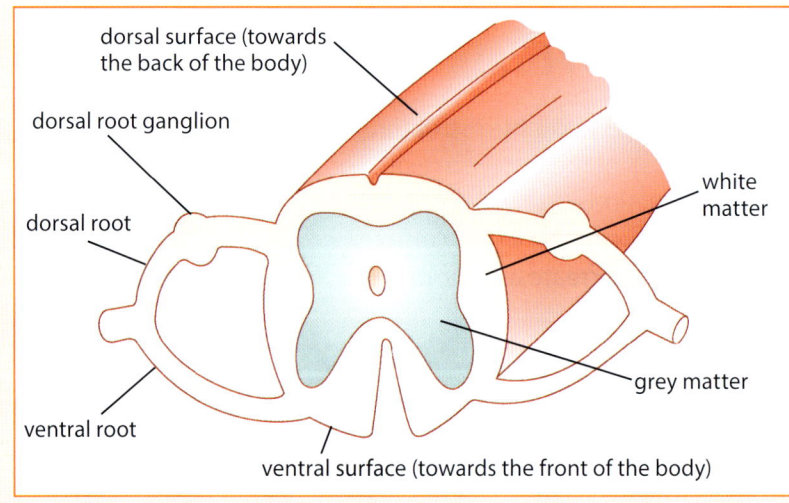

Figure 11.11 The structure of the spinal cord. If our spinal cord is damaged, we may lose all sensation below the damaged area, lose motor control or both.

Coordination and control

DID YOU KNOW?

Each side of your body is controlled by the opposite side of your brain – so what you see with your right eye goes to the left-hand side of your brain. The left-hand side of your brain is good at processing information and largely controls your speech, while the right-hand side is better at special awareness and recognising faces. If the right-hand side of your brain is damaged you may be unable to recognise even your closest family. It is often only when parts of the brain are damaged that we appreciate just what an amazing organ it is.

Checkpoint questions

5 Describe the role of a) the cerebrum b) the cerebellum and c) the medulla oblongata in a human brain.
6 Compare the structure and function of the spinal cord to that of the brain.

Voluntary and reflex control

Sense organs such as the eye, which we will look at in more detail later, respond to changes in the world around us and to changes in our internal environment. However, a change in a sense organ alone is simply not enough. The information needs to reach our central nervous system, and we need to process the information and then respond.

Voluntary actions

Many of the activities we carry out are voluntary – we choose to do them. So how do they come about? For example, if we see a mango that looks ripe at a market stall, we may well pick it up, feel it and smell it before deciding if it is as good as it looks, and buying it.

- When we see the mango, the nerve impulses are transmitted from our eyes along sensory neurons to our spinal cord.
- From the spinal synapses, the electrical signal continues up sensory neurons in the spinal cord until it reaches our brain.
- The information is analysed in our brain to place us in a market and to link back to memories of buying unripe or overripe mangos in the past as well, as well as the pleasure of eating a ripe mango.

Figure 11.12 Buying fresh food from a local stall is just one of the millions of voluntary actions we carry out each day.

- Nerve impulses are sent back from the brain along the motor nerves down our spinal cord.
- At more spinal synapses, the impulse is transmitted to motor nerves running to the effector muscles of our arm and hand that enable us to make a voluntary movement: reaching out to take hold of the mango, feeling it and bringing it to our nose to smell it.
- More information from stretch receptors in our muscles, olfactory (smell) receptors in our nose, pressure receptors in our skin and many others goes back to our brain very rapidly, to be processed again before we make our next voluntary movement.

Think of a different situation – crossing a road, greeting a friend or eating a meal – and work out what is happening in your nervous system.

In a conscious action, we are in control. We see a book, reach out and pick it up or hear an approaching car and stop moving. However, some of the actions of our body are so rapid that there is no time for conscious thought. Other actions take place without any awareness on our part. When was the last time you realised that your pancreas was producing insulin or that your gastric pits were filling up with digestive enzymes? So what is happening when these involuntary or reflex actions take place?

Activity 10.1

Investigating reaction times

The length of time it takes you to recognise a stimulus and react to it is your reaction time. These are very important in many situations – for example when you are driving, or at the start of a race. Some people react more quickly than others – and you can train yourself to speed up. In this investigation, you will be looking at reaction times by measuring how quickly your partner catches a metre ruler when you let it fall. If you collect all the data for everyone in the class, you can produce a graph to show the range of reaction times for your science group, and also do some statistical analysis to find the average median and mean reaction times for your class.

You will need:
- a metre ruler

Method

1. Work in pairs, with one partner holding the ruler and recording the distance and the other catching.
2. Hold the ruler so that your partner's hand is level with the 10 cm mark. They should be able to see the ruler and your hand.
3. Warn them that you will soon be dropping the ruler and after a few seconds, let go.
4. Repeat this three times and calculate the average distance the ruler travelled before your partner caught it.
5. Reverse roles.
6. Draw up a table and collect the results for the whole class Write up your investigation, including a graph to show the distribution of reaction rate across the class.
7. How could you develop or refine this investigation?

Coordination and control

Involuntary or reflex actions

Some of our responses to stimuli are automatic and very rapid – they do not involve conscious thought. These involuntary actions are known as **reflexes**.

Reflexes occur very rapidly and are usually involved in either of the following two responses and functions:

- Reflexes help us avoid danger or damage, for example, when we touch something very hot, we withdraw our hand before we are consciously aware of the sensation of pain. Another example is your knee jerk reflex. When you cross your legs, if someone hits below your kneecap gently but firmly, your leg jerks up into the air (see Figure 11.9).

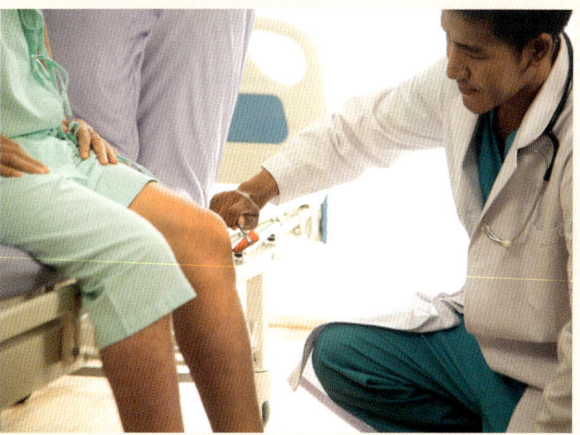

Figure 11.13 The knee jerk reflex is used by doctors to see how your nervous system responds. Try it yourself!

- Reflexes take care of basic bodily functions, leaving the brain free to coordinate other more complicated processes, for example, breathing – we don't consciously instruct our diaphragm to flatten and our intercostal muscles to contract, yet our breathing movements continue because breathing is a reflex action. Not only does gaseous exchange take place constantly, but we do not have to waste thinking time making sure it keeps going. The pupil reflex is another example of a protective reflex action that we do not have to think about. It protects our retinas from damage by too much light, and enables us to see as well as possible in low light levels.

The key point about a reflex action is that the incoming sensory messages do not reach a conscious area of the brain before instructions are sent out to take action.

- **Spinal reflexes** involve the spinal cord – the knee-jerk reaction when you are hit below the knee is a good example.
- **Cranial reflexes** involve the brain – the pupil reflex, where the pupils of your eyes get bigger or smaller depending on the level of light, is a good example.

Reflexes involve three types of neurons: sensory neurons, relay neurons and motor neurons. The receptors, neurons and effectors involved are called a **reflex arc**.

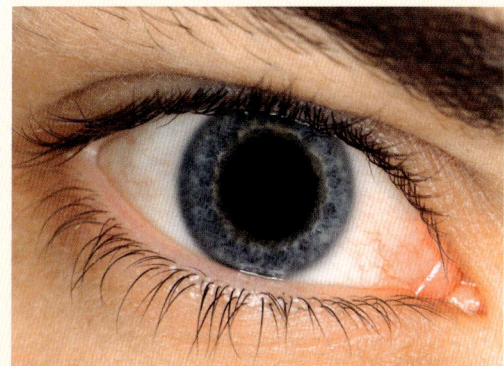

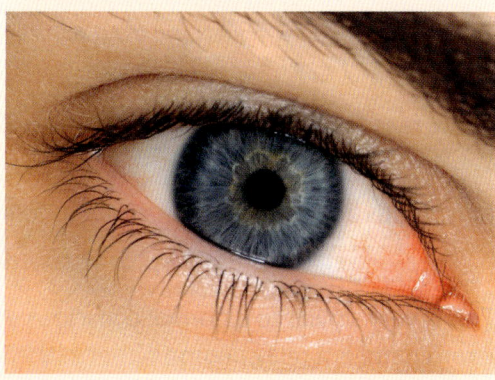

Figure 11.14 The pupils of the eye react to light in a cranial reflex. No conscious thought is involved. Pale blue eyes are shown in this picture because it makes it easier to see the changes in the pupil. The pupil gets bigger in dim light and gets smaller in bright light.

> **DID YOU KNOW?**
> Action potentials can travel along axons at speeds of 0.1–100 m/s. This means that nerve impulses can get from one part of a body to another in a few milliseconds. This is why reflexes can happen so quickly. It is the processing of information and decision making in your brain that takes the time.

Voluntary and reflex control

How spinal reflexes work in detail

Figure 11.15 shows what happens when you put your finger on a sharp pin.

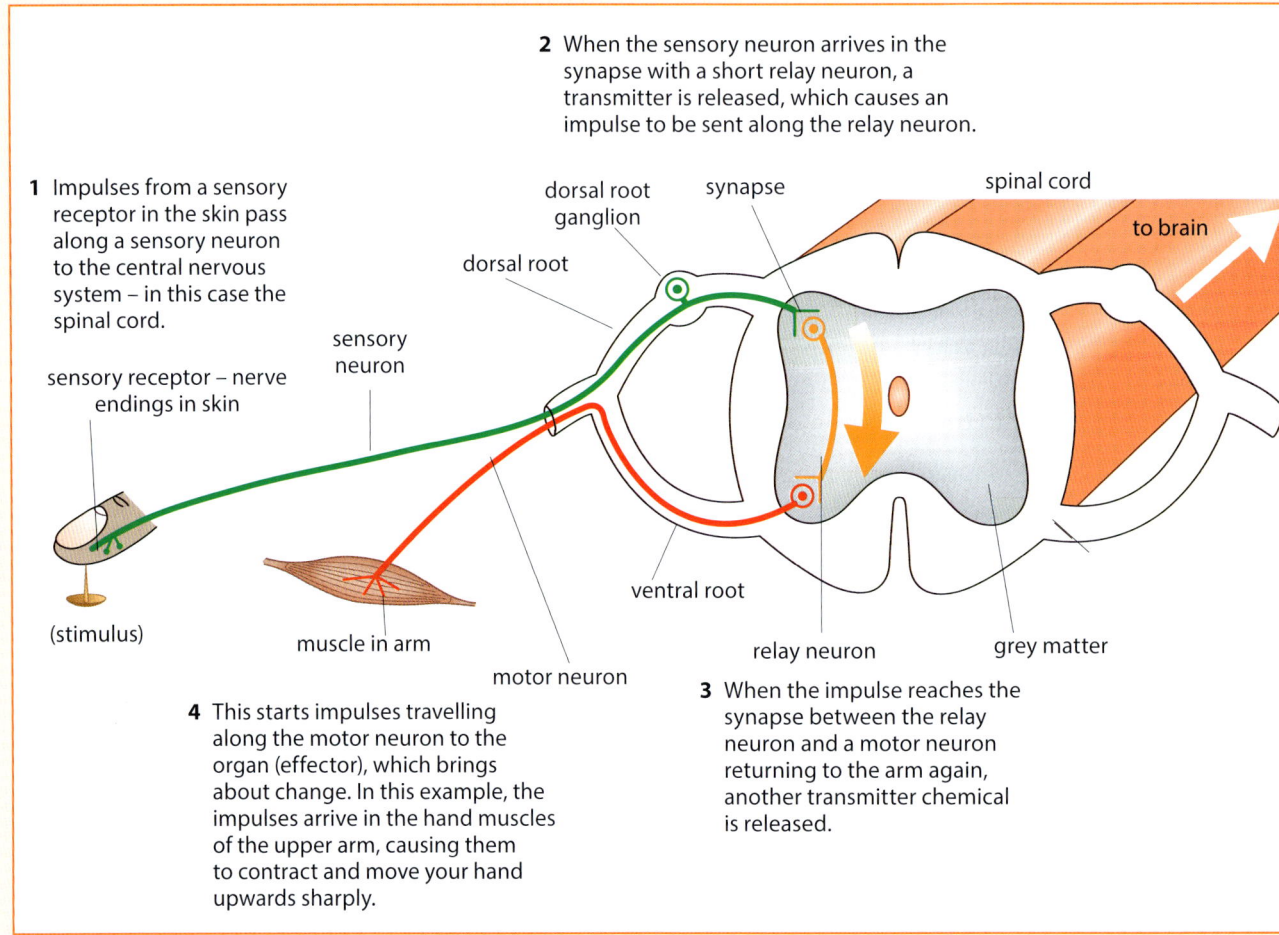

Figure 11.15 This spinal reflex action moves your hand away from the pin before you hurt yourself too much – and before your conscious brain realizes there is a problem.

Most reflex actions can be analysed as follows:

stimulus ➜ receptor ➜ coordinator ➜ effector ➜ response

This is very similar to a normal conscious action, except that in a reflex, the coordinator is a relay neuron either in the spinal cord or in the unconscious areas of the brain.

If you put your hand down on something hot, the reflex action would be:

hot plate on hand ➜ temperature and pain receptors in skin ➜ relay ➜
moving hand away from hot object ➜ muscles of arm ➜ neuron in spinal cord

By missing out the process of conscious thought, the whole action is speeded up. However, even as the impulse is moving through the reflex arc, other neurons have also been stimulated at the synapses in the CNS and these carry information up to the conscious brain so that you know what has happened, after the event.

Coordination and control

Checkpoint questions

7 Define a reflex action.
8 a Define a voluntary action
  b Compare and contrast a voluntary and a reflex action
9 If you are hit below the knee by a doctor with a medical hammer, your lower leg jerks upwards in a reflex action. Draw and label a diagram to show this reflex arc.

Conditioned reflexes

Some reflexes can be conditioned. A **conditioned reflex** is when a natural reflex becomes a response to a neutral stimulus that is different to the original stimulus. It is a type of simple learned behaviour. The classic example of a conditioned reflex was discovered by a Russian biologist called Ivan Pavlov.

- Pavlov noticed that dogs produce extra saliva when they see and smell food – this is a reflex action as the body prepares to eat and digest the food.
- He also noticed that the dogs salivated more when they saw the person who usually fed them, even if they were not carrying any food.
- Dogs do not salivate at a neutral stimulus such as a bell ringing.
- Pavlov began ringing a bell every time just before the dogs were fed.
- The dogs began to produce extra saliva when they heard the bell, before they saw or smelled food.
- The salivating reflex had become conditioned – the dogs had learned that the sound of the bell was associated with food.

The same thing happens in people – the lunch bell at school will make everyone salivate more! Other conditioned reflexes include livestock with an electric fence – the animals learn to avoid the fence because they get a mild shock. Eventually they do not go near the fence, even when it is switched off.

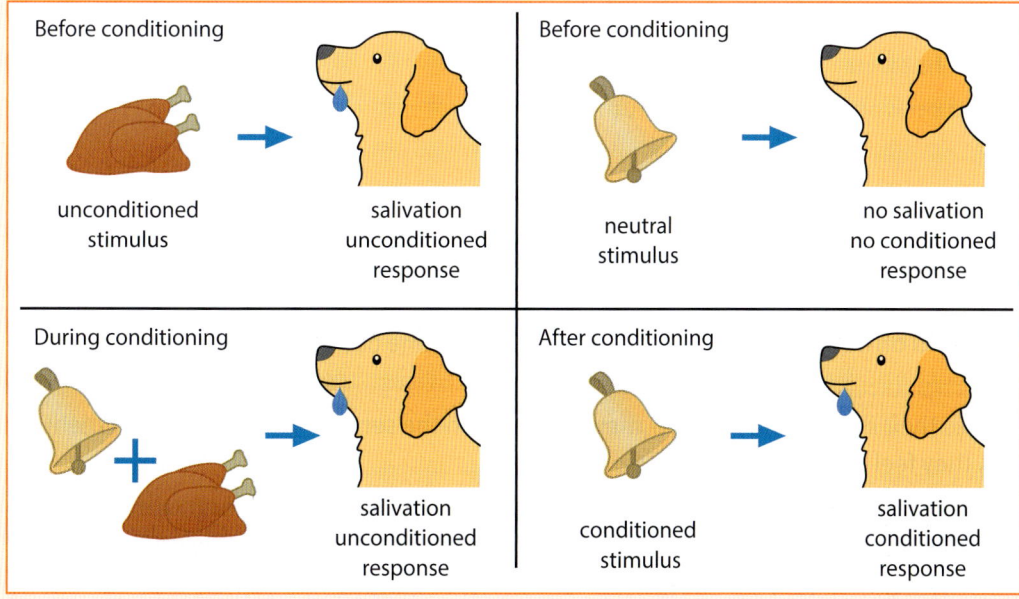

Figure 11.16 A conditioned reflex in dogs

The human sense organs

For any nervous system to work, there must be sensory receptors that respond to stimuli. In the human body, there are many different types of sensory receptors that respond to different stimuli. In every case, sensory receptors change the energy of the stimulus into electrical energy in a nerve impulse. Some of the most important are listed in Table 11.2. A sensory organ is an organ that contains a large number of sensory receptor cells.

Table 11.2 Sensory receptors in the human body

Receptor	Stimulus detected
Eye (retina, provides vision)	light
Ear (cochlea, organ of hearing)	sound
Ear (semi-circular canals, organ of balance)	movement
Tongue (taste buds, enable us to taste)	chemical
Nose (olfactory organ or organ of smell)	chemical
Skin (touch, pressure and pain	movement
Skin (temperature receptors)	heat
Muscles (stretch receptors)	movement (stretching)
Arteries and brain (chemoreceptors responding to pH and carbon dioxide level)	chemical

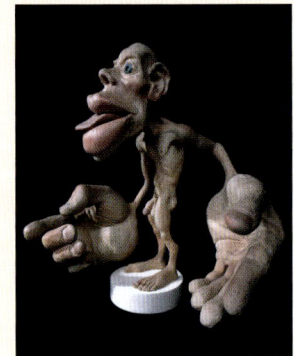

Figure 11.17 Different parts of the body contain different numbers of sensory receptors – this is what we would look like if our body parts reflected how sensitive they are.

Investigating different senses

1 TOUCH

You are going to investigate the sense of touch in different areas of skin.

You will need:
- a fine ballpoint or felt pen
- a bristle mounted on wooden holder or a blunt seeker

Method

1. Work in pairs. Take turns to carry out the investigation.
2. With the pen, draw a grid of 25 squares on the back of your partner's hand. Each square should be 2 mm x 2 mm
3. Draw an identical grid on a sheet of paper and label it with the name of the subject and the area of the body
4. Your partner closes their eyes or looks away – they must rely on the sense of touch alone. Ask them to say YES when they feel a touch.
5. Press the tip of the bristle against the skin in one of the squares until it just bends, or touch the skin with the blunt seeker as gently as possible. Touch each of the squares in turn, marking on the sheet of paper each one that gives a positive response.

Coordination and control

6   Now try other areas of the skin that you might expect to be more or less sensitive, for example, the palm of the hand, the arm, the leg, the foot, and so on.

7   Once you have tested three different areas, swap roles.

8   Are some parts of the skin more sensitive than others? Write up your experiment along with the results and explain your observations as well as you can.

2 TEMPERATURE

Is your sense of temperature absolute – or comparative? In other words, are your temperature receptors working like mini-thermometers or do they measure temperature relative to your body?

You will need:

- three bowls of water – one containing ice-cold water, one with hot water (but not too hot – you have to put your hand in it) and one with water at approximately room temperature.
- a watch, stopwatch or clock

Method

1   Place your left hand in the hot water and your right hand in the cold water for one minute.

2   Once the minute is up, place both hands in the water at room temperature.
    What does each hand feel like? What does this tell you about your sense of temperature? Write up your method, observations and explanations.

The human eye

The human eye is one of our most important sensory organs. Our eyes enable us to see in clear focus, in three dimensions (3D) and in colour. Not many other animals can manage all three.

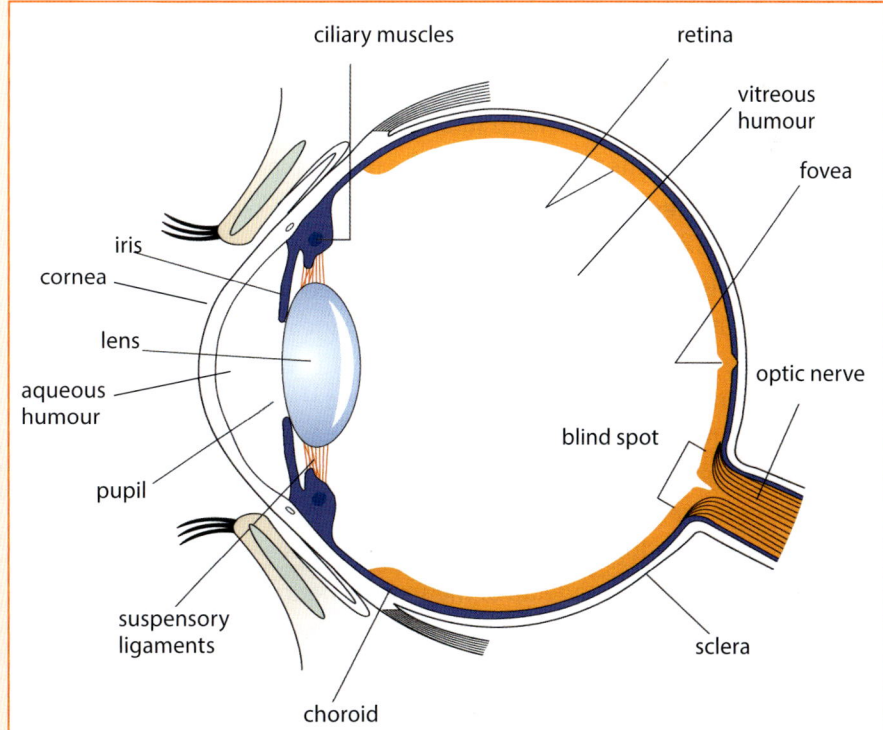

Figure 11.18 This horizontal section shows the main structures of the human eye – a very effective sense organ.

Our eyes are set in sockets in the skull that protect them. The main structures of the eye include the following (see Figure 11.18 on page 239):

- **Eyelids:** These close over our eyes to protect them from the entry of material like dust, sand and insects, which might injure or irritate them. The eyelids also sweep tear solution regularly over the surface of the eyes. This solution contains enzymes that destroy bacteria that might infect the eyes.
- **Sclera:** This is the white outer layer of the eye, very tough and strong so the eyeball is not easy to damage. The rest of the sclera has many blood vessels that supply the retina with food and oxygen.
- **Cornea:** This is the transparent area at the front of the sclera that lets light into the eye. The curved surface of the cornea is also very important for bending the light coming into the eye to make sure it enters the eye and is focused on the light-sensitive cells of the retina.
- **Choroid:** This is the dark layer containing pigmented cells that absorb light and stop it being reflected around the inside of the eye.
- **Aqueous humour:** This is liquid formed in the front part of the eye that helps to bend the light coming into the eye.
- **Vitreous humour:** This jelly-like fluid fills the eyeball and helps it to hold its shape.
- **Pupil:** This is the hole in the centre of the iris that lets light into the eye.
- **Iris:** This is the coloured part of the eye, made up of muscles that contract or relax to control the size of the pupil and so control the amount of light reaching the retina. The circular muscles run around the iris, while the radial muscles run across it like the spokes of a bicycle wheel. When the light is relatively dim, the radial muscles contract and the circular muscles relax, pulling the pupil wide open (it dilates). The change in the size of the pupil in response to light is a reflex action – you don't have to think about it (see Figure 11.14 on page 235).
- **Lens:** This is a clear disc made up largely of proteins that 'fine-tunes' the focusing of the light, bending it to make sure that it produces an image on the retina.
- **Suspensory ligaments:** These hold the lens in place.
- **Ciliary muscles:** These contract and relax, pulling on the suspensory ligaments to change the shape of the lens.
- **Retina:** This layer is made up of all the light-sensitive cells arranged together at the back of the eye. When an image is produced on the retina, the light-sensitive cells are stimulated. There are two types of light sensitive cells. **Rods** are found around the edge of the retina and they are sensitive to dim light and movement. **Cones** only work in bright light and give clear, defined images. They are found in the centre of the retina, especially the **fovea** (see Figure 11.18 on page 239). There are about 100 million light-sensitive cells in each of your retinas.
- **Optic nerve:** This is made up of the sensory neurons that carry impulses to the brain from the cells of the retina.
- **Blind spot:** This is where all the sensory neurons leave the eyeball.
- **Fovea:** This is the area of the retina where the focus is the sharpest – it contains only cone cells.

Figure 11.19 The structure of the human eye allows us to see in clear focus, in 3D and in colour.

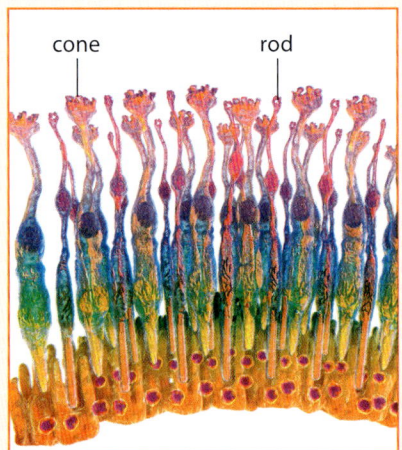

Figure 11.20 The retina contains the light-sensitive nerve cells: rods and cones.

Coordination and control

> **Checkpoint questions**
>
> 10 Define the term 'sense organs'.
> 11 Describe the role of the iris, the ciliary muscles and the retina in your eye.

DID YOU KNOW?

The pigment in the rods that responds to light is based on vitamin A. This is why a lack of vitamin A in your diet causes **night blindness**. You can't make the visual pigment you need to see in low light levels if you are short of vitamin A (see Section B Chapter 5).

Focusing the light

To see clearly, light from an object must be focused on your retina. For this to happen, the light must be bent or **refracted**. Light rays are refracted when they pass from one medium to another, for example, from air into water.

In our eyes, the light coming in is bent (refracted) twice – once as it passes from the air through the cornea and then again as it passes through the lens. As a result of this refraction, the image is focused onto the retina. The image also arrives on the retina upside down. The optical areas of our brain interpret this inverted image so that we are aware of the world the right way up.

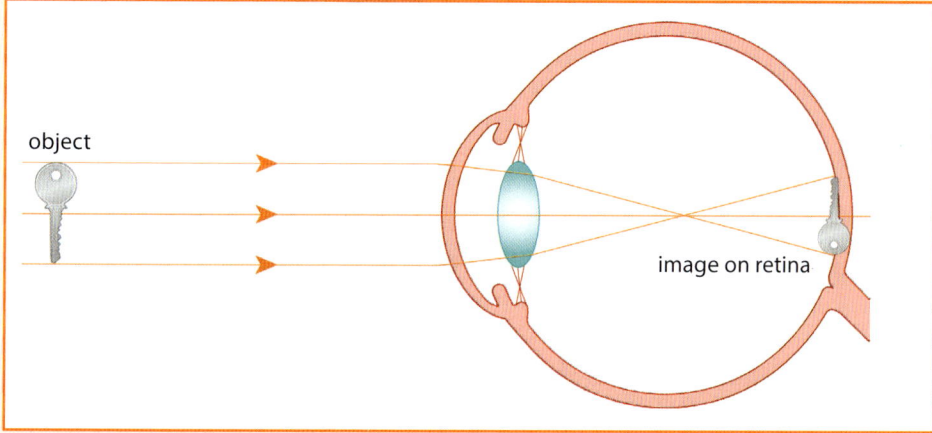

Figure 11.21 When an inverted image falls on the light-sensitive cells of the retina and a message is sent through the optic nerve to the brain, we can see.

Sometimes we look at objects close to us, for example, when we are reading or studying. At other times, we gaze into the distance looking at objects a long way away. The light arriving at our eyes in these circumstances travels differently. The light from a distant object reaching our eyes will travel in almost parallel rays, while the light from close objects will spread out or **diverge** very sharply.

How is the light focused? The cornea bends all the light entering the eyes towards the retina, but it is the lens that makes sure that we can see both close and distant objects equally well. It does this by changing shape. These changes in the shape of the lens are controlled by the contraction and relaxation of the ciliary muscles that surround it. In turn, the ciliary muscles pull, or do not pull, on the suspensory ligaments that hold the lens in place. The ability of the human eye to focus on objects at different distances is called **accommodation**.

The human eye

The process of accommodation works as follows:
- Light from distant objects travels in almost parallel rays that need relatively little bending. The ciliary muscles relax, so they pull on the suspensory ligaments. The ligaments in turn pull on the lens, stretching it relatively thin so the light is focused on the retina.
- Light from close objects is **diverging** and so needs more bending to bring it into focus on the retina. The ciliary muscles contract, so the suspensory muscles go slack and do not pull on the lens of the eye. The lens is fatter and more rounded, so it bends the light coming into the eye more, so that it is in focus on the retina.

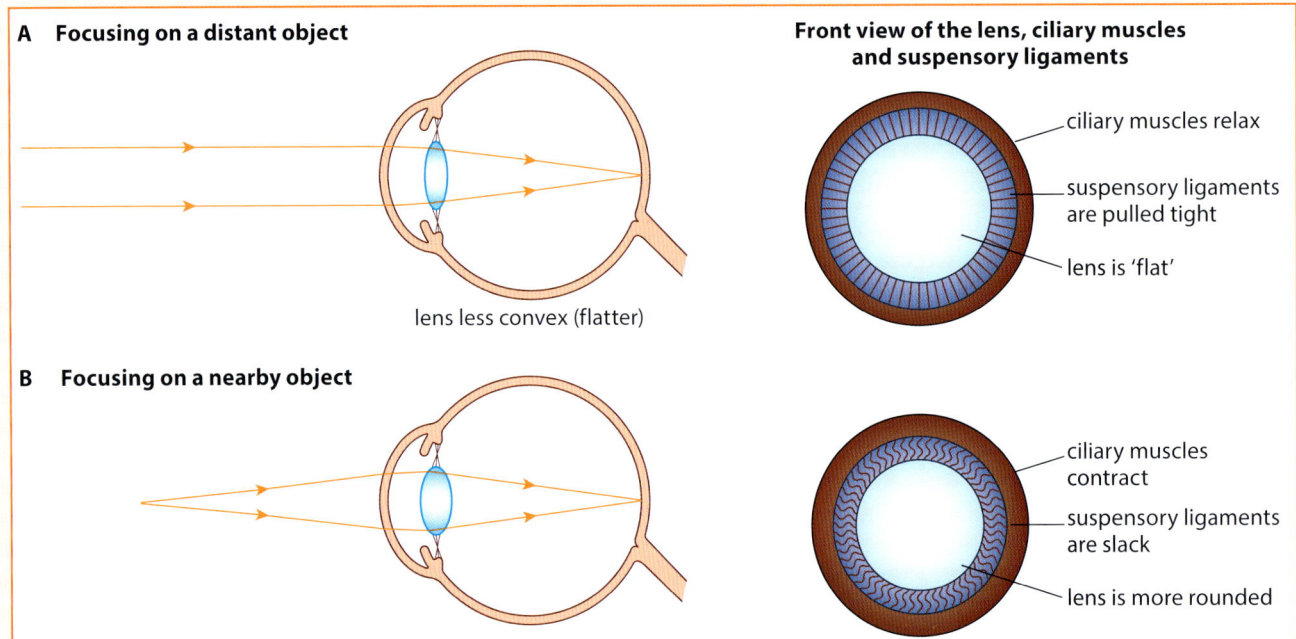

Figure 11.22 Accommodation in the eye is the relaxation and contraction of the ciliary muscles pulling on the suspensory ligaments, which changes the shape of the lens. This ensures both distant and close objects are focused equally clearly on the retina.

Common eye defects

There are a number of common eye defects that affect many people of all ages. The shape and ability of the lens to change shape, along with the size of the eyeball, affect how the eye focuses what we see. Here are some of the most common eye defects and diseases.

Myopia/short-sightedness

A **short-sighted** person focuses clearly on things that are close by, but objects in the distance appear blurred. There are two common causes for this:
- The lens is effectively 'too strong' – it is too curved even when the ciliary muscles are fully relaxed and so the light from distant objects is focused in front of the retina. This makes the image that actually lands on the retina out of focus and blurry.
- The lens is normal, but the eyeball is particularly long. Again, this means light is focused in front of the retina.

Coordination and control

Myopia can be corrected using concave lenses (diverging lenses) that spread the light out more before it gets into the eye. This means that the thicker lens can bring the rays of light into perfect focus on the retina, or there is room in the long eyeball for the light rays to be focused on the correct point (see Figure 11.23).

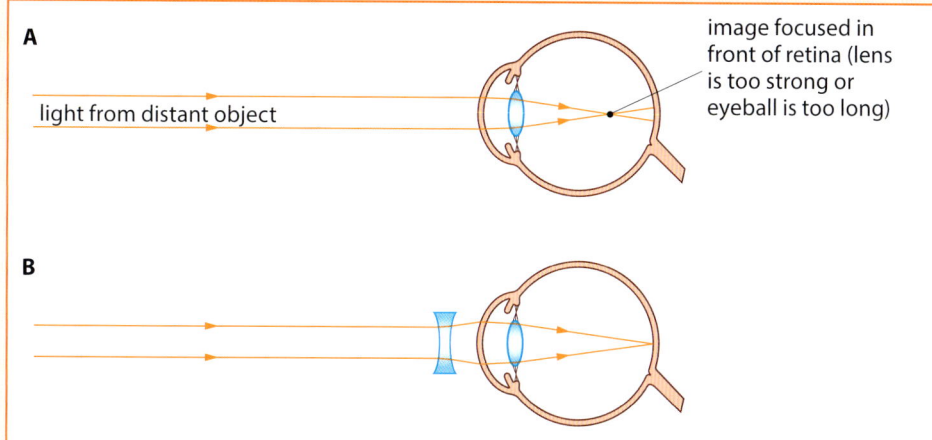

Figure 11.23 The eyes of short-sighted people focus light from distant objects in front of the retina, which makes them difficult to see clearly. A simple concave lens can make all the difference, as you can see clearly here.

Hyperopia/hypermetropia/long-sightedness

A **long-sighted** person can focus clearly on things at a distance, but objects close to them appear blurred. There are two common causes for this:

- The lens is effectively 'too weak' – it is too flat even when the ciliary muscles are fully contracted, and so the light from close objects is focused behind the retina. This means the image that actually lands on the retina is out of focus and blurred.
- The lens is normal, but the eyeball is particularly short. Again, this means light is focused behind the retina.

This problem is corrected using convex lenses (**converging** lenses) that bring the light rays together more before they reach the eye. Now the thinner lens can bring the rays of light into perfect focus on the retina or the short eyeball becomes the right length for the light rays to be focused on the correct point (see Figure 11.24).

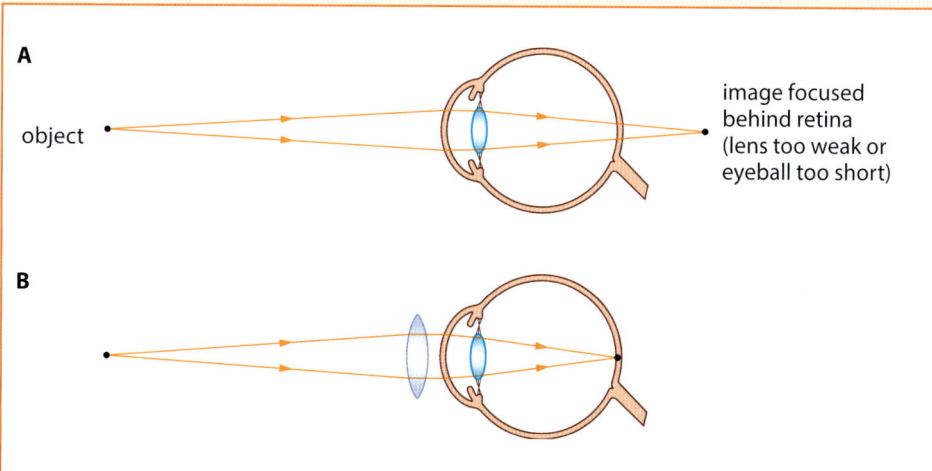

Figure 11.24 The eyes of long-sighted people focus light from close objects behind the retina, which makes them difficult to see clearly and makes tasks such as reading difficult. A simple convex lens can make all the difference, as you can see clearly here.

The human eye

Astigmatism

Astigmatism is another fairly common eye defect. The shape of the eye is irregular – more egg-shaped than round – so the cornea is curved asymmetrically, affecting how light is focused on the retina. In some people, it is the lens rather than the eyeball itself that is an unusual shape, but the end result is the same. Astigmatism can also be corrected by the use of lenses, but the situation is more complex than for long and short sightedness.

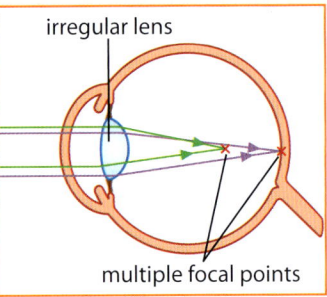

Figure 11.25 The lens in the eye is an usual shape, which causes the astigmatism.

Glaucoma

Glaucoma covers a number of eye conditions where the optic nerve carrying impulses from the eye to the brain is damaged where it leaves the eye. The risk of glaucoma increases as you get older, and it is particularly common in some families. The problem arises because pressure builds up inside the eye, which squeezes and damages the optic nerve. Why does the pressure increase? The eye constantly makes aqueous humour, and in a healthy eye, aqueous humour also drains away constantly through tiny ducts. If these ducts become narrow or blocked, the humour cannot drain away and pressure inside the eyeball rises. The problem is that this usually happens very slowly, with no symptoms, until it is too late and damage to the optic nerve becomes obvious. Without treatment, glaucoma eventually causes complete blindness. If glaucoma is detected early, eye drops or surgery can reduce the pressure in the eyeball and so protect the optic nerve, preventing further damage. This is another reason why regular eye tests are a valuable part of health care.

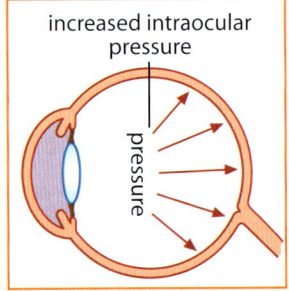

Figure 11.26 Glaucoma causes pressure in the eyeball that damages the optic nerve, eventually causing blindness. Regular eye tests can detect it early so that treatment can prevent further damage.

Cataracts

Cataracts are another potential cause of blindness – but again the good news is that the problem can be solved through relatively simple surgery. A cataract is a clouding of the lens of the eye. The vision becomes blurred, colours become washed out and faded, and eventually vision is lost as less and less light passes through the lens. Age is the most common cause of cataracts, but they are also more likely to develop in people with diabetes, and some children are born with **congenital cataracts**. As diabetes is relatively common in Caribbean countries, and as more people are living longer, cataracts are an increasing problem (look back to Chapter 10 for more about diabetes in the Caribbean). However, once cataracts start affecting your vision and interfering with your daily life, they can be removed and replaced with a clear plastic lens. This enables you to see clearly again, although you may need glasses for reading as the plastic lens is usually set up for distance vision.

Figure 11.27 Cataracts are caused by the lens of the eye becoming cloudy. An operation to replace the lens with a clear plastic one will enable you to see clearly again.

Checkpoint questions

12 Describe how lenses are used to correct sight defects.

13 Discuss the importance of regular eye tests for children and adults.

Hormonal control

As you learnt in Chapter 10, many processes in the body are coordinated by chemicals called **hormones**. Hormones act as chemical messages, produced in one part of the body but having an effect somewhere entirely different. There are two main types of glands in the body:

- **Exocrine glands:** These glands have a special tube or duct that carries the secretion from the gland where it is made to the place where it is needed. Sweat glands, salivary glands and mammary glands are all examples of exocrine glands.

- **Endocrine glands:** These glands produce hormones, and they have no ducts, so they are sometimes called **ductless glands**. They secrete the hormones they make directly into your bloodstream, to be carried all around your body. Most hormones only affect certain tissues or an organ – the **target organ** – and the hormone is picked up from the blood by receptors in the cell membranes of this organ.

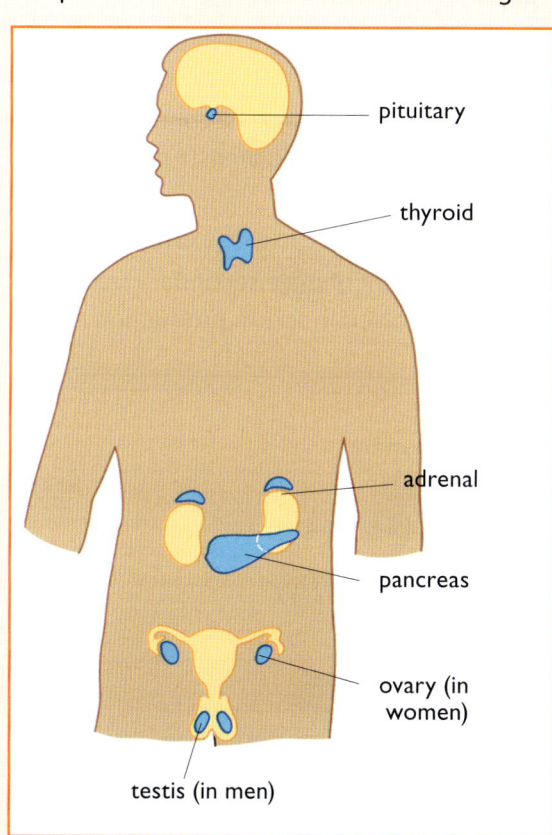

Figure 11.28 These endocrine glands control many things about us, including the way we grow, our sexual development, reproduction, and the sugar and water levels in our blood.

Some hormones, such as insulin, adrenalin and glucagon, act very quickly in the body rather like nervous control systems. Many others, such as thyroxine, the growth hormone and the sex hormones, have a much longer, slower impact on the body, but they are as important for coordination and control. Table 11.3 gives you a summary of the main endocrine glands in the human body, the hormones they produce and the effect that they have in the body. You will be learning much more about the actions of many of these hormones in the rest of this book.

Table 11.3 The main hormones produced in the body, the endocrine organ that produces them and their functions in the body.

Gland	Hormone	Functions of the hormone
Pituitary	Growth hormone	Controls the growth rate of children
	Thyroid-stimulating hormone (TSH)	Stimulates the thyroid gland to secrete thyroxine
	Antidiuretic hormone (ADH)	Controls the water content of the blood by its effect on the kidneys (see Chapter 10)
	Follicle-stimulating hormone (FSH)	Stimulates egg development and oestrogen production in women, and sperm production in men (see Chapter 12)
	Luteinising hormone (LH)	Stimulates egg release in women and testosterone production in men (see Chapter 12)
Thyroid	Thyroxine	Controls the metabolic rate of the body
Pancreas	Insulin	Lowers blood sugar level (see Chapter 10)
	Glucagon	Raises blood sugar level (see Chapter 10)
Adrenal	Adrenaline	Prepares the body for stressful situations – 'fight or flight'
Ovaries	Oestrogen	Controls development of female secondary sexual characteristics; involved in the menstrual cycle (see Chapter 12)
	Progesterone	Involved in the menstrual cycle (see Chapter 12)
Testes	Testosterone	Controls development of male secondary sexual characteristics; involved in sperm production (see Chapter 12)

Both the nervous system and the hormone system are important for coordination and control. They both have features in common, but in many ways they are very different. Table 11.4 compares these two systems.

The combination of nervous and hormonal control enables our body to work as a coordinated whole and plays a vital role in our life.

Table 11.4 Comparison of control features of the nervous and hormone systems

Nervous control	Hormonal control
• Electrical messages travel along neurons. • Chemical messages travel across synapses. • Messages travel fast. • Messages have a rapid effect. • Nerve impulse affects individual cells, for example muscle cells, and so has a very localised effect. • The response is usually short-lived.	• Messages are transported in the blood. • Only chemical messages are involved. • Messages travel more slowly – in minutes rather than milliseconds. • Messages often take longer to have an effect. • Effect is often widespread in the body – and will affect any organ or tissue with the correct receptors. • Effects are often long-lasting.

Coordination and control

Possible SBAs based on coordination and control

→ Investigate the sensitivity of sensory systems in different people/groups of people, for example:
  ⇨ visual field
  ⇨ colour perception
  ⇨ sense of smell
  ⇨ sense of touch for identifying things
  ⇨ sensitivity to pinprick touch.
→ Research synapses, or make a model synapse or an animation of a synapse.
→ Research chemical transmitters in the brain and their role in mental health.
→ Draw vertebrae from different vertebrates and measure/calculate the diameter of the different spinal cords.
→ Find out about the human ear and investigate different hearing ranges.
→ Investigate reaction times, for example:
  ⇨ effect of age
  ⇨ effect of drinking cola or coffee
  ⇨ effect of practising.
→ Investigate the link between the incidence of diabetes and cataracts in different Caribbean countries.
→ Research the regional incidence of different eye defects and rate of correction, and so on.

End-of-chapter summary

In this chapter, you have learnt that:

- living organisms need systems of coordination and control
- many multicellular organisms including human beings have both nervous and hormonal coordination and control systems
- the nervous system is the most rapid, and that nervous control involves:
  stimulus → receptor → co-ordinator → effector → response
- a nerve cell or neuron consists of a cell body, dendrites and an axon
- sensory neurons carry information from the sense organs to the central nervous system (CNS)
- motor neurons carry instructions from the CNS to the effector organs (muscles and glands)
- the central nervous system is the brain and spinal cord; information is assimilated and coordinated in the CNS
- neurons carry electrical impulses known as the action potential
- in any pathway, the junctions between neurons are called synapses; when an impulse arrives in one, neuron chemicals are released in the synapse to trigger an impulse in the next neuron

- a nerve contains many neurons; there are sensory nerves, motor nerves and mixed nerves
- the brain contains grey and white matter; it is the seat of conscious thought and much unconscious action
- the spinal cord carries information from all over the body to and from the brain
- cranial nerves come from the brain, while spinal nerves are from the spinal cord
- reflex actions avoid danger and run ordinary bodily functions – they avoid conscious thought
- reflex actions involve stimulus ➜ receptor ➜ coordinator ➜ effector ➜ response but the coordinator is the relay neuron in the spinal cord and there is no conscious thought involved
- the knee jerk reflex is a common example of a spinal reflex; it is used by doctors to test reflexes and in ordinary life to prevent stumbling
- sense organs detect changes in the internal or external environment
- the human eye includes the sclera, cornea, aqueous humour, vitreous humour, iris, pupil, lens, ciliary muscle, suspensory ligament, retina and optic nerve
- the light-sensitive cells – the rods and the cones – are found in the retina
- the iris controls the amount of light entering the eye
- the cornea and aqueous humour bend the light into the eye
- the lens controls the fine focus of the image onto the retina; the shape of the lens is changed by the contraction or relaxation of the ciliary muscles pulling on the suspensory ligaments
- when the ciliary muscles are relaxed, the suspensory ligaments are stretched and they pull the lens, thinner and flatter to focus light from distant objects on the retina
- short-sightedness, long-sightedness, astigmatism, glaucoma and cataracts are common defects of the eye; short-sightedness, long-sightedness and astigmatism can be corrected using lenses; the risk of cataracts is increased if you develop diabetes
- chemical coordination and control of the body is brought about by hormones secreted by special endocrine glands
- hormones are secreted directly into the blood and are carried around the body in the blood
- hormones may affect a single target organ or a range of organs and tissues; they have their effect through special receptor molecules on the cell membranes of the target organs and tissues
- hormonal control may be rapid, but is often relatively slow and long term
- Important endocrine organs include the pituitary gland, the thyroid gland, the adrenal glands, the pancreas, the ovaries and the testes.

End-of-chapter questions

1. Which one of these statements correctly explains the relationship between neurons and nerves?
   A   A neuron is a bundle of nerves.
   B   Neurons and nerves are the same thing.
   C   Neurons carry messages to the brain and nerves carry messages away from the brain.
   D   A nerve is a bundle of neurons.

2. Which one of the following is not part of a nerve cell?
   A   Cilia
   B   Dendrites
   C   Cell body
   D   Axon

3. What does a nerve impulse use to cross a synapse?
   A   Electricity
   B   Vibration
   C   Chemical transmitters
   D   Light rays

4. Imagine you have been out on the beach looking at some friends in the sea. You walk into the shade of a palm tree and begin to read a book. What changes would take place in your eyes?
   A   Your pupils would constrict, and your lenses would become flatter and less convex.
   B   Your pupils would constrict, and your lenses would become rounder and more convex.
   C   Your pupils would dilate, and your lenses would become flatter and less convex.
   D   Your pupils would dilate, and your lenses would become rounder and more convex.

5. Which part of the human brain is linked to basic balance and orientation?
   A   Cerebellum
   B   Cerebrum
   C   Hypothalamus
   D   Medulla oblongata

6. Which one of the following endocrine glands secretes a hormone that directly affects the metabolic rate of the body?
   A   Pituitary gland
   B   Ovary
   C   Thyroid
   D   Pancreas

7. Which one of these actions is not a reflex?
   A   Blinking
   B   Moving your foot away when you tread on a pin
   C   Driving
   D   A newborn baby gripping your finger

8 The nervous system is made up of a number of parts. Explain the function of each of these parts.
- a   The sense organs (2)
- b   The central nervous system (2)
- c   The sensory nerves (2)
- d   The motor nerves (2)

9
- a   What is the main difference between a voluntary action and a reflex action? (2)
- b   What is the value of reflex actions to the body? (2)
- c   Analyse the following reflex actions using the sequence:
  stimulus ➔ receptor ➔ coordinator ➔ effector ➔ response.
  - i   A doctor hits you just below the kneecap with a rubber hammer. (5)
  - ii You put your bare foot down on a drawing pin. (5)
  - iii Someone claps their hands near your face. (5)

10 The knee jerk reflex is a common example of a reflex. It is used by doctors to test reflexes and in ordinary life to prevent stumbling. A sharp tap just below the knee causes the leg to straighten.

- a   Write a description of what is happening at each of the numbered points 1–5 in the diagram. (5)
- b   This is a reflex action, yet you are aware that your leg has moved. Explain how you know consciously what has happened in a reflex action like this. (2)

11
- a   Describe, using diagrams if possible, how the eye works when you:
  - i   read this question.
  - ii look at a bird flying past the window.
- b   Describe the causes of short-sightedness, long-sightedness and cataracts, and explain how they can be corrected or treated.

12 The pupil of the eye is the hole through which light enters.
- a   Draw diagrams to show what the pupil and iris of the eye would look like in:
  - i   very bright light. (2)
  - ii ordinary light levels. (2)
  - iii very dim light. (2)
- b   Explain the pupil reflex and how the muscles of the iris change the size of the pupil. (6)

13 Here is some information about three eye problems. Use what you know about the way the eye works to explain why these conditions affect sight.
- **a** With cataracts, the lens of the eye goes cloudy or milky. (3)
- **b** Some people have an eyeball that is more egg-shaped than round. They are often short-sighted; they can see objects close to them, but not those at a distance. (3)
- **c** If the retina of the eye becomes detached from the back of the eye, people go completely blind in that eye. (3)
- **d** In glaucoma, the pressure in the eyeball increases, squeezing the optic nerve. (4)

14 Answer these questions. The use of diagrams will gain you marks.
- **a** How do we focus on objects that are near to us? (4)
- **b** How do we focus on distant objects? (4)
- **c** Show how the image of, for example, a tree that is formed on the retina of the eye is actually upside down. (4)
- **d** In an experiment, a scientist asked people to wear a special headset that seemed to turn everything upside down. They found it almost impossible to manage for several days as they saw an upside down world, but eventually they started seeing things the right way up again. This in turn caused problems when they stopped wearing the headset – everything now appeared upside down again. After a day or two, their perception of the world returned to normal. Can you work out what was happening at each stage of this experiment? (8)

15
- **a** Explain why certain actions such as breathing and the squeezing of the gut are maintained by reflex actions, while others such as speaking and putting food in the mouth are under voluntary control. (4)
- **b** Many bodily functions are under reflex control. In which other circumstances are reflex actions very important? Explain why. (3)

16 Write a short essay on the human brain and its role in controlling the body. (10)

17
- **a** What is a hormone? (1)
- **b** Where are the main human hormones produced? Make a table to show the main hormones produced in the body, where they are produced and what they do. (12)
- **c** How is control by hormones different from control by the nervous system? (9)

Chapter 12 The human reproductive system

When you have completed this chapter, you will be able to:

- distinguish between asexual and sexual reproduction
- describe the menstrual cycle in females
- explain the importance of family planning
- describe the structure and function of the human reproductive system
- evaluate different methods of birth control
- outline some of the issues related to abortion
- explain the importance of antenatal and postnatal care
- describe the process of giving birth
- explain the fertilisation of the ovum, the implantation of the embryo and the development of the embryo during pregnancy

What the examiners say

→ Candidates are mostly unable to differentiate between natural and artificial methods of birth control or to relate how artificial methods of birth control prevent pregnancy. Furthermore, candidates are unable to indicate safe and reliable methods of birth control, and tend to reference old wives' tales instead.

→ Candidates are unfamiliar with the stages of pregnancy and often inaccurately state where the baby develops or where fertilisation takes place. Additionally, candidates are unable to explain the changes that occur at the various stages of the menstrual cycle.

→ Generally, candidates do not always follow instructions when labelling diagrams. Additionally, candidates are often unable to accurately label the male and female reproductive system.

→ Candidates are challenged by the definitions of terms such as 'fertilisation' and 'ovulation', are often unaware of the functions of oestrogen, and are unfamiliar with the effects of the underproduction of other hormones such as FSH and LH on the body.

Revision tip

Make a note of things you do not understand and ask your teacher or a knowledgeable peer to explain them. Do not wait until the questions pile up – sort them out straight away!

Humans have babies, cats have kittens, and plants produce seedlings. For any species to survive, the individuals must be able to reproduce themselves. Without reproduction, species die out. In any form of reproduction, genetic information – the blueprint of life found in the nucleus of the cells – must be passed on from parents to offspring. There are two very different types of reproduction, both of which allow genetic information to be passed on from the parents and both of which result in new individuals.

Figure 12.1 All species must reproduce themselves to survive.

Asexual reproduction

Asexual reproduction involves only one parent. There are no special sex cells. Asexual reproduction gives rise to offspring that are genetically identical both to their parent and to each other. They are called **clones**. Many small, single-celled organisms reproduce asexually, and asexual reproduction is also widely found in plants. Asexual reproduction involves a form of cell division called **mitosis**. You will learn more about mitosis, genes and genetics in Section C of this book.

Asexual reproduction has many advantages.
- It is safe (there is no risk to the animal or plant).
- It is certain (there is no problem in finding a mate).
- It often gives rise to very large numbers of offspring.
- Successful genetic combinations can be passed on through the generations with no change. However, this strength is also a weakness. If conditions change – for example, if the climate warms up, a new predator or disease appears, or a competitor for the same food resource arrives – it can be devastating. A change may cause the loss of all the organisms, because if one can't cope with the new conditions, none of the other identical organisms can either.

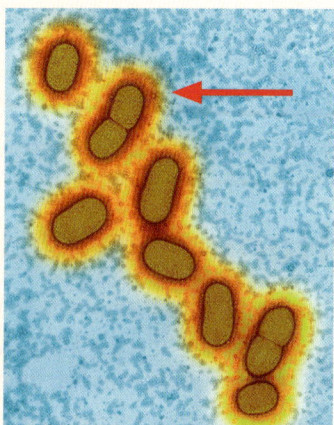

Figure 12.2 In asexual reproduction, there is only one parent. The two bacteria produced as some of these bacteria divide in two will be genetically identical clones.

Sexual reproduction

Sexual reproduction involves the joining of two special sex cells or **gametes**, which are usually, although not always, from two different individuals. Sexual reproduction gives rise to new individuals that have a mix of genetic information from both of their parents, so the offspring are always different from their parents. In animals, including people, the female gametes are called **ova** (singular **ovum**), sometimes referred to as eggs, while the male gametes are called **sperm**. The sex cells are produced by a special form of cell division called **meiosis**, which halves the number of chromosomes in the cells (you will learn more about meiosis in Section C).

Sexual reproduction is considerably more risky than asexual reproduction. The successful meeting of two different sex cells is by no means certain.

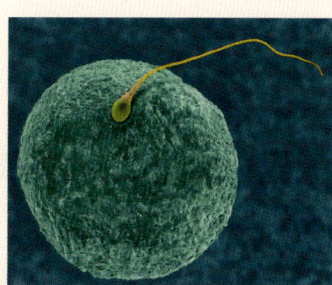

Figure 12.3 In sexual reproduction, there are two parents – the female gamete is fertilised by the male gamete, and the new individual has a mix of genetic material from both parents.

Asexual reproduction

However, the great advantage of sexual reproduction is that **genetic variety** is introduced. This is very valuable in any situation where the conditions are not stable. A change in conditions such as a new disease or a rise in temperature may be lethal for some organisms, but it is likely that others will contain a genetic combination that allows them to survive and take advantage of the new situation.

There are four main stages to sexual reproduction, and we will look at them in the context of human reproduction.

1. The sex cells are formed.
2. The male and female sex cells meet.
3. **Fertilisation** takes place – the male and female gametes must fuse.
4. The new cell (**zygote**), which is formed, develops into a new individual.

Figure 12.4 Flowers are very beautiful – but did you know that they are the sexual organs of the plant?

Checkpoint questions

1. Make a table to compare and contrast asexual and sexual reproduction.

Human reproduction

Human beings produce new human beings by a process called sexual reproduction. We use a process of **internal fertilisation**, and the embryo develops inside the body of the mother in a special organ called the **uterus** (**womb**). The gametes are produced in special sex organs. In the sexually mature male, sperm are produced in the **testes**. In the female, the ova are produced in the **ovaries** as a baby girl develops in her mother's uterus. When the girl becomes sexually mature, some of the ova mature in the ovaries each month, and one is released in each menstrual cycle (see pages 256–257).

The male reproductive system

The primary role of a man in human reproduction is to produce large numbers of gametes and position them in the female's reproductive system. Ideally, however, he should also invest time and care in supporting and bringing up the child once it is born. The structure of the human male reproductive system is closely related to the first two of these functions. The main structures of the male reproductive system are shown in Figure 12.5 and the function of each of the parts is explained in Table 12.1.

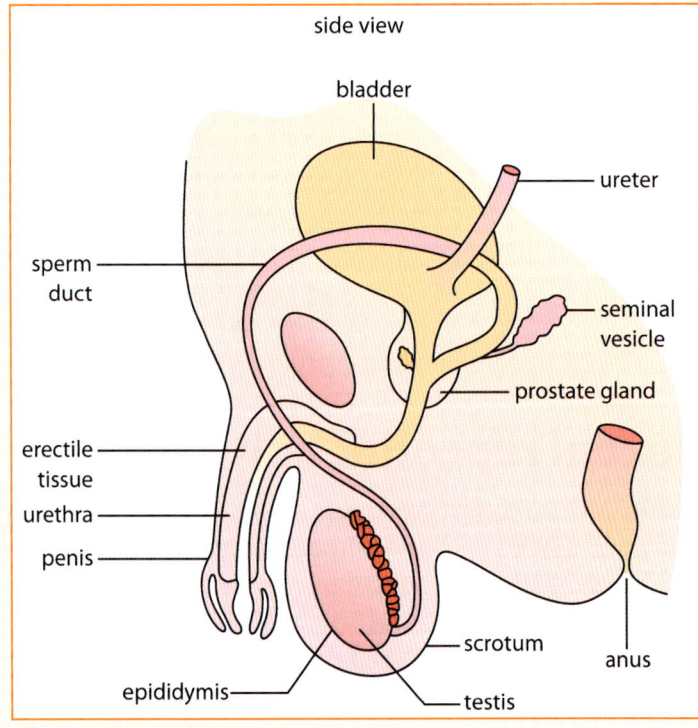

Figure 12.5 The human male reproductive system

Table 12.1 The male reproductive system

Structure	Function
Testes	These are responsible for producing sperm and testosterone.
Scrotum	The sac or pouch of skin that guards the testes. Since it is outside the main body cavity, the testes are usually 1–2 °C lower in temperature than the main body organs. The testes make sperm faster when they are cooler.
Epididymis	This is a long coiled tube that stores and transports sperm from the testes. The sperm matures during this process.
Epididymis	These are a pair of glands at the base of the bladder that secrete a thick fluid. This fluid is rich in sugar and provides sperm with energy.
Prostate gland	This is located at the base of the bladder and secretes a thin fluid that helps the sperm to move.
Erectile tissue	During an erection, the penis is flooded with blood. This makes it larger and firmer, and therefore possible to enter the female reproductive system.
Penis	At the peak of sexual arousal, muscular contractions in the penis cause **semen** to be ejaculated from the end. At other times, it carries urine from the bladder to the outside world.
Urethra	This tube allows semen or urine to be transported out of the body. The sphincter makes it impossible for both semen and urine to pass out of the penis at the same time.
Sperm duct	This is also called the vas deferens. This is a tube that contracts reflexively and moves the sperm from the testes to the urethra.

Sperm must move to the ovum in the female reproductive system. Once they reach the ovum, they must break down the protective layers to get into the egg and fertilise it. They are well adapted for the part they have to play in human reproduction. Figure 12.6 shows the different adaptations.

The female reproductive system

The role of a woman in human reproduction is to produce a relatively small number of large gametes or ova. She also provides the developing embryo with food and oxygen, and removes its waste products. Finally, after delivering the baby into the world, she provides it with a continued supply of food for a time.

The female sex organs are the ovaries, which are two walnut-sized organs found low in the abdomen. They are positioned close to the uterus and the Fallopian tubes, but are not actually attached to them.

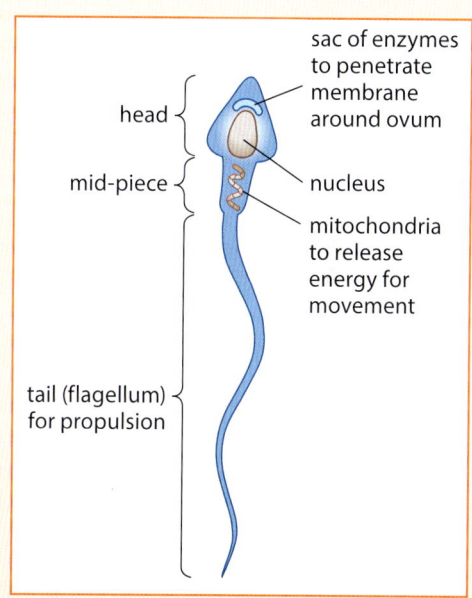

Figure 12.6 The adaptations of sperm to their role of reaching and fertilising the female ovum

The female reproductive system is adapted so that a mature ovum is released each month. It allows an embryo to develop and grow into a baby, which is then delivered. As you look at the structure of the female reproductive system and the way it works, you will see yet another example of an organ system that is well adapted to its function, as well as more hormonal coordination and control. Figure 12.7 shows the human female reproductive system.

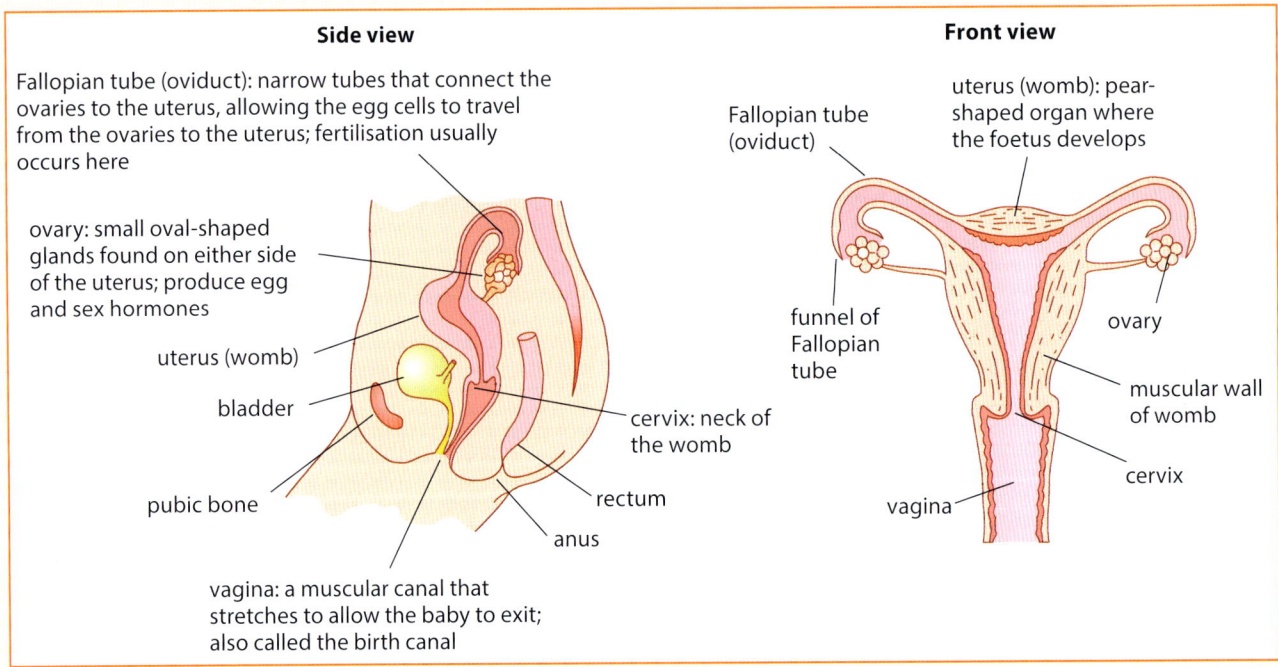

Figure 12.7 The human female reproductive system

The menstrual cycle

Chemical control by hormones is vital in the female reproductive system. Hormones control the whole process of **menstruation** and pregnancy. The menstrual cycle is a sequence of events that takes place approximately every four weeks throughout the fertile life of a woman, from the age of **puberty** to around 50 years of age (see Figure 12.7).

A baby girl has ovaries full of immature ova, but they do nothing until after puberty. Then, once a month, a surge of the hormones starts a few of the ova developing. The uterus builds up a thick, spongy lining with lots of blood vessels ready to support a pregnancy. About 14 days after the ova start ripening, one of them bursts out of its **follicle**. This is called **ovulation**. For a few days after ovulation the uterus lining stays thick and spongy, and stimulates the growth of more blood vessels ready to receive a fertilised ovum. If a pregnancy occurs, the embryo will immediately be richly supplied with food and oxygen. However in most months, the ovum is not fertilised and the woman does not become pregnant.

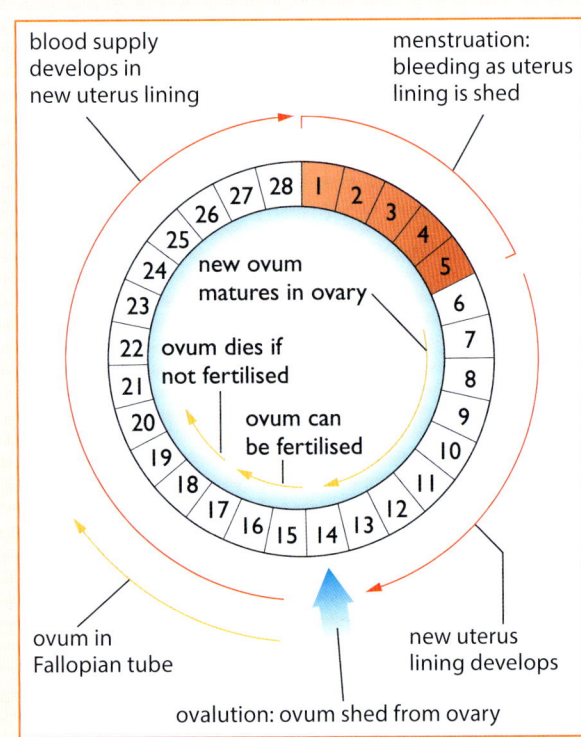

Figure 12.8 The menstrual cycle.

About ten days after ovulation (if no pregnancy has occurred), the blood vessels that are supplying the thick spongy lining of the uterus close down. The lining detaches from the wall of the uterus and is lost through the vagina as the monthly period or bleeding.

However, if the ovum has been fertilised, it will reach the uterus and sink into the thick, spongy lining, attach itself (**implant**) and start to develop. A pregnancy has begun.

The human reproductive system

The hormones of the menstrual cycle

All of the changes in the menstrual cycle are caused by the interactions of four main hormones that affect the female reproductive system. Between them, these four hormones control the menstrual cycle and female fertility. Here is a summary of these hormones, where they are produced and what they do.

These hormones are produced by the pituitary gland in the brain:

- **Follicle stimulating hormone (FSH)** stimulates the development of follicles in the ovary, where a number of ova mature. FSH also stimulates the ovaries to produce hormones, particularly **oestrogen**.
- **Luteinising hormone (LH)** stimulates the release of the egg from the ovary in the middle of the menstrual cycle and also affects the ovary so that it produces the hormone **progesterone** to keep the uterus lining in place.

These hormones are produced by the ovaries:

- **Oestrogen** stimulates the lining of the uterus to build up in preparation for pregnancy. It also affects the pituitary gland. As the oestrogen levels rise, the production of FSH by the pituitary gradually falls, which in turn means that the oestrogen levels fall. The rise in oestrogen levels has the opposite effect on the levels of the other pituitary hormone, LH. As oestrogen rises, the production of LH goes up. When LH reaches its peak in the middle of the menstrual cycle, it stimulates the release of a ripe egg from the ovary. This is when a woman is fertile and may become pregnant.
- **Progesterone** maintains the thickened lining of the uterus and stimulates the growth of blood vessels in the lining to prepare for a pregnancy. If a fertilised ovum arrives in the uterus, progesterone helps to maintain the pregnancy.

By the end of the cycle, when the menstrual bleeding is about to start, all the hormones are at a low level. Figure 12.9 shows the changes in the different hormone levels over 28 days and how they influence the menstrual cycle.

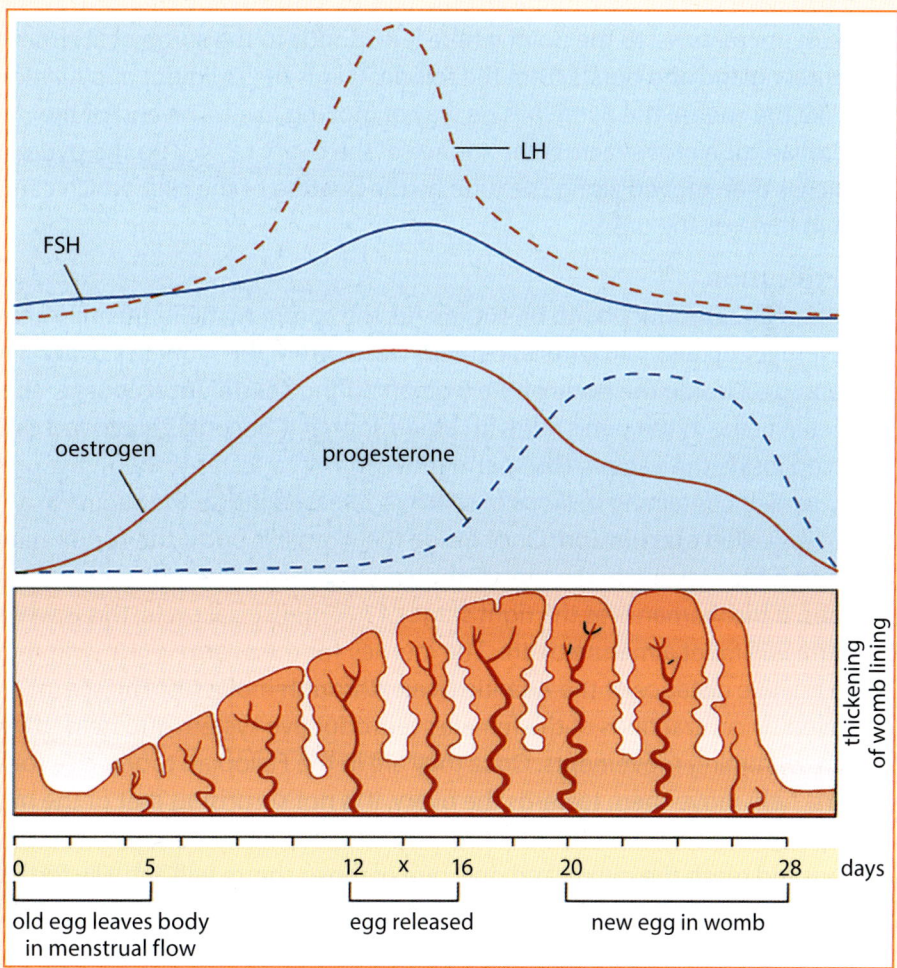

Figure 12.9 How the changes in the levels of different hormones influence the events of the menstrual cycle.

Human reproduction

The ovaries only contain a limited number of ova, so women do not have periods throughout their lives. Eventually, the ova in the ovaries run out. The hormone levels drop, the ovaries and uterus shrink, and the woman stops having periods. She is no longer fertile. This change, which takes place around the age of 50, is called the **menopause**.

Pregnancy

Once the body of a boy or girl has gone through puberty, it is physically prepared for sexual reproduction. However, all around the world, many people wait for adulthood before they have a child. For example, the mean age of childbearing in the Caribbean is 27.5 – this is relatively young compared to Australia and New Zealand where the mean age of child-bearing is 30.8. These countries have a lot of older mothers!

The main steps in a pregnancy are the following:

- Ovulation
- Fertilisation
- Implantation
- Development of the embryo and foetus

Ovulation

As you have seen, each month several follicles in the ovaries mature. Eventually one ovum matures to the point where it responds to the surge of LH from the pituitary gland and bursts from the follicle. This is the moment of ovulation. Unlike the sperm, the ovum has no way of moving itself. The end of the Fallopian tube moves across the surface of the ovary to pick up the ovum, which is then moved along the tube by the beating of the cilia, which carry the ovum towards the uterus.

Fertilisation

For human reproduction to be successful, the sperm made in the man's testes must meet up and join with an ovum released from the woman's ovary. The sperm gets inside the body of the woman during **sexual intercourse**. The erectile tissue in the penis fills with blood so that it becomes erect and can be placed inside the vagina. The sperm move from the testes through the urethra, and semen containing millions of sperm is released inside the vagina in a process called **ejaculation**. Once inside the woman's body, the sperm move through the cervix into the uterus, then through the uterus into the Fallopian tubes. If the woman is at the right stage of her menstrual cycle, this is where the sperm will meet a mature ovum. The journey for the sperm is like swimming the relative distance of the Atlantic Ocean through molasses! The sperm are moved by contractions of the female's reproductive system and helped by their own swimming movements. Once they are in the Fallopian tubes, cilia (beating hairs) help move them towards the ovary. It is not surprising that of the millions of sperm that set off on the journey, only between a few hundred to a few thousand reach the ovum, and only one of those sperm will actually fertilise it. Yet in spite of all the difficulties they face, sperm manage to reach the Fallopian tubes about half an hour after they are released.

> **DID YOU KNOW?**
>
> An average girl loses around 50 cm^3 of blood in each menstrual period, and will have approximately 450 periods during her fertile lifetime. This means she will lose, in total, around 22.5 litres (39.5 pints) of blood before she goes through the menopause.

Figure 12.10 Caribbean mothers, with a mean age of 27.5 years, are younger than those in Australia and New Zealand.

> **DID YOU KNOW?**
>
> Sperm live up to three days inside a woman's body waiting for an ovum. However, once an ovum is released from the ovary, it is fertile for only a few hours – 24 hours at most.

If the ovum and sperm meet in the Fallopian tube at the right time, the sperm cluster around the ovum and try to break through its protective layers. They use the enzymes in the acrosome in their front section (see Chapter 1 and Figure 12.6) to break down these barriers, until finally one sperm penetrates the ovum. This is the moment of **fertilisation**, which in humans is also called **conception**. At this point, the nucleus from the sperm, containing the chromosomes from the father, fuses with the nucleus from the ovum, containing chromosomes from the mother, and a potential new life begins. A tough **fertilisation membrane** forms around the fertilised ovum, which prevents any other sperm from penetrating it. The new cell is now called the **zygote** and it has a unique set of chromosomes. If all goes well, it will develop into a baby. (See Chapters 13 and 14 for more information about the way genetic information is passed from parents to their offspring.)

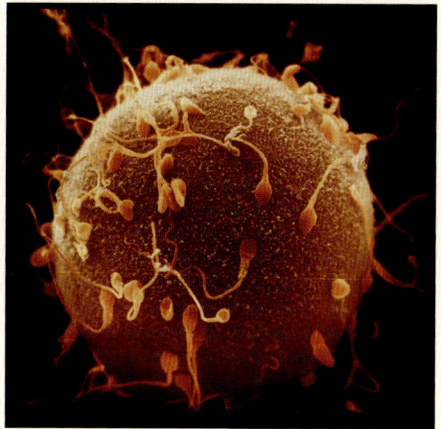

Figure 12.11 The moment of conception – when a sperm penetrates an ovum and the nuclei fuse, a unique new cell is formed.

DID YOU KNOW?

In most menstrual cycles, a single ovum is released, but sometimes two ova are released at the same time. If they are both fertilised by sperm, two babies may develop and the mother will deliver non-identical twins. The babies may be the same or different sexes – they are really normal siblings born at the same time.

More rarely, a fertilised ovum splits in two as the early embryo forms. Again two babies develop – but these are **identical twins**. Since they come from the same fertilized egg they are genetically identical – they are actually human clones!

Figure 12.12 Identical twins are also genetically identical.

When an ovum is fertilised, a complex series of events is set in motion, which in humans will lead to the birth of a fully formed new baby. The cell formed at fertilisation divides rapidly to form a hollow sphere (ball) made up of many small cells. This takes place as the zygote is moved along the Fallopian tube towards the uterus, which in humans takes about one week. By the time it arrives in the uterus, it is called an **embryo**.

Implantation

After several days in the uterus, the tiny embryo must attach itself if the pregnancy is to continue. Finger-like projections grow out and embed themselves into the lining of the uterus, usually at the top end. These eventually form the site of the **placenta**. This is called **implantation**. It is usually completed about 11 or 12 days after fertilisation, and from this time onwards, the woman is considered to be pregnant. The implanted embryo sends out chemical signals so the corpus luteum (the empty follicle in the ovary, now filled with yellow fat and producing pregnancy hormones) keeps producing progesterone. This in turn keeps the lining of the uterus thick and well supplied with blood vessels until the **placenta** takes over.

DID YOU KNOW?

60–70% of fertilised ova never reach the stage of being a baby. In fact, almost 50% of all fertilized eggs are lost before a woman even realizes she is pregnant. This is usually because there is something wrong with the embryo and it cannot develop normally.

Human reproduction

Development of the embryo and foetus

In the development of a human embryo, the major organ systems are completed by around the 12th week of the 40-week pregnancy. By the end of this time, the embryo is a tiny but recognisable human infant, although totally incapable of life outside of the uterus (see Figure 12.13). It is during this time that the embryo is particularly vulnerable to any harmful substances that the mother may take in. Eighty per cent of birth defects are caused by environmental influences that affect the development of the embryo, usually at the early stages when organ systems are forming. Drugs such as thalidomide (which affects the growth of the limbs), nicotine from cigarette smoking, alcohol, diseases such as rubella (German measles, which affects the ears and brain) and ionising radiation such as **X-rays** can all cause irreparable damage in the early stages of pregnancy.

From around 12 weeks onwards, the developing embryo is called a foetus. The remaining time in the uterus is used mainly for the tissues and organs to grow and mature. At this stage, the foetus is much less vulnerable to damage.

The role of the amniotic cavity and amniotic fluid

As the embryo grows, membranes form and fold around it to form the amniotic sac. This is filled with **amniotic fluid** in which the foetus continues to grow. The fluid supports the developing foetus, giving it the freedom to move about easily. It provides a medium in which the growing baby can practise swallowing and breathing movements, and holds the urine that is passed out of its body. The amniotic fluid also cushions and protects the foetus from damage by external injury or impact to the mother. Only a very violent blow would have any effect.

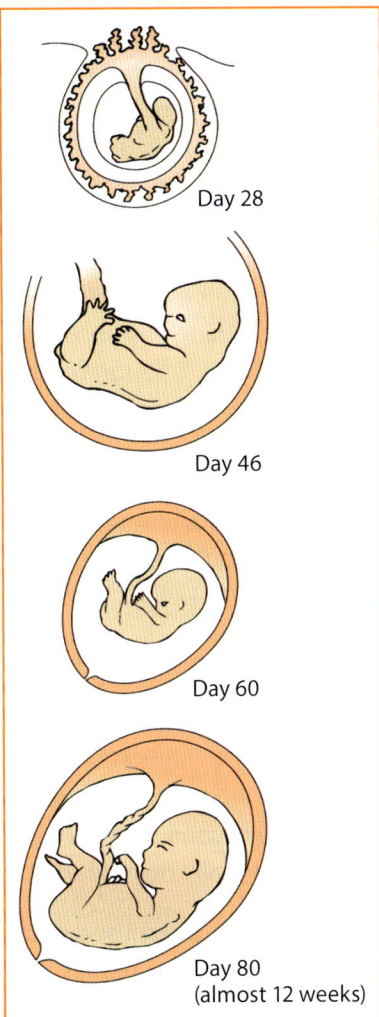

Figure 12.13 By 12 weeks, most of the organs and tissues have formed in the foetus. The rest of the pregnancy is spent growing and maturing.

The role of the placenta

The placenta develops at the same time as the embryo begins to form tissues and organs. The placenta contains blood vessels from both the mother and the embryo. This is where the exchange of food and oxygen takes place, as the mother provides the growing baby with the materials it needs. Carbon dioxide and urea produced by the foetus as it grows are also removed in the placenta and carried away in the mother's blood.

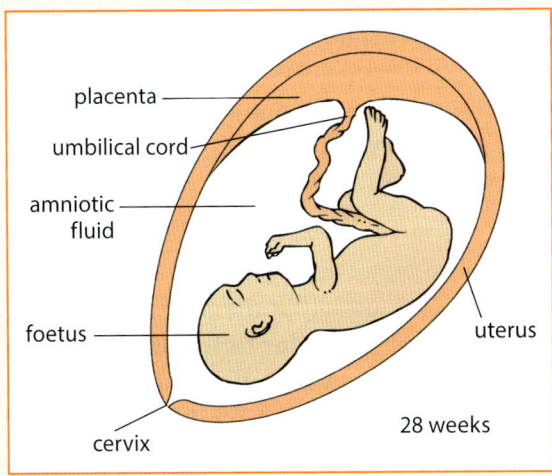

Figure 12.14 The amniotic fluid, contained in the amniotic sac, supports the developing foetus and protects it from harm – at this stage the mother will feel her developing baby as it kicks and stretches.

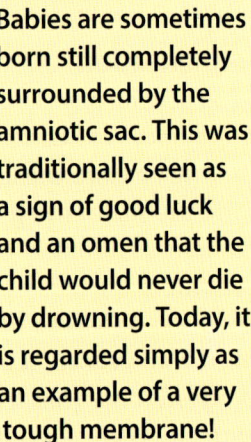

DID YOU KNOW?

Babies are sometimes born still completely surrounded by the amniotic sac. This was traditionally seen as a sign of good luck and an omen that the child would never die by drowning. Today, it is regarded simply as an example of a very tough membrane!

REMEMBER: The blood of the mother and the foetus DO NOT MIX!

Adaptations of the placenta include:
- lots of villi to give a big surface area for diffusion and exchange
- a good blood supply to maintain steep concentration gradients for substances moving both *from* the mother's blood *into* the foetal blood, and *from* the foetal blood into the mother's blood
- short distances between the mother's blood and the foetal blood to make diffusion easier.

The placenta also produces more and more progesterone, taking over from the ovary. It is vital that levels of progesterone stay high to maintain the pregnancy. If the levels fall, the mother may miscarry and lose the developing baby. The placenta also acts as a barrier, protecting the developing foetus from many microorganisms and drugs that cannot cross the placenta from the mother to the foetus. The foetus is connected to the placenta by the umbilical cord.

DID YOU KNOW?

A mature placenta towards the end of the pregnancy is a disc about 20 cm across and 6 cm thick. The surface area available for exchange between the mother and the foetus, made up of millions of microvilli, is about 16 m^2.

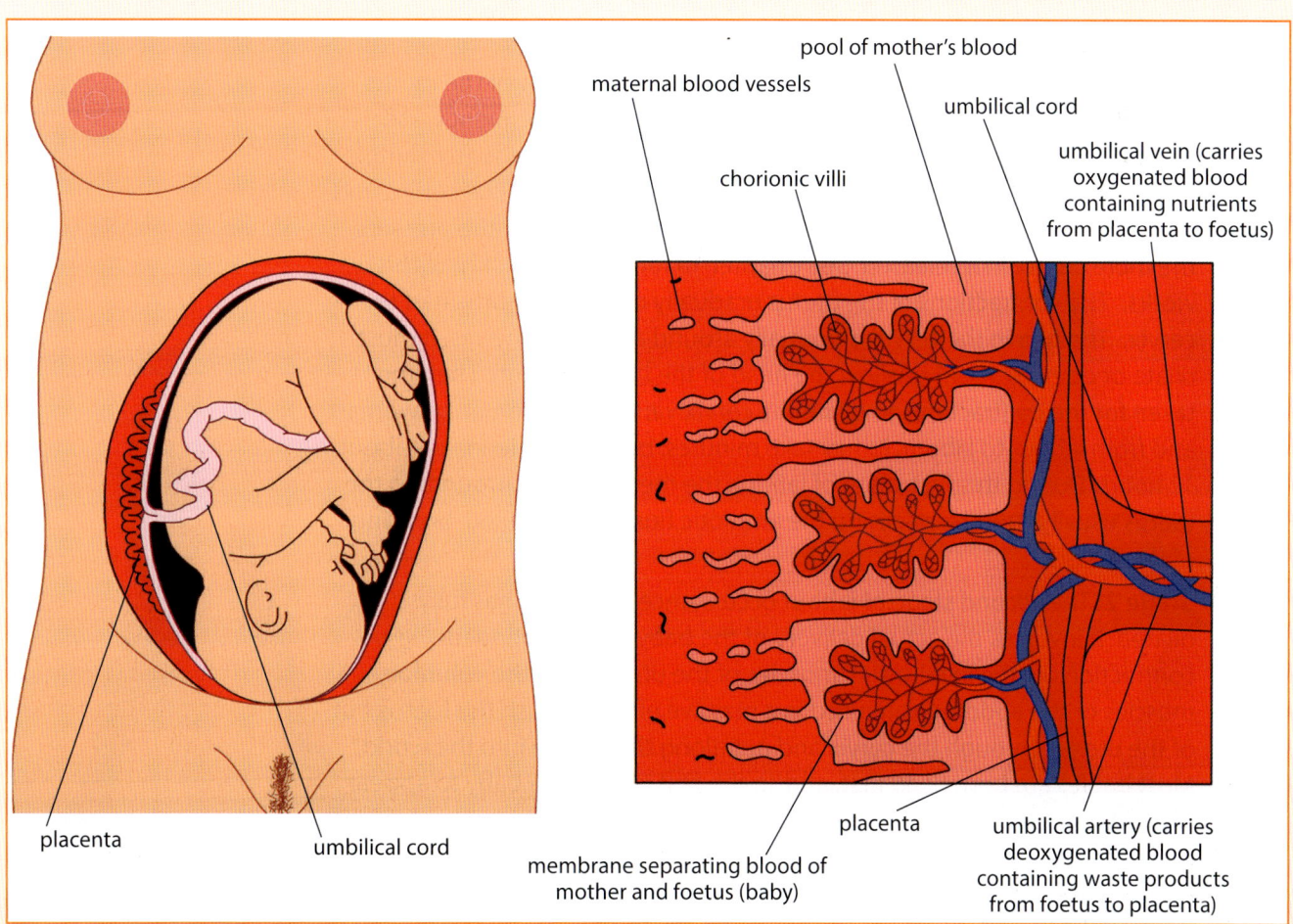

Figure 12.15 By the end of pregnancy, the placenta is supporting a fully grown baby. Adaptations such as the large surface area for diffusion and exchange make this possible.

Human reproduction

The birth process

The most dangerous journey most of us will ever undertake is the short distance from the uterus of our mother to the outside world. It is a passage which, even today, is full of difficulties and dangers for both mother and infant. At the end of pregnancy, which lasts approximately 40 weeks or 9 months, the foetus has outgrown the uterus. The placenta becomes less efficient and there just isn't enough space. It is time for the baby to be delivered. The signal for the delivery to start comes from the foetus rather than the mother. Chemicals produced by the foetus start the process of birth, known as **labour**.

To push a fully developed foetus from the uterus into the outside world is no easy task. Labour involves contraction of the uterine muscles, which are the strongest in the body. It is often lengthy and can be very painful. There are three main stages (see Figure 12.17):

Figure 12.16 At the end of the pregnancy, the foetus produces chemicals to start the delivery process and the mother's labour begins with contraction pains.

- **Stage 1:** During the first stage of labour, the muscular walls of the uterus contract, and the **contractions** steadily increase in both frequency and strength under the influence of the hormone **oxytocin**. This is an example of positive feedback – the head of the foetus pushes against the cervix, and the stretching stimulates the release of oxytocin. The greater the pressure, the more oxytocin is released. See Chapter 10 for the details of how this works. The contractions of the uterus gradually cause the **dilation** of the cervix. The gap in the cervix is normally around a few millimetres. It must dilate or open to 10 cm to allow the head of the baby through. The time taken for the contractions to dilate the cervix varies from woman to woman and from labour to labour. It may take minutes, but most often takes about 12 hours. During this first stage of labour, the membranes surrounding the foetus usually rupture, releasing the amniotic fluid. This is called 'breaking of the waters'.

- **Stage 2:** The second stage of labour occurs once the cervix has fully dilated. It involves the actual delivery of the baby. Both continued contractions of the involuntary muscles of the uterus, and the voluntary muscles of the abdomen are used in a massive effort to push the foetus out of the uterus, through the cervix and out along the vagina into the world. Most babies are born head first, and so the first part of the baby to appear at the entrance to the vagina is the top of the head. This moment, when the parents and healthcare professionals can see that the baby is about to be born, is called 'crowning'. The foetus is then expelled (pushed out) of the mother's body and is born.

- **Stage 3:** The third and final stage of labour is when contractions of the same uterine muscles, still stimulated by oxytocin, are used to expel the now useless placenta and membranes out of the body after they have peeled away from the uterine wall.

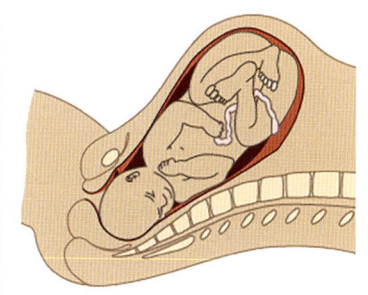 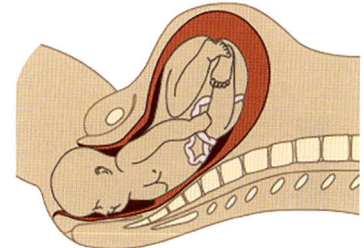 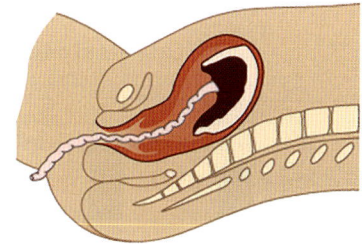

1. Contraction of the uterine muscles; dilation of the cervix.
2. Uterus contracts hard; crowning takes place; the baby is born.
3. Continued contractions expel the placenta and membranes.

Figure 12.17 The three stages of labour that result in the birth of the baby and the removal of the placenta from the mother's body.

As the foetus emerges at the moment of birth and the first breath is drawn into the lungs, a new, independent life begins. Up until now, the mother has done all the work for the baby, providing food, warmth, support and oxygen, and removing all the waste products. Now the baby has to manage these things itself and some very complex changes must occur so it can breathe and feed independently.

Figure 12.18 A newborn baby can now breathe, eat and sleep on its own, but will still need to be cared for by its mother and father in order to survive.

Checkpoint questions

3. Describe the role of the placenta and explain how is it adapted to perform its functions.

Antenatal/prenatal care

When a woman is pregnant, it is very important that she takes good care of her own health and that of her developing baby. **Antenatal care**, also known as **prenatal care**, from qualified health practitioners, such as doctors, nurses and midwives, has a big impact on reducing the problems that can develop during pregnancy and birth. Antenatal means 'before birth'. In the Caribbean, a very high proportion of women now have antenatal care. In some countries, such as Trinidad and Tobago or Barbados, almost all women get good antenatal care. In St Lucia and Jamaica, it is slightly lower – but still almost average for the region and better than many other countries in the world. The improvement in Caribbean antenatal care has played a major role in reducing the numbers of Caribbean mothers and babies who die during or immediately after birth (see Figure 12.19 on page 264).

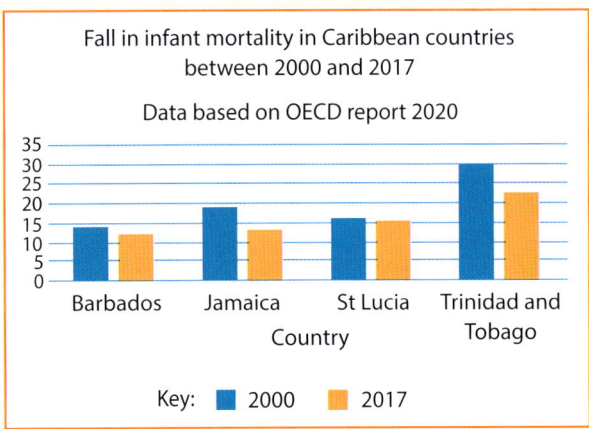

Figure 12.19 The rates of infant mortality have fallen across the Caribbean as antenatal care has improved over the last 20 years.

Adapted from "OECD/The World Bank (2020), Health at a Glance: Latin America and the Caribbean 2020, OECD Publishing, Paris, https://doi.org/10.1787/6089164f-en."

Good antenatal care involves at least four check-ups with a healthcare professional during pregnancy, which will include:

- advice about diet and health during pregnancy
- monitoring weight gain and blood pressure (as raised blood pressure can be dangerous for mother and foetus)
- checking on the growth of the foetus by external measurements of the uterus and by ultrasound scans
- monitoring the blood and urine for signs of problems.

It is very important that women do not smoke when they are pregnant, because this reduces the amount of oxygen they carry in their blood and therefore affects the amount of oxygen received by the developing foetus. Smokers generally have smaller babies and are at a much higher risk of having a stillborn child or a baby that dies in its first week of life (see Chapter 7). During antenatal care, mothers who smoke will be encouraged to give up smoking and be supported in their efforts.

Figure 12.20 Antenatal check-ups are important to monitor the health of the mother and the baby.

Some women develop high blood pressure while they are pregnant (see Chapter 8 to remind yourself about normal and high blood pressure). High blood pressure can put the life of both the mother and the foetus in danger – but careful monitoring through good antenatal care helps avoid these problems.

If a mother drinks alcohol while she is pregnant, she risks having a baby with a low birthweight OR a baby with foetal alcohol syndrome. Ethanol crosses the placenta, and if a developing foetus is exposed to high alcohol levels, it will not develop healthily and its brain will be affected. These changes are permanent. In the same way, many illegal drugs cross the placenta and affect the developing baby. These may cause problems in the development of the foetus, stillbirth (when a baby dies in the uterus) or the baby may be born addicted to the drug used by its mother, and suffer badly as it has the drugs withdrawn.

The human reproductive system

Good antenatal care will help support women to stop drinking or using drugs. It is best if women stop drinking and drug abuse before they become pregnant, but stopping at any stage will help their baby. You will learn more about the effects of alcohol and other drugs in later in the book.

The site where the placenta forms is of great importance to the progress of the pregnancy and will therefore be monitored during antenatal care. If the placenta forms lower down the uterus near the cervix (placenta praevia), then problems can arise at birth. Without medical intervention, the placenta would be delivered before the baby, which would then die from lack of oxygen. The mother would also be at risk from loss of blood from the site of the placenta. Ultrasound scans in pregnancy now pick up this condition so the pregnancy can be monitored and a Caesarean section carried out to avoid problems at birth.

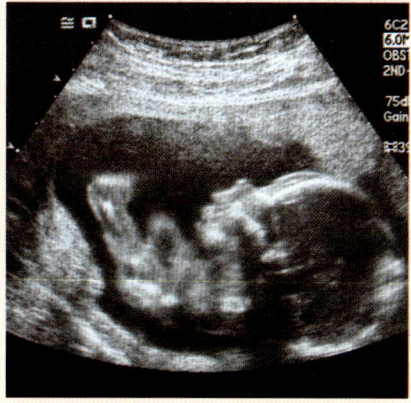

Figure 12.21 An ultrasound enables the healthcare professional to identify any risks with the pregnancy and avoid any problems at birth.

Placenta previa is just one condition that may mean a woman needs a Caesarean section to deliver her baby. Other complications, such as the shape of the woman's pelvis, the size of the foetus or the foetus becoming distressed and needing a fast delivery, are all complications that need a Caesarean section.

Postnatal care

Once the baby has arrived, successful **postnatal care** is vital. Postnatal means 'after birth'. A newborn baby is small and helpless. In fact, without lots of care from at least one and preferably both parents, human babies cannot survive. A baby needs to be kept warm and clean. It needs to be cuddled and loved, because human contact is very important for the proper development of a baby.

Feeding baby and feeding mother

Figure 12.22 The safe arrival of a healthy, full-term baby is the hoped-for outcome of every planned pregnancy.

Perhaps the most important aspect of caring for a new baby is to feed it successfully. Human beings, like all mammals, have evolved a method of providing the ideal food for their offspring. The mammary glands or breasts produce milk that is sterile and at the right temperature for babies to feed on whenever they need to.

Breastfeeding provides infants with the ideal food – a balance of fats, proteins, carbohydrates, minerals and vitamins, which exactly matches the needs of a growing human baby (see Chapter 5). The balance of the different components varies with the stage of the feed, the weather and the age of the baby to ensure that all the needs of the infant are met. For example, in hot weather, the milk is more dilute to make sure the baby doesn't dehydrate. Until the age of about six months, no additional feeding is necessary.

As well as providing a perfectly balanced diet, the major advantage of breast milk is that it provides a baby with antibodies, which gives it immunity to the same diseases as its mother has immunity to. To produce milk successfully, a mother needs to be well-fed herself, and to be well-rested so her body has the spare resources to make milk for her growing baby. Good postnatal care teams look after mother and baby, making sure both have a proper diet.

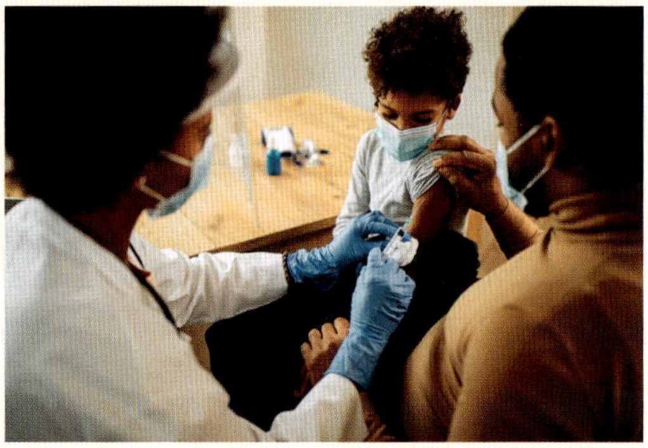

Medical visits

New mothers and their babies need regular visits from healthcare professionals to check that all is well and the baby is gaining weight. As the baby grows, mothers should take their growing babies regularly to medical clinics for check-ups to make sure they are making good progress and hitting their developmental milestones such as sitting, walking and talking. An important part of medical visits is to make sure the baby is up to date with **immunisation**. Vaccines against many deadly diseases protect children and save many hundreds of thousands of lives every year. Parents and healthcare teams in the Caribbean working together have eliminated polio in the area, measles and rubella are almost never seen, and other diseases are at all-time low levels. This has been a major success for our countries. You will learn more about **vaccination** in Chapter 16.

Figure 12.23 Medical checkups and vaccines start with new-born babies but carry on throughout a healthy childhood.

Birth control

In the 21st century, we have many methods of birth control (or **contraception**) that allow us to control our own fertility. Effective birth control lets us choose how many children we have, and when we have them. Most methods of birth control prevent conception while sexual activity continues whenever the couple desire it. The effectiveness of birth control methods is measured per '100 woman years' – in other words, if a hundred women used a method of birth control for a year, how many of them would end up pregnant?

Table 12.2 Controlling fertility

Method	How it works	Advantages	Disadvantages	Effectiveness
Natural methods	Natural methods of contraception are based on understanding the menstrual cycle and accurately predicting the moment of ovulation. Ovulation can be detected by the increase in temperature associated with it, by changes in the vaginal mucus or using a monitor that detects changes in the chemicals in the urine around ovulation. If sexual intercourse is avoided around the fertile time, pregnancy can be avoided.	There are no side effects and this method is permitted by, for example, the Catholic Church. Carried out with care and using scientific precision with recording techniques, it can be very effective.	It depends on full cooperation of both partners and it is not always easy to pinpoint ovulation, so pregnancy can result.	10 pregnancies per 100 woman years

The human reproductive system

Method	How it works	Advantages	Disadvantages	Effectiveness
Physical or barrier methods	These methods of contraception involve physical barriers that prevent the meeting of the ovum and the spermatozoa. A condom is a thin latex sheath that is placed over the penis during intercourse to collect the semen and so prevent the ovum and sperm from meeting. There is also a female condom for women to insert in the vagina before sexual intercourse.	No side effects; no medical advice needed; offers some protection against sexually transmitted infections such as syphilis and AIDS. The female condom gives women some control over contraception and exposure to sexually transmitted infections.	Can interrupt intercourse; sheath may tear or get damaged during intercourse allowing semen to get through; gives better protection when combined with a **spermicide** (a cream or gel that kills sperm).	Male condom – 2.5 pregnancies per 100 woman years. Female condom – 5 pregnancies per 100 woman years when used correctly.
	The diaphragm or cap is a thin rubber diaphragm that is inserted into the vagina before intercourse to cover the cervix and prevent the entry of sperm.	No side effects; offers some protection against cervical cancer.	Must be initially fitted by a doctor; may be incorrectly positioned or damaged and allow sperm past; gives better protection when combined with a spermicide.	
Hormonal methods	Hormonal methods use variations of the body's natural hormones to prevent conception. The pill is one of the most reliable methods of contraception. The mixed pill contains the female hormones oestrogen and progesterone. The raised level of oestrogen in the blood slows the production of FSH. Without rising FSH levels, no follicles develop in the ovary and no eggs mature to be released. Without mature ova there can be no pregnancy. The progesterone causes the mucus in the vagina and cervix to mimic that a pregnancy has occurred to stop more sperm getting in. The progesterone-only pill (mini-pill) doesn't contain oestrogen so it does not inhibit ovulation and needs to be taken at very precise time intervals to be effective.	The combined pill in particular is very effective at preventing pregnancy. The pill is taken at regular daily intervals and so does not interfere with intercourse. It may offer some protection against certain tumours.	The pill may increase the risk of certain tumours. It can cause raised blood pressure and an increased tendency for the blood to clot. The pill must be taken regularly. If the artificial hormonal level drops, the body's own hormones can take over and an egg can be released unexpectedly.	0.5 pregnancies per 100 woman years (due to human error in taking the pill)

Birth control

Method	How it works	Advantages	Disadvantages	Effectiveness
Sterilisation or surgical contraception	This is the ultimate form of contraception. By cutting or tying the tubes along which egg or sperm travel, conception is made almost impossible. This has the additional benefit of removing the human element associated with contraception, which is the major cause of failure in the other methods. In men, the sperm ducts (vas deferens) are cut and tied, preventing sperm from getting into the semen – this is called a vasectomy. In women, the Fallopian tubes are cut and/or tied to prevent the ovum reaching the uterus or the sperm reaching the ovum – this is called sterilisation or tubal ligation.	Almost 100% guaranteed to prevent pregnancy; permanent control of fertility; removes the problem of human error in contraception.	For women in particular, it involves a general anaesthetic; not easily reversible.	0.05 pregnancies per 100 woman years
Intrauterine devices (IUD)	This method does not prevent conception. The ovum and the sperm may meet, but the IUD interferes with and prevents the implantation of the early embryo. An IUD is a device made of plastic and a metal, frequently copper, which is inserted into the uterus by a doctor and remains there all the time.	Once inserted, no further steps need to be taken; relatively effective at preventing implantation and pregnancy.	Can cause pain and heavy periods; can cause uterine infections that may lead to **infertility**; if pregnancy does occur, it has a high chance of being in the Fallopian tubes (ectopic pregnancy).	2.5 pregnancies per 100 woman years

Effective family planning using a reliable method of birth control has a big impact on individuals and on society. If every child is a wanted child, born when the parents are feeling ready both financially and emotionally to support a baby, then everyone benefits. Couples who have children when they want them can enjoy those children. Having the number of children you choose, and spacing out the births to give the mother time to recover from one pregnancy before she is pregnant again, makes for healthier women, healthier children, less poverty and generally happier families. This in turn means society functions more efficiently. Less money needs to be spent on social services to support children who are unwanted or uncared for by their families, or on dealing with the health problems that result from too many pregnancies too close together. This frees up money for other things. Good family planning has a positive impact on everyone involved.

In the Caribbean, people often tend to start their families rather younger than in some parts of the world and teenage pregnancies are quite common. According to a 2018 PAHO report, the world teenage pregnancy average is 46 births per 1000 girls. For the Caribbean, this is estimated to be 66.5 per 1000 girls aged 15–19. This is on par with Latin America but lower than Africa. (Source: www.paho.org © Pan American Health Organization) Early pregnancies can lead to problems for both the young mother and the baby. Although the girl is fertile, her body has not finished growing and so may not be able to support pregnancy and birth very well. When you find out all that goes on during pregnancy and birth, you will be able to understand the stresses it places on the body of the mother. You will see why it is not ideal when the mother has growth demands of her own at the same time.

On the other hand, in some countries people are leaving it until their 30s to have a family, and this brings its own difficulties because people become less fertile as they get older. There is also a higher risk of the babies having genetic problems (see Chapter 14).

Using birth control wisely enables couples to have their families when they want them, at the time of life that suits them both.

Figure 12.24 Birth control allows couples to have their families when the time is right for them.

Checkpoint questions

2 Discuss the advantages and disadvantages of:
   a having a baby in your teens
   b having a baby in your 30s.
3 Summarise the advantages of disadvantages of three different methods of birth control.

Abortion: The choice

One of the issues that cannot be avoided when talking about methods of controlling family size, and which is also very important when looking at some of the issues linked to genetic diseases (see Chapter 14), is that of **abortion**.

Spontaneous abortion or miscarriage

Most abortions are spontaneous and natural, when for some reason an embryo or foetus does not continue to develop but dies and is lost. Natural abortion or **miscarriage** is very common in the first three months of pregnancy. Most women will experience at least one early miscarriage during their child-bearing years. Some of these are so early that a woman will not even have realised that she is pregnant; her period may simply be a bit late. In many cases, the embryo or foetus has some sort of genetic problem, but sometimes there is no apparent reason for the loss. A spontaneous abortion is often very distressing for both the mother and the father – the lost pregnancy was full of hopes and dreams. Fortunately, most people will eventually have another, successful pregnancy with a healthy baby at the end of it.

Medical abortion

There are circumstances where an apparently healthy pregnancy will be deliberately terminated (ended) in a medical abortion. Around the world, these are usually carried out when the woman is pregnant as a result of rape or incest, when there is a risk to either the physical or mental health of the mother, or if the foetus is severely handicapped in some way. Depending on the attitude of both the country and the doctors, the interpretation of these criteria varies greatly.

Most medical abortions are carried out in the first three months of pregnancy, when the embryo is still very small and a long way from being capable of independent life. The lining of the uterus, including the embryo, is removed either by vacuum extraction (where the contents of the uterus are removed by suction) or by scraping away the uterine lining (dilation and curettage). However, in many countries, abortion is legal up until 20 weeks or even later, but usually only when serious genetic handicaps are picked up by testing late in the pregnancy.

No one wants to have an abortion, but an unwanted or unhealthy pregnancy causes incredible suffering and sometimes an abortion is the best option for everyone.

The ethics of abortion

Abortion raises strong feelings in many people.

On one side are people who believe that women have a right to decide if and when they wish to have a baby, and that every child born has the right to be a wanted child. While everyone would prefer it if abortions were not necessary, these people feel that if a woman is pregnant and does not wish to be, if she has been raped or abused, or if the foetus has a genetic disorder or some other developmental problem, then a termination of that pregnancy is the right course to take.

Other people feel equally strongly that it is never right to take a human life, even that of an embryo that is only a few weeks old. They feel that the rights of the unborn child should be equal to or take priority over those of the parents and any other siblings who might already be in the family, and that abortion is always wrong.

These are issues of great sensitivity that need careful consideration, and each individual needs to reach their own conclusion. However, for individuals to be able to make a choice, abortion needs to be legal. When abortion is not legal, many women will still wish to terminate pregnancies that they do not want. In desperation, they may turn to illegal abortion. Often illegal abortion is also unsafe abortion. A legal early abortion carried out using medication or under sterile clinic conditions by a skilled doctor puts the mother at little or no risk – in fact less risk than carrying a baby to full term and giving birth. However, an unsafe abortion, carried out in unhygienic conditions by an unskilled practitioner, carries many health risks for the mother. These range from infection and future infertility to heavy bleeding and death. Many women in the Caribbean and other regions die every year as a result of unsafe abortions.

As the use of effective birth control continues to increase across the Caribbean, unwanted pregnancies should fall and with them the need for many abortions. Positive support for family planning is common in many Caribbean countries, and this is having a big impact on the numbers of abortions, and the numbers of women dying after illegal abortions.

Problems with the reproductive system

Infertility

Often, the first indication that there are problems with the reproductive system is when a couple decide they want to have a baby and the woman simply can't get pregnant. In this situation, a number of different tests will be carried out on both partners to try to find out the cause of the infertility, because different causes need different solutions. Worldwide, one couple in six have problems with fertility, and the problem is growing.

To find out why a woman is not conceiving, doctors look for both physical and chemical causes. One of the first checks will be to see if ovulation is occurring; no egg means no baby. There can be several explanations for a lack of ovulation. Sometimes the situation is very simple – there are no eggs in the ovaries. However, this is relatively rare and is the cause of infertility in only 1–2% of the women who have difficulty conceiving. If there are no eggs, then the woman will never conceive naturally. **IVF** (*in vitro* fertilisation) is the only chance of motherhood.

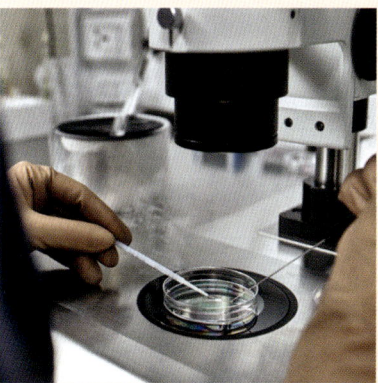

Figure 12.25 The IVF process involves mixing ova and sperm in a laboratory.

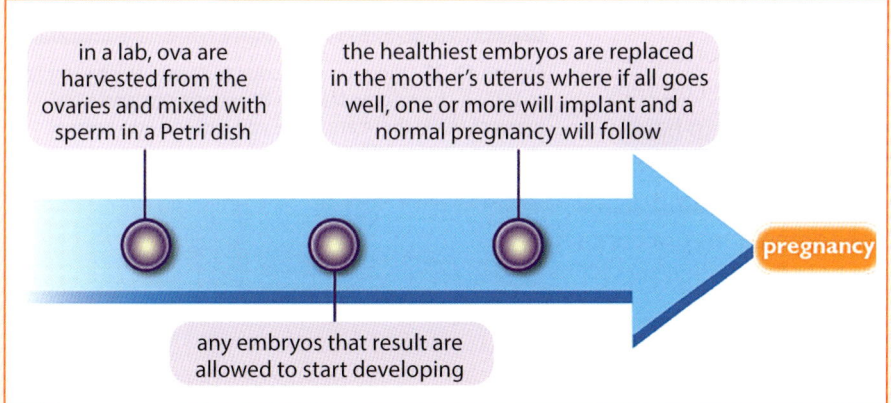

Figure 12.26 The process of IVF

If a woman has no ova, an egg donor is required. Some women are prepared to act as donors. They allow eggs to be collected from their ovaries and given to other, infertile women. A donated egg can be fertilised by sperm from the woman's partner and placed in the woman's body to develop.

However, in many women who do not ovulate, the cause is easier to deal with. Some women don't make enough FSH to stimulate the release of the ripe eggs from the ovary and others don't make any at all. By measuring the levels of different hormones in the blood, doctors can build up a fairly good picture of what is going on and whether ovulation is taking place. Synthetic hormones can be used to replace natural FSH, bringing about ovulation and so, hopefully, pregnancy.

One of the most common female physical problems preventing pregnancy is when the Fallopian tubes are twisted, scarred or blocked in some way. This prevents the sperm from meeting the egg, and even more importantly, stops the egg, fertilised or not, from travelling along to the uterus. Thirty per cent of female infertility is the result of damaged Fallopian tubes. Sexually transmitted infections (see Chapter 19 for more information about these) can often, but not always, be linked to damage in the Fallopian tubes.

The Fallopian tubes are inside the body, hidden to human eyes. About 11 cm long, they lead from the ovary to the uterus. Damage is revealed during an investigation called a **laparoscopy**.

During a laparoscopy, a fine telescope is inserted into the abdominal cavity, while the woman is under general anaesthetic, to look at the Fallopian tubes from the outside.

In the case of fertility of the male, there are two crucial factors: Is the man producing sperm in his semen, and are they normal, healthy and active? The answer to both these questions comes from a careful examination of a man's semen.

In normal, healthy semen, there should be 20 million sperm per 1 cm^3 of semen. Once the sperm count falls below this level, it begins to affect fertility. If the count is really low, things are more difficult.

The overall sperm count is important, but the ability of sperm to fertilise eggs successfully depends on more than numbers. The mobility of the sperm is very important too. In other words, how active are they and how well do they swim? For the man to be fertile, his sperm need to have actively lashing tails and around 50% of them must swim forward in straight lines rather than round and round in circles. It is also important that the semen does not contain too many abnormal sperm.

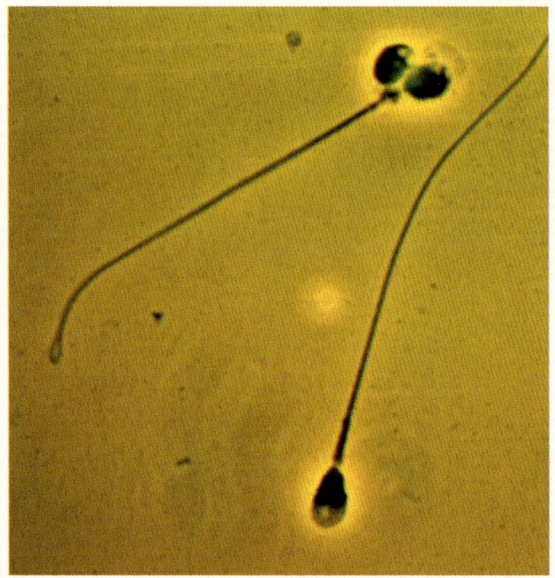

Figure 12.27 Using a microscope to view a fresh sample of semen shows how healthy the sperm are, and how many of them there are. Here you can see one normal sperm and one with two heads.

Cancers of the reproductive system

Infertility isn't the only problem that can affect the reproductive system. A number of common cancers can also affect certain parts. **Cancer** is one of the most common causes of death in the developed world. A cancer is a mass of abnormally growing cells (a tumour) and the cancer cells do not respond to the normal mechanisms that control the way the cells grow. They divide and invade the surrounding tissues. The tumours may split up, releasing small clumps of malignant cells into the bloodstream to be carried to other areas of the body where they start to divide again and form a secondary tumour. Cancer can be fatal as it interferes with the functioning of vital organs and even modern medical treatments cannot always guarantee a cure.

The human reproductive system

The biologist's toolkit: Research: reliable sources

When you research material to answer a question or for an assignment, how do you decide if the information you find is scientifically reliable? Social media is not the best place to go – that is often a source of opinion and not scientific fact.

Here are some pointers to help you find reliable sources:

- Look for international reputations, for example, the World Health Organisation, the Worldwide Fund for Nature, the Pan-Caribbean health and science organisations.
- Where was a study is carried out? If it is a well-known and respected institution such as the University of the West Indies (UWI), or Cambridge UK, or MIT USA, it is likely to be reliable. If it is a small organisation you have never heard of, then think carefully.
- Who funded the study? For example, cigarette manufacturers hid the risk of cigarette smoking for years and published evidence only showing benefits. So follow the money! Research funded by a neutral body is usually reliable.
- How big was the study and how long did it last? Generally, larger studies over long periods of time produce more reliable data than short studies using few people (like those used in cosmetic ads, for example).

In women, **cervical cancer** is a common problem which, if caught early, can easily be cured. There is a simple screening test for cancer of the cervix (the Pap smear test) but it is not easily available everywhere in the Caribbean and cancer of the cervix continues to be a big killer of women. Far too many women die of cervical cancer in the Caribbean and other regions every year.

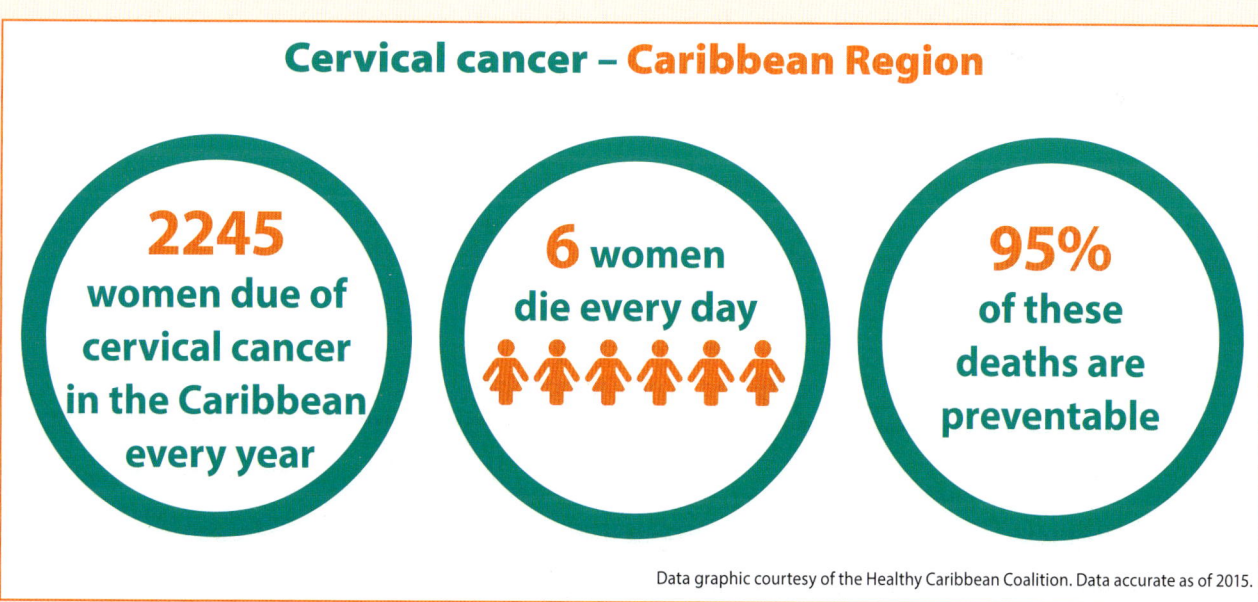

Figure 12.28 This data shows the impact of cervical cancers across all Caribbean countries. A total of 2245 women die of cervical cancer. Source: www.healthycaribbean.org/aus-dap-funding-the-cervical-cancer-prevention-initiative/

Problems with the reproductive system

95% of those deaths from cervical cancer could be prevented by a simple vaccine. Scientists have discovered that almost all cases of cervical cancer are the result of infection by the human papillomavirus (HPV). They have developed a vaccine against this virus that is given to teenagers before they are sexually active. The Caribbean countries are working together to try to eliminate this terrible disease.

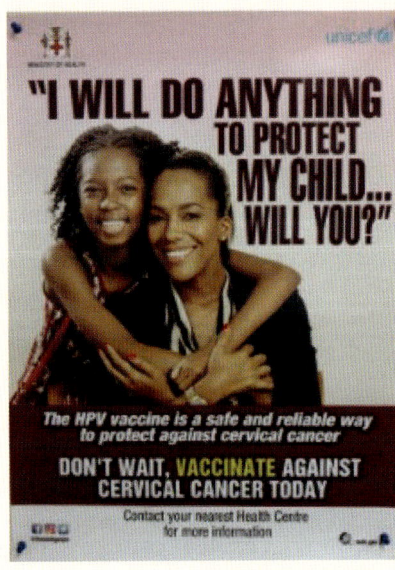

Figure 12.29 We have the chance to wipe out cervical cancer with a simple vaccine. Now the challenge for the Caribbean is to get the message across and get vaccines into our young people.

Ovarian cancer is another cancer found in relatively young women. No one knows why the ovary changes in this way. Scientists think that it is at risk because it is a very active tissue. There is also a genetic link – ovarian cancer can run in families. It is a very dangerous cancer because it causes few symptoms and grows rapidly in the abdomen. If caught early, it can easily be cured. The problem is that it is rarely caught early, so it spreads and is often fatal.

Uterine cancer is more common in older women after the menopause. Again it can be dangerous because the tumour grows in the uterus and produces few symptoms. If it is found, it can be treated.

In young men, **testicular cancer** is one of the most common tumours. Like the ovaries, the testes contain very active tissue that is constantly dividing and making sperm. Tumours can usually be detected as new lumps on the testes. If testicular cancer is caught early, the affected testis can be removed and the cancer cured. The biggest problem is that young men do not always examine their body carefully, and if they do discover a lump, they may be too embarrassed to visit their doctors – an embarrassment that may cost them their lives.

In older men, the most common cancer of the reproductive system is **prostate cancer**. Here, the prostate gland becomes cancerous. As it grows, it can cause problems with urination. Caught early, it can be cured, but prostate cancer still claims many lives. Caribbean men are particularly prone to aggressive prostate cancers.

Human reproduction is an amazing process. The more you learn about it, the more amazing it is that any baby is born safe and well. However, in the Caribbean, as in almost every country in the world, there is still a long way to go before every baby born is wanted and cared for.

Checkpoint questions

4. How are hormones used to help treat infertility problems?
5. What are the main cancers that affect the reproductive organs?
6. Why do you think cancers of the reproductive system are often ignored until it is too late?

Possible SBAs based on the reproductive system

→ Investigate and draw male and female gametes and early embryos using prepared microscope slides.
→ Explore and describe sexual reproduction in different plants, comparing for example flower structures, number and size of gametes and seed dispersal mechanisms.
→ Produce an animation of the human female menstrual cycle to use in teaching.
→ Investigate the reproductive cycles and pregnancies of different mammals, for example, cats/dogs/farm animals/other primates.
→ Make a model of different stages of human pregnancy to use with expecting parents in antenatal classes.
→ Research aspects of reproductive health across Caribbean countries or within your own country and produce some statistical analysis, for example, on antenatal care; on teenage pregnancies; on postnatal care; on maternal or baby deaths, and so on.
→ Research the background to the HPV vaccine/incidence of cervical cancer across the Caribbean or in your own country/impact of HPV vaccine globally/plan an education campaign to increase uptake of this life-saving vaccine, and so on.

End-of-chapter summary

In this chapter, you have learnt that:

- asexual reproduction involves one individual and the production of identical offspring known as clones, using the form of cell division known as mitosis
- Sexual reproduction involves two gametes from different parents; when these sex cells fuse, a new, genetically unique individual is formed

- the sex organs in human beings mature and become active at puberty
- the human male reproductive system produces sperm in the testes; the sperm travel from the testes to the penis in the sperm duct, and secretions are added from the seminal vesicle and the prostate gland to produce semen
- in the human female reproductive system, the ovaries release mature ova from developing follicles once a month in the menstrual cycle; the uterus develops a blood-rich lining each month to prepare for a pregnancy; if an ovum is fertilised, it will implant in the lining; if not, the lining is shed as the monthly period
- the menstrual cycle in women is controlled by hormones released from the pituitary gland (FSH and LH) and by the ovary (oestrogen and progesterone)
- sperm are deposited in the female reproductive system during sexual intercourse
- sperm move up through the cervix and the uterus to reach the Fallopian tubes
- the ovum released at ovulation is moved along the Fallopian tube towards the uterus by the beating of cilia
- if the sperm and the ovum meet in the Fallopian tubes, one sperm may penetrate the ovum; this is the moment of fertilisation or conception
- it takes around 40 weeks (nine months) for the fertilised ovum to grow and mature into a fully developed fetus ready for birth; during pregnancy, the mother supplies food and oxygen to the fetus though the placenta, and removes the waste products the same way
- there are three stages to the birth process: the contraction of the muscles of the uterus to bring about the dilation of the cervix; more contractions, crowning and the delivery of the baby; contractions and the delivery of the afterbirth; the contractions are controlled by oxytocin in a positive feedback loop
- the number and timing of pregnancies can be planned using a variety of birth control methods; there are a number of types of birth control – natural, barrier, hormonal and surgical – and each has advantages and disadvantages
- some pregnancies end early in a spontaneous abortion or miscarriage, often because there is a problem with the embryo or foetus
- some unwanted pregnancies are terminated by medical abortion; the health or even the life of the mother may be at risk, the pregnancy may be the result of rape or incest or the foetus may have a disability; the reasons for which abortion is legal vary from country to country; in some places illegal abortions are a big problem and many women die as a result
- there are health and ethical issues linked to the use of both birth control and abortion
- there are a number of cancers that affect the reproductive systems including cervical cancer and ovarian cancer in women and prostate cancer in men; cervical cancer can be almost completely eliminated using the HPV vaccine.

End-of-chapter questions

1. Which one of the following is not an advantage of asexual reproduction?
    - A Successful characteristics are passed on to offspring.
    - B It often results in large numbers of offspring.
    - C There is need to find a mate.
    - D Genetic variety is introduced.

2. What is the type of cell division involved in the formation of gametes called?
    - A Meiosis
    - B Mitosis
    - C Metaphase
    - D Metatarsal

3. Which one of the following reproductive hormones is produced by the pituitary gland?
    - A Oestrogen
    - B Testosterone
    - C Follicle stimulating hormone
    - D Progesterone

4. In which part of the female reproductive system does fertilisation of the ovum take place?
    - A Uterus
    - B Cervix
    - C Fallopian tube
    - D Vagina

5. How long is the average human pregnancy?
    - A 30 weeks
    - B 35 weeks
    - C 40 weeks
    - D 45 weeks

6.
    a. With reference to the different parts of the diagram labelled on the right, explain the events of the female menstrual cycle. (10)
    b. Describe the differences you would expect to see if the woman became pregnant during this cycle. (4)
    c. Suggest the differences you would expect to see if the woman was infertile because she did not produce FSH in her pituitary gland. (4)
    (Use diagrams to help you answer parts b and c if you wish to.)

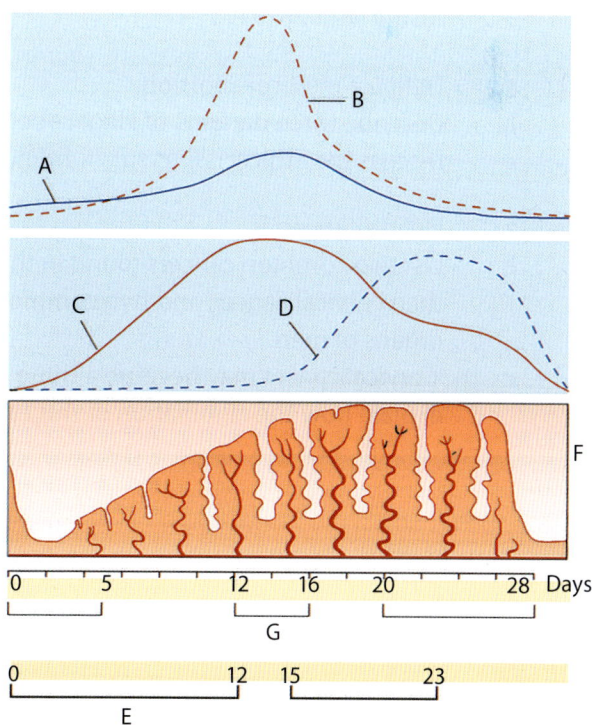

End-of-chapter questions 277

7 a Explain why it is important that there are very large numbers of sperm present in the semen. (3)

    b Give one advantage and one disadvantage of the human testes being situated in the scrotum. (2)

    c Describe the sequence of events from the arrival of sperm in the vagina to the fertilisation of an ovum. (6)

8 The birth of a baby has been described as 'an everyday miracle'. Write an essay describing the events of a normal human pregnancy from conception to birth. (10)

9 a Describe how the nutritional needs of a pregnant woman differ from the nutritional needs of a woman who is not pregnant. (3)

    b Define antenatal care and explain why it is important. (3)

    c Give three advantages of breastfeeding an infant. (4)

10 a Define the term 'birth control'. (1)

    b Describe the main forms of available birth control. (10)

    c The biggest cause of failure of birth control is usually human error. What does this mean? (2)

11 a Hormones can be used to control fertility artificially. Explain:

        i how hormones can be used to prevent pregnancy in the **contraceptive** pill. (3)

        ii how hormones can be used to help infertile couples become pregnant. (3)

    b Describe the advantages of family planning to:

        i the individual. (3)

        ii society. (3)

    c In your opinion, do the benefits of family planning outweigh the problems? Justify your answer. (3)

12 a Define a medical abortion. (3)

    b Describe three dangers of illegal abortions. (3)

    c Discuss the ethical issues surrounding legal and illegal abortions. (4)

13 a Define cancer. (3)

    b List two common cancers found in the reproductive organs of women other than cervical cancer, and two common cancers found in the reproductive organs of men. (4)

    c 'Cervical cancer may become a thing of the past'. Discuss this statement and explain how the elimination of this type of cancer might be possible. (5)

End-of-section questions

1 **a** Name the part of the skin where new cells are made. (1)
   b State what melanin is and explain its importance in the skin. (3)
   c When you feel too warm, you sweat and your skin become flushed. Explain how these changes cool you down. (3)
   d The diagram below shows a vertical section through the human eye. Write the correct name of each part labelled by the letters A to H. (4)
   e Define short-sightedness and describe how to correct it. (3)

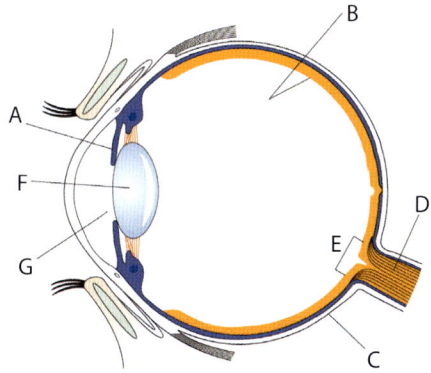

2 **a** Write the numbers 1 to 18 in your notebook or answer booklet, and write the name of the corresponding part in the diagrams below. (6)

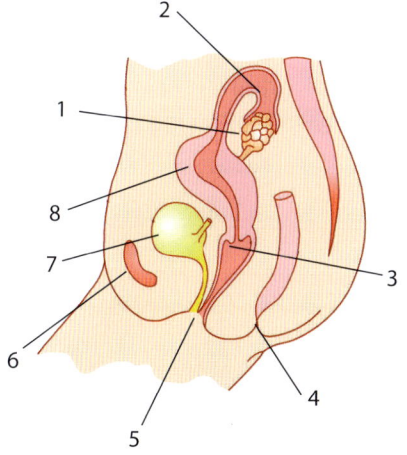
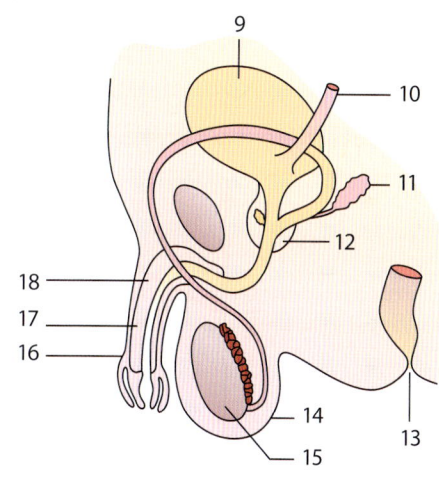

   b Name and briefly describe two methods of birth control. (6)
   c Name one negative side effect of using birth control. (3)

3 **a** Draw an outline of the human body. Insert and label the endocrine glands found in human beings. (6)
   b For any of the hormones produced by the endocrine gland, describe the results of its:
      i) overproduction.
      ii) underproduction. (3)
   c Name two hormones that affect the menstrual cycle in a fertile female. (2)
   d With the help of a graph, explain how the hormone levels change within one menstrual cycle in a typical fertile female. (4)

4 a Name the types of teeth found in mammals. (3)
    b State four ways we can help keep our teeth healthy. (4)
    c Label the parts indicated by letters T, V, W, X, Y and Z in the diagram below. (3)

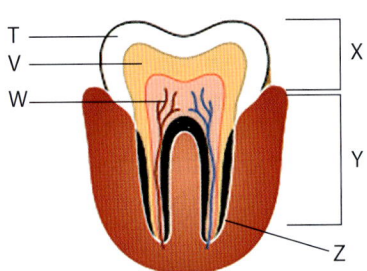

    d State the main differences between the teeth of a carnivore and those of a herbivore. (2)
    e Draw a timeline to show the changes in human teeth from birth to adulthood. (3)

5 a Define respiration. (1)
    b When we breathe, oxygen and carbon dioxide move across the surface of alveoli in the lungs. Explain how this process works. You can use a diagram to help you explain. (5)
    c Describe how osmosis differs from the process you named in part (b)? (2)
    d In a diagram of the heart and related blood vessels in a human, show how oxygenated blood flows using arrows. (4)
    e Compare and contrast a large vein and a large artery. Your answer can be in the form of a table. (4)

Data analysis

In an experiment simulating the digestion of food in the stomach, graph A shows the rate of reaction of pepsin at different pH levels. Use the graph to answer the following questions.

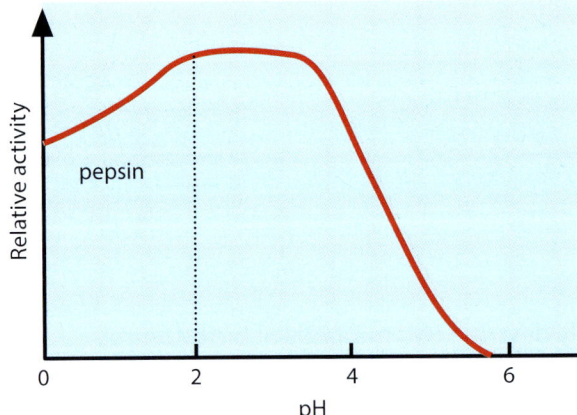

1 State the pH at which pepsin work best? (2)

2 a Name the food type broken down by pepsin. (1)
    b Explain how the pH needed by pepsin is produced in the body. (2)

3 The enzyme amylase is used in a similar experiment. Using the graph in your answer book, draw the rate of reaction of amylase at different pH levels. Explain why this graph is different to the one for pepsin. (3)

Case study

A top Jamaican athlete has just finished running a 200 m race, which he won. He is sweating heavily, breathing hard and his skin looks flushed. He is sipping from a bottle of room-temperature water as he does a TV interview.

1 Describe the four life processes that were speeded up in his body in the course of the race. (4)

2 Explain what has happened in two of these life processes. (8)

3 Explain why the athlete sips room temperature water instead of drinking large amounts of cold water. (3)

Revision tool kit 2

Chapter:				
Topic & sub-topic	Very comfortable	Average	Need more work	Do not know at all

Use this inventory as a guide, so you know which aspects in a topic you need to spend more time on. As you master a sub-topic, review the inventory.

Downloadable inventories with the topics and sub-topics filled in are on the website.

Section C: Heredity and variation

Learning outcomes

At the end of Section C, you will:

- understand the concept of a gene as it pertains to DNA, chromosomes and alleles
- understand the role of genes and heredity in determining how traits can be altered and inherited by asexual and sexual means through the process of mitosis and meiosis respectively
- develop an awareness of the importance of genetic variation and its role in natural selection
- appreciate the social and ethical implications of genetic engineering.

Take a look around your classroom. Everyone has many things in common, yet each of you (even if there are identical twins in your class) can be identified and is different. Now think about your family – parents, brothers and sisters, grandparents, aunts and uncles, and cousins. At a big family gathering, people will look more alike than you and your classmates. This is because you are related and will share similar features. Yet you are still all unique and different. This is because there is variety in your genetic make-up and variety in the way you have been brought up.

In the nuclei of the cells of our body are the chromosomes that carry all the details that make up our basic appearance and physiology. This information is held within the structure of a very special chemical – DNA (deoxyribose nucleic acid) – which makes up those chromosomes. The details of this amazing molecule will be described here, along with the impact that it has on the inheritance of everything from the shape of our nose to the chemistry of our liver.

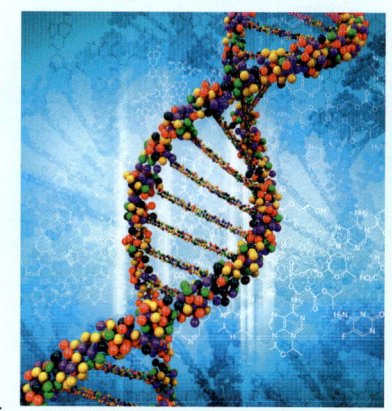

Figure C1 DNA – the molecule of life

The whole of life as we know it depends on reproduction. Every time we fall over or graze our skin, every time cells in our body wear out and need to be replaced, every time we grow a little taller, we rely on the ability of our cells to reproduce themselves by mitosis. We rely on a special form of cell division called meiosis to make the sex cells or gametes in our sex organs needed to conceive a new baby. However, the growth and development of the foetus from that point onwards depends completely on mitosis. These two forms of cell division are vital for the continuation of life, and you will be learning about their similarities and differences in this section of the book.

Think back to your family gathering. Family likenesses are often a topic of conversation. People might say 'I see you've got the family ears' or 'I'm glad to see you've inherited my intelligence'. Characteristics certainly are inherited, passed on from parents to their children. In this section, you will explore just how this happens, how it is possible to predict some of the possible combinations of offspring that may be produced, and how sometimes a genetic inheritance will include diseases, as well as other characteristics.

Figure C2 The people in this family group are clearly related to one another, yet none of them are identical.

> **Revision tip**
>
> *Summarise the information you are trying to learn in different ways. For example, make a concept map, which might help you to see a pattern you did not notice before.*

Chapter 13: Cell reproduction and variation

When you have completed this chapter, you will be able to:
- explain the importance of mitosis in asexual reproduction
- describe the process of mitosis
- describe the role of meiosis and sexual reproduction in producing variation
- understand the role of environmental influences in producing variation
- define the terms 'DNA', 'chromosome', 'gene' and 'allele', and explain the difference between a gene and an allele
- understand and explain the importance of meiosis in sexual reproduction
- distinguish between genetic and environmental variation
- describe the process of meiosis and how it differs from mitosis

What the examiners say

→ Candidates typically appear intimidated by this aspect of the syllabus; they often confuse key terms and provide incomplete responses to the questions.

→ Specifically, candidates often confuse the relationship between mitosis/meiosis and asexual/sexual reproduction; find it difficult to differentiate between genetic and environmental variation; they are unable to discuss the gametes derived from males and females, and how this determines the gender of offspring; they are unable to discuss the negative effects of genetic engineering in food production; they are unable to differentiate between continuous variation and discontinuous variation; they are limited in their ability to differentiate between phenotype and genotype, and often mistake genotype for gametes; and are unfamiliar with the crossover concept as applied to the relationship between DNA and chromosomes.

Your body is made up of millions and millions of cells. Throughout your life, you need to make more cells. You need cells to grow, cells to replace cells that age and die, cells to repair the damage that happens in everyday life and cells to be the gametes that will start a new life. For all these things to happen, your cells must reproduce themselves. In this chapter, you will learn the amazing processes that your cells use to divide. You will discover how every body cell has the same information, yet every gamete is different.

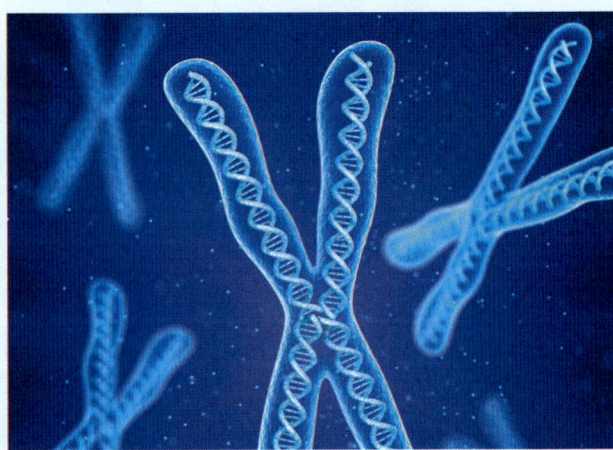

Figure 13.1 Chromosomes are made up of DNA, which stores all the genetic information passed on to us by our parents.

Chromosomes

Inside the nucleus of every cell (see Chapter 1) there are thread-like structures called **chromosomes**. This is where the genetic information passed on from parent to child is stored (see Figure 13.1 on page 284). The chromosomes are made up of **DNA (deoxyribose nucleic acid)** and this amazing chemical carries the instructions to make all the proteins in our cells. Many of these proteins are enzymes, which in turn control the production of all the other chemicals that make up our body and affect what we look like and who we are. The **genetic information** is stored in small sections of the DNA called **genes** (see Figure 13.2). Each chromosome contains thousands of genes joined together.

Each gene has at least two different forms – and the different forms of a gene are called **alleles**. The easiest way to understand this is with an example. In a certain position on a chromosome, there will be a gene that determines whether your thumb is straight or curved.

- Gene = shape of your thumb
- One allele = curved
- Another allele = straight

Whether your thumb ends up straight or curved will depend on the combination of alleles you inherit from your parents – you will learn more about this in Chapter 14.

Each different type of organism has a different number of chromosomes in the cells – people have 46 chromosomes and tomatoes have 24, while turkeys have 82! Chromosomes come in **homologous pairs** (matching pairs), so people have 23 pairs of chromosomes.

Scientists can photograph the chromosomes in human cells when they are dividing and arrange them in pairs to make a special picture called a **karyotype** (Figure 13.3).

Human karyotypes show 23 pairs of chromosomes. In 22 of them, both the chromosomes in each pair are the same size and shape, regardless of whether we are male or female. These 22 pairs of chromosomes are called the **autosomes**. They control almost everything about the way we look and the way our body works. The remaining pair of chromosomes is different for males and females. A girl has a pair of two similar **X chromosomes**, but a boy has one X chromosome and another, much smaller, **Y chromosome**. These are called the **sex chromosomes** because they decide whether we are biologically male or female. Everyone inherits an X chromosome from their mother. If this joins with a sperm carrying another X chromosome, we will be female. If it is fertilised by a sperm carrying a Y chromosome, we will be male.

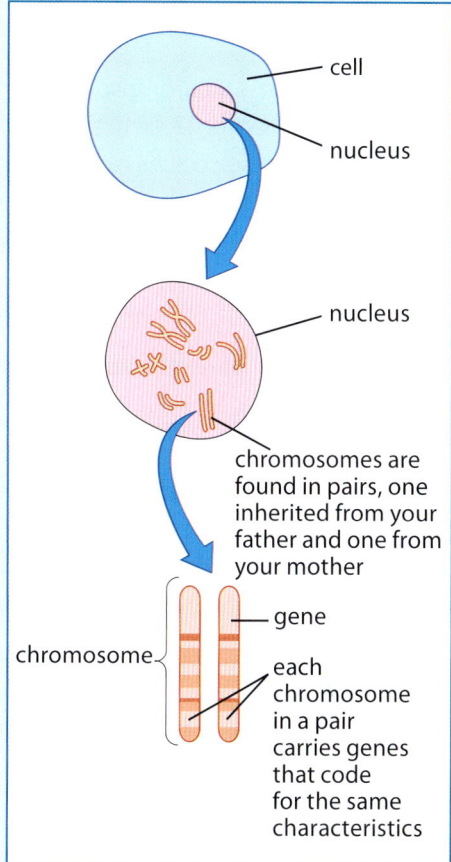

Figure 13.2 The nuclei of our cells contain the chromosomes that carry the genes that control the characteristics of the whole body.

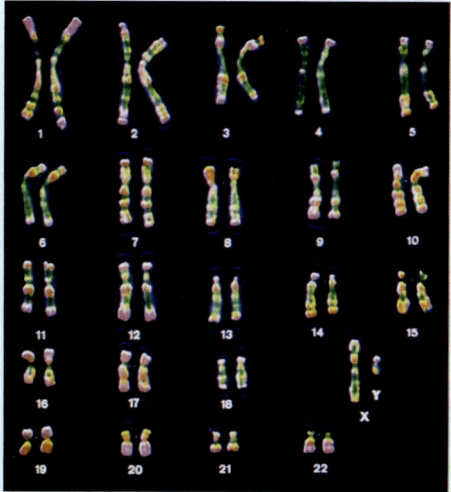

Figure 13.3 Karyotypes like this one of a healthy man have helped scientists find out more about the mysteries of inheritance.

Haploid and diploid cells

We inherit half of our chromosomes from our mother and half from our father. As you saw in Chapter 12, fertilisation takes place when the gametes (ovum and sperm) fuse. Each sex cell has a single set of chromosomes, so when two sex cells join during fertilisation, the new cell formed has a full double set of chromosomes.

- In humans, the egg cell has 23 chromosomes and so does the sperm. They are **haploid** cells, with chromosome number **n**.
- At fertilisation they fuse to form a new cell with the full human complement of 46 chromosomes. This is a **diploid** cell with a chromosome number of **2n**. Now look at Figure 13.4.

The combination of genes on the chromosomes of every newly fertilised ovum is completely unique. Once fertilisation is complete, the unique new cell begins to divide by a process called **mitosis**. This will continue long after the foetus is fully developed and the baby is born. (See Chapter 12 for more details on fertilisation and the development of the foetus.)

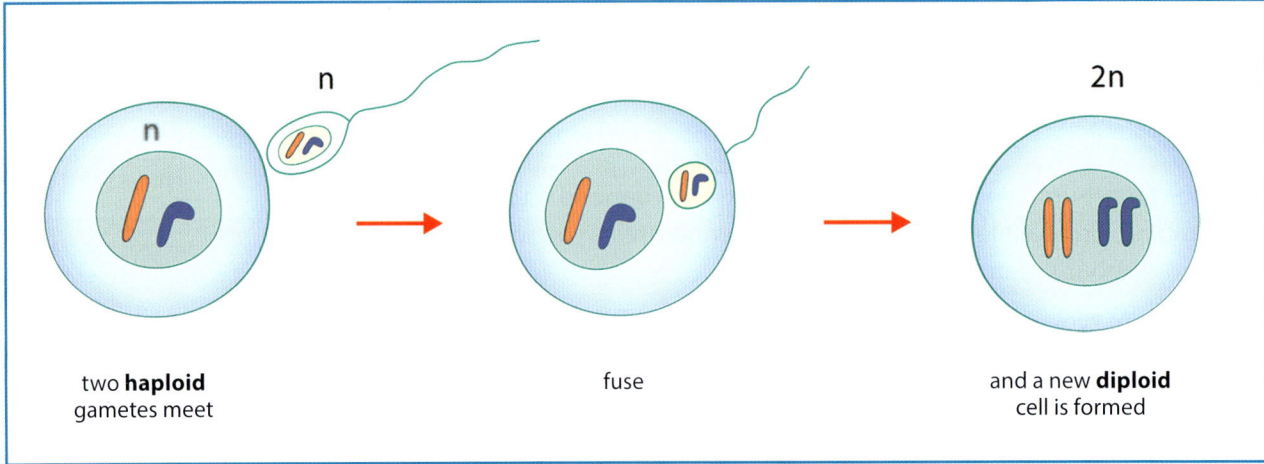

Figure 13.4 Gametes are haploid cells (n). At the moment of fertilisation, they fuse to form a new diploid cell (2n). This will grow and reproduce by mitosis to form a new individual.

Mitosis

Mitosis is the process of cell division that produces identical diploid **daughter cells**. As a result of mitosis, every body cell has the same genetic information. In asexual reproduction, the cells of the offspring are produced by mitosis from cells of their parent. This is why they all contain exactly the same chromosomes with no variation.

How does mitosis work?

Mitosis is actually one continuous process, but to make it easier to understand, we divide the process into stages, as shown in Figure 13.5.

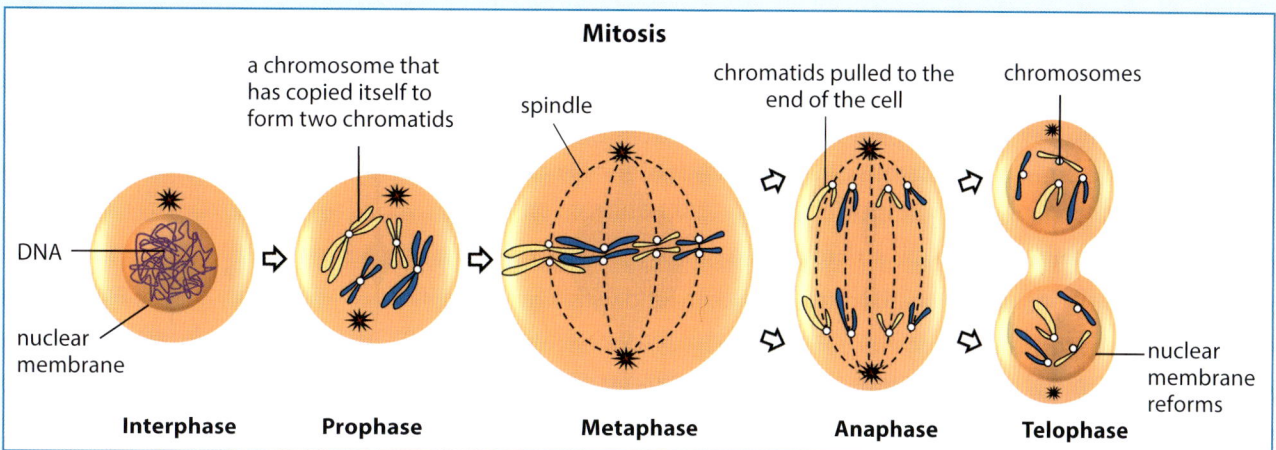

Figure 13.5 The formation of identical daughter cells by simple division takes place during mitosis. It supplies all the new cells needed in our body for growth, replacement and repair. Our cells really have 23 pairs of chromosomes, but for simplicity, this cell is shown with only two pairs.

Here are the stages in the mitosis process, as shown in Figure 13.5:

- **Interphase:** The cell makes copies of the DNA, ready for division.
- **Prophase:** The DNA **condenses** – it becomes visible. The chromosomes have already divided, so each one appears as two **chromatids** joined in the middle. The membrane round the nucleus breaks down and disappears.
- **Metaphase:** A protein spindle forms. The chromatids line up around the middle.
- **Anaphase:** The chromatids are pulled to different ends of the cell. Now they are called chromosomes again. See Figure 13.6.
- **Telophase:** New nuclear membranes form around the chromosomes and the cytoplasm starts to pinch in.
- Finally, two identical daughter cells are formed through cytokinesis. This is the process of dividing the cytoplasm into two separate daughter cells, each with its own nucleus.

> **DID YOU KNOW?**
> We lose our body cells at an amazing rate – 300 million cells die every minute. Fortunately mitosis takes place all the time to replace them.

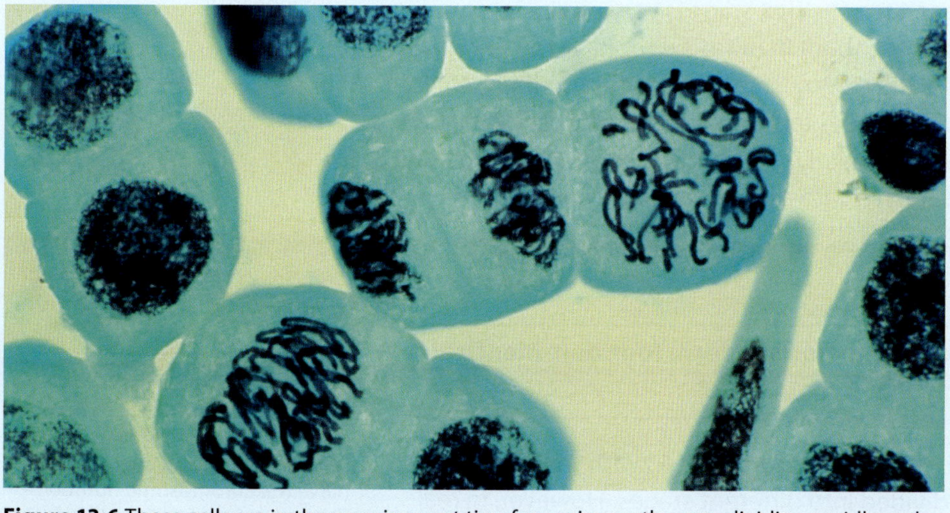

Figure 13.6 These cells are in the growing root tip of an onion so they are dividing rapidly and the chromosomes have taken a dark stain. You can see mitosis taking place, with the chromatids in different positions as the cells divide.

Why is mitosis so important?

Mitosis takes place all the time, in tissues all over the body. Mitosis always produces identical, diploid daughter cells that have the same number (2n) and type of chromosomes as the parent cell. These daughter cells are also known as **clones**. Mitosis is incredibly important in living organisms for:

- **growth** – making more cells so organisms get bigger
- **repair** – replacing cells that are worn out or damaged with identical cells
- **asexual reproduction** – the production of new, identical individuals that are identical to their parent organism, for example, strawberry plants and bacteria.

However, mitosis is not the only type of cell division. Another type is called meiosis, which takes place only in the reproductive organs.

> **Checkpoint questions**
>
> 1 Explain why new cells are needed continuously in our body.
> 2 Produce a flow diagram to show how a cell divides by mitosis.

Meiosis

Meiosis is a special form of cell division where the chromosome number is reduced by half to produce haploid cells that are NOT identical to the parent cell. These are the sex cells or gametes. Meiosis only takes place in the reproductive organs where the gametes are made.

The reproductive organs in humans, like most animals, are the ovaries and the testes. This is where meiosis takes place in the production of the ova and sperm.

When a cell divides to form gametes, the first stage is very similar to normal body cell division, which you learnt about in mitosis (Prophase I to Telophase I and cytokinesis in Figure 13.7). The chromosomes are copied so there are four sets of chromatids. The cell then divides to form two identical daughter cells. These cells then divide again immediately, without the chromatids doubling again. This forms four gametes, each with a single set of chromosomes.

The names of the stages of meiosis are important, but more critical for understanding is the principle:

> **A single diploid cell divides by meiosis to form four non-identical haploid cells.**

The details of this process are shown in Figure 13.7. As with mitosis, the diagram shows the process divided into different stages. However, in real life it is a single flowing process, which has been described, rather poetically, as the 'dance of the chromosomes'.

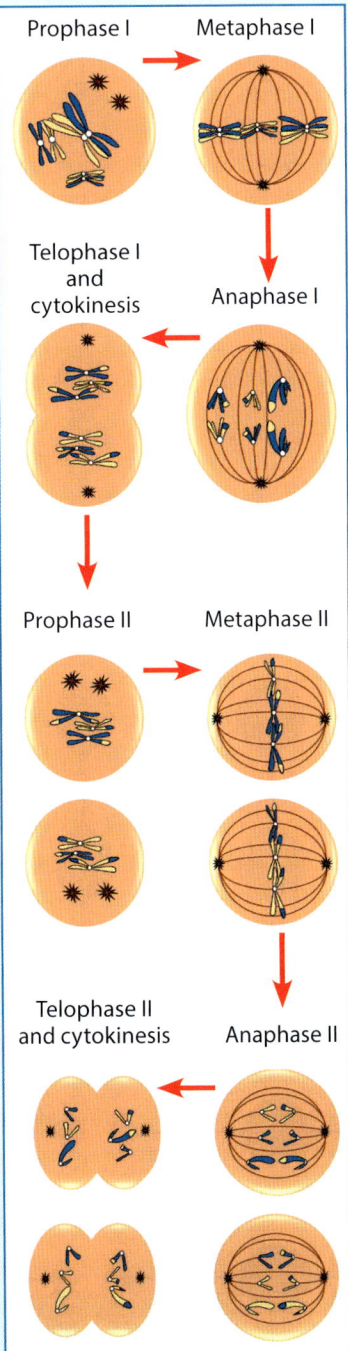

Figure 13.7 The formation of unique haploid sex cells in the ovaries and testes involves a special type of cell division, meiosis, to halve the chromosome number. The cell is shown with only two pairs of chromosomes to make it easier to follow what is happening.

Cell reproduction and variation

Why is meiosis so important?

There are two main reasons why meiosis is so important:

- **It halves the number of chromosomes in the gametes, making sexual reproduction possible:** Our normal body cells have 46 chromosomes in two matching sets or homologous pairs, 23 from our mother and 23 from our father. If two body cells joined together in sexual reproduction, the new cell would have 92 chromosomes, which simply would not work! Fortunately, as a result of meiosis, our sex cells contain only one set of 23 chromosomes, exactly half of the full chromosome number. So when the gametes join together at fertilisation, the new cell formed contains the normal number of 46 chromosomes.

- **It introduces variation into the gametes:** Each gamete we produce is slightly different from all the others. This variation is key for the survival of the species – you will discover why later in this chapter and in Chapter 14. The combination of chromosomes that come from the mother and the father is different in each gamete. What's more, there is some exchange of genes between the chromosomes in a process called **crossover** (see Figure 13.8). As a result, no two ova or sperm are the same. In these two ways, lots of variation is introduced into the genetic mix of the offspring.

> **DID YOU KNOW?**
>
> One testis produces over 200 million sperm each day by meiosis. As most males have two working testes, that gives a total of 400 million sperm produced every 24 hours! Only one sperm is needed to fertilise an ovum. However, as each tiny sperm has to travel 100 000 times its own length to reach the ovum, less than one in a million ever complete the journey, so it's a good thing that plenty are made!

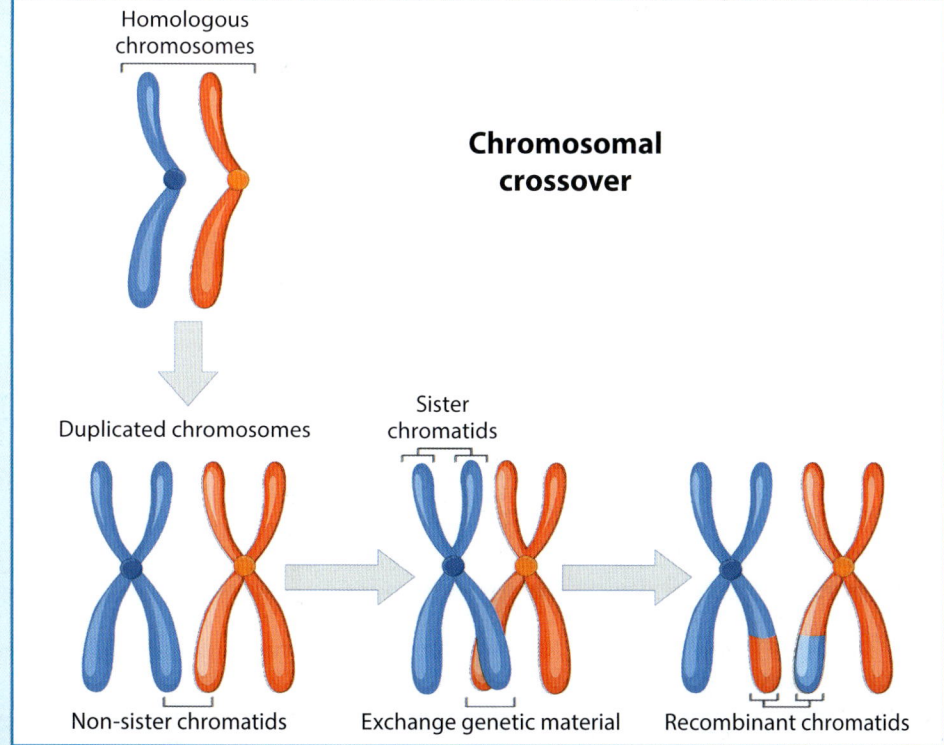

Figure 13.8 During meiosis, the original pair of chromosomes (one from the mother and one from the father) are copied to form pairs of sister chromatids. When the chromatids crowd together and line up on the spindle at metaphase 1, the inside non-sister chromatids break and rejoin with each other. This is called crossover and you can see how it results in all four of the chromatids being different.

Meiosis

> **Checkpoint questions**
>
> 3 Name the male and female gametes, and describe how they differ from normal body cells.
> 4 State the type of cell division needed to produce the gametes.
> 5 Explain why meiosis is so important in the human body.
> 6 Make a table to compare and contrast the processes of mitosis and meiosis.

Genetic variation

The differences between asexual and sexual reproduction (see Chapter 12) are the result of the different types of cell division involved in the two processes.

Asexual reproduction

In asexual reproduction (Figure 13.9), the offspring are produced from the cells of a single parent as a result of mitosis, so they contain exactly the same chromosomes and the same genes as their parent. There is no variation in the genetic material.

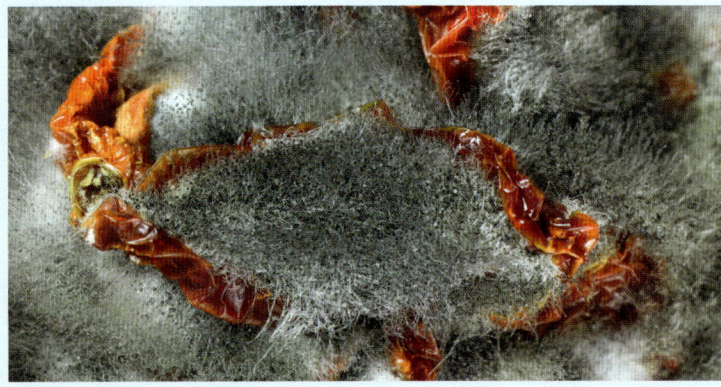

Figure 13.9 An example of an organism that reproduces asexually is mould.

Sexual reproduction and genetic variation

In sexual reproduction, the gametes are produced by meiosis in the sex organs of two parents. This introduces variation, as each gamete is different. When the gametes fuse, one of each pair of chromosomes, so one of each pair of genes, comes from each parent. The combination of genes in the new pair will contain alleles (different forms of the gene) from each parent. This also helps to produce different characteristics in the offspring.

Another big source of genetic variation is **mutation**. A mutation is a change in the genetic material. A mutation may affect a single gene, several genes or a whole chromosome.

Most mutations do not have any effect on us at all. Some mutations make things better, as you will see. But sometimes a mutation means systems in the cell or the body do not work as they should, and this may cause real problems.

For example, if a mutation involves a whole chromosome, it can result in a condition such as Down syndrome. In this condition, the baby has an extra copy of chromosome 21 (Figure 13.10). This results in many different problems in the child's mental and physical development. With a lot of love and support, people with Down syndrome can lead happy lives (Figure 13.11), but they often have health problems as a result of their extra chromosome. According to a 2022 United Nations estimate, Down Syndrome affects 1 in 1000 live births worldwide. The risk of it happening increases with the age of the parents, particularly the mother.

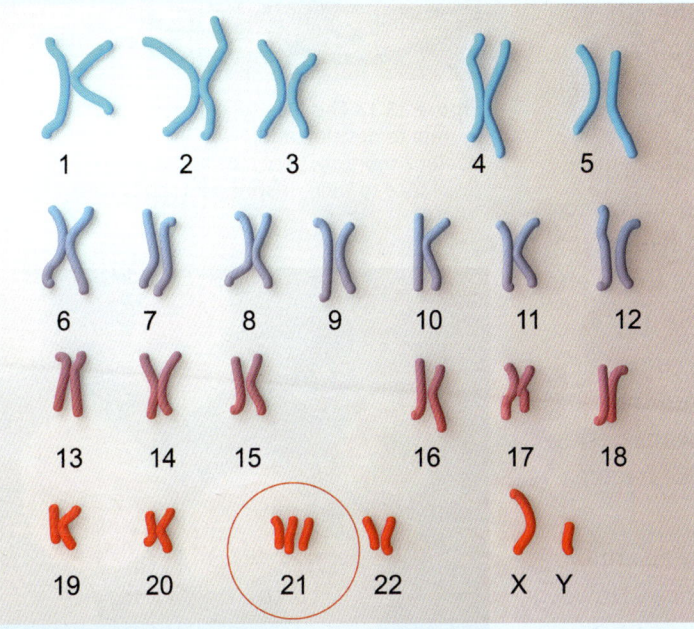

Figure 13.10 A karyotype showing the arrangement of chromosomes in a male with Down syndrome. Chromosome 21 has an extra copy.

Figure 13.11 Although their extra chromosome causes health problems for people with Down syndrome, with support they can lead happy and productive lives.

Why genetic variation is important

Genetic variation is extremely important in living organisms. The identical clones that result from asexual reproduction based on mitosis do very well when the conditions suit them. But if there is a change in the environment there may be problems. This could be a change in anything from the average temperature to more rain, or from a new disease to a new predator appearing (see Figure 13.12). If all the organisms in a population, whether animals or plants, have the same genetic make-up, then they are all vulnerable to change, because if one cannot cope with it, none of them can.

This is the great advantage of the genetic variation that results from sexual reproduction based on meiosis, and also the effect of mutations. Each individual has a different genetic make-up. So if conditions change, although some individuals will struggle to survive, others will cope well and the genetic make-up of the whole population will gradually change. This is called natural selection or 'survival of the fittest'.

> **Natural selection:** The organisms best suited by their genes to a particular environment are most likely to survive and breed, passing on those useful alleles.

Figure 13.12 The brown tree snake was introduced into Guam in the 20th century. There were plenty of animals like birds and rodents for the snakes to hunt and eat. With no predators to hunt them, the snake numbers grew quickly. These snakes have caused the extinction of most of the native bird population in Guam, who had no defence against this new predator.

Genetic variation

The development of antibiotic-resistant bacteria is causing many problems around the world – and gives us a good example of both natural selection in action and how important genetic variation is to a population. Many bacteria are useful, but some cause terrible diseases – you will learn more about these in Section D. Most bacteria reproduce asexually, but mutations still take place, so there is some genetic variation in a population of bacteria.

- Antibiotics are chemicals that kill bacteria. When they were first developed, their effect was almost miraculous and they have saved millions of lives. They still save many, many lives, but some problems have arisen.
- Each time an antibiotic is used, most of the bacteria are wiped out. However, there are almost always a few bacteria that have a different genetic variation as a result of a mutation. If this mutation makes them resistant to the effect of the antibiotic (Figure 13.13) – they are not killed by it.
- The few bacteria that remain by the end of a course of treatment will be the fittest for the conditions, so they will survive and reproduce their **antibiotic-resistant genes**. This means that an antibiotic-resistant strain of bacteria **evolves**.
- The small number of resistant bacteria left by an antibiotic can usually be dealt with by the body's defences. However, if someone stops taking their antibiotic medicine too soon, or if antibiotics are used too widely, resistant strains develop so that the antibiotic no longer works.
- This has already happened with several different antibiotics – we now have a few strains of bacteria that are resistant to almost all known antibiotics (Figures 13.14 and 13.15).

Figure 13.13 This illustration shows how bacteria mutate to develop antibiotic resistance. As the bacteria multiply, they pass on genes for antibiotic resistance to their offspring (green circles).

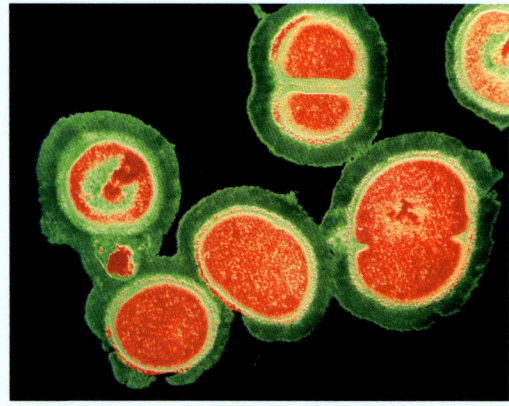

Figure 13.14 A coloured transmission electron micrograph (TEM) of a deadly cluster of MRSA Staphylococcus aureus bacteria. The MRSA bacteria are resistant to most antibiotics and are commonly found in hospitals, which makes them very dangerous.

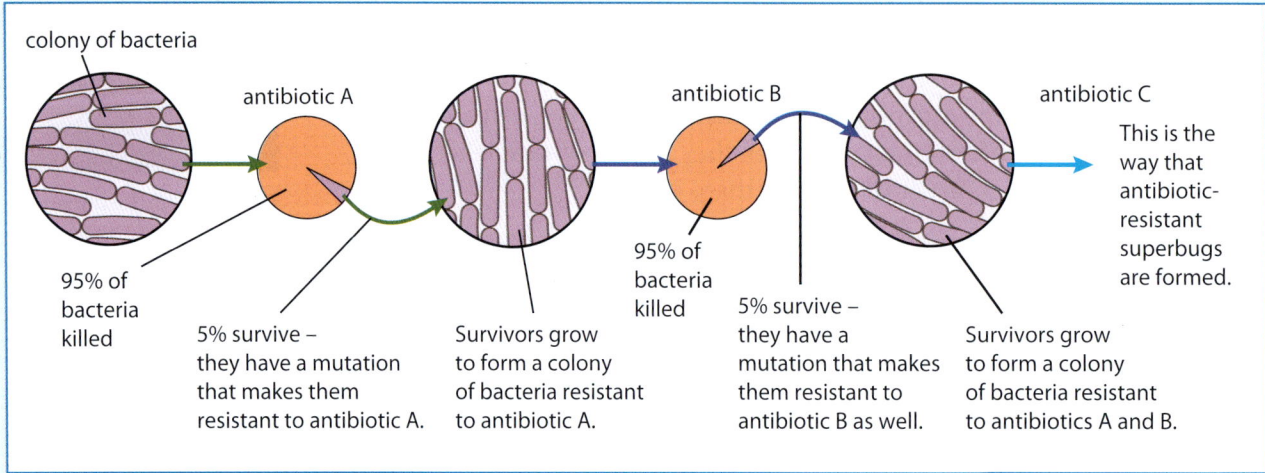

Figure 13.15 It is very important to complete a course of antibiotics, even if you feel better. The bacteria with some resistance to the medicine will be the last ones to be killed by the drug. If they are not destroyed, they may go on to reproduce, passing on the genetic variation and forming an antibiotic-resistant strain.

Cell reproduction and variation

Genetic variation in humans

Genetic variation is important in human beings too. Examples of genetic variation include the determination of sex – whether you inherit two X chromosomes (XX) and so are biologically female or an X and a Y chromosome (XY), making you biologically male. In Chapter 8, you discovered human blood groups – in the ABO system, you will be either blood group A, B, AB or O, and this is directly inherited. And you inherit the ability to roll your tongue or not to roll it. These are examples of genetic variation in humans.

For example, for the Maasai people in Africa, a very tall, thin body shape is an advantage as it allows them to lose heat more easily. Over time, tall, thin people have survived more successfully so this is the body shape selected through natural selection. Now the Maasai are famous for their height. Similarly, the Inuit people tend to be relatively short and rounded because this shape reduces heat loss. When you live in some of the coldest places on the Earth, this is a definite survival advantage and natural selection has done its work. See Figure 13.16.

Figure 13.16 Genetic variation and natural selection has enabled people to adapt successfully to very different environments.

Environmental variation

The genes we inherit certainly determine much of what we are. For example, a manchineel tree will never grow into a breadfruit tree, regardless of the soil or the weather conditions it grows in! Our basic characteristics are inherited from our parents and certain aspects of our features may be very similar to theirs. Eye colour, hair colour and texture, the shape of our nose and earlobes, our sex and skin colour – all of these features are determined by our genes and are the result of genetic variation.

However, while our genes undoubtedly play a major part in deciding how we will look, the conditions in which we grow and develop are also very important. This is easy to see in plants, because you can produce large numbers of genetically identical plants and change the conditions in which they grow. By changing the amount of light and soil nutrients available, genetically identical plants end up looking far from identical. Plants deprived of light or nutrients, or kept in the cold, do not make as much food as plants with plenty of everything they need, so the deprived plants are smaller and weaker – they are not able to fulfil their genetic potential. See Figure 13.17.

It is not ethical to take genetically identical people and put them in different conditions to see how the environment affects them. However, we can observe the effect of the environment on the appearance of people. We can look at easily measured characteristics such as height, weight and life expectancy, and measure how they change over time. The genetic make-up of a population won't change much over a hundred years, so if a feature shows a shift, it can usually be accounted for by a change in the environmental conditions. Here is a Caribbean example:

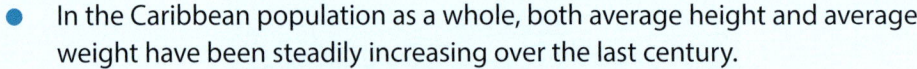

Figure 13.17 These two genetically identical cress plants show the effect of changing environmental conditions. The cress plant on the left was placed in sunlight, while the cress plant on the right was kept in a cupboard.

- In the Caribbean population as a whole, both average height and average weight have been steadily increasing over the last century.
- There has been no genetic pressure to get bigger; taller or heavier people have not had an advantage in the population so it is highly unlikely that these changes are due to genetic differences.
- Over the same period of time, the average diet has changed dramatically. Not only does almost everyone in the Caribbean now have enough food to eat, but the food available has become much higher in fats and sugars (see Figure 13.18).
- Adequate food for everyone means that the genetic height potential for almost every individual can be fulfilled, whereas in years gone by, the growth of many children was stunted by malnutrition.
- A higher energy intake than necessary means that people are also gaining more fat.

This gives us evidence that environmental causes and genetic causes affect the appearance of individuals of the same species. As it is extremely rare for any organism to grow and develop in absolutely ideal conditions, for most organisms a combination of genetic and environmental variation determines what they look like.

Figure 13.18 The weight of the Caribbean population is increasing because almost everyone now has enough food to eat, but this food is often high in fats and sugars.

Checkpoint questions

7. Explain the importance of genetic variation and natural selection.
8. Compare and contrast continuous and discontinuous variation in people.

Cell reproduction and variation

The only genetically identical human beings are identical twins (see Figure 13.19). It would be unethical to take such twins away from their parents by force and have them brought up in very different families to see if it made any difference to how they turned out. However, over the years, some identical twins have been adopted by different families. Some of them did not even realise that they had a twin until they were quite elderly. Scientists have researched this, tracing the twins through their very different lives, and looking at the similarities and differences between them. The similarities in appearance at the first meeting of these twins, in their hairstyles and dress, were often quite astonishing. But the measurable facts – things like height, weight and IQ – showed that just like other organisms, some of the differences between human beings are due to genetics and some are due to environment.

To see the effect of being brought up in different environments, scientists compared the results from the separated twins with identical twins brought up together, and with brothers and sisters (siblings) who were not twins.

- When the results for twins brought up separately and apart are very similar, it suggests that genetics is the strongest factor affecting the appearance.
- When the results for the identical twins brought up together are very different from the twins brought up apart, then it appears that environmental factors also have a strong impact on determining the appearance.

This evidence suggests that some factors, such as height, are strongly affected by our genes, whereas weight is much more the result of the home environment (see Table 13.1).

DID YOU KNOW?

Everyone has a unique set of fingerprints. Even identical twins have different fingerprint patterns. Something in the environment before you are even born is involved in making your unique fingerprint patterns.

Figure 13.19 In spite of their identical genetic makeup, identical twins become more different as they grow older and the environment affects them more. In this image, the young girls are more similar than the older women, who have had a longer time to experience environmental variation.

Table 13.1 Comparison of variations in identical twins brought up together and apart, and non-twin siblings

Measured difference in	Identical twins brought up together	Identical twins brought up apart	Non-twin siblings
Height	1.7 cm	1.8 cm	4.5 cm
Weight	1.9 kg	4.5 kg	4.7 kg
IQ (a measure of intelligence)	5.9	8.2	9.8

Environmental variation

Continuous and discontinuous variation

Take a look around your classroom, at the people in the local market or at a crowd watching cricket. You can compare many different features, such as height, hair colour and skin colour, which vary across all the individuals. There will be some very short people, some who are very tall, and many different heights in between. The same is true of foot size, weight, intelligence, and so on. These characteristics are said to show **continuous variation** because there is a gradual transition between the two extremes. Features that show continuous variation are usually determined by a number of different genes and are also affected by the environment – the availability of food, the impact of disease, and so on.

However, not all features exist in a wide variety of forms. Some characteristics are either present or they are not – we are male or female; we can roll our tongue or we can't; we have blood group A, B, AB or O; or we have dangly or attached earlobes. These features are said to show **discontinuous variation**. Characteristics that show discontinuous variation are usually determined by a single gene with little or no environmental impact. See Figure 13.20.

We human beings have relatively few characteristics that are determined by single genes. The great majority of our characteristics are the result of the interaction of many genes, often with the environment in which we live playing a part as well. However, the only way we can hope to understand the way in which we inherit genetic information from our parents is to look at how the simple features that show discontinuous variation are inherited. We will look at this in Chapter 14.

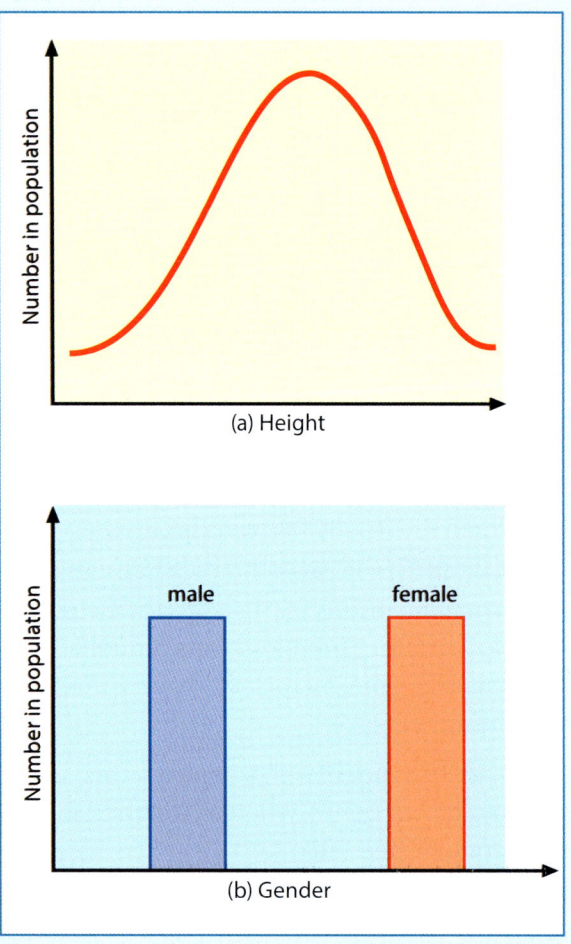

Figure 13.20 The difference between (a) continuous and (b) discontinuous variation is plain to see when you look at different features shown by individuals.

Possible SBAs based on cell reproduction and variation

→ Investigate how our knowledge and understanding of DNA and chromosomes have developed over time, and produce a timeline for the classroom wall.

→ Make models of haploid gametes and diploid cells and use them to explain the events of fertilisation.

→ Make posters to compare the stages in mitosis and meiosis.

→ Make your own slides of growing onion root tip cells and draw examples of the different stages you can see. Keep a record of your technique and any problems you overcome.

→ Make models to demonstrate crossover, for example, with modelling clay, and use them to make a podcast explaining their importance in introducing variation.

→ Investigate examples of continuous and/or discontinuous variation in your class/school/family/community, and produce graphs of your data along with explanations of your findings.

End-of-chapter summary

In this chapter, you have learnt that:

- in body cells, chromosomes are found in homologous pairs
- chromosomes contain the genes that carry the genetic information about an individual that is passed on from one generation to another
- chromosomes are made up of DNA (deoxyribose nucleic acid)
- different forms of the same gene are called alleles
- humans have 46 chromosomes arranged in 23 pairs
- 22 pairs of chromosomes carry information about the body generally; the final pair are the sex chromosomes and these determine whether you are female (XX) or male (XY)
- in asexual reproduction, the cells divide by mitosis; the chromosomes duplicate themselves and then separate to form two identical daughter cells (clones)
- cells produced by mitosis have identical genetic information
- body cells divide by mitosis to produce more identical cells for growth, repair, replacement or, in some cases, asexual reproduction
- cells in the reproductive organs divide to form the gametes (sex cells)
- body cells have two sets of chromosomes – they are diploid with the chromosome number 2n; gametes have only one set of chromosomes – they are haploid with the chromosome number n
- gametes are formed from body cells by meiosis
- the process of meiosis introduces variation because no two gametes are ever the same – different combinations of chromosomes form the different gametes, and crossing over takes place during meiosis
- sexual reproduction gives rise to variation because genetic information from two parents is combined
- genetic variation is very important for helping species to survive when conditions change
- natural selection describes the process where those organisms best adapted for a particular environment are most likely to survive and breed, passing on their useful genetic make-up
- variation is not all down to genetics – the environment also affects variety
- two main kinds of variation are seen in organisms; in continuous variation, are there is a gradual transition between the two extremes (for example, height and weight); in discontinuous variation, a feature is either present or it is not (for example, male or female, dimples or no dimples).

End-of-chapter questions

1. What are chromosomes made up of?
   A RNA
   B Protein
   C DNA
   D Acetic orcein (1)

2. How many chromosomes would you expect to find in a normal human body cell?
   A 23
   B 50
   C 84
   D 46 (1)

3. Which combination of chromosomes would result in a male human being?
   A XXX
   B XY
   C XX
   D YY (1)

4. Which **one** of the following statements is not true of mitosis?
   A In the initial stages of cell division, the chromosomes divide to form daughter chromatids.
   B Mitosis is used to replace old, worn-out cells.
   C Two identical daughter cells called clones are formed.
   D Genetic variety is introduced during the process. (1)

5. Which of the following processes occurs during meiosis?
   i One cell division
   ii Creation of a spindle
   iii Production of four haploid cells
   iv Doubling of the chromosomes to form pairs of chromatids

   A i, ii and iv
   B ii, iii and iv
   C i, iii and iv
   D i and iii only (1)

6. Which **one** of the following statements is true of meiosis?
   A Genetic variation is introduced during the process.
   B Meiosis is used to replace old, worn-out cells.
   C Two identical daughter cells called clones are formed.
   D Meiosis is involved in asexual reproduction. (1)

7 Which **one** of the following is not an example of discontinuous variation?
   A   Sex
   B   Blood group
   C   Height
   D   Ability to roll your tongue (1)

8 Copy this statement and fill in the gaps with the appropriate terms.

New cells are needed for _____ and to _____ worn out cells. The new cells must have the same _____ in them as the originals. Each cell has a _____ containing the _____ grouped together on _____. The type of cell division that produces identical cells is called _____. (7)

9 Division of the body cells takes place all the time in living organisms.
   a   Why is mitosis so important? (3)
   b   Explain why the chromosome number must stay the same when the cells divide to make other normal body cells. (3)
   c   Describe the process of mitosis and explain the experimental procedure that enables us to see this process taking place. (10)

10 a   State the number of pairs of chromosomes in a normal human body cell. (1)
   b   State the number of chromosomes in a human ovum. (1)
   c   Give the number of chromosomes in a fertilised human ovum. (1)
   d   Describe what happens to the chromosomes when an ovum and a sperm meet at fertilisation. (2)
   e   Explain how sex is determined in a human at the moment of fertilisation. (2)

11 a   Name the special type of cell division that produces gametes from ordinary body cells. (1)
   b   State where in the body this type of cell division takes place. (2)
   c   Explain why this type of cell division so important in sexual reproduction. (2)
   d   Describe the process of cell division by which gametes are formed – draw diagrams if necessary. (8)

12 The variation between individuals does not depend only on genetics. Write an essay explaining how the variation between people arises, giving evidence where possible for the effects you describe. (15)

Chapter 14: Genetics

When you have completed this chapter, you will be able to:

- describe genetic conditions resulting from mutations including Down syndrome, Turner's syndrome, Klinefelter's syndrome, and albinism
- explain the inheritance of a single pair of characteristics (monohybrid inheritance)
- explain the concept of genetic engineering and discuss the advantages and disadvantages of this technology
- discuss natural selection including the prevalence of sickle cell anaemia in people of African descent
- carry out simple genetic crosses using the Punnett square
- define the terms 'dominant', 'recessive', 'homozygous', 'heterozygous', 'genotype' and 'phenotype'
- describe the inheritance of sex-linked conditions such as haemophilia and colour blindness
- explain the inheritance of genetic traits including albinism, tongue rolling and sickle cell anaemia
- explain the term 'mutation' and develop the idea of genetic diseases

What the examiners say

- Candidates are unable to convincingly explain the terms related to the study of genetics and lack a basic understanding of the topic.
- Additionally, candidates are unfamiliar with the methods of treating conditions such as haemophilia, and sometimes seem unfamiliar with the relationship between the recessive allele in the X chromosome and sex-linked diseases.

Revision tip

Discuss the answers to questions and tests with your teacher. This will help you to understand how to use the information

In the previous chapter, we looked at the processes of mitosis and meiosis, and considered why there is so much variety in the human race. In this chapter, we are going to look in more detail at genetics and inheritance – the science of how information is passed from parents to their children.

Ideas about genetics, chromosomes and genes are everywhere in the 21st century. We read about them in the papers, see them on TV and learn about them in science lessons. Yet strange as it may seem, for hundreds of years people had no idea about how information was passed from one generation to the next.

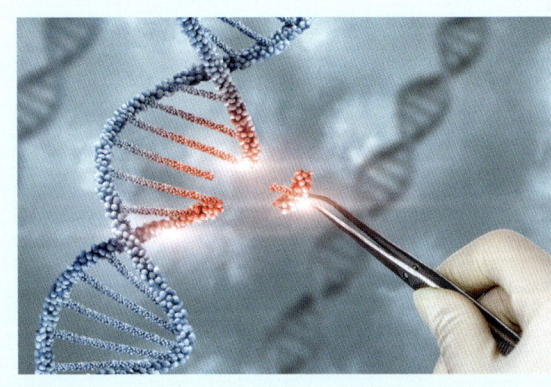

Figure 14.1 Genetic engineering may help to cure some diseases in the future by replacing the faulty genes with healthy ones.

Discovering genetics

For centuries, people thought that the characteristics of the parents blended together so that the distinct characteristics of each parent were lost. In other words, a cross between a black dog and a white dog would produce grey puppies. The birth of Gregor Mendel in 1822 was the beginning of the end for those theories. He carefully bred different pure strains of peas – round peas, wrinkled peas, green peas and yellow peas – and then carried out breeding experiments with them. See Figure 14.2. He developed a theory to explain his observations, based on independent particles of hereditary material, some of which were dominant over others but which never mixed together.

Mendel kept precise records of everything he did and made a statistical analysis of his results – something almost unheard of in those times. Finally, in 1866, when he was 44 years old, Mendel published his findings. They explained some of the basic laws of genetics in a way we still refer to today. At the time, his work was ignored, because no one had seen chromosomes in the nucleus of a cell – but later scientists realised that Mendel's findings were correct! See Figure 14.3.

Figure 14.2 Gregor Mendel's experiments and observations with pure strains of peas explained some basic laws of genetics that still apply today.

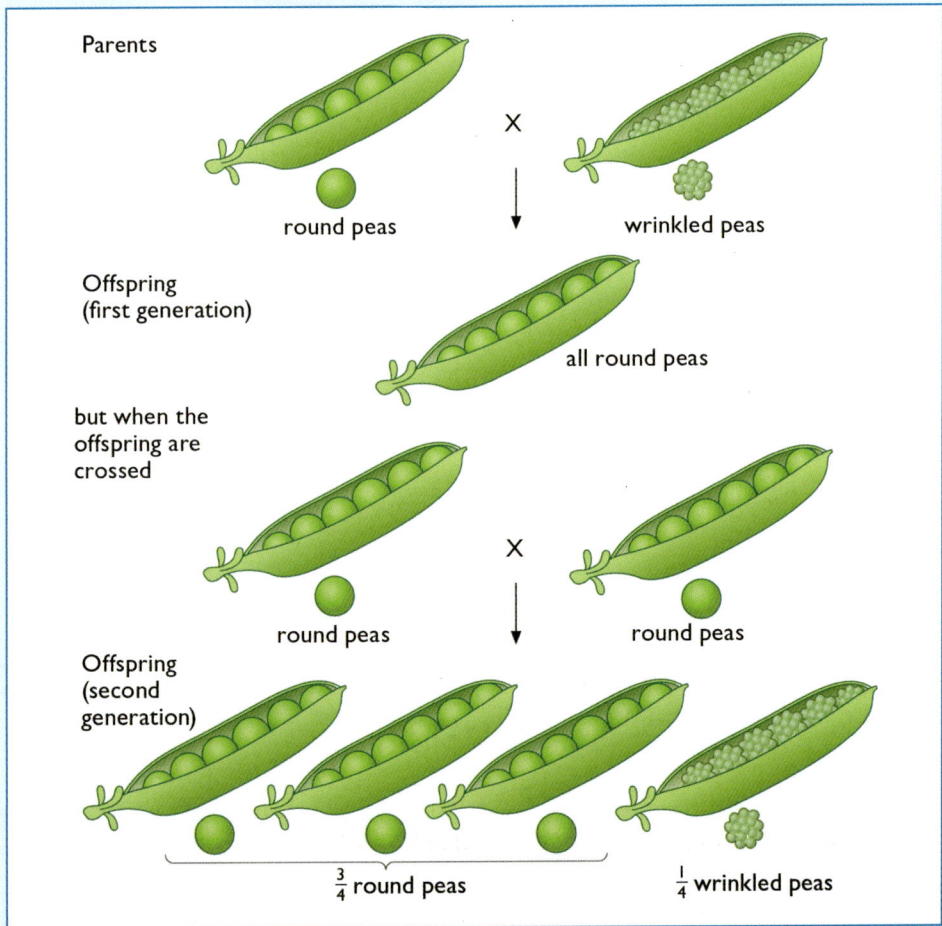

Figure 14.3 Gregor Mendel observed that the round shape of peas seemed to dominate the wrinkled shape. However, information for a wrinkled shape was carried and could emerge again in later generations. In other words, there were unique units of inheritance that were not mixed together.

How inheritance works

As a result of combining genetic information from two parents, the offspring that result from sexual reproduction show much more variation than the offspring from asexual reproduction. As you saw in Chapter 13, this is a great advantage in making sure the species survives. The more variation there is in a group of individuals, the more likely it is that at least a few of them will have the ability to survive difficult conditions. If we take a closer look at how sexual reproduction works, it becomes clear how variation appears in the offspring.

Let's look at what we have learnt so far:

- The chromosomes we inherit from our parents carry our genetic information in the form of genes.
- Many of these genes have different forms called alleles.
- A gene can be modelled as a position on a chromosome.
- An allele is the particular form of information in that position on an individual chromosome, for example, the gene for dimples may have the dimple or the no-dimple allele in place.
- Most of our characteristics, like our eye colour and nose shape, are controlled by a number of genes (Figure 14.4).
- Some characteristics are controlled by a single gene.
- Often there are only two possible alleles for a particular feature, but sometimes we inherit one from several different possibilities.

Figure 14.4 Genes control different characteristics such as eye colour, nose and ear shapes, and if we have dimples. Some characteristics are controlled by several genes, while other characteristics like dimples are controlled by a single pair of genes.

Monohybrid inheritance

Characteristics inherited through a single pair of genes are examples of **monohybrid inheritance**. Almost every example we consider in this book will be a case of monohybrid inheritance.

There are genes that decide whether (see Figure 14.5):

- our earlobes are attached closely to the side of our head or hang freely
- our thumbs are straight or curved
- we can roll our tongue (although there is some evidence that two genes are involved in this)
- we have dimples when we smile
- we have hair on the second segment of our ring finger.

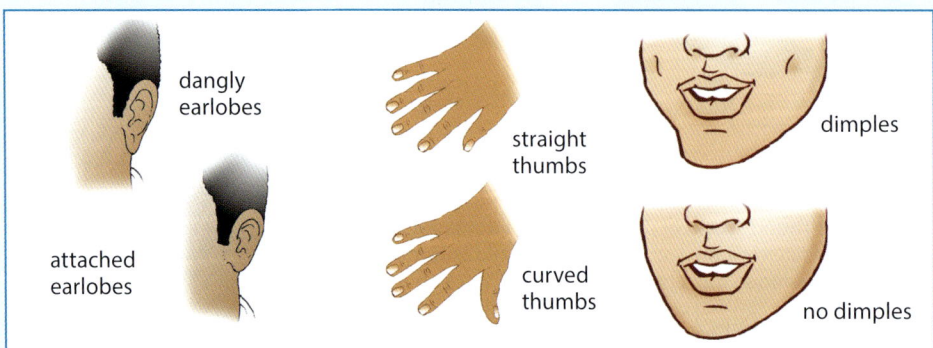

Figure 14.5 These are all human characteristics controlled by a single pair of genes, so they are very useful in helping us understand how inheritance works.

We use genes that control a single characteristic to help us understand how inheritance works.

- **Dominant** alleles control the development of a characteristic even when they are only present on one of our chromosomes, for example, dimples and dangly earlobes. When we write a symbol for a dominant allele, we use a capital letter such as **D** for dimples or **E** for dangly earlobes.
- **Recessive** alleles only control the development of a characteristic if they are present on both chromosomes – in other words, no dominant allele is present, for example, no dimples or attached earlobes. When we write a symbol for a recessive allele, we use a lower case letter such as **d** for no dimples (see Figure 14.6 and Figure 14.7) or **e** for attached earlobes (see Figure 14.9 on the next page).

Your genetic make up

You get a random mixture of alleles from your parents, which is why you don't look exactly like either of them. So how does this work? Here is an example.

The gene that controls dimples has two possible forms – an allele for dimples, which is dominant **D**, and an allele for no dimples which is recessive **d**.

- If you inherit two identical alleles, whether they are dominant or recessive, you are **homozygous** for that characteristic, for example, **DD** or **dd**
- If you inherit one of each type of allele, you are **heterozygous** for that characteristic, for example, **Dd**.

Your **genotype** describes your genetic make-up. Your **phenotype** describes your physical appearance or observable characteristics. We often use these terms relating to the inheritance of a specific characteristic, for example:

- His genotype was **Ee**. His phenotype was dangly earlobes.
- Her genotype was **ee**. Her phenotype was attached earlobes.

Look at Figure 14.7. In this case, on the pair of homologous chromosomes that the child inherits, the chromosome inherited from the father carries the dominant allele for dimples. The chromosome inherited from the mother carries the recessive allele for no dimples. The child is heterozygous – their genotype is **Dd** – and their phenotype is to have dimples, because dimples is a dominant characteristic.

Figure 14.6 This child is heterozygous for the dimple characteristic. His genotype is **Dd** and his phenotype is dimples.

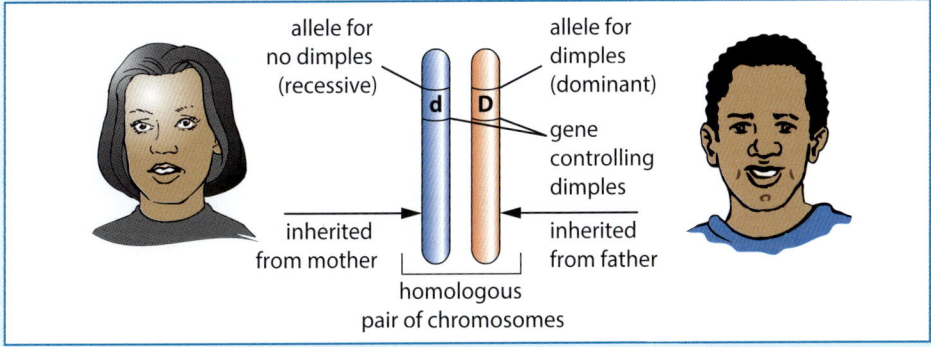

Figure 14.7 The different alleles that you inherit result in the development of very different characteristics. Here, the child inherits one allele for dimples from their father and one allele for no dimples from their mother. The child is heterozygous, and will have dimples, because having dimples is a dominant characteristic.

How inheritance works

How does it work?

If you have brothers and sisters, they could inherit a different combination of alleles to you, and so have different characteristics. This is why family members look similar, but different! See Figure 14.8.

Figure 14.8 Children inherit different allele combinations from their parents, which is why they have some features that are the same and others that are different.

We will look in detail at the way we inherit the shape of our earlobes. There are two possible alleles controlling this characteristic:

- the allele to have dangly earlobes (represented by a capital **E** because it is a dominant allele)
- the allele to have attached earlobes (represented by a small **e** because it is the recessive allele).

The characteristic that we inherit (dangly or attached earlobes), as shown in Figure 14.9, depends on the following:

- If we inherit a dangly allele from both our parents **EE** (**homozygous dominant**) or only from one **Ee** (**heterozygous**), we have dangly earlobes. (Gentotye **EE** or **Ee**, phenotype dangly earlobes.)
- If we inherit an attached allele from both our parents **ee** (**homozygous recessive**), we have attached earlobes. (Genotype **ee**, phenotype attached earlobes.)

We can use this type of information to predict the characteristics of the offspring a couple might have. Using simple crosses where a characteristic is inherited by a single pair of alleles, we can start to build up a way of predicting which offspring might result from a particular cross. This becomes very useful when we begin to look at genetic diseases.

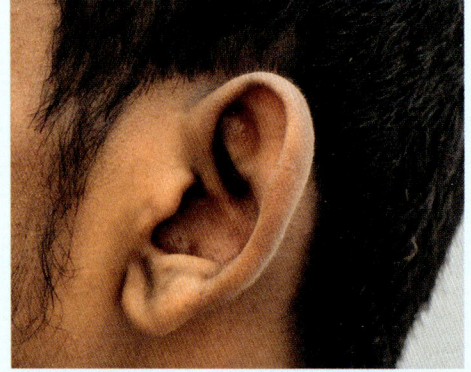

Dangly earlobes (**E**) dominant

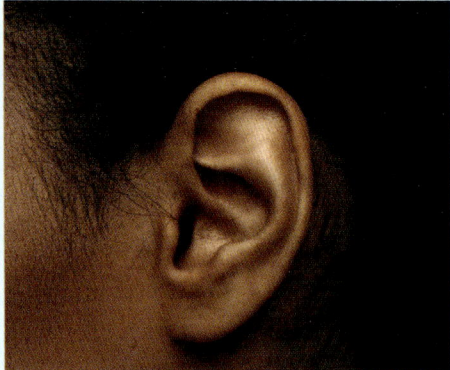

Attached earlobes (**e**) recessive

Figure 14.9 Whether our earlobes dangle or are attached directly to our head is determined by the alleles we inherit from our parents.

Genetics

The biologist's toolkit: Genetic diagrams – the Punnett square

We model genetic crosses using simple genetic diagrams. A genetic diagram shows you:
- the alleles for a characteristic carried by the two parents
- the possible gametes that can be formed from these
- the ways the alleles could combine to form the characteristic in their offspring *and* the expected ratios of the different possible offspring; the offspring are shown as the first filial generation (F_1).

The most widely used genetic diagram is a Punnett square, named after the geneticist who developed it. Punnett squares are compact and easy to use as long as you work through them methodically.

There are a number of steps to follow:
- Take the parental genotype and chose a letter to represent the characteristic. Sometimes it is obvious – in this case D or d. If you choose your own letter, make sure there is a clear difference between the capital letter and the lowercase letter, for example, Aa, Bb, Gg – this avoids confusion later!
- Provide a key to explain which symbol is which.
- Split the genotype of the parents to find the genotype of the gametes. For example, if the genotype of the parent is DD, both gametes will carry the allele D. If the genotype of the parent is Dd, half the gametes will carry the allele D and half will carry the allele d.
- Draw and use a Punnett square to work out the possible offspring.

Here is an example of a Punnett square showing the cross shown in Figure 14.7 on page 303.

Parental genotypes: mother dd, father Dd

D = dominant dimples

d = recessive dimples

Gametes from mother: d and d   Gametes from father: D and d

	D	d
d	Dd	dd
d	Dd	dd

Genotype of possible offspring: Dd or dd

Phenotype of possible offspring: dimples or no dimples

Ratio of dimples to no dimples: 1 : 1 or 50% of each

To work out the possible gametes, you need to look at the genotypes of each of the parents. So, to use our dangly earlobes example again, if you:
- have the genotype **EE**, both of the possible alleles you could pass on in your gametes are **E**
- attached earlobes, you have genotype **ee** and both of the possible gametes you might produce would carry the recessive allele **e**
- are heterozygous, **Ee**, your gametes may contain either the dominant allele **E** or the recessive allele **e** – and it is completely a matter of chance which one will meet up with another gamete.

In a genetic diagram, you will end up with all the possible genotypes of the offspring – that is, the alleles they might inherit. From this you can work out their possible phenotypes, which are the physical characteristics that they will have as a result of their genotype. For example, someone with a genotype **EE** or **Ee** will have dangly earlobes as their phenotype. Only someone with the genotype **ee** will have attached earlobes as their phenotype.

You can use simple diagrams to work out genetic crosses. Figure 14.10 provides you with some clear examples to follow. You need to indicate clearly the genotypes of the parents, the possible gametes, the possible genotypes of the offspring and also the ratio of the different possible phenotypes of the offspring.

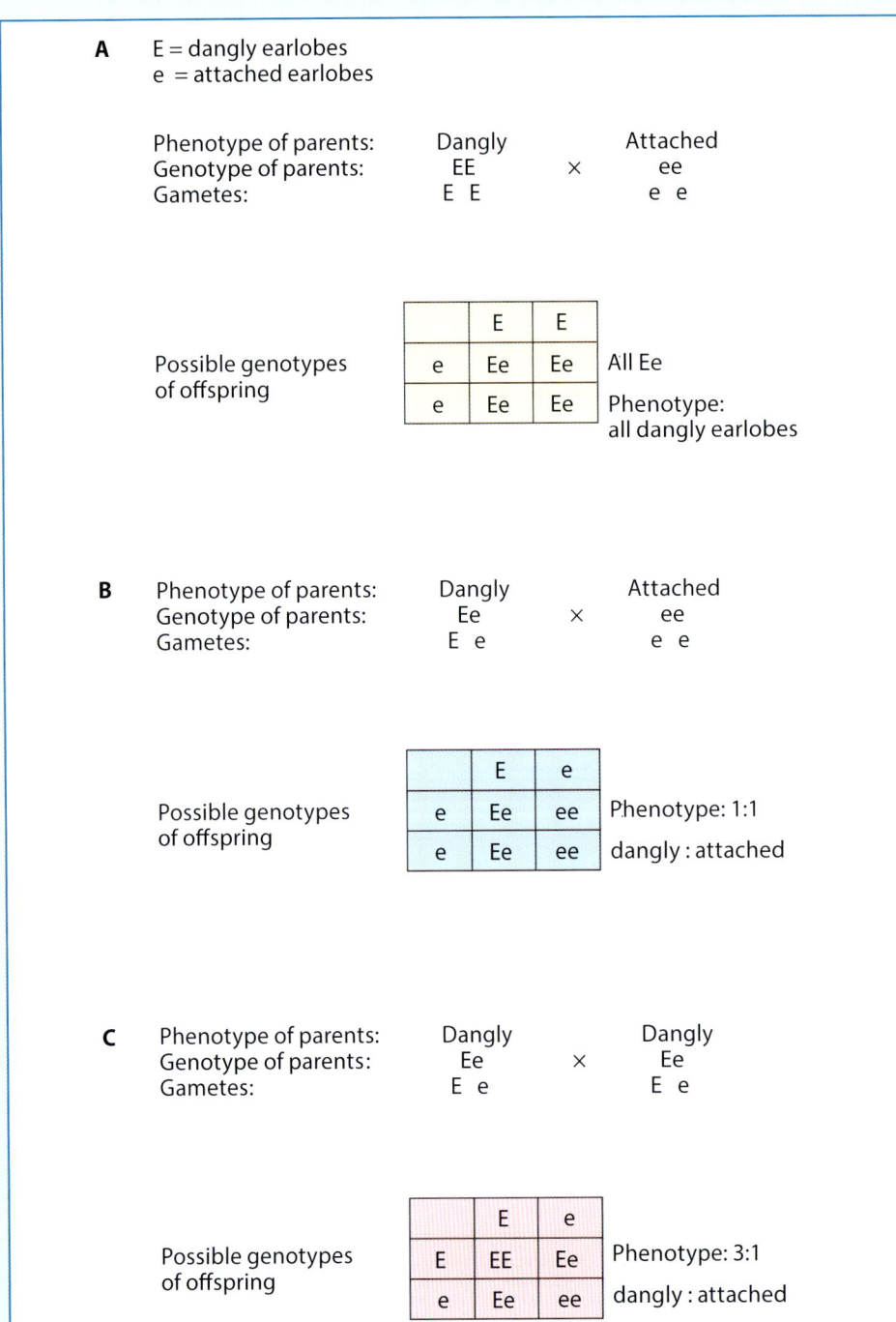

Figure 14.10 Punnet squares like these show us the parents, possible gametes and possible offspring of a monohybrid genetic cross and let us work out the likelihood of particular combinations of alleles being passed on.

Another example of monohybrid inheritance in people that is often quoted is the ability to roll your tongue or not – see Figure 14.11. For many years it has been thought that the ability to roll your tongue is inherited through a dominant allele in the way shown in the Punnett square in Figure 14.12. However, it has been observed that in some pairs of identical twins, one can roll their tongue and the other cannot, and some people teach themselves to roll their tongue. Now geneticists are left trying to work out exactly what happens, and tongue rolling is not always seen as a simple example.

Figure 14.11 The inheritance of tongue rolling is still often shown like this, but now scientists are not so sure. This reminds us that scientific ideas are always changing.

T = roller
t = non-roller
offspring phenotype: 3 rollers, 1 non-roller

	T	t
T	TT	Tt
t	Tt	tt

Figure 14.12 Punnett square to show roller and non-roller

Checkpoint questions

1 Describe how the inheritance of dominant traits differs from the inheritance of recessive traits.

2 Look at your thumbs. Are they straight or curved? This is a characteristic inherited on a single gene. The allele for straight thumbs is dominant over the allele for curved thumbs. Choose suitable symbols and, using Punnett squares, draw a genetic cross between:
   a two heterozygotes
   b a heterozygote and a homozygous recessive individual.

Inherited conditions in humans

Our genetic inheritance from our parents often contains nothing worse than a big nose or a tendency to gain weight easily. However, sometimes the genetic combination we receive has more noticeable and dramatic effects. A mutation – a change in the DNA – may cause problems. A number of factors increase the risk of mutations in the gametes:

- Older parents, especially women – the ova are present all through life, so mutations may build up. Older men are also more likely to experience mutations as their testes make sperm.
- Ionising radiation such as X-rays increase the risk of mutations in the gametes
- Certain substances, including some medicines, increase the risk of mutations.

In the inherited condition called **albinism**, where a mutation affects a single gene, the melanin pigment in the skin, hair and eyes fails to develop. The standard allele for pigment to develop is represented by A and it is dominant. The allele for albinism, a, is recessive.

Albinism is found throughout the animal kingdom and people are no exception. People with albinism are very vulnerable to sun damage to their skin, so they have a greatly increased risk of developing skin cancer (Figure 14.13). They have to take great care to protect their vulnerable skin from sunlight. Their eyes are also very sensitive to light and they often have problems with their vision, but apart from this they are typical, healthy individuals.

Figure 14.13 Albinism is particularly noticeable in regions where most people have a high level of protective melanin in their skin, such as the Caribbean. The birth of a child with albinism often appears quite random, but it is the result of hidden recessive alleles on both sides of the family.

If people do not understand genetics, the arrival of a baby with albinism can cause great distress. In the past, people with albinism often suffered discrimination as a result of their unusual appearance. However, for albinism, and indeed any other genetic traits, looking at the family history can show exactly how a characteristic has been passed on (see *The biologist's toolkit: Genetic diagrams – family trees* on the next page). See Figure 14.14 below.

A   A = melanin production
a = albinism – no melanin production

Phenotype of parents: melanin produced   melanin produced
Genotype of parents: Aa   ×   AA
Gametes: A a   A A

Possible genotypes of offspring

	A	a
A	AA	Aa
A	AA	Aa

Phenotype: all produce melanin

B   Phenotype of parents: melanin produced   melanin produced
Genotype of parents: Aa   ×   Aa
Gametes: A a   A a

Possible genotypes of offspring

	A	a
A	AA	Aa
a	Aa	aa

Phenotype: 1:1 melanin produced : albinism

Figure 14.14 Children with albinism are usually born to parents who produce melanin. A parent with albinism will not have a child with albinism unless their partner also carries the allele for albinism.

Genetics

The biologist's toolkit: Genetic diagrams – family trees

A pedigree or family tree models how a particular genetic characteristic is passed through a family.

Recessive characteristics such as albinism often show up relatively rarely, as the recessive allele must be inherited from both parents. Dominant characteristics are seen more often, as an individual only needs one allele to show the characteristic.

On a family tree:
- men are usually shown as squares and women as circles
- individuals affected by a particular trait are shaded.

Once a pedigree is built up, it is possible to see when a mutation may have taken place, and to work out the possible genotypes of many of the individuals. Figure 14.15 below shows the pedigree for albinism in a family.

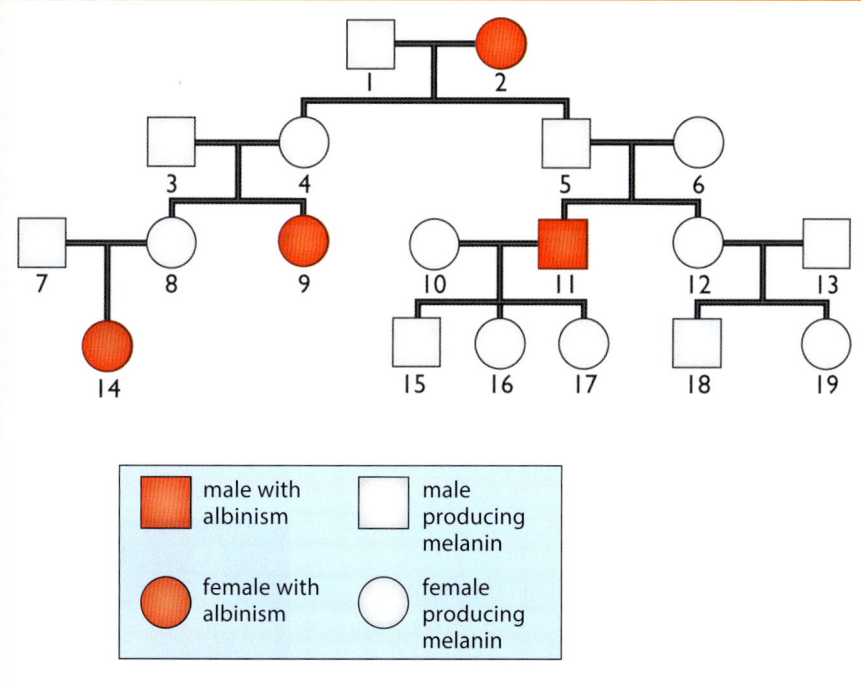

DID YOU KNOW?

Some genetic disorders cause such chaos in the cells that the foetus cannot develop properly and miscarries. But other genetic disorders, such as colour blindness, have little effect on people's lives.

Figure 14.15 In this family tree, you see that albinism is relatively rare, because it depends on both partners carrying the recessive albino allele and by chance passing it on to their child. The chances of two **carriers** meeting, having children AND both passing on the recessive gene is relatively small (see Figure 14.14).

Genetic diseases

Sometimes conditions passed on in our genes can be harmful or even lethal. They are called **genetic diseases** or **genetic disorders**. When a baby is born with a genetic disease, it is an example of a **congenital defect** – damage that is present at birth. In fact, 2–3% of all babies are born with congenital defects, and 25% of those defects are due to problems in the genes.

As you saw in Chapter 13, sometimes the genetic material changes – we call this a **mutation**. Some of these genetic disorders, like Down syndrome, are the result of a mutation that results in a extra copy of a whole chromosomes, but most are the result of single gene defects. We can use our knowledge of dominant and recessive alleles to work out the risk of inheriting one of these genetic diseases. A very common genetic disease that particularly affects people of African descent is **sickle cell anaemia**.

Sickle cell anaemia

It is easy when looking at genetic diseases to think of all mutations as bad and harmful. However, it is very important to remember that mutations are a source of variation. Something that is a disadvantage under some circumstances may become an advantage in other circumstances. Some mutations are so severe that it is impossible to imagine that they could ever be of benefit, but life can be full of surprises. A good example of this is the genetic disease called **sickle cell anaemia**, which affects many people of African descent across the Caribbean and other areas of the world. The inheritance of sickle cell anaemia is an example of natural selection in action.

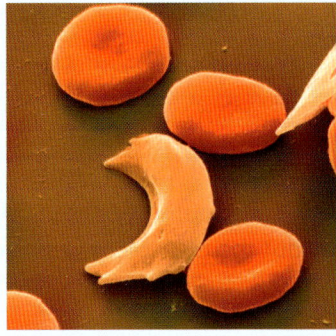

Figure 14.16 This simple change in the shape of the red blood cells – a sickle shape – gave sickle cell anaemia its name.

Sickle cell anaemia, also known as sickle cell disease, is the result of a single mutation that affects our haemoglobin. The haemoglobin fills our **red blood cells** and carries oxygen around the body (see Chapter 8 to remind you of the importance of the red blood cells). This disease changes the shape of the red blood cells from the usual biconcave discs to shrunken sickle shapes (see Figure 14.16). These sickle-shaped red blood cells give the disease its name. The problem is that these cells cannot carry oxygen properly. They are stiff and do not move through the blood vessels easily. They can even cause blockages in the blood vessels.

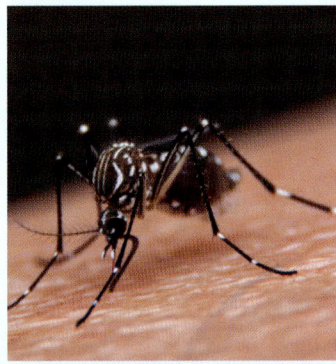

However, the mutated allele responsible for sickle cell anaemia has another effect. Malaria is an infectious disease carried by the bite of mosquitoes, and it causes death and illness to millions of people every year across the world (see Figure 14.17 and Chapter 18). Sickle cell anaemia is particularly common in those areas of the world badly affected by malaria, such as Africa.

Around the world, offspring who are homozygous for sickle cell anaemia may die of the disease without regular, intensive treatment including blood transfusions. Many people who are homozygous for the allele for normal haemoglobin will become ill or die of malaria. Those who are heterozygous for the sickle cell allele will be the healthiest. Although they will be slightly anaemic, this will not have a big effect on their quality of life, and they will not be struck down by malaria. The sickle cell allele causes the death of many of those unfortunate enough to be born with two copies, but it saves the lives of many thousands more who are carriers of just one copy of the allele. This explains the prevalence of the sickle cell allele in people of African descent.

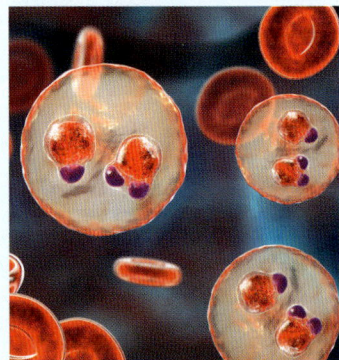

Figure 14.17 A mosquito bites a person and injects them with the plasmodium parasite, which causes malaria. The sickle-shaped red blood cells may prevent some of the plasmodium parasites from infecting the red blood cells. This reduces the number of parasites in the red blood cells and provides some protection against malaria.

In terms of natural selection, heterozygotes for sickle cell anaemia had a better chance of surviving to reproduce than homozygotes for either the dominant or the recessive allele – see Figure 14.18.

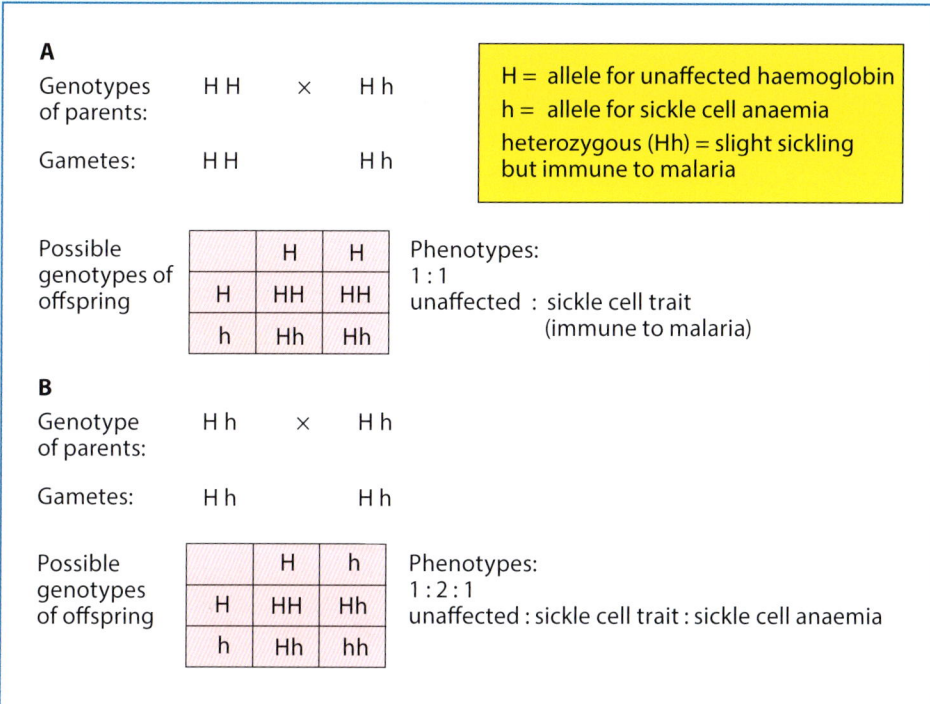

Figure 14.18 It is such an advantage to inherit just one sickle cell allele that natural selection has resulted in a high prevalence of sickle cell anaemia and sickle cell trait in people of African descent, wherever they now live in the world.

Sex-linked diseases

All of the examples of monohybrid inheritance we have looked at so far involve genes found on the autosomes – the ordinary chromosomes that control most of our body. However, some gene mutations that cause problems are found on the sex chromosomes. You learnt in the previous chapter how human beings inherit either two X chromosomes (female) or an X and a Y chromosome (male) (see Figure 14.19 and Figure 14.20).

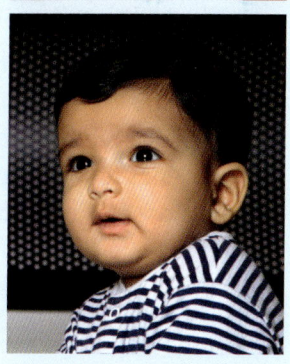

Figure 14.19 A baby girl inherits two X chromosomes (XX) and a baby boy inherits an X and a Y chromosome (XY).

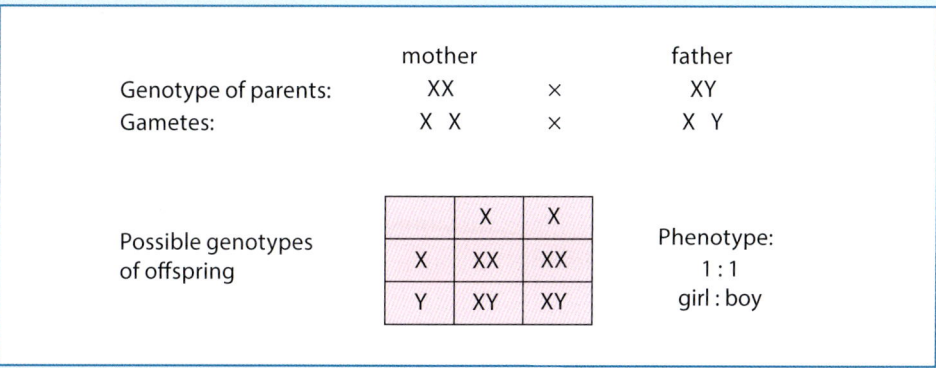

Figure 14.20 You use Punnett squares to model the inheritance of the sex chromosomes – every time a baby is conceived there is a 50% chance it will be a boy – and a 50% chance it will be a girl!

Sex-linked diseases 311

Some problems are linked to the inheritance of the sex chromosomes themselves. Just as Down syndrome occurs when a mutation leads to three copies of chromosome 21 in the cells, if meiosis goes wrong and there is a mutation as the gametes form, we may end up with ova or sperm with an extra X chromosome. This may cause gametes that contain XX or XY instead of being X or Y only, or gametes that do not contain a sex chromosome at all. Examples of the problems that may arise include the following:

- **Klinefelter's syndrome:** This affects only males, who are born with an extra copy of the X chromosome so they are XXY. The males affected are still genetically male and they often do not realise there is a problem. It affects around 1 in 660 males. They are likely to be infertile and have a low sex drive, and there are some linked health problems but many people who are affected never know about it. See Figure 14.21.

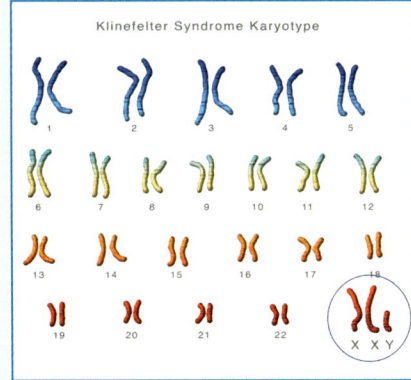

Figure 14.21 Klinefelter's syndrome karyotype showing the extra X chromosome

- **Turner's syndrome:** This affects only females, who are born with a single X chromosome instead of the usual two XX chromosomes. Females who are affected by Turner's syndrome are usually shorter than average and have underdeveloped ovaries, so they are infertile. It is often not diagnosed until puberty and has some linked health problems, but most affected individuals lead healthy, active lives. See Figure 14.22.

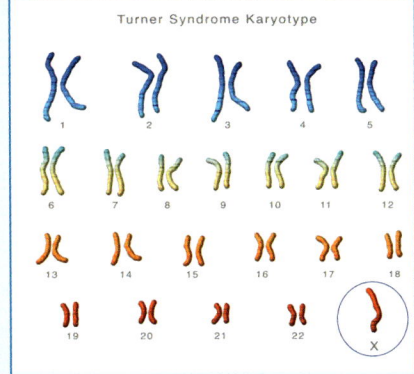

Figure 14.22 Turner's syndrome karyotype showing the single X chromosome

Some genetic diseases are carried on the sex chromosomes – usually the X chromosome because the Y chromosome is small and carries almost exclusively information about maleness. Genetic disorders that involve genes found on the sex chromosomes are called **sex-linked diseases** or **sex-linked conditions**.

Colour-blindness

One of the best known and most common sex-linked conditions is **red–green colour blindness**. A mutation on the X chromosome results in a recessive allele that affects the ability to see in full colour. The mutation affects the production of visual pigments in the retina of the eye (see Chapter 11). The condition is not serious, and if affected, people manage to work their way around it – even traffic lights are not a major problem. However, there are certain jobs, such as being a commercial pilot, which people with colour blindness are not allowed to do. Have a look at Figure 14.23.

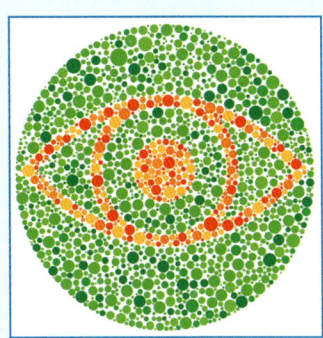

Figure 14.23 If you can see the eye shown in this circle then you are not colour blind. If not, you may have one of the most common sex-linked mutations.

Females have two X chromosomes. They are only red–green colour blind if they are homozygous for the recessive allele, so colour blindness is not very common in females. Since males only have one sex chromosome, they only have to inherit one faulty allele to be affected, so colour blindness is much more common in males than females – see Figure 14.24.

The percentage of people affected by colour blindness also varies in different ethnic groups. In fact, around 8% of Caucasian men and 1% of women are red–green colour blind, while 4% of black American men and 0.8% of women are affected by colour blindness.

When you draw out a genetic cross for a sex-linked condition, you have to show the chromosomes as well as the alleles, and you need to indicate whether offspring are male or female (see Figure 14.24).

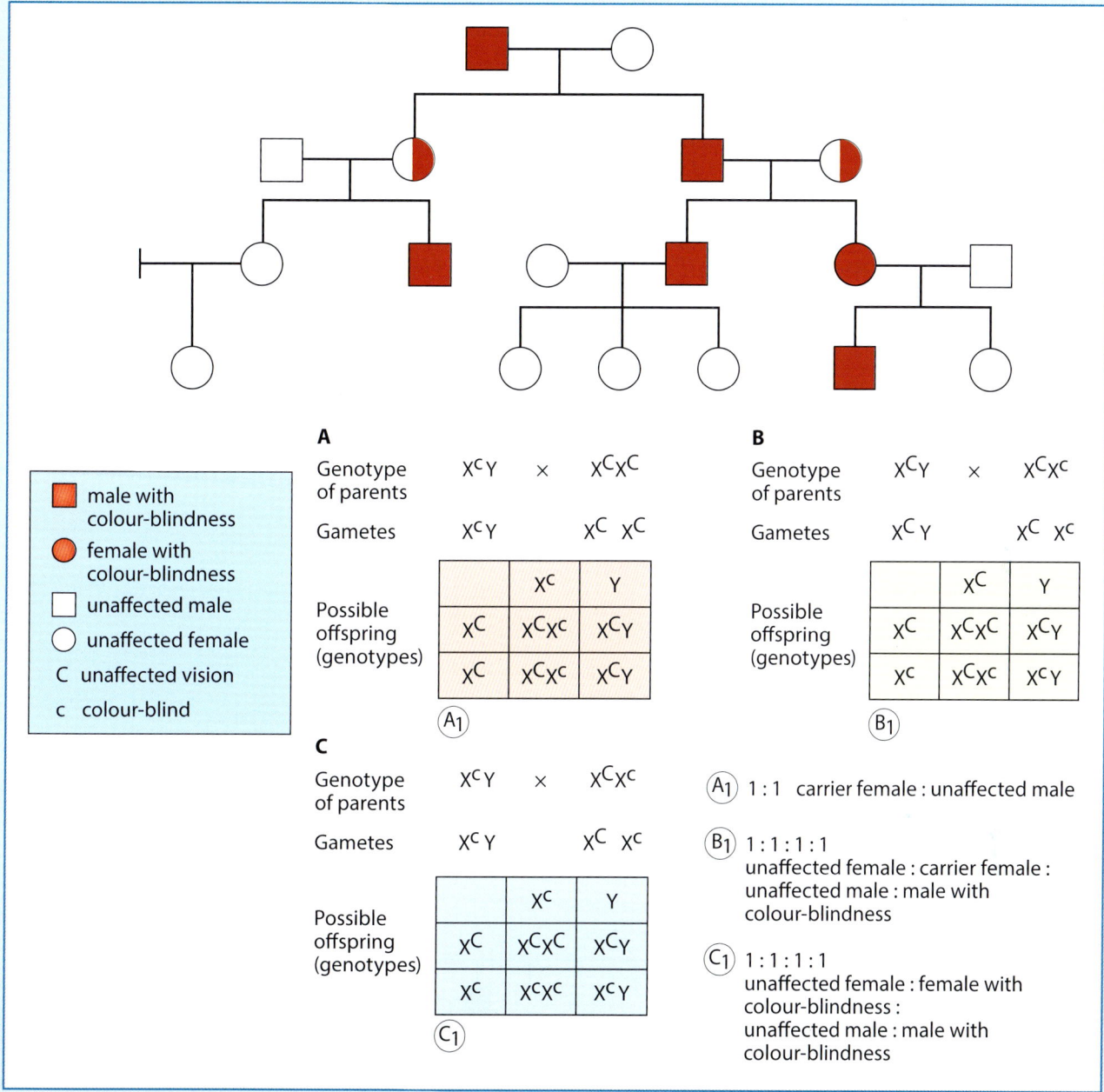

Figure 14.24 Colour blindness in a family tree and Punnett squares to show how this sex-linked condition is inherited.

Haemophilia

Colour blindness is an inconvenience but nothing more. However, some sex-linked conditions are much more serious. Human males are always affected much more by sex-linked conditions than females, as they only have one X chromosome. Therefore, they only have to inherit one faulty allele to be affected. Their mothers are usually heterozygous carriers who are often quite unaware that they are carrying a potentially damaging recessive allele.

Sex-linked diseases

Haemophilia is a sex-linked condition in which one of the proteins needed for the clotting of the blood (see Chapter 8) is missing. As a result, the blood does not clot properly and without treatment, the slightest injury or bruise can result in heavy bleeding either externally or internally – so much so that the affected person may die. Fortunately there is an effective treatment; the missing clotting factor can be injected regularly. The treatment has to continue for life as there is no cure for this condition as yet, although genetic engineering offers real hope.

The gene for the clotting protein is carried on the X chromosome and the homozygous form is almost always lethal. Haemophilia can occur anywhere – the mutation favours no one. As the future Queen Victoria of Britain (see Figure 14.25) developed in her mother's womb, the mutation took place in her embryonic ovaries. She had eight children, most of whom married into other royal families. As a result, haemophilia plagued the royal families of Europe for many years (see Figure 14.26).

Figure 14.25 Queen Victoria of Britain passed on the mutated gene for haemophilia to her children, and through them to her grandchildren and great-grandchildren.

Genotype of parents: $X^H X^h$ × $X^H Y$
Gametes: X^H X^h X^H Y

Possible offspring genotypes

	X^H	X^h
X^H	$X^H X^H$	$X^H X^h$
Y	$X^H Y$	$X^h Y$

Offspring phenotypes:
$X^H X^H$ female with all clotting factors
$X^H X^h$ carrier female with all clotting factors
$X^H Y$ male with all clotting factors
$X^h Y$ male with haemophilia

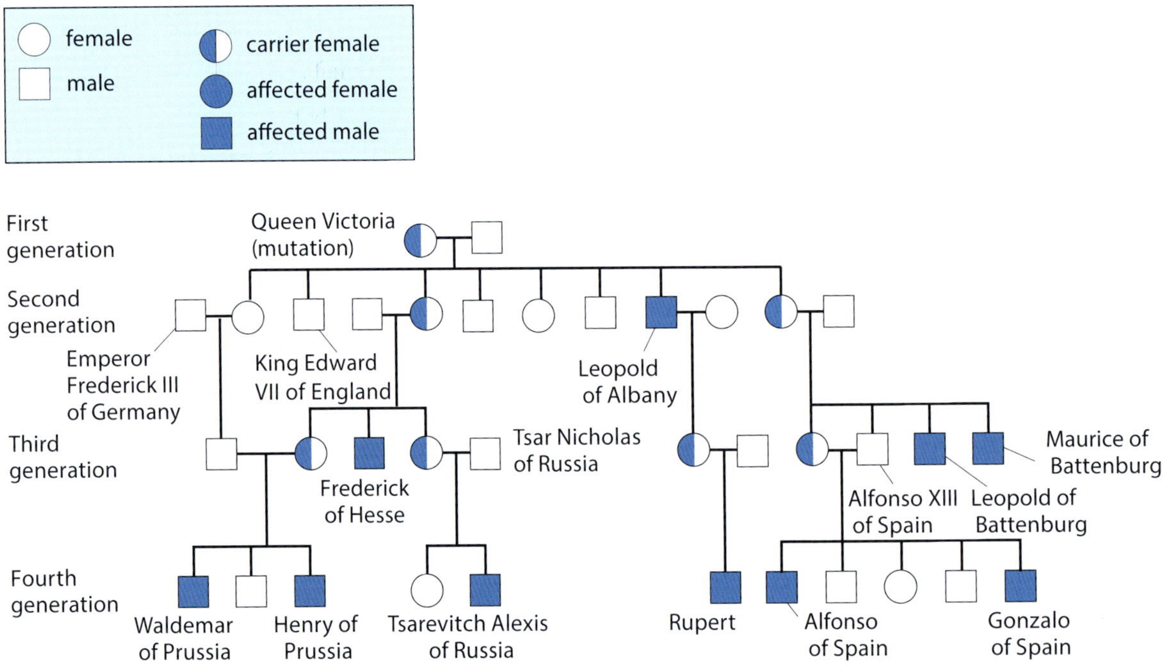

○ female   ◐ carrier female
□ male   ● affected female
        ■ affected male

Figure 14.26 Genetic disorders such as haemophilia can affect a family for many years, as this Punnett square and British royal family tree clearly shows.

Curing genetic diseases

So far, we have no way of curing genetic diseases. Scientists hope that genetic engineering (genetic modification) will enable them to cut out faulty alleles and replace them with healthy ones. They have tried this in affected people, but so far it is proving more difficult than they thought.

There are genetic tests that can show people in affected families whether or not they carry the faulty allele. This, in turn, allows them to make choices as to whether or not to have a family. It is also possible to screen embryos for the alleles that cause these and other genetic disorders. These tests are very useful but raise many ethical issues (see Chapter 12). The challenge of treating genetic diseases is one that has not yet been solved, but maybe in your lifetime, diseases such as sickle cell anaemia and haemophilia will become things of the past.

Genetic engineering

In recent years, scientists have discovered ways of changing the genetic material of an organism in a process called **genetic engineering** (also called **genetic modification** or **gene editing**). Genetic engineering is used to give an organism new characteristics that we want to see or that are desirable.

The concept of genetic engineering is as follows:

- Identify a gene that is faulty, or which could be improved, in an organism.
- Identify a healthy or better gene for this function in another organism, which may be the same species, a closely related species or a very different species.
- Using special molecular tools (see Figure 14.27), cut out the failed gene and replace it with another small piece of DNA from another organism.
- The new DNA, containing the new or modified gene, works in the organism and overcomes the problem first identified.

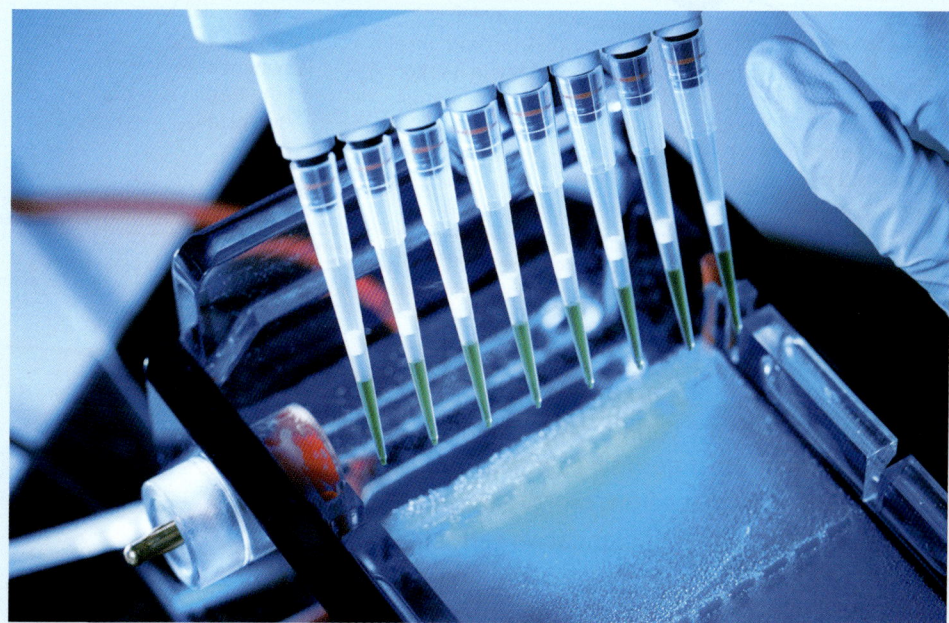

Figure 14.27 Gel electrophoresis is one of the molecular tools that scientists use in genetic engineering. They use it to view cut-out and inserted DNA.

The new DNA, which contains genetic material from two different organisms (often two different species), is called **recombinant DNA** (see Figure 14.28). So, for example, genes from the chromosomes of one of your human cells can be 'cut out' using enzymes and transferred to the cell of a bacterium. Your gene carries on coding for a human protein, even though it is now in a bacterium. The organisms that contain changed or edited DNA are known as **genetically modified organisms** or **GMOs**.

Figure 14.28 Synthetic insulin is genetically engineered using recombinant DNA, which is then cultured in bioreactors like these, before being harvested and prepared for use.

If bacteria are cultured on a large scale, they make huge quantities of protein. If genetically modified bacteria are cultured, they will make huge quantities of proteins from another organism. Scientists are making use of this ability. We now use GMOs to make a number of drugs and hormones for medicines.

One of the best examples of how we use GMOs is the production of human insulin using genetically engineered bacteria. Figure 14.29 shows you how this works. People with diabetes need supplies of the hormone insulin (see Chapter 10). This insulin used to be extracted from the pancreas of pigs and cattle but it wasn't quite the same as human insulin, and the supply was variable. Both of those problems have been solved by the introduction of genetically engineered human insulin. Using recombinant DNA, with a human insulin-producing gene inserted into the bacterial genetic material, this piece of genetic engineering means that people with diabetes are guaranteed a regular supply of pure, human insulin to keep them healthy.

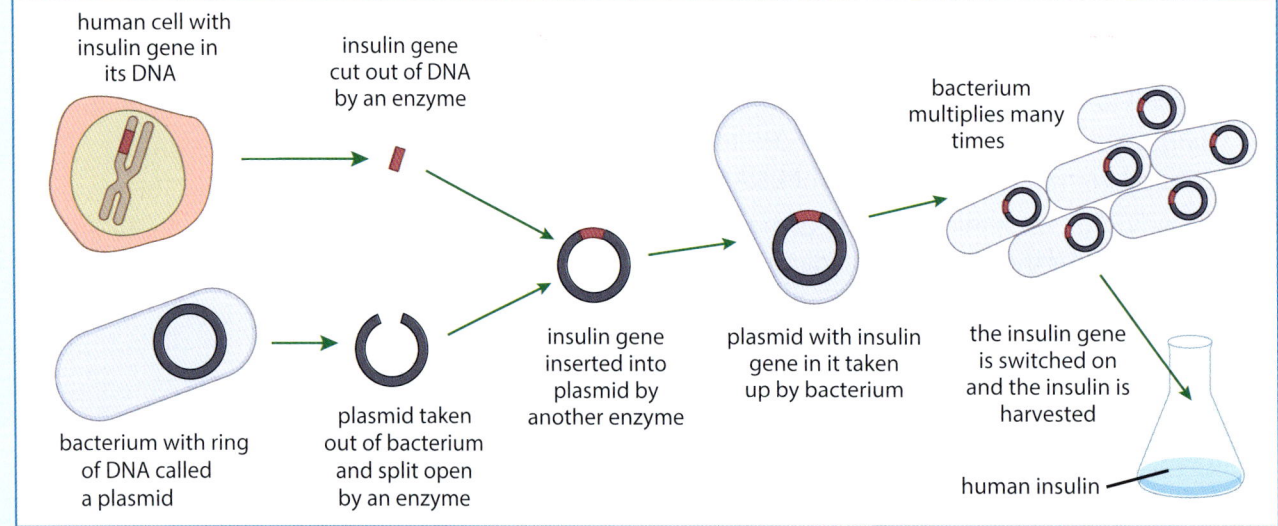

Figure 14.29 The principles of genetic engineering: In this case, bacterial cells receive a gene for the production of insulin from a human being and go on to make huge quantities of the insulin needed by people with diabetes all over the world.

Transferring genes to animal and plant cells

There is a limit to the types of proteins bacteria are capable of making. As a result, genetic engineering has moved on to insert genes into many different types of organisms. Scientists have found that genes from one organism can be transferred to the cells of another type of animal or plant at an early stage of their development. As the animal or plant grows, it develops with the new desired characteristics from the other organism.

Genetics

Advantages and disadvantages of genetic engineering

There are many advantages to using genetic engineering. Here are some examples:

- Crop plants and farm animals have improved growth rates, helping to provide more food for all.
- Genetically modified (GM) crops usually have bigger yields or higher nutritional values than traditional crops, for example, golden rice with increased vitamin A (Figure 14.30), so the crops have increased food value.
- Crops can be modified to grow well in particularly hot and dry or wet conditions, to help overcome problems of climate change, for example, rice that can withstand being submerged for up to three weeks during flooding.
- Crop plants are being engineered to produce their own pesticide or are being made resistant to the herbicides used to kill weeds, for example, GM soy beans used globally contain herbicide resistant genes.
- Some exciting work involving producing bananas that contain human vaccines and other fruits containing human medicines is being done. This would make it much easier to supply vaccines to people in many countries across the developing world.

Figure 14.30 Genetically modified golden rice crop with increased vitamin A for added nutrition.

However, there have been some concerns about possible disadvantages to genetical engineering, such as the long-term issues GMOs. As evidence builds, many of the concerns, so far, do not apply.

- Some people are concerned about possible long-term issues arising – but as the evidence builds, no problems have been identified.
- Some people were concerned about eating GM crops – but everyone eats a wide range of DNA every day and we have the enzymes we need to break it down, so this has not proved to be a problem.
- There have been concerns about genes from GM crops spreading into wild populations – so far this has not proved a major problem and in many cases hasn't happened at all.
- GM crops cost more than ordinary crops, and there are concerns that this will make them unavailable to poorer countries who need the advantages they bring the most.

The biggest ethical issue comes with the use of genetic modification on people to try to cure genetic diseases. Most people would like to see an end to genetic diseases, but the concern is that people would try to use GM to have more intelligent children, or prettier children, or taller offspring with darker or lighter skin. There are many issues for us all to think about, but also enormous benefits.

Possible SBAs based on cell reproduction and variation

→ Using beads, modelling clay, paper chains or other media, make models of genetic crosses.
→ Investigate the incidence of a specific genetic disease, for example, sickle cell anaemia in your school/community/country, and do some statistical analysis to see which age groups/ethnic groups, and so on, it appears in most frequently.
→ Look at patterns in genetic diseases across the Caribbean.
→ Investigate the incidence of Down syndrome across the Caribbean compared to a country where women tend to have children when they are older. Consider different factors that will affect the incidence including maternal age, prenatal testing for genetic abnormalities, and so on.
→ Find and draw out/make karyotypes to illustrate different human genetic conditions, for example, normal male and female, Down syndrome male and female, Klinefelter syndrome, Turner syndrome, and so on.
→ Develop a game where students have to work out genetic crosses to score points and win prizes.
→ Produce an animation to show how genetic engineering is carried out.
→ Make a video presentation of the issues – advantages and disadvantages – surrounding the use of GM technology.

End-of-chapter summary

In this chapter, you have learnt that:

- Gregor Mendel was the first person to suggest separately inherited factors, which we now call genes
- genes have different forms called alleles
- genes come in pairs; a pair of genes controls a particular characteristic or set of characteristics
- each member of a pair of genes may have a different allele
- the term 'genotype' describes the genetic make-up of an individual, while their phenotype describes their physical characteristics
- a characteristic controlled by a dominant allele will be present even if only one of these alleles is inherited (present on only one chromosome)
- a characteristic controlled by a recessive allele is only present if the allele is inherited from both parents (present on both chromosomes)
- if an individual is homozygous for a genetic trait, they have two identical alleles for that characteristic
- if an individual is heterozygous for a genetic trait, they have two different alleles for that characteristic
- in human body cells, the sex chromosomes determine whether you are female (XX) or male (XY)
- you can represent genetic crosses – the parents, the gametes and the possible offspring – using simple genetic diagrams known as Punnett squares
- some features or characteristics are controlled by a single gene and you can use these to look at monohybrid inheritance; examples include dimples, dangly or attached earlobes, straight or curved thumbs, and tongue rolling

- new alleles (forms of genes) result from mutations (changes) in existing genes
- the chance of mutations occurring is increased by ionising radiation and certain chemicals
- mutations are often neutral, but they can be harmful – causing abnormalities in babies and cancer – or beneficial, giving an individual an advantage in the race for survival
- the individuals with the genetic make-up that makes them best suited to their environment are most likely to survive and produce offspring – this is natural selection
- some diseases and disorders are inherited; there are a number of conditions that are inherited – some of them have little impact on health and life expectancy, while others have a very severe effect
- healthy people who have a disease-causing recessive allele are known as carriers
- albinism is an inherited condition where melanin pigment is not produced in the body
- sickle cell anaemia is a genetic disease that causes severe illness in those people who inherit two sickle cell alleles; however, it gives resistance to malaria to people who inherit one sickle cell allele; this is an example of monohybrid inheritance where being heterozygote brings great benefit to the carriers of the disease; this is the cause of the high prevalence of sickle cell anaemia in individuals of African descent
- some genetic conditions are sex-linked – in other words, the gene is on the X chromosome; sex-linked conditions are more common in males as they only have one X chromosome
- in Klinefelter's syndrome, the male has an extra X chromosome (XXY), and in Turner's syndrome, the female has a single X chromosome
- red-green colour blindness is sex-linked and much more common in males than females, but it has little or no effect on health
- haemophilia is a sex-linked condition where the blood does not clot; it is often fatal if it is not treated, but modern clotting factors allow haemophiliacs to lead relatively normal lives
- genetic engineering – also known as genetic modification – involves changing the genetic make-up of an organism by removing and/or inserting the DNA
- bacteria containing recombinant DNA are used to produce human insulin for people with diabetes
- genetically modified organisms (GMOs) are being used in the production of food and medicines
- there are many advantages and some disadvantages and ethical issues linked to the use of genetic engineering.

End-of-chapter questions

1. What is the basic unit of inheritance?
   A   DNA
   B   A chromosome
   C   The nucleus
   D   A gene (1)

2. Winston has dangly earlobes. He has inherited one allele for dangly lobes from his mother and one for attached lobes from his father. Which **one** of the following terms best describes Winston's genotype for his earlobes?
   A   Homozygous
   B   Heterozygous
   C   Homologous
   D   Autosomal (1)

3. Which **one** of the following statements is a definition of a recessive allele?
   A   An allele that controls the development of a characteristic even when it is only present on one of the chromosomes
   B   An allele that controls the development of characteristics alongside another different allele that is also expressed in the phenotype
   C   An allele that only controls the development of a characteristic if it is present on both chromosomes
   D   An allele that only occurs on the sex chromosomes (1)

4. Which **one** of the following can cause mutation and so variation in the DNA of your cells?
   A   Ionising radiation
   B   Excess salt
   C   Eating too much
   D   Nitrogenous waste (1)

5. Which **one** of the following genetic conditions is sex-linked?
   A   Sickle cell anaemia
   B   Haemophilia
   C   Albinism
   D   Cystic fibrosis (1)

6. a   Describe how Mendel's experiments with peas is evidence that there are distinct 'units of inheritance' that are not blended together in the offspring. (4)
   b   Suggest a reason why people didn't accept Mendel's ideas. (1)
   c   The development of the microscope played an important part in helping to convince people that Mendel was right. Suggest how. (1)
   d   Gregor Mendel did his work on peas. Use Punnett squares to help you work out the genotypes and phenotypes of the possible offspring of a cross between:
       i   pure breeding round peas (RR) with pure breeding wrinkled peas (rr) (4)
       ii   a round pea carrying a wrinkled gene (Rr) and a wrinkled pea (rr).
       Show all your workings. (4)

Genetics

7 Define the following words: 'gene', 'allele', 'dominant allele', 'recessive allele', 'homozygous', 'heterozygous'. (6)

8 The allele that gives a person dimples is dominant over the allele for no dimples. One partner in a couple has dimples and the other has no dimples. Would you expect their children to have dimples? Explain as fully as you can, using genetic diagrams to explain your answer. (8)

9 Whether or not you are a person with albinism decided by a single gene with two alleles. The typical allele, A, is dominant over the albinism allele, a. Use this information to help you answer the following questions. Show any working.

Tom is a person with albinism but Sandy is not. They are expecting a baby.
  a   We know exactly what Tom's alleles are. What are they and how do you know? (2)
  b   If the baby has normal colouring, what does this tell us about Sandy's possible genotype? (2)
  c   If the baby has albinism, what does this tell us about Sandy's possible genotype? (3)

10 a   Define a mutation. (2)
  b   Describe the potential effects of a mutation. (2)
  c   If you have a medical X-ray, you may have a lead apron placed over your abdomen. Suggest a reason for this. (3)

11 Achondroplastic dwarfism is a genetic condition that affects the long bones of the body, which do not grow to typical size, although in every other way affected individuals are quite typical. It is inherited as a dominant gene. Embryos that are homozygous for achondroplastic dwarfism die before birth. Use the family tree to help you answer the questions on the next page.

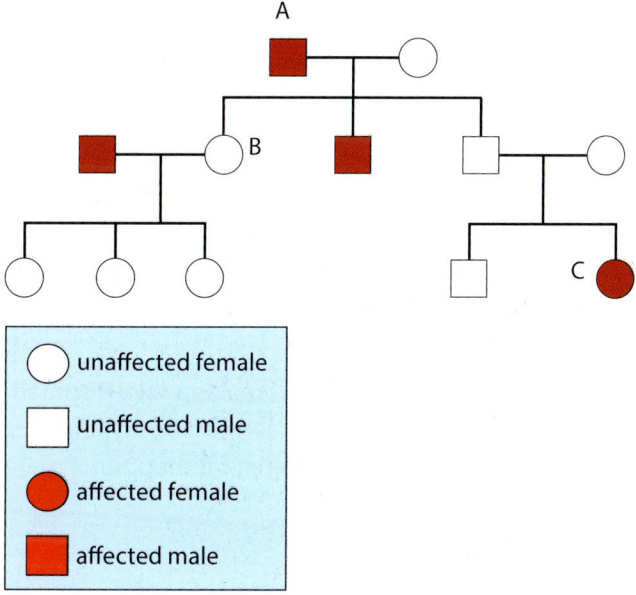

- a Chose a suitable capital and lower case letter to represent the two alleles. Give the genotype you would expect for individuals A, B and C on the family tree and explain your answers. (8)
- b In the family where two people with achondroplastic dwarfism married, one pregnancy ended in miscarriage. What might be the explanation for this? Use a genetic diagram to help you explain. (5)
- c Huntington's disease is a fatal genetic disease. It is inherited in the same way as achondroplastic dwarfism, but affected individuals look unaffected until the symptoms of the disease start to appear in middle age. It is now possible to test people in an affected family for the condition, although there is no treatment available. Do you think this is a useful test to take? What difficulties might arise from it? (5)

12
- a What is sickle cell anaemia? (2)
- b Two people who are heterozygous for sickle cell anaemia have four children. What are the chances of them having:
  - i an unaffected child?
  - ii a child with sickle cell anaemia?
  - iii a child who is resistant to malaria? (5)

  Produce a genetic diagram to show this cross and help to explain your answer.
- c People who are heterozygous for the sickle cell trait have a greatly improved immunity to malaria, a tropical disease that kills about two million people a year in the developing world. How does this affect which of the children in the family above is/are most likely to live to adulthood and why?
  - i If the family lives in Africa (2)
  - ii If the family lives in Eastern Europe (a poor area with little or no malaria) (2)

13
- a Draw what you would expect the red blood cells of a normal person, a carrier of sickle cell and a person with sickle cell anaemia to look like. (3)
- b The red blood cells of children who inherit two sickle cell genes do not carry oxygen very efficiently. Huge numbers are destroyed in the spleen so there are relatively few red blood cells. Their spleens become distended and painful. The strange-shaped red blood cells do not flow easily through blood capillaries, so they often block them, cutting off the blood supply to areas of tissue. One gene seems to have a lot of effects. Explain how a gene that causes a change in the shape of the haemoglobin molecule can cause symptoms of severe anaemia, a swollen painful spleen and painful joints. (6)
- c With global warming, it seems possible that malaria might spread into Europe, where at present the allele for sickle cell anaemia is rarely found, except in immigrant populations. What would you expect to happen if malaria spreads? Explain your answer. (4)

14
- a What is a sex-linked condition? (1)
- b Explain why sex-linked conditions are more common in males than females. (2)

c   Look at the diagram. It shows a family tree affected by the sex-linked condition haemophilia.
   i   Choose suitable letters to represent the alleles. (2)
   ii What are the possible genotypes of individuals A, B and C? (3)
   iii Produce genetic diagrams to explain your answers. (12)

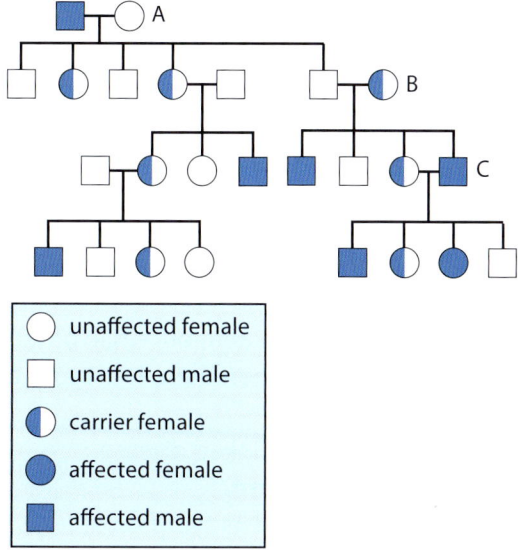

○ unaffected female
□ unaffected male
◐ carrier female
● affected female
■ affected male

15 Write an essay on the genetic disease haemophilia and discuss its impact on people's lives both in the past and in the present. Include genetic diagrams to explain how the disease is spread. (15)

16 a   Define genetic engineering. (3)
   b   State why genes are inserted into animals and plants, as well as bacteria. (3)
   c   Discuss how genetically modified organisms might reduce the amount of pesticides that farmers spray on their fields. (3)

17 a   Draw a flow diagram to show the stages of producing human insulin from genetically modified bacteria. (3)
   b   Make two lists, one to show the possible advantages of genetic engineering and the other to show the possible disadvantages. (3)

End-of-section questions

1. **a** Name the types of cell division that occur in the human body. (2)
   b Name **one** difference between the terms in each of the following pairs:
      i   Sperm and egg
      ii   Genotype and phenotype
      iii Chromosome and gene
      iv Genetic variation and environmental variation (8)
   c A father has dangly earlobes and is heterozygous for this condition. His wife has attached earlobes. Draw a diagram to work out the possible genotypes and phenotypes of their offspring. Use D and d for the dominant and recessive alleles respectively. (5)

2. **a** Describe albinism. (2)
   b Describe how a baby with Down syndrome differs from a baby with albinism by referring to their genetics. (4)
   c Some characteristics show continuous variation, while others show discontinuous variation. Describe the difference between these two types of variation and list two examples of each. (6)
   d i   Describe what is meant by a sex-linked disease. (2)
      ii   Define a carrier of a genetic disease. (1)
      iii Name **one** sex-linked disease and its symptoms. (2)

3. **a** Two parents showing normal characteristics have a child who bleeds easily and the bleeding is difficult to stop when he cuts himself. What sex-linked disease does this child possibly have? (2)
   b Use B and b to represent the genes for normal and abnormal blood clotting, respectively. Copy and complete the table below to show how the child may have ended up with the characteristic described. (6)

   c Name an effective treatment for the disease in part **a**. (1)
   d i   Define genetic engineering. (2)
      ii   Suggest **four** ways in which genetic engineering is useful. (4)

4. **a** Briefly describe how the genetic disease sickle cell anaemia affects human beings. (4)
   b i   A person is heterozygous for the sickle cell trait. Explain this statement. (1)
      ii   Suggest how this person might be affected if bitten by a malaria-carrying mosquito and explain the benefits of this response compared to a homozygote for sickle cell disease or a homozygote for normal haemoglobin. (2)

5. **a** State the role of Gregor Mendel in the study of genetics. (2)
   b Explain why identical twins show very similar features compared to **non-identical twins**. (4)
   c Draw a labelled diagram to show what happens to an egg or ovum when fertilisation occurs. (3)

Heredity and variation

d   Copy numbers 1–9 in the genetic cross diagram below and write the labels. (5)

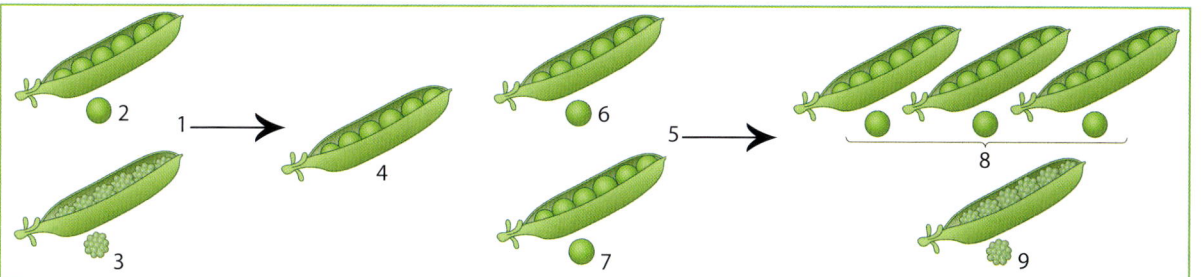

Data analysis

1   Use the data in the table below to draw a histogram showing the heights of males and females in the population of an unspecified Caribbean country. (7)

Height ranges (metres)	No. of males	No. of females
1.2–1.39	100	500
1.4–1.59	600	600
1.6–1.79	400	300
1.8–1.99	200	100
2.0–2.19	50	0

2   State the type of variation shown on your histogram. (1)

3   Would you expect a similar type of graph of the different blood types in this same population? Explain your answer. (2)

Case study

Mr and Mrs Jay were called to their children's school by a worried principal. Their son seemed to be having a lot of problems in art class, as he was not able to use colours as instructed by his teacher. He was painting trees in red and flowers in green, and did not seem to understand that his teacher had told him to do the opposite. His twin sister did not have this problem and was a very promising art student. Mrs Jay mentioned that her mother also mixes up the colours red and green.

1   Suggest a genetic condition that may be the cause of their son's problem. (1)

2   Explain why his twin sister may not have this problem. (4)

3   Draw a diagram to show how the son may have inherited his illness from his grandmother. (7)

Revision tool kit 3

This activity requires you to put in some serious effort!

1   Choose your least favourite Human and Social Biology topic. Depending on how much time you have, go back to that chapter in your textbook and read through it again.

2   Find a peer who is very comfortable with the topic. Try to 'teach' him or her the topic. Your partner will be able to tell if you are truly the expert! This can be a very efficient strategy if you are working in a dedicated study group. Each person becomes the expert for a topic and teaches the group.

Section D: Diseases and their impact on humans

Learning outcomes

At the end of Section D, you will:
- understand the basic concepts of human well-being and disease
- appreciate the social and economic importance of disease control.

Our health is very important to us, but it can be threatened by disease. Disease has an enormous impact on humans. It affects us personally, in our everyday life, and it also affects the productivity and economic situation of the countries around the world. But what is disease, and how does it affect our body? These are the questions you are going to answer as you work through this section of the book.

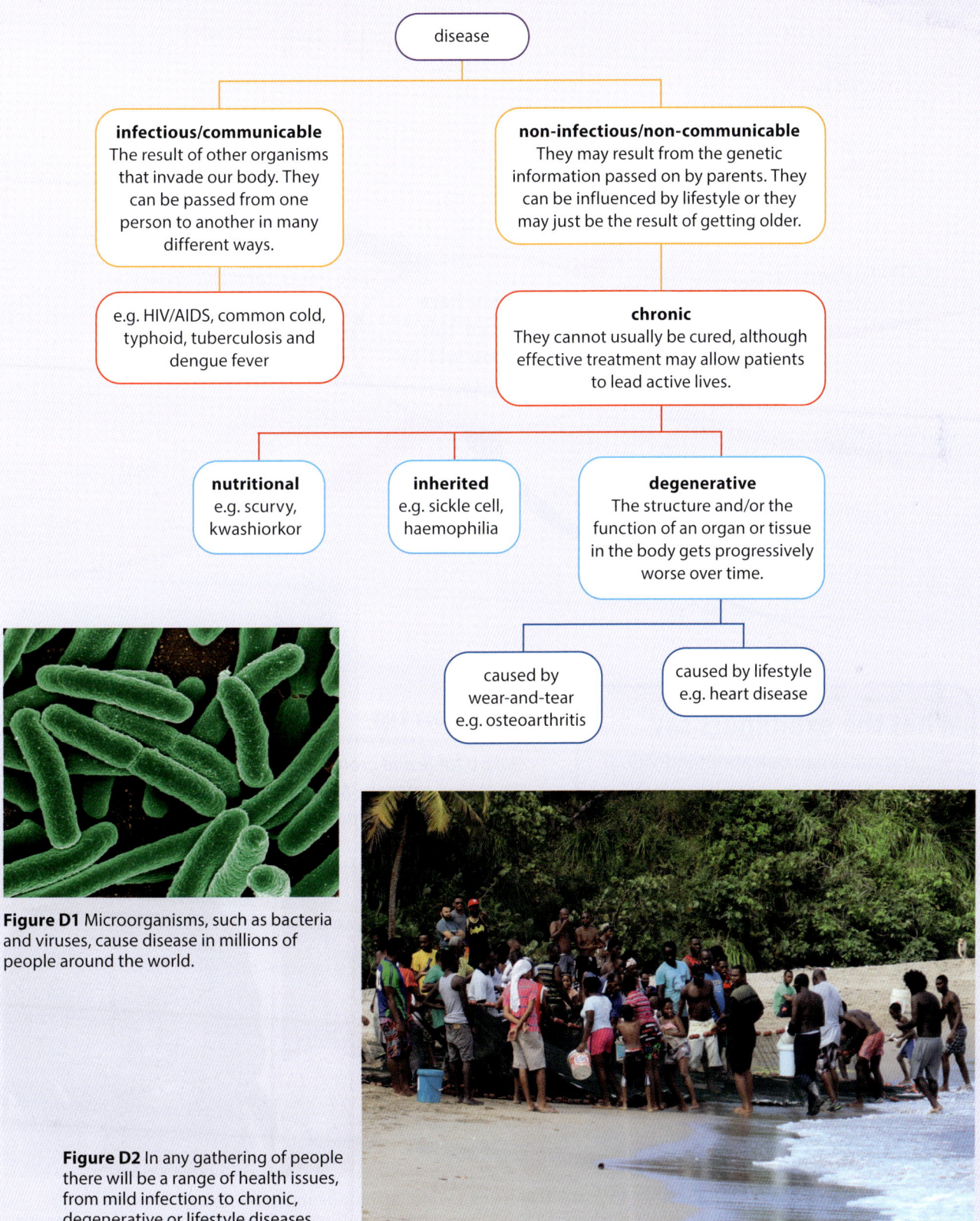

Figure D1 Microorganisms, such as bacteria and viruses, cause disease in millions of people around the world.

Figure D2 In any gathering of people there will be a range of health issues, from mild infections to chronic, degenerative or lifestyle diseases.

Chapter 15
Health and diseases

When you have completed this chapter, you will be able to:

- describe how changes to lifestyle, diet and exercise can be used to treat obesity, Type II diabetes and secondary hypertension
- classify communicable/infectious diseases and non-communicable diseases
- differentiate between the signs and symptoms of diseases
- define the terms 'infectious disease', 'chronic diseases', 'nutritional deficiency diseases', 'inherited disorders', 'lifestyle diseases' and 'mental health problems'
- define the terms 'health' and 'disease'
- state the main causes, primary symptoms a possible treatment of asthma
- explain diabetes mellitus (Type II) and secondary hypertension as complications of obesity

What the examiners say

→ Candidates can talk about individual diseases, but appear to be unfamiliar with the relationship between some chronic diseases and changes in the body such as weight gain.
→ Candidates are unable to adequately explain the role played by the pancreas in the development of diabetes.

Revision tip

Make a table and group all the diseases you come across in the correct category. Make each group a different colour – it will help you to remember them!

Health is very difficult to define and even more difficult to measure. If you wake up with a headache on a school day you may feel so unwell that you stay in bed. But if it is carnival time, you might completely ignore a similar headache!

Figure 15.1 Which of these people looks well, and which looks ill? Research shows that people generally associate happiness with health.

What is health?

Health is more than a set of symptoms. If we feel well and happy, we are unlikely to look for the signs of disease (see Figure 15.1). On the other hand, if we feel unwell, it is difficult to convince ourselves that there is nothing medically wrong with us. The World Health Organization (WHO) has defined health and disease as follows: 'Health is a state of complete physical, mental and social well-being, and not merely the absence of disease or infirmity. Disease is the state in which a part, organ or system of the body is damaged, or does not function properly.'

When we have a disease, we usually feel unwell in some way. When we go to the doctor, they identify the disease by symptoms and signs.

- **Symptoms:** These are what a person feels and cannot be seen by the healthcare professionals, for example, a sore throat or feeling tired. They are difficult to measure, as people experience things differently. See Figure 15.2.
- **Signs:** Clear evidence of a disease that doctors can get from examining a patient or from tests, for example, raised blood sugar level, rash on the skin or an enlarged heart.

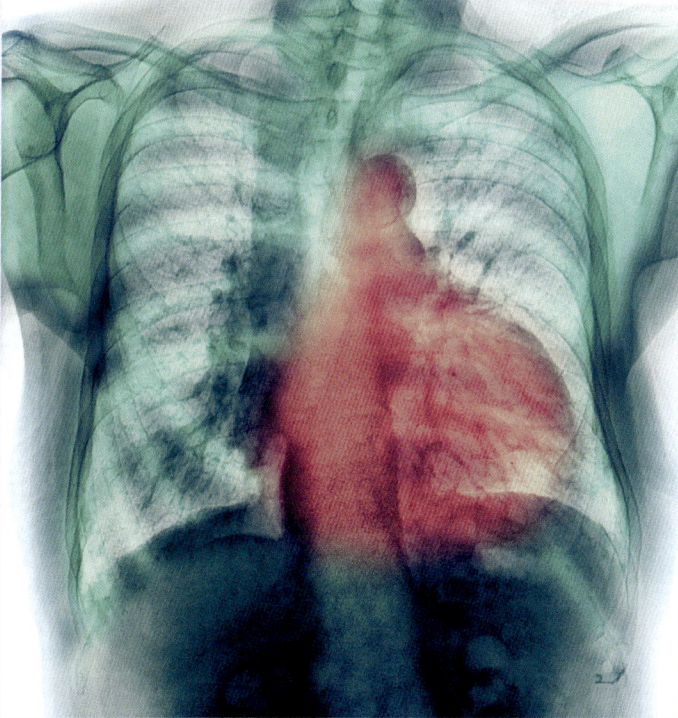

Figure 15.2 A sore throat is a symptom of a disease and is difficult to measure because people feel things differently. An X-ray showing an enlarged heart is a sign of disease, because it shows clear evidence of the problem.

Classifying diseases

We classify (group) diseases based both on the type and cause of the disease, and by the part of the body that is affected. The main categories of diseases are:
- Communicable/infectious/pathogenic diseases
- Non-communicable diseases.

Communicable/infectious/pathogenic diseases

These diseases are passed from one person to another – they are **communicable** or **infectious**. They are caused by microorganisms known as **pathogens** (see Chapter 1 for more information on microorganisms). **Infectious diseases** are the ones most likely to have affected you personally – the common cold, influenza and COVID-19 are typical examples. Other examples of infectious diseases that we will look at in this section include the following:

- **Sexually transmitted infections (STIs)** are infectious diseases that are passed from person to person through sexual contact, for example, gonorrhoea and HIV/AIDS.
- **Vector-borne diseases** are infectious diseases where the disease-causing organism is carried from one person to another by a vector, for example, malaria where mosquitoes are the vectors.

Non-communicable diseases

Diseases that cannot be passed from one person to another are called non-communicable diseases. In this section, we will be looking at several different types of non-communicable diseases including the following:

- **Chronic** diseases are long-term illnesses that are not infectious. They are often **physiological**, such as diabetes, or **degenerative**, where the function of a system gets worse over time, for example, osteoarthritis (see Figure 15.3 and Chapter 9).
- **Nutritional deficiency diseases** are diseases caused by a lack of particular nutrients in the diet, for example, night blindness, scurvy (see Figure 15.4), rickets or marasmus. Look back to Chapter 5 to remind yourself about this type of non-communicable disease.
- **Inherited disorders** are the result of mutations in the genetic material, for example, Down syndrome and sickle cell anaemia. You learned about these non-communicable diseases in Chapter 13 and Chapter 14.
- **Lifestyle diseases** are diseases that we develop as a result of lifestyle choices we have made. Some of these, including sexually transmitted infections, are communicable, but many of them are not, for example, smoking-related diseases such as lung cancers and heart disease, obesity (see Figure 15.5) or diseases related to taking illegal drugs. You have already met some of these diseases in earlier chapters.
- **Cancers** are mainly non-communicable diseases that result from the uncontrolled growth of cells in an organ or tissue, for example, breast cancer, ovarian cancer and prostate cancer. Some cancers such as cervical cancers have been found to be communicable. You met some of these cancers in Chapter 12 on the reproductive system.
- **Mental health problems** affect our central nervous system and in turn our whole sense of well-being, for example, anxiety and depression (Figure 15.6).

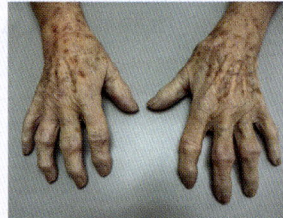

Figure 15.3 Osteoarthritis of the hands

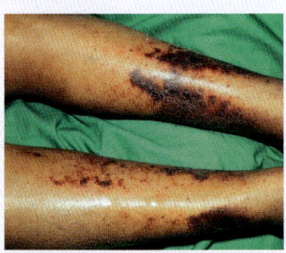

Figure 15.4 Scurvy causes bleeding of the gums and around the lower legs and shins.

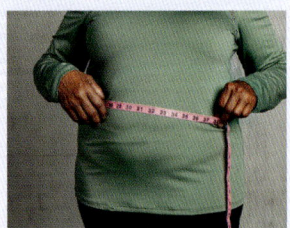

Figure 15.5 Obesity can lead to diabetes, heart disease and even a stroke.

Figure 15.6 Depression can cause fatigue, insomnia, obesity and an increased risk of heart disease.

Another way of categorising diseases is to focus on particular body systems. For example, **respiratory diseases** are diseases that affect the respiratory system, such as **asthma**, influenza, COVID-19 and bronchitis. Some respiratory diseases are chronic, while others are infectious.

We **treat** most diseases – we can relieve the symptoms to make people feel better, at least for a time. For example, the pain relief tablets we use if we have a headache treat the symptom – they get rid of the headache – but they don't cure it. The headache will come back, or go away, in its own good time.

There are some diseases we can **cure**. This means we can completely get rid of the disease and make people better. For example, if you get a serious bacterial infection such as pneumonia, drugs called **antibiotics** are used to destroy the bacteria and cure the disease.

The social and economic impact of disease

Understanding the causes of different types of disease and how to treat, cure or prevent them is important for several different reasons.

Individuals, families and society

If a person is unwell, they cannot function properly. They will not learn well at school, they cannot go to work and may feel that they are a burden on their family. In a short infectious disease like a cold, these effects don't last long and are not a problem. But if an infection makes us seriously ill, or has long-term effects (like long COVID), or if we have a chronic or long-term condition, disease has a major impact and makes life very difficult for the individual affected, as well as their family and friends (see Figure 15.7).

Figure 15.7 Disease can have a significant effect on the lives of the people affected, as well as their family, friends and even society as a whole.

If one individual has a long-term illness, it has a considerable effect on them and their social group. They need care and support, and cannot work to support their family – it affects their personal productivity.

A long-term non-communicable disease has the following impact on an individual:

- It affects the goals they set in life – can they get married? Will they be able to work?
- It affects the occupation they do or aspire to do, which means that it also has an effect on their earning ability.
- It affects their ability to exercise – and exercise has been show to be of benefit to overall physical and mental well-being.

Figure 15.8 If an individual goes to hospital, it is a concern for them and their family. If too many people need hospital treatment, it becomes a social and economic problem for the whole community.

Map data © Google Nov 2020

If many people have the same condition, it puts pressure on society as a whole, because healthy people have to work and support those who are unable to work and support themselves. Chronic ill-health in a country may have the following wide-ranging impact on the economy (see Figure 15.8):

- There will not be enough people who are fit for work.
- Many people will need financial support for themselves and their families.

Eventually, society could collapse under the burden of disease.

Epidemics and pandemics

Sometimes there are major outbreaks of disease – often these are infectious diseases, but sometimes they are non-communicable diseases.

- An **epidemic** is an outbreak of disease that affects a large proportion of a community, region or the population of a country
- A **pandemic** is an epidemic that spreads to many different countries or even over different continents. The global spread of COVID-19 in the early 2020s is an example of a pandemic.

Epidemics and pandemics – especially pandemics – have major socio-economic implications. Huge numbers of people may be away from work, causing factories and other places of work to close down. See Figure 15.9. It may be necessary to lock down whole countries, affecting travel and tourism. People cannot work when isolated, but still need to buy food and other necessities. The impact of major disease outbreaks has been made very clear through the effects of the global COVID-19 pandemic on the physical and mental health of individuals, on the structure of communities and on the economies of countries around the world.

Figure 15.9 During the COVID-19 pandemic lockdown, people had to stay home sometimes for weeks or months. Cities and towns were deserted, and many businesses had to close down.

Health and diseases

Chronic and lifestyle-related diseases

We will look at a number of chronic and lifestyle-related diseases that have a particular impact in the Caribbean. Some of them you have met before in earlier chapters – but here we will bring together the whole picture.

Obesity

Obesity basically means being very overweight. It can be diagnosed using a number of methods, and in the Caribbean, the most common ones are the Body Mass Index (BMI) and waist/hip ratio. To find your BMI, you divide your weight in kilograms by your height in metres squared (kg/m^2). A BMI of greater than 30 is obese. (See Chapter 5 for more information on the causes of obesity and the BMI.)

Doctors have also found that the ratio of your waist to your hips is a good guide to obesity, and also to your risk of linked diseases such as Type II diabetes and heart disease. A waist-to-hip ratio in women of 0.8 or more and in men of 1.0 or more is regarded as obese. Obesity has been increasing steadily in the Caribbean – see Figure 15.10 – and so have the many diseases linked to it.

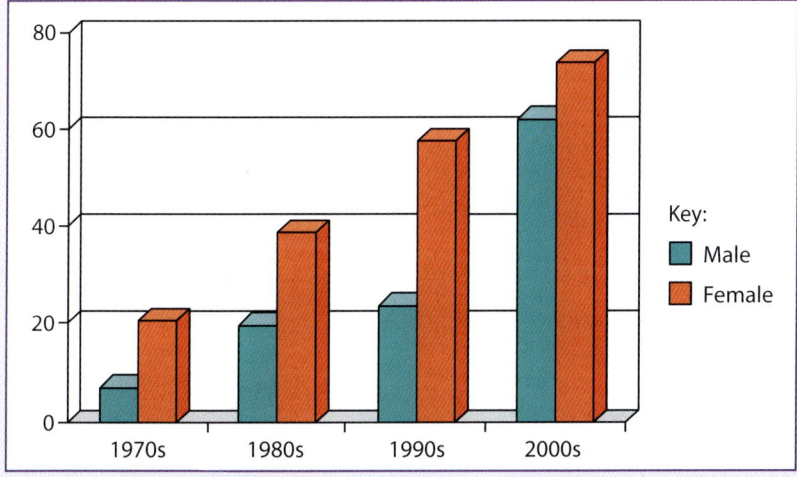

Figure 15.10 This data shows how the percentage of the Caribbean population who are overweight or obese changed over four decades. Look back at *The biologist's toolkit* on page 100 in Chapter 5 to remind you how to interpret a bar chart.

As you saw in Chapter 5, there are several reasons for obesity, including the following:

- Consuming more food than we use – the extra energy is stored as fat.
- A lack of exercise – this can be by choice or following surgery that restricts movement.
- Steroid-induced (rare) – if steroid drugs are needed as part of the treatment for a medical condition, it can lead to excessive weight gain.
- Hormone-induced (rare) – sometimes if certain hormones are over- or under-produced in the body, it can cause obesity.
- Genetics – we don't inherit obesity, but some families certainly seem to gain weight very easily. This may be due to differences in the chemical reactions in the body (the metabolism).

DID YOU KNOW?

Your fat cells are formed as you develop in your mother's uterus. It doesn't matter how fat or thin you are, you don't make any more fat cells. The ones you have just get bigger as you get fatter!

The effects of obesity

Obesity is a problem because it is linked to many diseases – (see Figure 15.11).

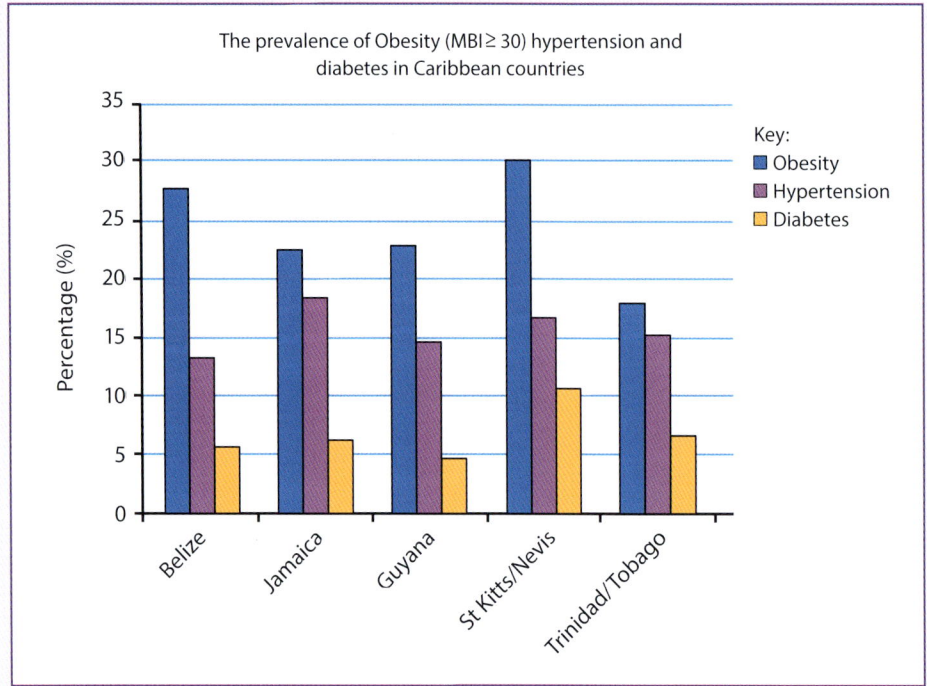

Figure 15.11 Levels of obesity and two associated diseases in five Caribbean countries. Look back at *The biologist's toolkit* on page 100 in Chapter 5 to remind you how to interpret a bar chart.

- Obesity increases the risk of developing Type II diabetes.
- Obesity is linked to an increased risk of heart diseases and hypertension.
- Obesity contributes to high blood cholesterol levels – not least because obese people often take in a lot of fat in their diets. Strokes are more likely, often linked to high blood pressure, and so is gall bladder disease as a result of fatty deposits. This is very painful and often involves surgery to cure it. See Figure 15.12.
- Data suggests that certain cancers, particularly cancer of the breast, prostate and colon, are linked to obesity. The increased risk of colon cancer is probably linked to the diet of many obese people. The effect on breast and prostate cancers seems to come from the effect of fat cells on hormone levels.
- Obesity often leads to poor self esteem and quality of life.
- Obese people tend to be unhappier and feel themselves to be socially unacceptable. There is also much prejudice against overweight people; people often wrongly believe that being overweight or obese means that the person is lazy. This can lead to job discrimination and discrimination in general.

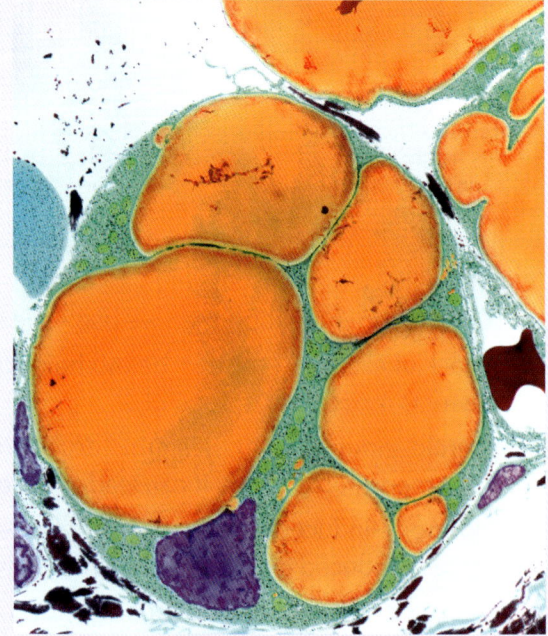

Figure 15.12 Fat cells use very little energy. Eighty-five per cent of the contents of a fat cell is a globule of fat!

In the Caribbean, obesity is a very real health issue. For example, the Jamaican Health and Lifestyle Survey (2016–17) showed that 54% of Jamaicans over 25 years old were overweight or obese. Part of the problem in the Caribbean is that being overweight has traditionally been viewed as healthy, and fat women are considered to be more attractive than those who are thin. See Figure 15.13.

Fast food that is high in fats and salt is fine as an occasional treat, but for many people it has become a regular part of their diet. This type of food is one factor that has caused the increase in obesity in the Caribbean.

Figure 15.13 Obesity and the diseases often linked to it are increasing across the world – and the Caribbean is no exception.

Obesity should be treated by aiming for substantial weight loss over a long period. In one sense, dealing with obesity is simple: we simply need to eat less food, eat more healthily and do more exercise. Making some healthy lifestyle changes will give us guaranteed results without needing any medicines.

However, in the case of obesity, there is an alternative to treatment. Prevention is better than cure; the elements of a healthy lifestyle should be adopted and maintained as early as possible in life and then obesity and the health problems it brings will not be an issue.

Diabetes mellitus

Diabetes mellitus (commonly called diabetes) is a condition where we lose control of our blood sugar and there is a constantly high level of glucose in the blood. This is called **hyperglycaemia**. The classic sign of diabetes that doctors look for is the presence of sugar in the urine. See Chapter 10 to refresh your memory on the role of the pancreas and insulin in the control of our blood sugar level. In diabetes, this delicate balance is lost.

Types of diabetes

There are two types of diabetes:
- Type I diabetes (see Figure 15.14):
  - 10% of people with diabetes have Type I diabetes.
  - The pancreas makes little or no insulin.
  - It often develops during childhood or teenage years.
  - **Signs and symptoms** include feeling thirsty, urinating a lot, sugar in the urine, weight loss, hunger and tiredness.
  - **Treatment** includes insulin injections, managing the diet carefully, and balancing insulin injections and food intake (see Figure 15.15).

Figure 15.14 Type 1 diabetes often develops during childhood, and children need to learn how to use devices to check their blood glucose level regularly.

Chronic and lifestyle-related diseases

- Type II diabetes:
  - 90% of people with diabetes have Type II diabetes
  - It is linked strongly to obesity; insulin is produced by the pancreas but the cells of the body stop responding to it.
  - There may be a genetic link.
  - It often develops in adults who become obese and is being seen increasingly in obese children and teenagers.
  - **Signs and symptoms** are similar to those for Type I diabetes but associated with obesity.
  - **Treatment** includes losing weight, increasing exercise levels and managing the diet. If this isn't enough, there are drugs that increase insulin production in the body or that help your cells respond more effectively to the insulin produced.

The effects of diabetes

When diabetes is well managed, the person affected lives a normal, active life. However, if diabetes is not managed properly, the results can be very serious.

When diabetes is poorly managed, the raised blood sugar level often damages the blood vessels. This leads to a number of problems such as blindness, kidney disease, nerve disease, heart disease and strokes.

People with badly managed diabetes are also at risk of infections; if the feet are seriously affected, they may need to be removed (amputated) (see Figure 15.16). A UWI study in Barbados found that 1 in 100 people with diabetes has had an **amputation**. This in turn increased the risk of dying relatively quickly. Another study found that people wearing poor footwear – thongs, very high heels or no shoes at all – were 3.5 times more likely to need to have their foot or leg amputated than people who wore comfortable, protective shoes.

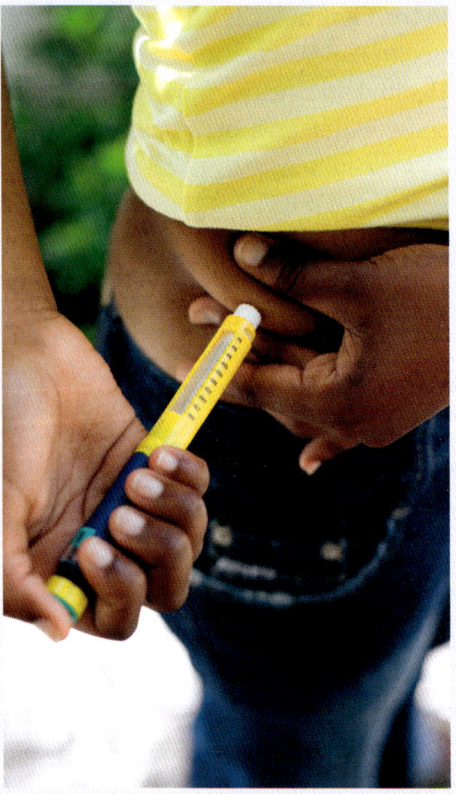

Figure 15.15 Injecting insulin becomes part of normal life for anyone with Type I diabetes.

Figure 15.16 Serious infections and amputations are a common side effect of badly managed diabetes.

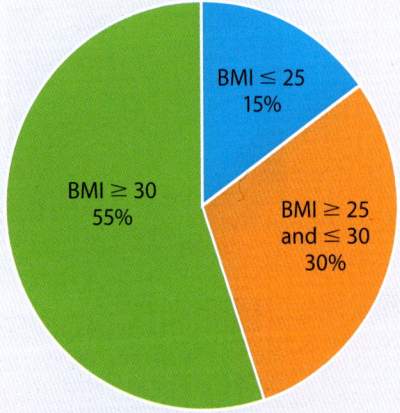

Body Mass Index (BMI) of people diagnosed with Type II diabetes

- BMI ≤ 25: 15%
- BMI ≥ 25 and ≤ 30: 30%
- BMI ≥ 30: 55%

Figure 15.17 This data shows that 85% of people diagnosed with Type II diabetes in a year were overweight or obese. Obesity and Type II diabetes are very closely linked, giving people an opportunity to really help their own health by controlling their weight gain. Look back at *The biologist's toolkit* in Chapter 5 on page 102 to remind you how to use a pie chart.

Gatineau M, Hancock C, Holman N, Outhwaite H, Oldridge L, Christie A, Ells L. Adult obesity and type 2 diabetes. Oxford: Public Health England, 2014.

Health and diseases

Cardiovascular diseases

Cardiovascular diseases affect the heart and the circulatory system. Refer back to Chapter 8 to remind yourself of how the heart and circulatory system work. A number of chronic diseases affect the cardiovascular system, including hypertension and coronary heart disease.

Hypertension

Hypertension is the medical name for high blood pressure. As you learnt in Chapter 8, blood pressure is the force exerted on the walls of the arteries when blood is pumped around the body by the heart. Blood pressure is considered high if the systolic pressure is greater than 140 mmHg or the diastolic pressure is greater than 90 mmHg.

About 90% of the cases of hypertension in the Caribbean are what is known as **primary hypertension**. This is high blood pressure that is not the result of a known medical condition. Over 60% of the cases of primary hypertension in the Caribbean are the result of obesity. The other cases are the result of factors such as high-salt diets, lack of exercise and excess alcohol.

The remaining 10% of cases are **secondary hypertension**. This is high blood pressure that is a symptom of another disease such as chronic kidney diseases, hormonal disturbances or tumours.

Hypertension in the Caribbean

There are record levels of high blood pressure in the Caribbean; four or more in every ten adults have the disorder. In fact, hypertension is a leading listed cause of death in the Caribbean because of the strokes and heart disease it causes.

When we have high blood pressure, our heart enlarges over a period of time, much like when we go to the gym and lift weights, and the muscles we exercise grow in size. As the heart increases in size, it requires more food and fuel to function. After a while, the body simply cannot supply enough oxygen for the heart muscles and the heart begins to fail. It can no longer pump the blood effectively. This can eventually lead to a heart attack.

Another problem with high blood pressure is the stress it puts on the blood vessels. If the pressure of the blood is too high, it can burst a delicate blood vessel in the head, in the same way as a balloon bursts if you try and blow it up too much. When a blood vessel is broken in the brain, blood bleeds into the brain tissue. This is called a stroke and it can cause serious damage to the brain and even death.

Coronary heart disease

In coronary heart disease, the coronary arteries feeding the heart muscle become narrowed by fatty plaques (see Figure 15.18). Eventually they become blocked by the fat or by a blood clot and this results in a heart attack, when the blood supply to part of the heart muscle fails. The muscle cannot work and the heart stops pumping (see Chapter 8).

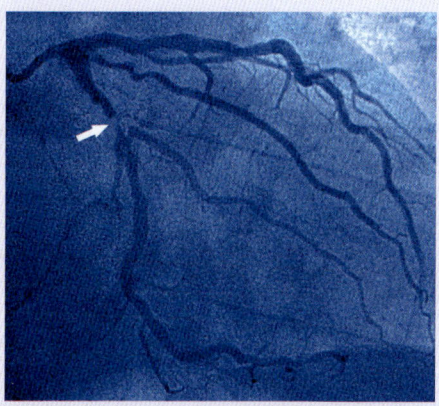

Figure 15.18 A special type of X-ray called an angiogram shows up blockages in the coronary artery. Problems like this are usually solved by surgery.

Heart attacks may be silent and without any symptoms (asymptomatic). However, they are often a frightening experience for both the patient and onlookers. The patient usually feels a crushing pain in the chest, accompanied by sweating, breathlessness and heaviness of the feet. Factors that increase the risk of a heart attack are:

- age – they are more common as you get older
- smoking (see Chapter 8)
- obesity
- high-salt or high-fat diets
- lack of exercise
- genetic tendency – some families have a genetic increased risk of heart disease.

Treatment of a heart attack

The key to saving a patient who has had a heart attack is speed.

- Call an ambulance immediately and give them aspirin to reduce the clotting of their blood.
- If the heart stops, give CPR until help arrives (see Chapter 8).
- In hospital, patients will be given:
  - nitro-glycerine to dilate the blood vessels, get more oxygen to the heart and so ease the pain
  - clot-busting drugs that break down the clots in the coronary arteries, allowing blood to flow back to the heart muscle
- Also in hospital, if the heart loses its rhythm, healthcare workers may use a **defibrillator** to give the heart a shock and get it back into a healthy rhythm.

Preventing further heart attacks

Simple measures to prevent more heart attacks include:

- losing weight
- eating a healthy, low-fat, low-salt diet
- exercising regularly.

Doctors may also need to operate to treat the coronary arteries.

- One option is to replace the blocked arteries with healthy blood vessels from the patient's own leg. This is called **bypass surgery** and it is often very successful combined with the simple measures given above.
- Another option is to insert a stent into the coronary arteries – a mesh tube that holds the blood vessels open so they can supply the heart with blood. This process is called an **angioplasty** (see Figure 15.19).

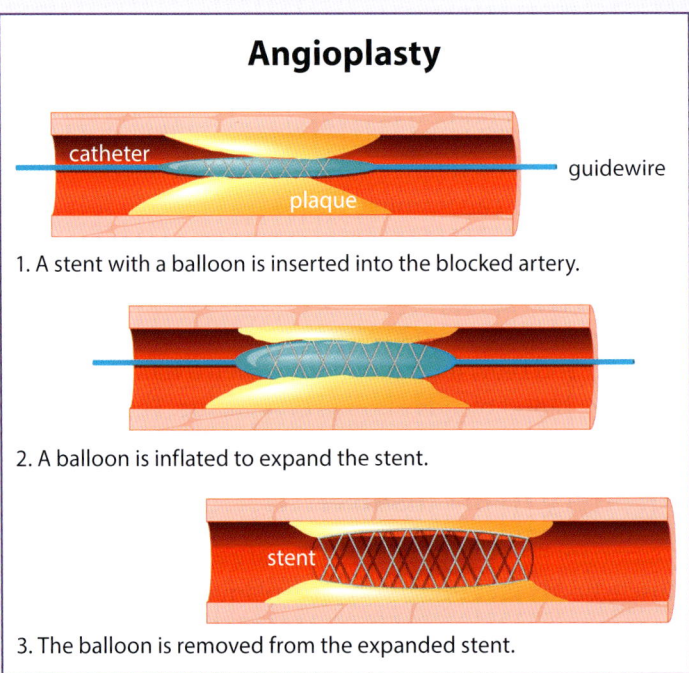

Figure 15.19 Inserting a stent into a blocked coronary artery restores blood flow to the heart muscle and helps prevent another heart attack.

Health and diseases

Heart attacks in the Caribbean

In the Caribbean, there has been a marked increase in heart attacks over the last half century. In fact, it is one of the most common cause of death across the region – 11.82% of deaths in Jamaica in 2018 were due to coronary heart disease (see Figure 15.20). Heart problems are a complication of hypertension, obesity and diabetes mellitus, as discussed earlier.

Adapted from "OECD/The World Bank (2020), Health at a Glance: Latin America and the Caribbean 2020, OECD Publishing, Paris, https://doi.org/10.1787/6089164f-en."

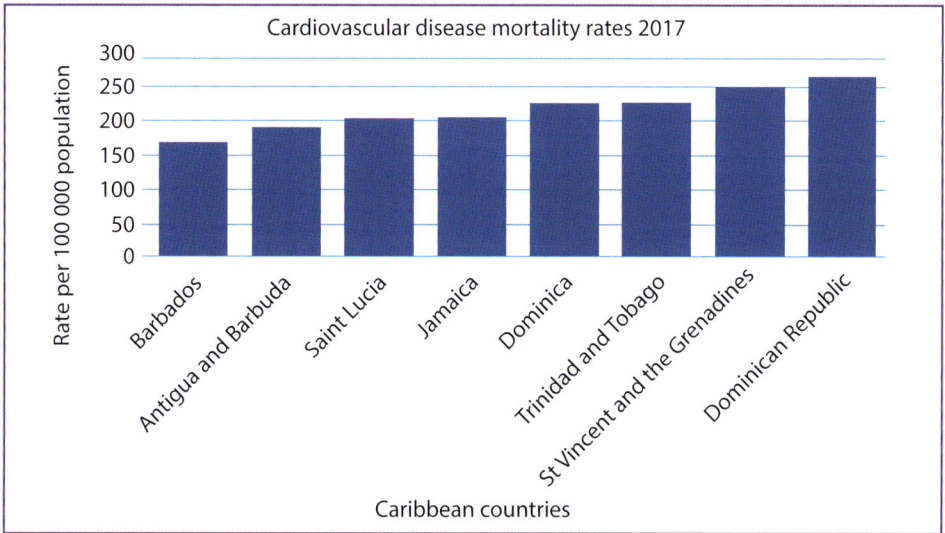

Figure 15.20 These data show how deaths from cardiovascular disease vary across Caribbean countries. Thery are all comparatively high – the European average is 119.4 per 100 000, with only 46.6 per 100 000 dying of cardiovascular disease in France.

Several other factors have contributed to the rise in heart disease in the Caribbean, including:
- cigarette smoking
- increased life expectancy (resulting from overcoming many of the common infectious diseases)
- physical inactivity
- excessive alcohol consumption
- genetic predisposition.

However, in the Caribbean, probably the largest factor contributing to heart attacks is poor diet. The Caribbean diet is high in starchy carbohydrates such as green bananas, potatoes, wheat-flour bread and cornmeal porridge. The excessive use of cooking oil in the preparation of food also contributes to increased blood lipids.

Thousands of people in the Caribbean are dying at a relatively young age each year as a result of high blood pressure and coronary heart disease. To avoid cardiovascular disease and its complications in the future, we must look carefully at our lifestyle. If we, the young generation, choose the healthy options described on the previous pages as preventative measures, we can look forward to a long, healthy life. See Figure 15.21.

Figure 15.21 A healthy lifestyle with plenty of exercise will prevent cardiovascular diseases such as hypertension and coronary heart disease.

Chronic and lifestyle-related diseases

Asthma

Asthma is another chronic disease, and one that affects young and old alike. It is a disease of the respiratory system (see Chapter 7 to remind you of how a healthy respiratory system works). People with asthma have difficulty breathing. The lining of the airways swell up, narrowing the tubes and making it difficult to move air into and out of the lungs (see Figure 15.22).

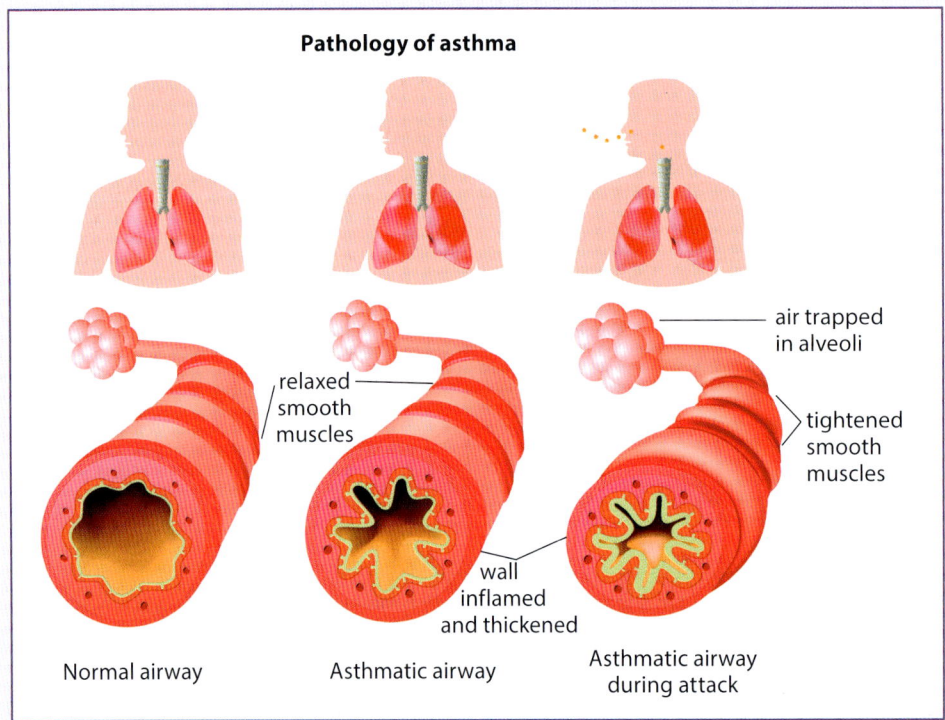

Figure 15.22 Asthma is a chronic disease of the respiratory system.

Someone who has asthma isn't affected all the time. Some people have attacks several times a day, others only a few times a year – perhaps if they get an infection such as a cold. Asthma attacks range from mild, where the sufferer feels uncomfortably out of breath, to life-threatening where they gasp frantically for air.

What causes asthma?

People with asthma have over-sensitive airways that become irritated when exposed to triggers such as:
- respiratory infections
- flower or grass pollen
- mould
- animal fur
- dust mites
- smoke
- cold air
- exercise.

Figure 15.23 Common triggers that irritate the sensitive airways of people with asthma

Asthma can also be triggered by stress. We don't really understand why some people have asthma and others don't. Children are more likely than adults to become asthmatic, possibly because their airways are small and so easily become narrowed. A 2017 study showed that 19.6 Jamaican children aged 2–17 were wheezy, while 16.7% had doctor-diagnosed asthma.

Types of asthma attacks

An asthma attack can be divided into two major components: the hyper-reactive response and the inflammatory response.

- **The hyper-reactive response:** When an irritant (pollen, dust mites (Figure 15.24), and so on) is breathed in, the tubes leading into the lungs respond by narrowing. The muscles surrounding the tubes spasm, narrowing the airways. This is called bronchospasm. The linings swell, narrowing them even more. The airways are unable to relax and as a result, an asthmatic person wheezes, as they are unable to get enough air.

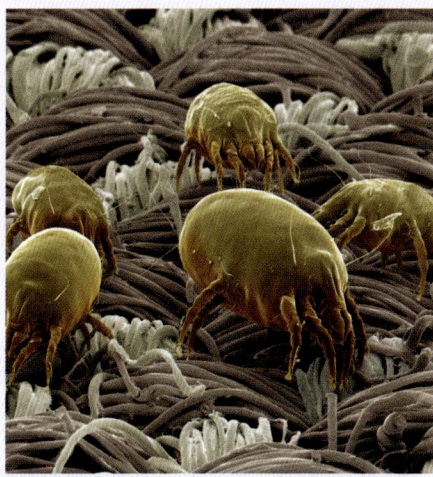

Figure 15.24 Dust mites are found in furniture, mattresses, pillows and carpets, however well we clean. They, and their faeces, are a common trigger of asthma.

- **The inflammatory response:** The immune system responds to inhaled irritants by sending white blood cells and other immune factors to the lungs where they create an excess of airway-clogging mucus. This brings on a bout of coughing and further hampers breathing.

Treating asthma

There are two main methods of treating asthma. The medicine is delivered through an inhaler (see Figure 15.25), so it is breathed straight into the areas where it is needed. To give immediate relief of the symptoms, many people use an inhaler to deliver measured doses of a chemical like adrenaline (see Chapter 11), which dilates (opens up) the airways. This is only used when it is needed, in other words, when having an asthma attack.

The inhaler used to relieve symptoms is often combined with a preventative treatment. This is usually a steroid that is also taken by an inhaler and used regularly every day. The steroid reduces the sensitivity of the lining of the tubes, making an asthma attack less likely.

A peak flow meter, though not very widely used in the Caribbean, is helpful for people who must take asthma medication daily. It is a device that measures how well air moves out of the lungs. The peak flow meter can indicate if there is any narrowing in the airways, hours, and sometimes even days, before any asthma symptoms appear.

In addition to medication, there are many steps that can reduce the symptoms and improve the quality of life of asthma sufferers. They mainly involve avoiding potential triggers.

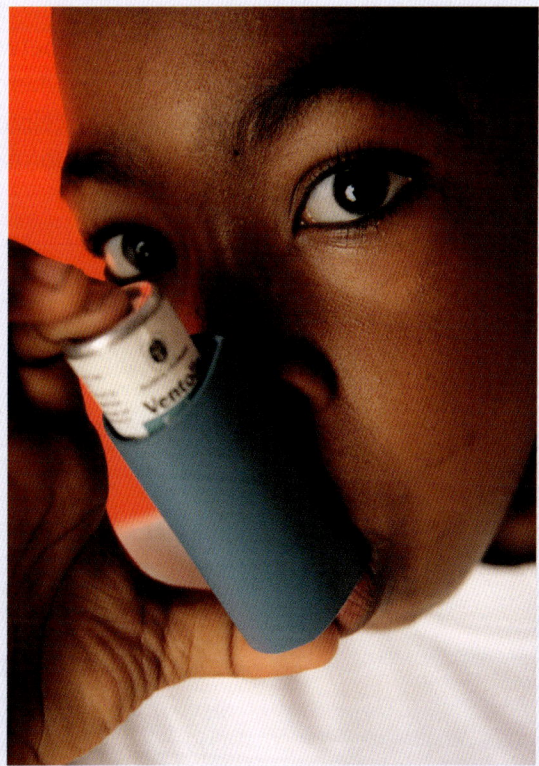

Figure 15.25 The use of inhalers to relieve symptoms and reduce the sensitivity of the airways allows most people with asthma to lead active lives.

Chronic and lifestyle-related diseases

Here are some steps to follow to reduce asthma symptoms:

- Keep furry pets away from living and sleeping areas, and brush and wash them frequently to reduce hair shedding into the surroundings.
- Wash and vacuum carpets, mattresses and pillows frequently to pick up any hair, mould and dust.
- Avoid inhaling vehicle and other fumes.
- Avoid areas where there are likely to be plants, flowers or grass with pollen.
- Encourage asthmatics not to smoke and to avoid smoky places.
- Encourage asthmatics to take up an exercise routine to improve their lung capacity and ability to breathe deeply – swimming has been shown to be effective, and singing can also be very useful. See Figure 15.26.

Figure 15.26 Obadele Thompson is a Bajan Olympian who competed in three Olympic games and won a bronze medal in the 100 metres sprint in the 2000 Sydney Olympics. He also had asthma from the age of three years old – proving that well-managed asthma may be a chronic disease – but it needn't stop you succeeding in life!

Cancers

Cancers affect many people, in many families, in every country of the world. It causes 20% of the deaths seen in the Caribbean countries. You have already met a number of cancers that affect different organs of the body:

- cancer of the colon (bowel cancer) in Chapter 6
- lung cancer and other cancers of the respiratory system in Chapter 7
- cancers of the reproductive system, especially of the cervix, uterus and prostate in Chapter 12.

So what are cancers? Healthy body cells grow, divide and then grow again before the next time they divide. The gap between the divisions may be days, months or even years. A cancer develops when the normal control of cell growth and division is lost (see Figure 15.27). The cells divide rapidly, and do not grow properly between divisions. A mass of abnormal cells develops – this is a **tumour**.

There are two types of tumours:

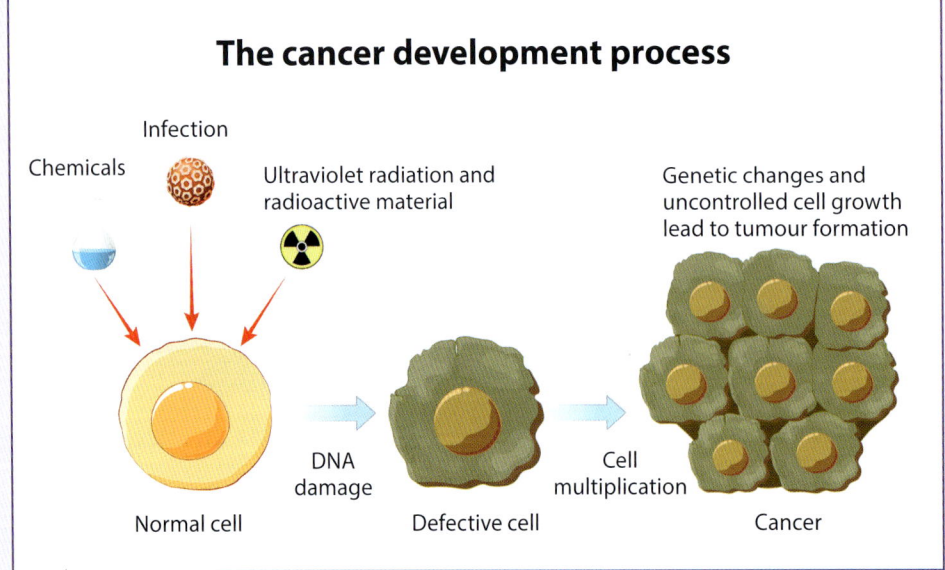

Figure 15.27 Cancer development following DNA damage.

- **Benign tumours** are collections of abnormal cells contained in a membrane. They do not spread to different parts of the body, but they may grow very quickly and get very large. This can be dangerous and even life-threatening if a tumour forms somewhere with little space, for example, in the brain inside the skull.
- **Malignant tumours** (cancers) have cells that break off and spread around the body in the blood or lymph. They lodge somewhere and begin to divide and invade new tissues (see Figure 15.28). The uncontrolled growth continues and secondary tumours form. A cancer often stops the normal functions of the tissues where it is growing and if left untreated, most cancers will kill the person affected sooner or later. Because of the way cancers spread around the body, it can be very difficult to track them down and treat them.

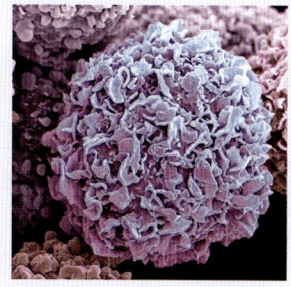

Figure 15.28 A scanning electron micrograph (SEM) of breast cancer cells. Their fine surface folds break off and can spread throughout the body.

What causes cancer?

We don't have the full answer to what causes cancer. Scientists do not fully understand how cancers form. However, we do know that the causes of some types of cancers usually involves some damage to the DNA of the cells.

- **Age:** The risk of developing most cancers increases with age.
- **Genetic factors:** We know there are genetic links to some cancers including early onset breast cancer and ovarian cancer.
- **Mutations:** These are changes in the genetic material. These may be random, but are often caused by external factors such as cancer-causing chemicals called **carcinogens**, for example, the tar in tobacco smoke and asbestos.
- **Ionising radiation:** These include UV light from the Sun and X-rays. This form of radiation interrupts the normal cell cycle of growth and division, and causes cancers such as melanomas in the skin.
- **Virus infections:** These cause about 15% of human cancers. For example, cervical cancer is usually caused by infection with the human papillomavirus Virus (HPV).

How do we treat cancer?

There are several methods of treatment for cancer – also known as modalities. They include the following:

- **Surgery:** The tumour and any affected tissue is removed.
- **Radiation therapy:** Targeted radiation is used to destroy cancer cells and stop them growing further. See Figure 15.29.
- **Chemotherapy:** Drugs that destroy actively dividing cells are used to kill cancer cells. They will also have some affect on healthy cells, so people having chemotherapy may feel unwell for a time, but this happens in order to destroy the cancer cells.
- **Targeted therapies:** These are new therapies using genetic tools, alterations to the immune system, hormone therapies and so on, which are helping to increase our success with treating cancer.

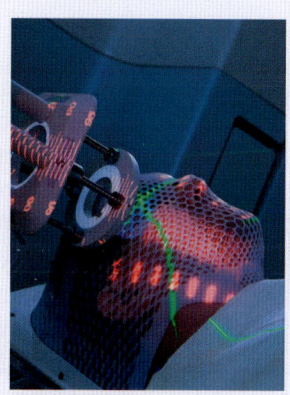

Figure 15.29 Radiation therapy to destroy skin cancer on the face

As well as treating cancer, we can do our best to avoid it. This isn't always possible, but there are a number of things we can do to reduce our risk of developing cancers. Some are general good sense:

- Eat healthily (see Figure 15.30).
- Do not become obese.
- Take lots of exercise.
- Don't smoke or drink heavily.

Others are specific to particular cancers – see Table 15.1.

Figure 15.30 Eating healthily can reduce the risk of developing cancer.

Table 15.1: A summary of common cancers

Type of cancer	Causes if known	Symptoms/signs	Type of treatment	How to reduce risk
Respiratory tract (lung)	Up to 90% caused by smoking tobacco	Persistent cough, breathlessness, coughing up blood, tiredness	Depending on the stage of disease: surgery to remove the lung, radiotherapy or chemotherapy	Do not smoke or give up smoking tobacco.
Breast	Causes not completely understood, but include age, faulty genes (especially early breast cancer), being overweight or obese and alcohol	Most common: a breast lump. Other changes in the size or shape of the breasts	Depending on the stage of disease: surgery to remove the lump or the whole breast, radiotherapy, chemotherapy and hormone therapy	Maintain a healthy weight, exercise regularly, don't drink alcohol, check breasts regularly – early treatment most likely to cure the disease.
Cervical	Almost all result from infection by HPV – human papillomavirus	Vaginal bleeding that is unusual for you – NOT a period; changes to vaginal discharge; pain during sex; pain in lower back, pelvis or lower tummy	If caught early after a smear test: removal of precancerous cells. Otherwise, depending on the stage of the disease, surgery of part or all of the reproductive system, radiotherapy and chemotherapy	Have an HPV vaccine when young, which prevents infection and removes the risk of cervical cancer.
Ovarian	Not completely understood but linked to: age; faulty genes; having breast or bowel cancer; starting periods young or menopause late; not having children; being overweight/obese; smoking	Regular swollen tummy/feeling bloated; pain or tenderness in tummy; no appetite; needing to urinate more often; back pain; tired; bleeding after the menopause	Depending on the stage of disease: surgery; chemotherapy	Maintain a healthy weight – lose weight if you are overweight or obese. Stop smoking. Discuss with your doctor if ovarian cancer runs in your family.

Health and diseases

Type of cancer	Causes if known	Symptoms/signs	Type of treatment	How to reduce risk
Uterine	Not well understood – linked to high levels of the female hormone oestrogen which, in turn, is linked to obesity	Bleeding after the menopause; bleeding between periods or after sex; changes to vaginal discharge; a lump in your tummy, pain in your lower back or pelvis	Depending on the stage of disease: surgery to remove the uterus and other organs; radiotherapy; chemotherapy; hormone therapy if the cancer has spread too far for surgery or radiotherapy	Maintain a healthy body weight. Do regular exercise. Follow a healthy diet with little alcohol.
Colon	Not fully understand but linked to: • age – 90% over 60 years old • diet high in red meat and low in fibre • being overweight or obese • lack of exercise • family history • alcohol use • smoking.	Blood in the faeces; a change in bowel habits; persistent lower abdominal pain and bloating caused by eating	Depending on the stage of disease, surgery to remove the affected bowel (often the only treatment needed); radiotherapy; chemotherapy; new targeted therapy which makes chemotherapy more effective	Maintain a healthy body weight. Reduce the amount of red meat eaten. Eat a diet high in fibre. Exercise regularly. Don't smoke/stop smoking. Reduce alcohol intake.
Prostate	Largely unknown; risks are increased by: • age • ethnicity – most common in Afro-Caribbean men and least common in Asian men • slight genetic link • obesity.	Increased need to urinate; straining to urinate; a feeling that the bladder has not fully emptied	Early, slow-growing prostate cancer does not need treating If it is more advanced or faster growing, surgery to remove the tumour; radiotherapy to destroy cancer cells; hormone therapy to slow growth of any cancer cells	Maintain a healthy body weight/ lose weight if overweight or obese.

Cancers

Figure 15.31 shows data on common cancer incidence and mortality in the Caribbean based on an article in a medical journal.

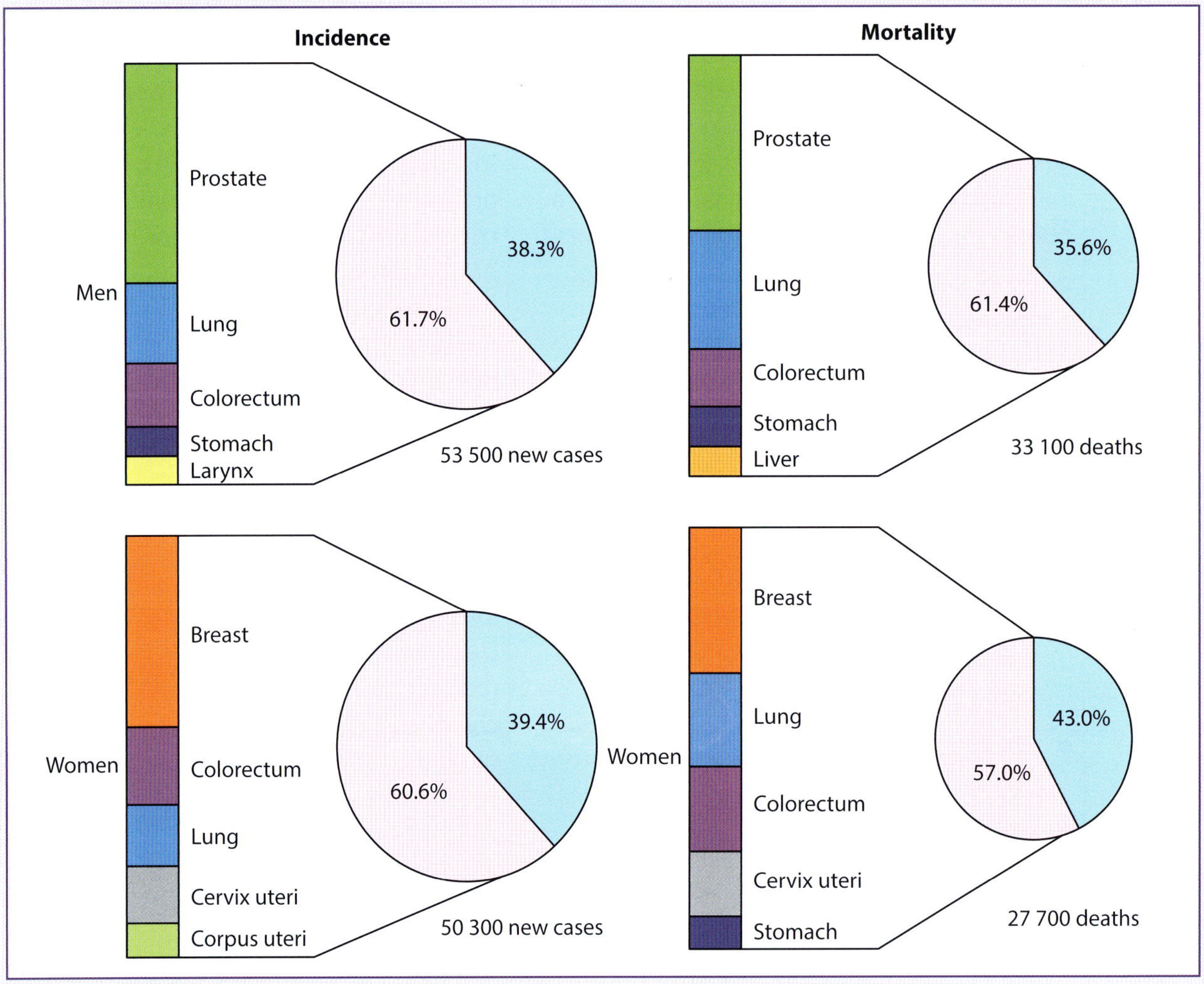

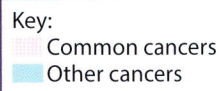

Author: Dingle Spence,Rachel Dyer,Glennis Andall-Brereton,Michael Barton,Susannah Stanway,M Austin Argentieri,Freddie Bray,Shamir Cawich,Sophia Edwards-Bennett,Christopher Fosker,Owen Gabriel,Natalie Greaves,Barrie Hanchard,James Hospedales,Silvana Luciani et al. Publication: The Lancet Oncology. Publisher: Elsevier. Date: September 2019. © 2019 Elsevier Ltd. All rights reserved.

Figure 15.31 Data on common cancer incidence and mortality in the Caribbean based on an article in the highly respected medical journal 'The Lancet' in 2019. Look back to *The biologist's toolkit* (pages 67, 75 and 107 could help) to remind yourself of the importance of reliable sources.

Mental health problems

Many of the diseases we will be studying in this, Section D, are physical diseases that focus on particular systems of your body, such as your respiratory system or your digestive system. Some diseases, however, affect our mind – the way we feel, think and behave. Good mental health is as important as good physical health and the two are closely linked. Mental health problems were historically often referred to as **neurosis**. This term is not widely used now but you may still come across it.

Some mental health problems, such as schizophrenia and bipolar disorder, are chronic and need long-term management and medication. Many common mental health problems, however, come and go, and affect most people at some stage in their lives. Three of the most common mental health problems are **anxiety**, **stress** and **depression**.

Health and diseases

Anxiety

Everyone feels uneasy, fearful or anxious sometimes – before an exam or an interview, for example. This is a normal response to life. Anxiety only becomes a mental health problem when it gets out of control, when it may be anything from mild to severe. Look at Figure 15.32.

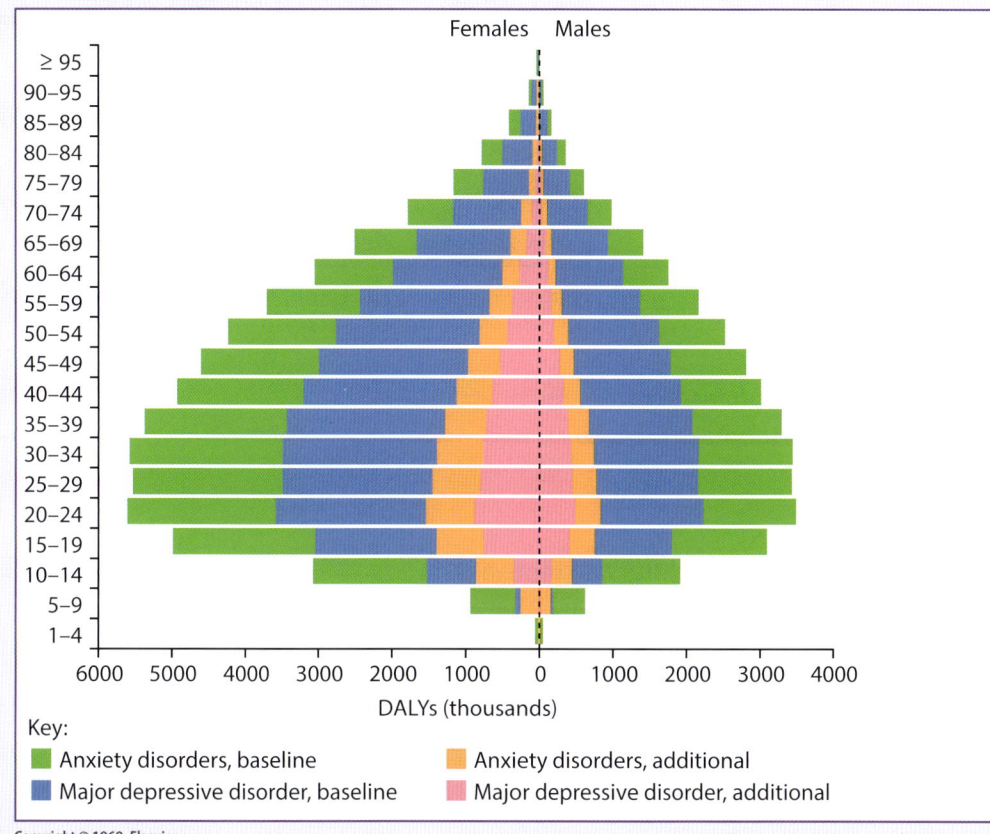

Figure 15.32 The World Health Organization has calculated the impact of anxiety around the world in DALYs (Disability-adjusted life years) in 2020, during the global COVID pandemic. The good news is that all types of anxiety reduce as you get older.

Symptoms and causes of anxiety

Symptoms of anxiety range from a constant sense of anxiety to panic attacks, phobias and social anxiety, which makes mixing with others difficult.

The causes of anxiety vary and different doctors have different ideas, but the main causes seem to be:
- overactivity in areas of the brain involved with emotion and behaviour (see Figure 15.33)
- imbalance of the neurotransmitters serotonin and noradrenaline in the brain, which are linked to mood changes (see Chapter 11 to remind yourself about the role of neurotransmitters at synapses in the nervous system)
- genetics, as the tendency to develop anxiety seems to be inherited – you are five times more likely to experience anxiety if a relative has been affected
- traumatic experiences or stress as a child
- long-term health conditions
- alcohol or drug misuse.

Figure 15.33 People with anxiety find mixing with others very difficult.

Mental health problems

Treatment of anxiety

People can help themselves to recover from mild and moderate anxiety in several ways. These include:

- exercise
- self-help courses
- stopping smoking (the drugs in cigarettes may increase anxiety
- reducing alcohol and/or caffeine intake – both of these can increase feelings of anxiety.

Health professionals have other treatments that can help to overcome anxiety if it becomes severe:

- psychological or talking therapies, such as, cognitive behavioural therapy, which helps to 'retrain' your brain (see Figure 15.34)
- medicines that help to balance the neurotransmitters in the brain, for example, SSRIs (selective serotonin reuptake inhibitors).

Figure 15.34 Talking to a professional therapist can help with recovering from anxiety.

Stress

Most people feel stressed at times, and it is often helpful and motivating. When we are stressed, we release hormones such as adrenaline that cause changes such as an increased heart rate. These changes are useful if you are in physical danger – for example, you need to run away from a bull – but less so if you are feeling stressed about work or a relationship.

Stress becomes a mental health problem if you feel stressed all the time, and everything feels too much for you. Stress is often a reaction to mental or emotional pressure or feeling out of control. DON'T feel alone – most people feel stressed at some point in their lives and there are many ways to help overcome both the stress and the symptoms it brings.

Symptoms of stress

Stress produces a range of symptoms that may be physical, mental or behavioural:

- **Physical symptoms** include headaches, dizziness, muscle tension or pain, stomach problems, chest pain, rapid heartbeat and sexual problems.
- **Mental symptoms** include difficulty concentrating, worrying constantly, becoming forgetful, feeling overwhelmed and finding it hard to make decisions.
- **Behavioural symptoms** include sleeping a lot or not being able to sleep, being more irritable than usual, avoiding places or people, and drinking, smoking or eating more (see Figure 15.35).

Figure 15.35 Overeating on unhealthy foods is a behavioural symptom of stress.

Treatment of stress

There are many ways to treat or overcome stress. It is important to find the main cause of our stress – for example, work, family problems, financial problems, health, exams, appearance – and to try to reduce it in that area of life. This in turn will reduce overall stress levels. It is not always easy – but there are a number of treatments that help:

- **Self-help techniques:** there are many ways for an individual to reduce their stress levels – sometimes on their own and sometimes involving family, friends or professionals. Examples include:
  - Talk to friends, family or a counsellor.
  - Discover natural stress-busters, especially exercise (see Figure 15.36).
  - Practise calming breathing exercises.
  - Plan ahead for stressful events.
  - Investigate time-management techniques to help you take back control of your life.

- **Things to avoid:**
  - DON'T try to do everything at once; set small, achievable targets.
  - Focus your time and energy ONLY on things you can actually do something about.
  - DON'T use alcohol, food or drugs to relieve stress, as they can end up adding to the problem.

Figure 15.36 Exercise is a natural stress-buster, and it is good for you too.

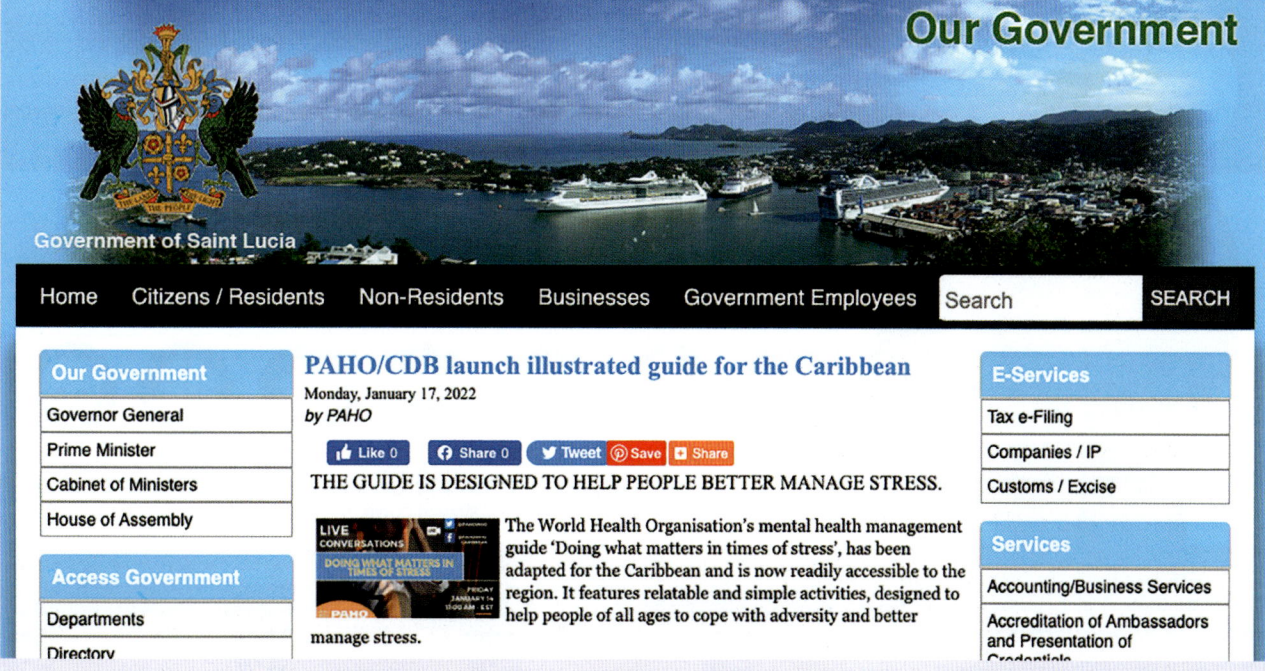

Figure 15.37 Governments in the Caribbean recognise the difficulties that mental health problems such as stress cause and they are working to support everyone affected – such as this Saint Lucia project from 2022.

Depression

Almost everyone feels down at times – it is a normal reaction when things are tough or sad. But depression that lasts weeks, and certainly depression that lasts months, is not healthy and has a serious effect on overall well-being. However, with the right treatment, most people with depression make a full recovery.

Symptoms of depression

As with most mental health problems, the symptoms of depression (Figure 15.38) vary from person to person and depend on the severity of the illness. They include:

- unhappiness and hopelessness
- loss of pleasure in life
- anxiety (see pages 347–348)
- sleeping badly
- sleeping all the time
- feeling tired however much sleep you get
- loss of interest in sex.

Figure 15.38 Treatment can help people with depression, who are feeling unhappy and anxious.

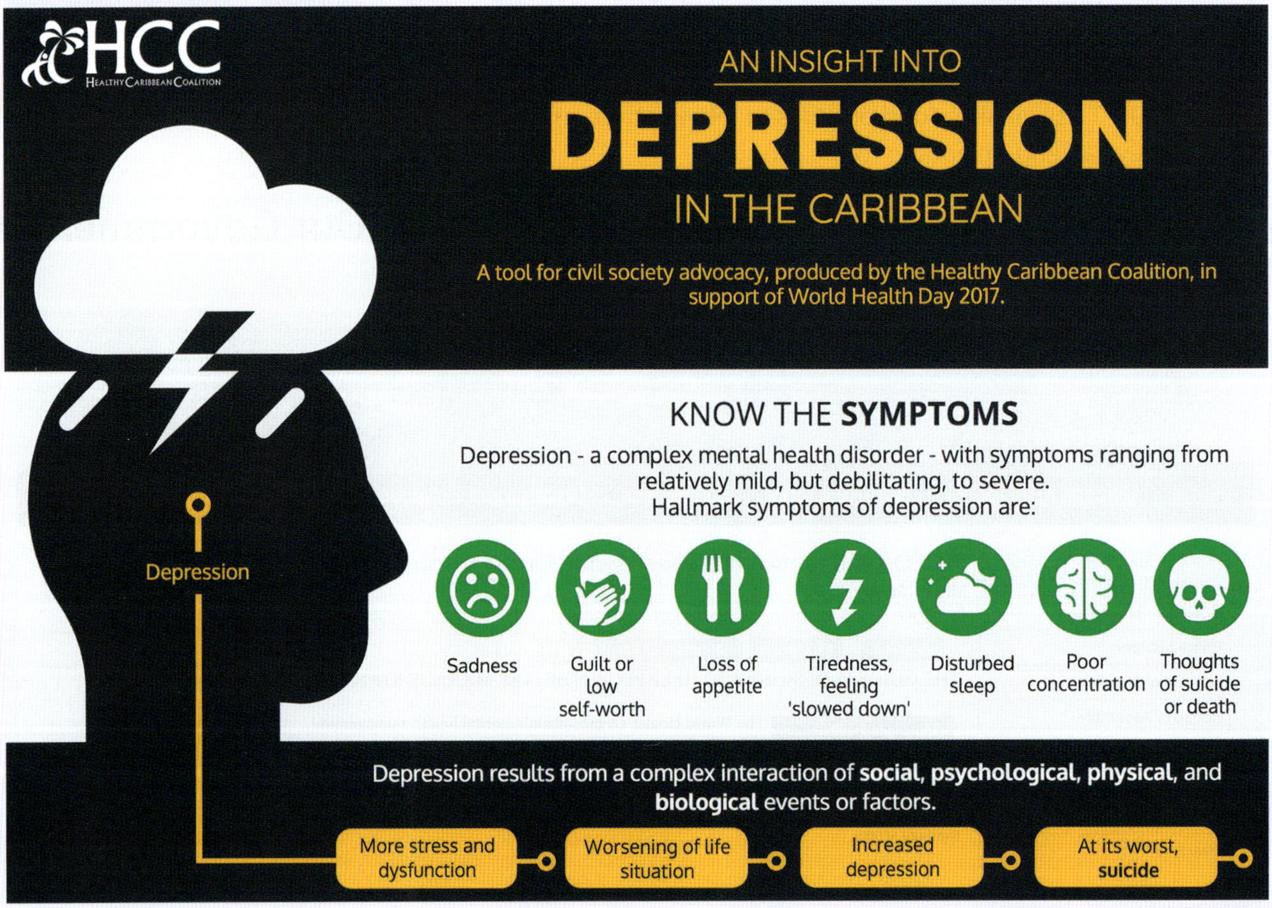

Figure 15.39 Raising awareness of depression is important so people can look out for each other. Graphic courtesy of The Healthy Caribbean Coalition.

Health and diseases

Causes of depression

There seem to be a number of different causes of depression and doctors and scientists do not fully understand them all. They include:

- a life event such as bereavement, family breakdown or job loss that triggers a sense of sadness, which does not grow less or go away
- long-term stress
- genetics, as the tendency to become depressed may run in some families
- giving birth, as the hormone changes around giving birth may cause **post-natal depression** in some women
- loneliness
- illness, because diseases such as coronary heart disease, cancer or arthritis increase the risk of developing depression
- alcohol and drugs, which may trigger a spiral of depression.

Treatment of depression

There are a number of treatment options for depression, depending on how severe the condition is. These treatment options include the following:

- Self-help techniques; as with other mental health problems, there are many ways for an individual to help themself overcome their depression, especially if it is relatively mild. Examples include:
  - talking things through with friends, family or a counsellor
  - doing natural mood-lifting activities, especially exercise
  - joining a self-help group – talking to others in the same situation and helping each other may bring significant benefits. See Figure 15.40.
- Health professionals have other treatments that help to overcome depression if it becomes severe and overwhelming:
  - psychological or talking therapies such as cognitive behavioural therapy
  - counselling services
  - medication, such as anti-depressants, which are drugs that affect the chemistry of the brain and help to restore balance, enabling the patient to function again until they naturally find pleasure in life.

The diseases discussed in this chapter so far are all examples of non-communicable diseases. Some are chronic, whereas others are curable. They all have a large impact on the quality of life of an individual and their family. However, with careful treatment and a sensible choice of lifestyle, the effect on the sufferers can be kept to a minimum. In fact, in some cases, the right lifestyle choices will allow the person to avoid the diseases altogether. For example, plenty of exercise helps to reduce, prevent or cure diseases ranging from diabetes and cardiovascular problems to some cancers and mental health problems such as stress and depression.

You will examine the influence of lifestyle on your health throughout this section of the book.

Figure 15.40 Joining a self-help group and talking to others in the same situation can help with depression.

Possible SBAs based on health and well-being

→ Produce a leaflet/podcast/poster series on 'Good health' covering all aspects of health in the Caribbean.

→ Chose ONE of the main types of non-communicable diseases and carry out a project to discover the incidence of these diseases and the social and economic impact they have on either the local community (using surveys and so on) OR on your country as a whole.

→ Carry out a survey in your school/ local community to discover how aware people are of the health problems associated with obesity. Use your results to develop resources to help get the message across about the importance of maintaining a healthy weight to the people who need it most.

→ Carry out a survey in your school/ local community to discover how aware people are of the health benefits associated with regular exercise. Use your results to develop resources to help deliver the message about the importance of regular exercise in reducing the risk of many different diseases to the people who your survey shows need it the most.

→ Investigate the impact of diabetes within your country – and possibly do a comparison with the Caribbean as a whole.

→ Survey your classmates/school to discover the incidence of asthma and do background research to see if numbers in the community/country are increasing or decreasing. Present your evidence and suggested strategies to reduce the health burden with regard to asthma going forward.

→ Investigate the screening policies of your country for different types of cancers/investigate the success of screening policies in other parts of the world, for example, for cervical cancer, or breast cancer and discuss the economic balance to be made.

→ Develop a 'mental wellness' pack for all students, to help them cope with the pressure of school, exams, and so on. This must contain some information about mental health and some evidence about the impact it has on young people in the Caribbean.

End-of-chapter summary

In this chapter you have learnt that:

- health is a state of complete physical, mental, and social well-being and not merely the absence of disease or infirmity; disease is the state in which a part, organ or system of your body is damaged or does not function properly
- communicable or infectious diseases are passed from one person to another; they are caused by microorganisms know as pathogens
- non-communicable diseases are not passed from one person to another; they include chronic diseases, which are long-term illnesses that may be physiological (diabetes, asthma) or degenerative, where the function of a system gets worse over time (osteoarthritis); nutritional deficiency diseases caused by a lack of particular nutrients in the diet (night blindness, scurvy, or marasmus); inherited disorders (Down syndrome, sickle cell anaemia); lifestyle diseases result from lifestyle choices we have made (smoking-related diseases such as lung cancers and heart disease); cancers that result from the uncontrolled growth of cells in an organ or tissue (breast cancer, ovarian cancer, prostate cancer); and mental health problems (anxiety, depression)
- both communicable and non-communicable diseases have an impact on the individual affected, and often on their family as well; many diseases also have a social and economic impact on the whole of society

- an epidemic is an outbreak of disease that affects a large proportion of a community, region or the population of a country; a pandemic is an epidemic that spreads to many different countries or even over different continents
- chronic diseases can be treated but not cured
- obesity or being overweight occurs when we eat more food than our body can use; obesity is linked to a number of health problems including hypertension, coronary heart disease and Type II diabetes
- diabetes mellitus is the condition that causes persistently high blood sugar levels, and major symptoms include increased frequency of urination, increased thirst, increased hunger and tiredness; the main clinical sign is sugar in the urine
- hypertension is the medical name for high blood pressure; blood pressure is considered high if the systolic pressure is greater than 140 mm Hg or diastolic is greater than 90 mm Hg
- coronary heart disease results when the heart fails to pump; usually is accompanied by a crushing pain in the chest, but it can also be painless
- asthma is shortness of breath due to constricted airways; it is triggered by certain irritants in the environment, or stress; it requires medication to relieve the symptoms and reduce the chance of an attack; it is helpful to avoid of potential triggers
- cancers result from the uncontrolled growth and division of cells; this results in tumours that may break off and spread around the body; left untreated, most cancers cause death
- the causes of cancer are many and varied, and scientists do not understand all of them, but they include carcinogenic chemicals such as tar in cigarette smoke, genetic factors and mutations, age, ionising radiation such as UV light and X-rays, and some viruses
- common cancers in the Caribbean and globally include respiratory tract cancers, especially lung cancers, breast cancers, cancers of the cervix, uterus and ovary in the female reproductive system, and colon cancers and prostate cancer in the male reproductive system
- ways of treating cancers include surgery to remove the tumour, radiotherapy, chemotherapy and modern therapies such as hormone therapy or immunotherapy
- mental health problems affect our mind – the way we feel, think and behave; good mental health is as important as good physical health, and the two are closely linked
- common mental health problems in the Caribbean include anxiety, stress and depression; there are a number of possible treatments for all of these diseases
- the above diseases are all common in the Caribbean community and are among its major causes of death
- the major risk factors for obesity, diabetes, hypertension and coronary heart disease include poor dietary habits and lack of physical activity; there is also evidence to suggest genetic factors
- the treatment of these conditions involves making sensible lifestyle choices, targeting the risk factors, and use of appropriate medication; eaing healthily, maintaining a healthy weight or losing weight if necessary, and doing regular exercise reduces the risk of developing many of the diseases in this chapter, and also helps you to recover if you are affected.

End-of-chapter questions

1. What are the **two** types of diabetes mellitus?
   - A I and II
   - B A and B
   - C Glucose and fructose
   - D Insulin and tablets (1)

2. Which of the following is not a possible cause of obesity?
   - A Overeating
   - B Immobility
   - C Steroids
   - D Running (1)

3. What is hypertension a major risk factor for?
   - A Strokes
   - B Epilepsy
   - C Red eyes
   - D Wet feet (1)

4.
   - a Describe how the coronary arteries narrow in coronary heart disease and explain why this is potentially dangerous to health. (10)
   - b Name the main signs and symptoms of a heart attack. (6)
   - c State how a patient with a heart attack can be treated. (4)

5.
   - a Name **three** 'triggers' of asthma. (3)
   - b Name and discuss the components of an asthma attack. (10)
   - c Describe the different types of asthma medication. (7)

6.
   - a Define the term 'hypertension'. (1)
   - b Explain the concept of blood pressure. (4)
   - c Give **five** major risk factors for hypertension. (5)

7.
   - a Describe the **two** most common ways used in the Caribbean for diagnosing obesity. (6)
   - b Explain how culture is influencing the increase in obesity in the Caribbean. (4)
   - c Using the data in the diagrams below and on the next page, explain how the evidence suggests strong links between obesity and other common diseases. (3)

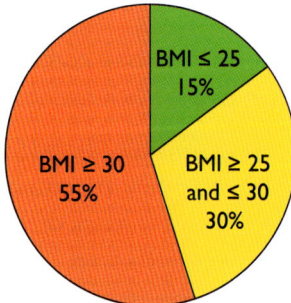

Body Mass Index (BMI) of people diagnosed with Type II diabetes

Gatineau M, Hancock C, Holman N, Outhwaite H, Oldridge L, Christie A, Ells L. Adult obesity and type 2 diabetes. Oxford: Public Health England, 2014.

Health and diseases

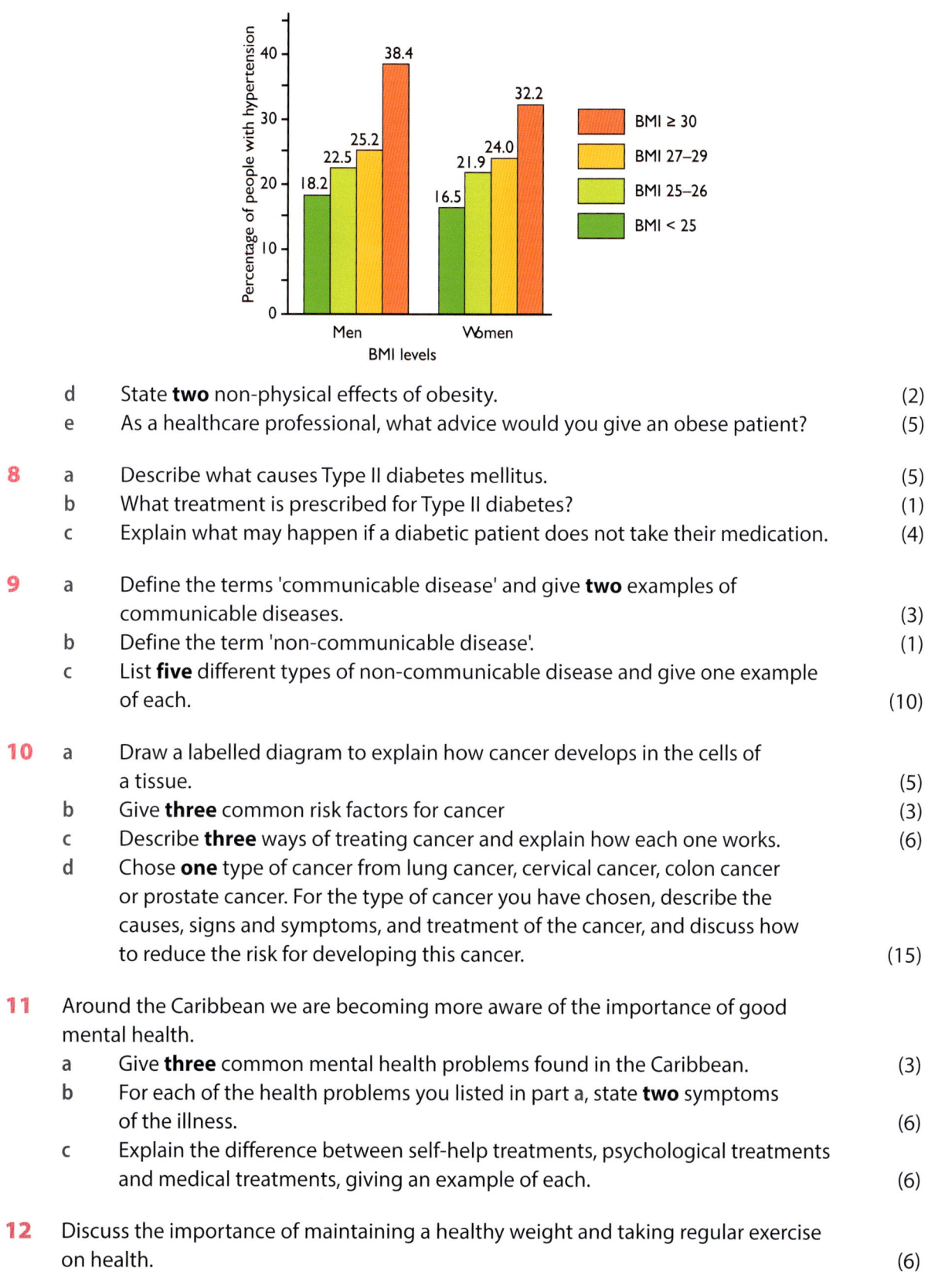

    d   State **two** non-physical effects of obesity. (2)

    e   As a healthcare professional, what advice would you give an obese patient? (5)

8   **a**   Describe what causes Type II diabetes mellitus. (5)

    b   What treatment is prescribed for Type II diabetes? (1)

    c   Explain what may happen if a diabetic patient does not take their medication. (4)

9   **a**   Define the terms 'communicable disease' and give **two** examples of communicable diseases. (3)

    b   Define the term 'non-communicable disease'. (1)

    c   List **five** different types of non-communicable disease and give one example of each. (10)

10   **a**   Draw a labelled diagram to explain how cancer develops in the cells of a tissue. (5)

    b   Give **three** common risk factors for cancer (3)

    c   Describe **three** ways of treating cancer and explain how each one works. (6)

    d   Chose **one** type of cancer from lung cancer, cervical cancer, colon cancer or prostate cancer. For the type of cancer you have chosen, describe the causes, signs and symptoms, and treatment of the cancer, and discuss how to reduce the risk for developing this cancer. (15)

11   Around the Caribbean we are becoming more aware of the importance of good mental health.

    a   Give **three** common mental health problems found in the Caribbean. (3)

    b   For each of the health problems you listed in part a, state **two** symptoms of the illness. (6)

    c   Explain the difference between self-help treatments, psychological treatments and medical treatments, giving an example of each. (6)

12   Discuss the importance of maintaining a healthy weight and taking regular exercise on health. (6)

Chapter 16: Infectious diseases caused by viruses and bacteria

When you have completed this chapter, you will be able to:

- describe different ways in which infectious diseases are transmitted from one person to another
- describe different types of immunity
- list the main types of infective organisms
- describe ways to limit the spread of infectious diseases or prevent them completely
- describe a variety of ways of treating infectious diseases
- distinguish between disinfectants, antiseptics and antibiotics
- define the term 'vector'
- explain the methods used to control the growth of microorganisms
- distinguish between immunity and immunisation; vaccine and vaccination

What the examiners say

- Too many candidates cannot differentiate between bacteria and viruses.
- Candidates are uncertain about the categories of diseases, are sometimes unable to discuss the types of immunity, and in many instances, talk about preventative measures instead.
- Candidates often distort their responses by using terms, such as 'antibodies' and 'antibiotics', incorrectly and interchangeably. There are also candidates who are often unable to accurately define antibiotics.

Revision tip

When you are revising for exams, eat healthily, take lots of short breaks and make sure you get plenty of sleep.

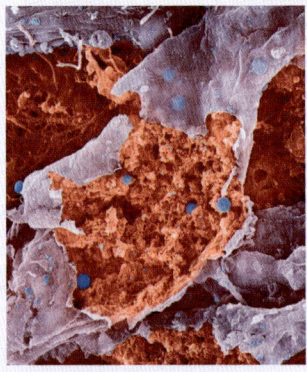

Figure 16.1 Scanning electron micrograph (SEM) of flu viruses (blue) budding from a burst epithelial cell in lungs.

In this chapter, we are going to look at the principles of infectious diseases (also called **communicable diseases** or **transmissible diseases**). This is the type of disease most likely to have affected you directly. Infectious diseases are caused by microorganisms and **parasites**, and cause illness all over the world, in every country. They attack rich and poor alike, and there are infectious diseases that affect every organ system in the body. The diseases range from relatively mild, such as the common cold and tonsillitis, through to known killers, such as tetanus, influenza (see Figure 16.1), COVID-19, HIV/AIDS, pneumonia, the plague and malaria. Once you understand the principles of infections disease, you can apply them to any example you are given.

Infectious diseases

Infectious diseases are often **acute** – they come on abruptly (rapidly) with distinct signs and symptoms that last for a limited time.

Unlike the non-communicable diseases we met in the last chapter, infectious diseases are spread from one infected person or animal to another.

The main disease-causing agents are viruses, bacteria and fungi, also called **pathogens** (see Chapter 1 for details of the structure of these organisms). However, not all infectious diseases are caused by microorganisms. Some of the parasites that infect the human body and cause disease can grow to very large sizes (you will be looking at these in more detail in Chapter 18).

> **DID YOU KNOW?**
>
> The largest human tapeworm, *Diphyllobothrium latum*, can grow up to 15 m in length inside your gut – so it certainly isn't a microorganism!

Microorganisms and disease

Bacteria and viruses cause disease because they attack the cells of our body. Once they get inside our body, they take advantage of the warmth, food and oxygen, and reproduce rapidly.

- Bacteria reproduce by simply splitting in two. They often produce toxins (poisons) that affect the body and cause the symptoms of disease. Sometimes they damage the cells directly.
- Viruses take over the cells of the body as they reproduce, damaging and destroying them (see Figure 16.2). They very rarely produce toxins.

Common disease symptoms are:
- a high temperature
- headaches
- rashes.

These symptoms are caused by the damage to the cells and the toxins produced by the pathogens. Symptoms also appear as a result of the way the body responds to the damage and toxins.

Checkpoint question

1. State the most common disease-causing organisms.

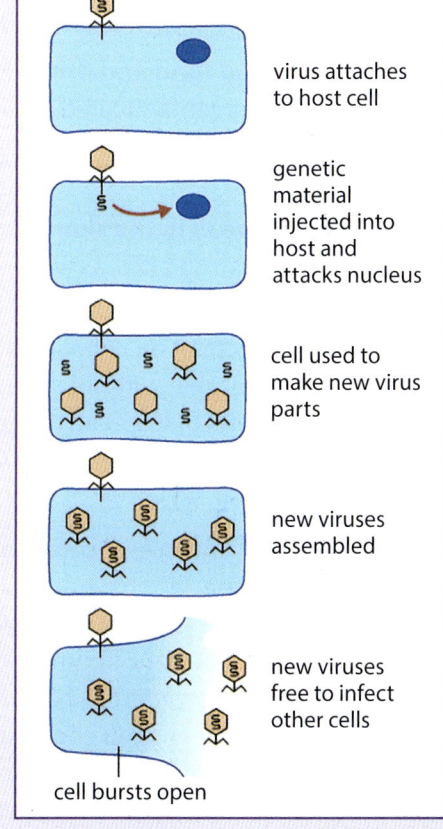

Figure 16.2 Viruses cannot reproduce without destroying their host cells. Every time the cells burst open, more viruses are released to attack more cells. Each time the body comes under attack by a mass of viruses, the temperature spikes.

How do infectious diseases spread?

The key point about infectious diseases is that the organisms that cause the diseases are spread from one person to another. The more microorganisms that can get into our body, the more likely it is that we will get a disease. The spread of a disease is also called the **transmission** of a disease.

Some common ways in which infections can be spread are described below.

- **Droplet infection:** When we cough, sneeze or talk, tiny droplets are expelled from our breathing system. If we have an infection, those droplets contain microorganisms. Other people breathe in the droplets, along with the viruses or bacteria they contain, and so pick up the infection, for example, **flu (influenza), COVID-19, tuberculosis** and the **common cold**.
- **Direct contact:** Some diseases are spread by direct contact of the skin, e.g. **ringworm** (a fungal disease) and some **sexually transmitted infections** like genital herpes.
- **Contaminated surfaces:** Many diseases are spread when we get pathogens on our hands from, for example, coughing into a hand, and then touching surfaces. The next person who touches that surface/object transfers the pathogens to their skin. If they then touch their mouth or nose – the pathogens have a way into the body. See Figure 16.3.
- **Contaminated food and drink:** Eating raw or undercooked food, or drinking water contaminated by sewage, means that we take large numbers of microorganisms, such as **gastroenteritis** and **cholera**, straight into our gut.
- **Through a break in the skin:** Microorganisms enter the body through cuts, scratches and bites, as well as through needle punctures, for example, **HIV/AIDS** and **hepatitis**.
- **Vectors:** A vector is an animal that spreads disease-causing organisms from one host to another without suffering any harm itself. In this textbook, we consider vectors that spread diseases between people. There are a number of common animal vectors in the Caribbean, including rats, mosquitoes, and houseflies.

Figure 16.3 Droplets and aerosols filled with microorganisms fly into the air at up to 100 miles an hour when you sneeze.

Overcrowded and unhygienic conditions encourage the spread of disease. When people live in overcrowded conditions, with no sewage or clean water systems, infectious diseases are very common. Overcrowding helps microorganisms to spread from person to person in every way, from droplet infection to contaminated water vector animals. See Figure 16.4. When living conditions improve, the incidence of infectious diseases falls dramatically.

It is as important to look at how diseases are spread as it is to find treatments for specific diseases. Understanding the way diseases spread allows us to prevent them from being passed on from person to person. We will look at this in more detail later in this chapter and also when we consider the different effects of lifestyle on disease.

Figure 16.4 Overcrowded and unhygienic conditions encourage the spread of many infectious diseases.

Checkpoint questions

2. Describe how microorganisms cause disease.
3. Make a table to summarise the ways in which diseases may be spread.

Infectious diseases caused by viruses and bacteria

Preventing the spread of infectious diseases

We can use our knowledge of how diseases are spread to reduce the risk of them spreading. By using strict hygiene measures, and preparing and storing food carefully, we can avoid many diseases. Here are some key ways to prevent the spread of infectious diseases.

Hygiene

Good hygiene means keeping yourself and your surroundings clean. Follow these guidelines:

- Wash your hands with soap and water after using the toilet, before cooking and after touching any potential sources of harmful microorganisms. Sanitise your hands with an alcohol-based hand rub (ABHR), if you cannot wash your hands.
- Cough or sneeze into your elbow or into a tissue, and then wash your hands. This prevents too many pathogens getting into the air and removes them from your hands.
- Cover your mouth and nose with a mask in public if you have a disease of the respiratory system to prevent spreading pathogens into the air as you speak, cough and sneeze (see Figure 16.5 and Figure 16.6).
- Use **disinfectants** on kitchen work surfaces, toilets and so on to reduce the number of pathogens.
- Keep raw meat away from food that is eaten uncooked to prevent the spread of pathogens.
- Cover any cuts and apply **antiseptic creams** to stop pathogens from getting directly into the body.

Figure 16.5 Wearing a mask in public if you have a respiratory illness protects people around you.

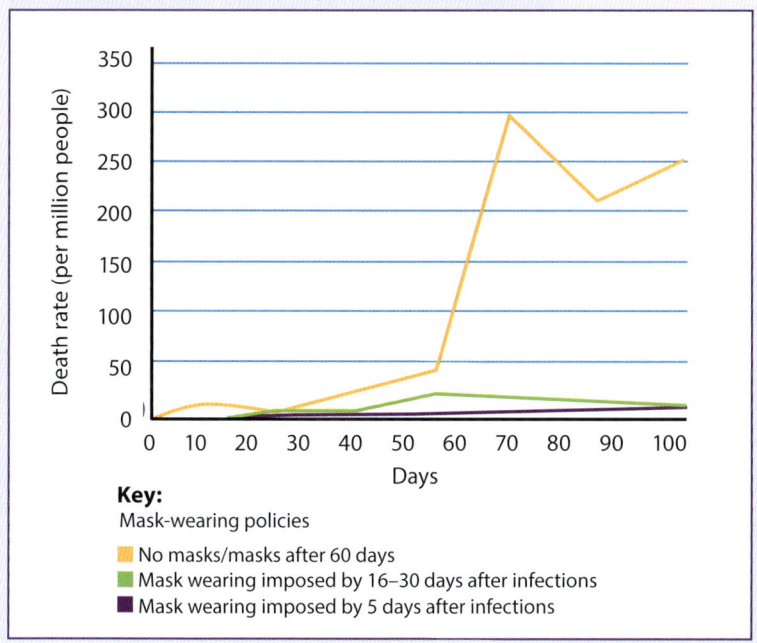

Key:
Mask-wearing policies
- No masks/masks after 60 days
- Mask wearing imposed by 16–30 days after infections
- Mask wearing imposed by 5 days after infections

Figure 16.6 In the COVID-19 pandemic, scientists collected clear evidence showing that wearing masks reduced both the number of cases of COVID-19 in the population and, as a result, the number of deaths was also lower.

Isolation

If someone has an infectious disease, the fewer people they see, the less likely they are to pass on their disease. Keeping people in isolation (Figure 16.7) is very important in diseases such as cholera and ebola – and it is also important in respiratory diseases such as COVID-19 and the flu.

Destroying or controlling vectors

Rats, mosquitoes and houseflies are examples of animals that act as vectors of human diseases. They spread diseases from the plague to malaria to gastroenteritis. If the vectors are destroyed, the spread of the diseases they carry will be prevented. If the number of vectors is controlled, the spread of the diseases they cause will be reduced. You will learn more about vectors in Chapter 18.

Vaccination

Vaccination is a process that has saved millions of lives around the world since it was first developed in the 19th century. Doctors use an element of a pathogen to help your immune system learn how to deal with it. Then if you meet the living microorganism, your immune system deals with it immediately and you do not become ill. You will learn more about vaccinations later in this chapter.

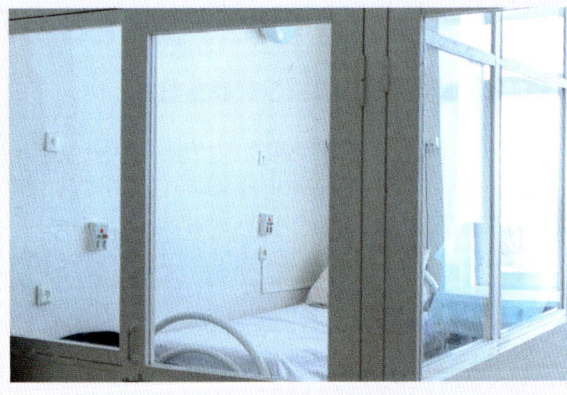

Figure 16.7 Many hospitals have special isolation units for treating people with infectious diseases.

Defences against disease

If we fail to prevent the spread of pathogens in our environment, our body can defend itself in several ways. This prevents the pathogens from getting inside the body and invading our cells.

- Our skin covers our body and acts as a barrier, preventing microorganisms from reaching the vulnerable tissues underneath.
- If our skin is damaged or broken in any way, the blood forms clots, which dry into scabs and seal over the cut (see Figure 16.8 and Chapter 8).
- The acid in the stomach destroys the majority of microorganisms taken into the digestive system (see Chapter 6).
- The breathing system is vulnerable, because air is drawn right inside our body every time we breathe. However, the whole system produces a sticky liquid called mucus that covers the lining of the organs and the tubes, and traps microorganisms. The mucus is then moved out of the body or swallowed down into the gut where the microbes are destroyed by the acid in the stomach.

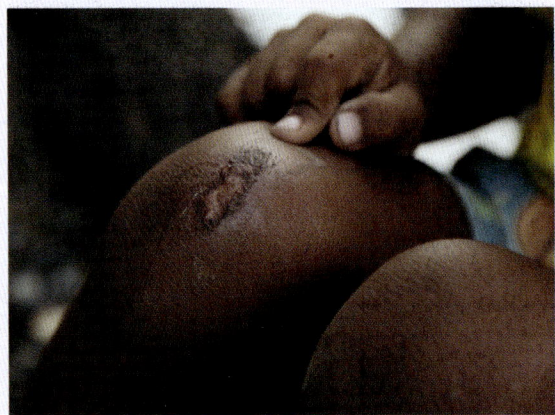

Figure 16.8 If our skin is damaged, the scabs that form protect us against the entry of microorganisms while new skin grows. This is particularly important in children!

The immune system

If microorganisms do manage to get through the body's defence system, we have more ways of protecting ourselves from attack. Our white blood cells help to defend us against pathogens and protect us from the worst effects of disease. They make up the immune system. (See Chapter 8 for details of how the white blood cells are produced in the lymph glands.)

Natural immunity

Every cell has protein markers on the outside of it – including bacteria and viruses. These proteins are called antigens. This is how cells recognise each other – and also recognise 'strangers'. As you can see in Table 16.1, our white blood cells make antibodies against any 'stranger' antigens such as those on bacterial or viral pathogens (see Figure 16.9). When the antibodies stick to the antigens, they disable or destroy the pathogen, which is then removed by the phagocytes.

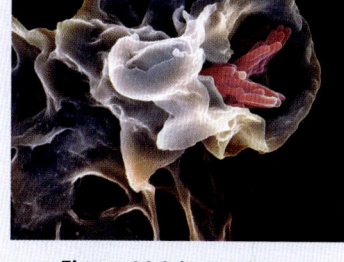

Figure 16.9 A scanning electron micrograph (SEM) of a white blood cell antibody (purple) surrounding and destroying a tuberculosis bacterium (pink).

Table 16.1 The different types of white blood cells, their roles and how they defend the body against disease

Role of the white blood cell	How it defends against disease
Lymphocytes – produce antibodies 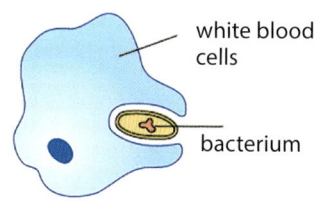	Some types of these white blood cells produce special chemicals called antibodies. The antibody attaches to a special marker protein (antigen) on the surface of the microorganism. Each type of microorganism has unique markers, so unique antibodies are needed for each different type of pathogen. Once they have produced antibodies against a particular bacterium or virus, the white blood cells can produce them again very rapidly so the person is immune to that particular disease.
Phagocytes – ingest microorganisms	These white blood cells ingest (take in) microorganisms once antibodies have bound to their antigens. They digest the microorganisms within the cell, destroying them and preventing them from causing disease.
Lymphocytes – produce antitoxins 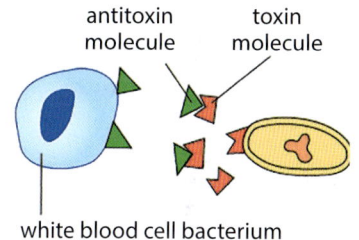	Some types of these white blood cells produce antitoxins that counteract the toxins (poisons) released by some microorganisms.

Once we have had a disease, our immune system 'remembers' the antigens from the pathogen and the correct antibodies to deal with it. If we meet the pathogen again, our white blood cells produce antibodies immediately. These antibodies destroy the pathogen before it can cause the symptoms of disease. In other words, we are immune to the illness. This is called **natural active immunity** because we have developed it actively for ourselves.

There is another type of natural immunity – this is the **natural passive immunity** that a mother gives to her unborn baby via the placenta. Antibodies pass from the mother to her foetus. If a mother breastfeeds her baby once it is born, she gives it even more natural passive immunity. This is because breast milk – and in particular the colostrum that is produced in the first few days after the baby is born – is very rich in antibodies. These antibodies protect the baby against many diseases until its own immune system gets going and starts to develop natural active immunity against pathogens. Breast milk will protect the child for as long as he/she continues to breastfeed. It is called passive immunity because the baby does not develop the immunity itself – it gets the antibodies from the mother.

Artificial immunity

Some diseases cause permanent damage, or even kill us, before our body manages to produce the right antibodies. Examples include tetanus, diphtheria, polio and even measles. Ideally, we will not come into contact with the pathogens that cause these diseases, but if we do, we can now prevent ourselves from falling ill in the first place. We have developed ways to protect ourselves against some of the most dangerous diseases by developing **artificial active immunity**.

Immunisation is the process by which someone is protected from a disease through **vaccination**. It is one of the greatest achievements of modern medicine because it protects us from a disease without us experiencing the serious effects of the illness. Immunisation has spared millions of people from the effects of devastating diseases, most recently from COVID-19 in the global pandemic of the early 2020s.

How does immunisation work?

1. First, create a **vaccine**. A vaccine contains either a weakened or dead strain of the pathogen, or part of the pathogen DNA or part of its structure, such as the protein spikes of the SARS-CoV-2 virus that causes COVID-19.

2. Vaccinate people. This gets the vaccine inside the body by injecting it or – for some vaccines – through drops in the mouth.

3. The vaccine inside the body triggers our immune response and our white blood cells develop the antibodies to the disease (see Figure 16.10).

Then if, in future, we meet the live pathogen, our body can destroy it before we become ill. This form of artificial immunity is called **artificial active immunity**. The antigen is introduced artificially, but our body actively makes its own antibodies.

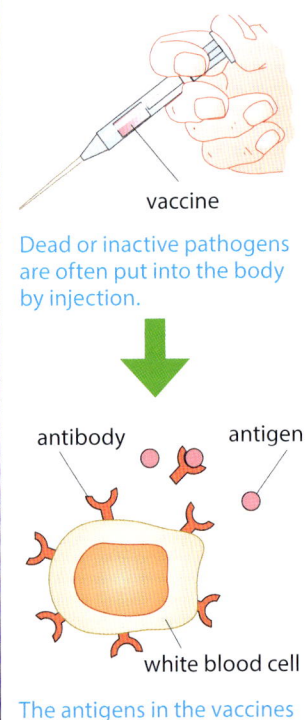

Dead or inactive pathogens are often put into the body by injection.

The antigens in the vaccines stimulate your white blood cells into making antibodies. The antibodies destroy the antigens without any risk of you getting the disease.

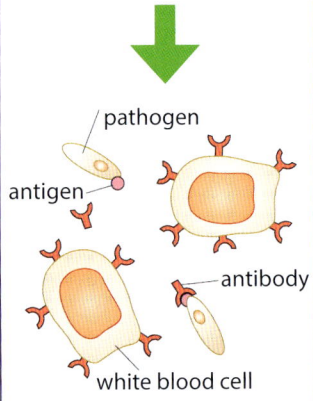

You are immune to future infections by the pathogen. That's because your body can respond rapidly and make the correct antibody as if you had already had the disease.

Figure 16.10 Immunisation is used to give us artificial active immunity to a number of dangerous diseases.

Infectious diseases caused by viruses and bacteria

Sometimes it is too late to give a normal vaccine. If you step on a rusty nail, you are at risk of developing tetanus. In some countries, if you are bitten by a dog, there is a risk of rabies. If you have not already been vaccinated, all is not lost. You can be given a vaccine that contains the antibodies you need to combat the specific pathogen. This is called **artificial passive immunity**. For example, you may have a shot of tetanus antibody after treading on the rusty nail, or a series of rabies antibody injections if you are bitten by a possibly rabid dog. However, passive immunity is short-lived. You need a normal vaccination once the initial danger is over so that you can develop your own artificial active immunity.

Vaccination and preventing the spread of disease

Over the last 50 years, there has been a tremendous move in the Caribbean to improve the uptake of vaccines against measles and other diseases. The number of people in the Caribbean being vaccinated has gone up from around 37% in 1970 to over 90% in the 21st century. As a result, many diseases have almost disappeared from the region. What is more, as a result of these vaccination programmes, the use of antibiotics and a rising standard of living, the number of children in the Caribbean who die each year of infectious diseases has fallen dramatically.

The impact of immunisation programmes has been immense. Smallpox, for centuries a killer, is no longer in existence thanks to worldwide vaccination programmes. Polio is going the same way, and as you can see in Figure 16.11, the Caribbean has been a very successful part of that immunisation programme.

> **DID YOU KNOW?**
>
> Much of the early work on immunisation was done by a Frenchman called Louis Pasteur. He was a family man and was broken-hearted when three of his five children died at a young age of infectious diseases. He was determined to do something about it. He convinced people by his experiments that diseases were caused by microorganisms. He developed vaccines against some of them, including anthrax and rabies. By the end of his life, he was close to a vaccine against diphtheria, the disease that had killed his little girls. Most of us are still vaccinated against this disease today.

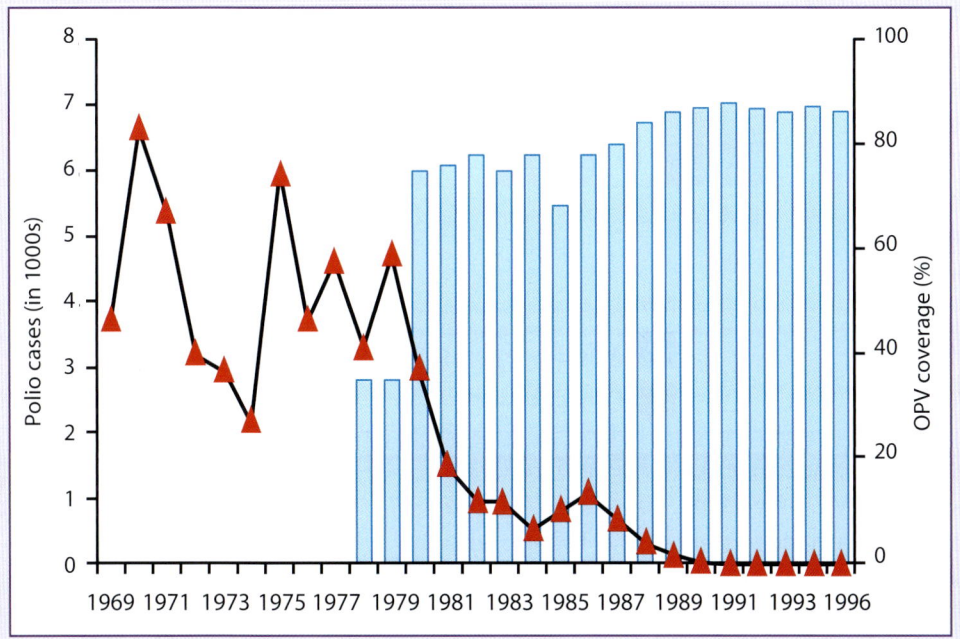

Eliminating Polio in Latin America and the Caribbean, Author Molly Kinder, 2004.

Figure 16.11 This graph shows the dramatic fall in polio cases in the Caribbean as the uptake of the oral polio vaccine (OPV) increased to almost 90% of the population. This is part of a global success story of using immunisation to eliminate polio, a virus infection that paralysed and even killed hundreds of thousands of children around the world.

The immune system

The Caribbean has also been extremely successful at vaccinating against and building immunity to measles, which is virtually never seen in the region (see Figure 16.12).

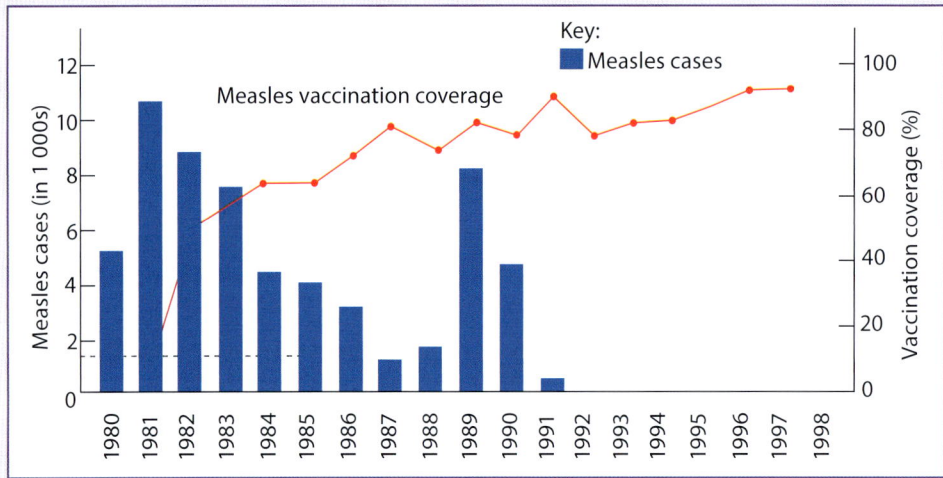

Karen N. Lewis-Bell, Beryl Irons, Elizabeth Ferdinand, Laura L. Jackson, J. Peter Figueroa "The Expanded Program on Immunization in the English- and Dutch-speaking Caribbean (1977–2016): reasons for its success" https://www.scielosp.org/article/rpsp/2017.v41/e127/.

Figure 16.12 This graph shows clearly how effective measles vaccinations have been at reducing the numbers of people in the Caribbean who suffer from this disease and its side effects, which can include blindness and death.

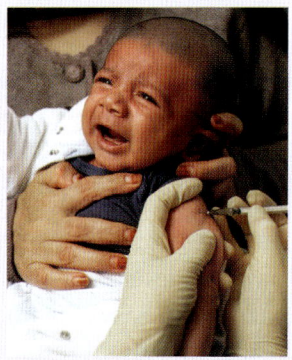

Figure 16.13 No one likes having an injection, but a quick jab when you have a vaccination is well worth it to avoid some terrible diseases, such as polio, diphtheria, measles or COVID-19.

Within a year of the outbreak of the global COVID-19 pandemic in the 2020s, vaccines were developed to control the disease, save lives and end the pandemic. These vaccines have been used in their billions globally and have saved millions of people from serious illness and death. In many countries, 85–95% of the population are fully vaccinated against COVID-19, and serious illness and deaths are falling fast. Unfortunately, in spite of an excellent history of successful immunisation programmes, we here in the Caribbean have been relatively slow at taking up the vaccines (see Figure 16.14). The reasons for this include the following:

- There have been problems with getting access to vaccines in the Caribbean.
- Vaccine hesitancy has been an issue. Social media spread opinions from many persons on the safety of vaccines, all of which are not based on scientific facts. As a result, some persons have been influenced against taking vaccines like those for polio and measles, which have saved numerous lives worldwide (see Figure 16.13).

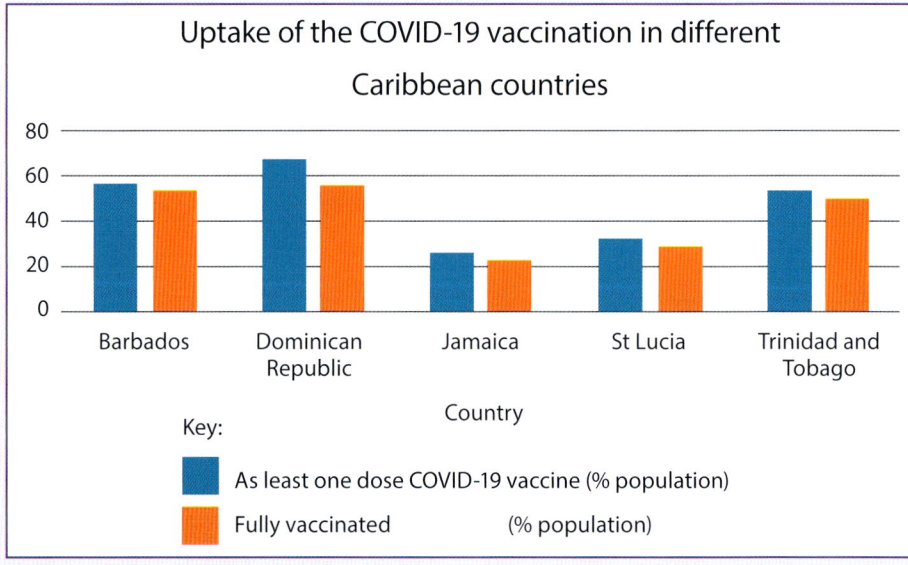

© 2022 Reuters. All rights reserved

Figure 16.14 The uptake of COVID-19 vaccines varies widely across different Caribbean countries.

Fortunately, many people follow the existing science and make an informed choice to have the COVID-19 vaccine. For example, more than 50% of the population of Barbados, Trinidad and Tobago, and the Dominican Republic are now double vaccinated (see Figure 16.14). This is excellent news and progress continues across the Caribbean.

Infectious diseases caused by viruses and bacteria

A chemical approach to controlling microorganisms

One way to lower the risk of spreading infectious diseases is to reduce the number of microorganisms around us. This helps to limit the number of pathogens that get into our body. We try to produce a **sterile** environment. **Sterilisation** involves killing all the microorganisms in a material or on the surface of an object, making it safe to handle without fear of contamination. A surface or object is either **sterile** (it contains no microorganisms) or it is not. Pathogens in our environment can be attacked chemically in different ways, such as, with **antiseptics** and **disinfectants**.

Disinfectants

A disinfectant (Figure 16.15) is a chemical substance applied to an inanimate object to kill microorganisms. This means that disinfectants are used on floors and surfaces, but not on people! Disinfectants do not usually make a surface completely sterile – they simply reduce the numbers of microorganisms present. The ideal disinfectant should be:

Figure 16.15 Disinfectants like these allow us to keep our surroundings safe from microorganisms.

- fast-acting
- effective against all types of infectious agents without destroying tissues or acting as a poison on the person using it
- able to easily penetrate the material to be disinfected without damaging or discolouring it
- easy to prepare and stable on exposure to heat, light or other environmental factors
- not unpleasant to work with, either in terms of its smell or its feel.

Two common and widely used disinfectants are:

- chlorine, which is used to disinfect our drinking water in the Caribbean
- household **bleach**, which we use in our home to keep sinks, toilets and surfaces germ-free.

Antiseptics

Antiseptics (Figure 16.16) are chemical substances applied to living tissue to kill microorganisms – disinfectants for the skin! Think back to when you last cut yourself. You probably applied an antiseptic. An open cut is an invitation to microorganisms as they have easy access to the inside of your body. An antiseptic will kill all the microbes on your skin and over the cut, preventing infection from getting into your tissues. Antiseptics often sting your skin. This is because they not only kill bacteria, they also damage human tissue slightly.

Different disinfectants and antiseptics have different ways of getting rid of microorganisms. Some of them kill the microorganisms, while others simply stop them growing, but they all make life safer for us.

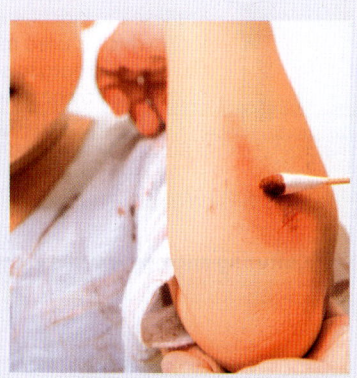

Figure 16.16 Antiseptics kill the microbes on your skin around a wound to prevent them getting into the tissues.

Investigating disinfectants and antiseptics

Many microorganisms can be grown in the laboratory. This allows us to learn a lot more about them and which chemicals will kill them. If we want to find out more about microorganisms, we need to culture them. This means that we need to grow very large numbers of them so that we can see the colony as whole.

If we culture bacteria in Petri dishes (see Figure 16.17) we can add samples of different antiseptics or disinfectants and see how they affect the growth of the microorganisms.

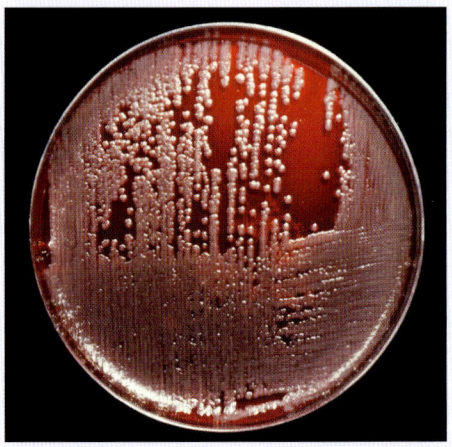

Figure 16.17 Culturing microorganisms like these bacteria makes it possible for us to see them and investigate how antiseptics and disinfectants affect the way they grow.

The biologist's toolkit: Principles of culturing microorganisms

If we give bacteria the right conditions, they grow and divide very rapidly, which is why it is relatively easy to culture them in the lab. Some bacteria divide every 20 minutes. This is a basic technique widely used for culturing bacteria in the laboratory.

- To culture microorganisms, we must provide them with all they need. This usually involves providing a culture medium containing carbohydrate to act as an energy source, along with various mineral ions, and in some cases, extra protein and vitamins. The culture medium will be either **agar** jelly or a sterile broth.
- Everything must be **sterile** when we culture microorganisms, so we ONLY grow the organisms we are investigating – not any dangerous ones from our skin or our environment.
- An appropriate temperature and oxygen are also usually provided.

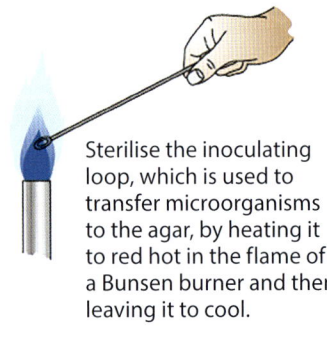

Sterilise the inoculating loop, which is used to transfer microorganisms to the agar, by heating it to red hot in the flame of a Bunsen burner and then leaving it to cool.

Culturing microorganisms in the lab

Dip the sterilised loop in a suspension of the bacteria you want to grow and then use it to make zigzag streaks across the surface of the agar.
Tilt the lid of the Petri dish to keep out unwanted microbes and close the lid as quickly as possible to avoid contamination.

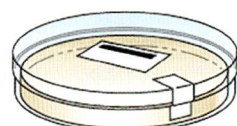

Secure the lid of the Petri dish with short pieces of adhesive tape to prevent microorganisms from the air contaminating the culture – or microbes from the culture escaping. Do not seal all the way around the edge to allow air in. More dangerous microorganisms grow when there is no oxygen.

Infectious diseases caused by viruses and bacteria

Antibiotics

Sometimes we can't keep microorganisms out of our body and we get ill. Fortunately, we now have a way to attack bacteria inside our body – the medicines known as **antibiotics**. Antibiotics are compounds that kill bacteria but do not damage human cells. This is the big difference between antibiotics, antiseptics and disinfectants.

If we swallow disinfectants or antiseptics, we will harm or even kill ourselves as well as the bacteria that are making us ill. However, we can take antibiotics into our body safely. They circulate in the blood and travel to all the body tissues, including the site of infection (see Chapter 8). Antibiotics only damage bacteria – our body cells are safe from them. **Penicillin** was the first antibiotic to be discovered, and it is still in use today. More antibiotics have been discovered since, but we really need more to be developed.

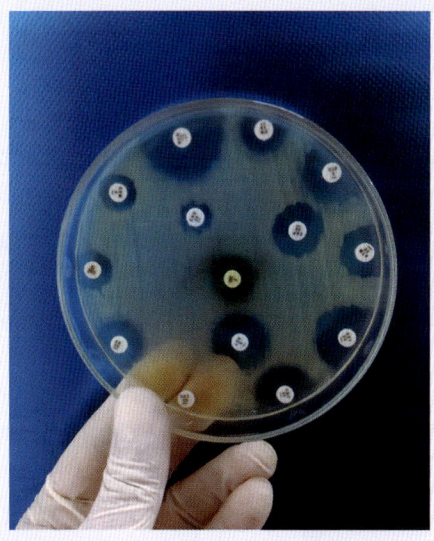

Figure 16.18 We use Petri dish cultures of bacteria to investigate the effectiveness of both different types of antibiotics and different strengths of the same antibiotic. If an antibiotic is effective, then the bacteria cannot grow close to them.

Antibiotics work in the following ways:
- Some interrupt the metabolic pathways of bacteria so they die.
- Some weaken the cell walls of bacteria so that they burst. Penicillin works in this way.
- Some interfere with protein synthesis in the bacteria so that they cannot reproduce.
- Some damage the bacterial DNA so that they cannot reproduce.

Fungal infections such as **ringworm** are cured using **anti-fungal agents**. These are similar to antibiotics but attack fungal cells. If you have ever suffered from athlete's foot between your toes, or ringworm on your skin, you will probably have used an anti-fungal cream to help you get rid of it.

Many antibiotics kill many different types of bacteria. They are used against common illnesses such as tonsillitis. Other antibiotics only work against a very specific type of bacteria. When doctors choose the right antibiotic to use, they consider the pathogen that is most likely to be involved and the cheapest antibiotic that will deal with it.

Antibiotics cure many killer diseases, even the plague. They have revolutionised medicine but we are left with two big problems:

1. Many microorganisms are becoming resistant to antibiotics, as you saw in Chapter 13. We must be very careful to use antibiotics only when they are really needed, and finish every course of antibiotics we are given, to help make sure that these amazing medicines continue to work for us.

2. We do not yet have a solution for diseases caused by viruses. Antibiotics have no effect on them. We do have a few drugs that are effective against viruses (see Chapter 19 for more about HIV/AIDS and the success of antiretrovirals in treating this disease), but for most viral diseases we have no real treatment or cure. This is something scientists and doctors are working on all the time. See Figure 16.18.

A chemical approach to controlling microorganisms

Checkpoint questions

4. During the hurricane season, people are often asked to add a capful of bleach to a large bucket of drinking water. Suggest a reason for this.
5. Describe how you might investigate whether or not a particular medicine is effective against bacteria.
6. If you go to the doctor with a sore throat or an ear infection, sometimes you will be given antibiotics and sometimes you won't. Explain why the treatment you are given varies.

Using heat to control microorganisms

As you have seen, we use antiseptics and disinfectants to reduce the microorganisms in our environment. However, they both have limitations – for example, they are usually toxic to both people and microorganisms, so we cannot use them to control microorganisms in our food.

Heat is a highly efficient means of sterilisation, provided that the material to be sterilised is **heat-resistant**. There are different ways of using heat to control microorganisms. Some of them produce a completely sterile environment and others substantially reduce the numbers of microorganisms.

- The simplest and best known method of sterilising using heat involves boiling. The objects to be sterilised are placed in boiling water (at 100 °C) and kept there for some time. Ten minutes will kill most cells, but some viruses and bacteria (for example, hepatitis viruses and *Clostridium* bacteria) take several hours of boiling to kill.

- **Autoclaving** is the method most commonly used to sterilise materials that are not damaged by heat. This method is very important for producing sterile equipment in hospitals and laboratories. An autoclave is very similar to a pressure cooker, which we might use in our kitchen (see Figure 16.19). The autoclave is used at 15 pounds per square inch of pressure, which raises the boiling point of water to 121 °C. Fifteen to 45 minutes of 'cooking' at these temperatures is enough to kill all microorganisms and sterilise the equipment.

Figure 16.19 An autoclave for sterilising equipment.

- **Ultra-high temperature (UHT)** is a way of treating food to kill all the microorganisms on or in it. The temperatures used range from around 135 °C to 150 °C. The food is only heated to these extreme temperatures for two to six seconds, but that is long enough to kill any microorganisms present and completely sterilise it. Since UHT treatment sterilises food, it not only gets rid of any disease-causing microorganisms but also destroys the organisms that cause food to go bad (see Figure 16.20). As a result, UHT food or milk will last for years if no air is allowed to get to it.

Figure 16.20 UHT technology removes the need to refrigerate milk or food until the packaging is opened.

- **Pasteurisation** is another technique widely used to treat milk, beer and other food stuffs, and make them safe to take into the body. Pasteurisation is not strictly speaking a method of sterilisation, because it does not kill all the microorganisms in the food. To pasteurise food, it is heated to either 71.6 °C for at least 15 seconds or 62.9 °C for 30 minutes (see Figure 16.21). Either way, most pathogens and most of the microorganisms that make food go bad are destroyed, so the food is much safer to eat and lasts longer. However, because it is not sterile, it will eventually go bad.

Figure 16.21 Milk being pasteurised

- Canning food (Figure 16.22) prevents the growth of microorganisms that would make the food go bad and cause disease. The food is heated to boiling point and sealed into the can, which kills microorganisms in the food and prevents the entry of new organisms from the air. The contents of the can remain sterile until it is opened.

Figure 16.22 Canning prevents microorganisms getting into the food from the air, and destroys any microorganisms already in the food. Canned food will last for years without going bad.

Possible SBAs based on infectious diseases caused by viruses and bacteria

→ Create an activity to use with primary age children to show them how pathogens are spread on dirty hands, and to demonstrate the importance of hygiene in reducing the spread of disease.

→ Produce a set of resources – posters/podcasts/blogs/films/and so on – to explain the different ways in which infectious diseases are spread and how we can use this knowledge to reduce the spread of these diseases.

→ Make a model/animation of the human immune system and how it works.

→ Investigate the incidence of infectious diseases in babies who are breast fed and so are given natural passive immunity from their mothers, and babies who are bottle fed.

→ Investigate the development of vaccinations and produce a timeline of the major steps in their development and major successes in reducing or eliminating disease.

→ Investigate the childhood vaccination programme in your country and compare it to other countries in the region and/or countries further afield.

→ Produce resources to encourage people in the Caribbean/your own country to take up the COVID-19 vaccination – use international data and local comparisons and highlight the importance of protection in the long term.

→ Carry out a practical investigation into the effectiveness of different disinfectants/different concentrations of the same disinfectant/antiseptics/different concentrations of the same antiseptic on the growth of bacterial cultures. Follow all safety precautions when growing bacterial cultures.

→ Carry out a practical investigation into the effectiveness of heating objects to different temperatures to sterilise them/the effectiveness of different temperatures of washing up water on sterilising crockery, and so on.

→ Design and produce a board game or a simple computer game Microorganisms vs People to give other students/children/everyone as much information as possible in a fun way about how infectious diseases are spread and what we can do to prevent them.

End-of-chapter summary

In this chapter, you have learnt that:

- infectious diseases can be passed from one person to another, either directly or indirectly
- infectious diseases are caused by microorganisms such as bacteria, viruses and fungi; they can also be caused by parasitic organisms such as tapeworms
- the signs and symptoms of disease are the result of the effect of the microorganisms and the toxins they produce on the cells in your body, and the reaction of your body to the invading pathogens
- understanding how diseases are spread can help us to prevent the spread of disease
- transmissible diseases are spread from person to person by droplet infection, direct contact, contaminated surfaces, contaminated food and drink, through a break in the skin, and by animal vectors
- overcrowded and unhygienic conditions increase the spread of disease in all of these ways
- the spread of infectious diseases is reduced by good hygiene, isolating those infected, destroying or controlling vectors, and vaccination
- the body's defences against the entry of microorganisms include the skin as a barrier, the blood clotting mechanism, the stomach acid, and the mucus and cilia of the respiratory system
- the lymphocytes of your immune system produce antibodies in response to the antigens on the surface of the pathogens
- antibodies disable the pathogens, which are then engulfed and digested by phagocytes
- some lymphocytes produce antitoxins
- when our immune response meets and overcomes a pathogen, we acquire natural active immunity and should not get the same disease again
- a baby gets natural passive immunity from its mother; antibodies from the mother pass into her foetus across the placenta; many more antibodies are passed to the baby through breast milk, protecting the baby in its early months of life
- immunisation protects us against diseases that can cause serious damage and death
- dead or weakened strains of bacteria are put into your body in a vaccination; your immune system responds, so you are protected if you meet the live pathogen; you have artificial active immunity
- if you are vaccinated with antibodies against a disease, you will have temporary artificial passive immunity
- disinfectants are chemicals used to kill microorganisms or stop them from growing on inanimate objects and surfaces
- antiseptics are chemicals used to kill microorganisms or stop them growing on your skin, cuts, and so on
- antibiotics are chemicals that are safe to use inside your body; they destroy bacteria or stop them growing but do not harm human cells
- antifungal drugs are similar to antibiotics, but they destroy fungal cells
- antibiotics do not affect viruses
- heat is important in sterilising both equipment and food.

End-of-chapter questions

1. What are infectious diseases also called?
   A   Crippling diseases
   B   Casualty diseases
   C   Communicable diseases
   D   Chronic diseases (1)

2. What is **one** major difference between viruses and bacteria?
   A   Viruses have no nucleus, whereas bacteria do.
   B   Viruses are living organisms and bacteria are not.
   C   Viruses have membranes, whereas bacteria do not.
   D   Viruses can only reproduce in living cells, whereas bacteria split in two. (1)

3. What is an acute illness?
   A   It is of longer duration.
   B   It is abrupt in onset.
   C   It is of indefinite duration.
   D   It is uncertain in onset. (1)

4. Which **one** of the following would you use to clean up if your puppy was sick on the floor?
   A   Antiseptic
   B   Antibiotic
   C   Disinfectant
   D   Antidiuretic (1)

5. What is the immunity called that is passed from mother to foetus by way of the placenta?
   A   Natural active
   B   Artificial active
   C   Natural passive
   D   Artificial passive (1)

6. What is the immunity acquired by administering a tetanus shot called?
   A   Natural active
   B   Artificial active
   C   Natural passive
   D   Artificial passive (1)

7. a   Describe **five** ways in which infectious diseases are passed from one person to another. (5)
   b   Explain how understanding the ways in which infections are spread helps us to prevent disease. (3)

8. Explain how the following help to defend our bodies against disease:
   a   Skin (2)
   b   Blood clots (3)
   c   Mucus (3)

9 **a** Define the terms 'immune response', 'antigen', 'antibody' and 'antitoxin'. (4)
    b Describe how you would develop natural active immunity to chickenpox. (2)

10 Antibiotics have played a very important role in successfully treating many infectious diseases. Write an essay on antibiotics and the different ways in which they can destroy pathogens. (10)

11 Doctors always try to use the lowest effective dose of an antibiotic. This makes the treatment as cheap as possible, and means there are higher doses to try if the disease persists. How could you investigate the lowest possible effective dose to use against a particular strain of bacteria? (10)

12 **a** Define immunisation. (2)
    b Describe the difference between a vaccine and vaccination. (2)
    c Describe how a vaccine produces artificial active immunity to a disease such as diphtheria or measles. (4)
    d Discuss the importance of vaccines in reducing the incidence of serious diseases such as polio, measles, tetanus and COVID-19. (4)

13 **a** Compare and contrast disinfectants and antiseptics. (4)
    b Define antibiotics and explain how they differ from disinfectants and antiseptics. (4)
    c List **four** different ways in which antibiotics have an effect. (4)

14 **a** Define sterilisation. (2)
    b Suggest **one** advantage and one disadvantage of using heat for sterilisation. (2)
    c Describe **four** ways in which we use heat to reduce the spread of disease. (4)

Chapter 17: Infectious diseases in action

When you have completed this chapter, you will be able to:

- list the causes of a range of infectious diseases, including acute respiratory infections such as influenza, bronchitis and COVID-19, cholera and gastroenteritis
- suggest ways in which the spread of these named infectious diseases may be prevented
- discuss possible treatments of a number of common infectious diseases
- describe the signs and symptoms of some common infectious diseases
- explain how each of these named infectious diseases are spread

What the examiners say

→ Candidates are unable to accurately identify the microorganisms that cause the various infectious diseases or explain how these diseases can be treated.

→ Candidates are unfamiliar with the **causative agents** of most infectious diseases, their signs and symptoms, as well as effective treatment methods. In some instances, candidates repeat home remedies as treatments for infectious diseases rather than the scientifically proven ones.

→ Candidates continue to misspell key terms.

Revision tip

Revise for set periods of time and then take a break to rest and refresh. You'll remember better if you are relaxed! Planning your own revision timetable can really help.

There are hundreds of different bacterial and viral infections. As you learned in Chapters 15 and 16, the microorganisms that cause an infection are called pathogens. Bacterial infections can usually be treated using antibiotics, but we don't have many effective drugs against viruses yet. When you are studying any disease in detail, keep in mind the general principles of infectious diseases you learnt in the last chapter.

In this chapter, we will consider some common infectious diseases that affect two body systems you have already studied in detail – our respiratory system and our digestive system. See Figure 17.1.

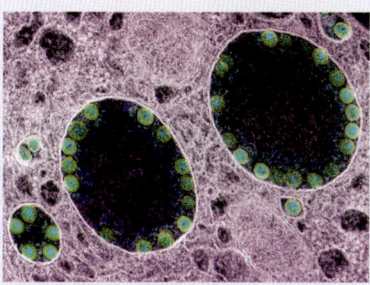

Figure 17.1 A transmission electron micrograph (TEM) of virus particles (green) infecting epithelial cells.

Infections of the respiratory system

In Chapter 15, you learnt about asthma, which is a non-communicable disease that affects the respiratory system. Here, we will look at respiratory infections that affect the airways and the lungs (see Chapter 7). The pathogens that cause respiratory infections are usually spread via droplets in the air. The most common acute respiratory infections covered in this section include influenza, bronchitis and COVID-19. Almost everyone will have suffered from at least one of these infections during their lifetime!

Influenza

Influenza, better known as the flu, is a common acute infection of the respiratory system. An acute infection is one that comes on suddenly, has clearly defined symptoms and doesn't last very long. The flu may be common, but it is also sometimes deadly, especially for the old and the very young. It is estimated that around 36 500 deaths every year in the Caribbean are the result of flu, and globally up to 650 000 people die as a result of a flu infection. Some strains of flu are particularly dangerous. They can spread across a whole country (an epidemic), several countries or even across the world (a pandemic). Over the last few centuries, there has been a major outbreak of flu approximately every 30 years. As people travel more and more freely from country to country, the risks of a flu outbreak becoming a pandemic are always increasing. Look at Table 17.1.

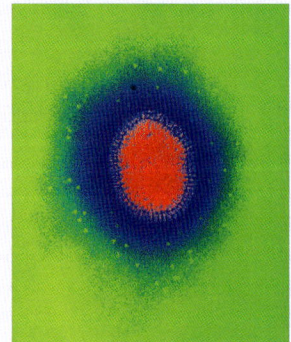

Figure 17.2 The influenza virus is incredibly small but has caused the deaths of millions of people around the world.

Table 17.1 The causes, effects and treatment of influenza (flu)

Cause	How is it spread?	Signs/Symptoms	Treatment	Prevention
Influenza viruses, called type A, type B and type C	Droplet infection, for example, sneezing, coughing and direct contact	• A high fever, which makes you feel hot and cold in turns; you will probably sweat a lot when hot, and then feel shivery when cold; the fever peaks each time the viruses burst out of your cells • Aching muscles and painful joints • Severe headaches • Sore throat and swollen glands • Breathlessness – the flu attacks the lungs' deeper respiratory. • Exhaustion	• Flu is caused by a virus and there is no cure. • Treat the symptoms. • Rest. • Drink lots of fluids. • Take pain relievers.	• Wash your hands every time you sneeze or cough into your hand to wash away the droplets that carry the virus. • Throw out those damp, germ-laden, used tissues as soon as you use them. • Stay away from school or work until you feel better. • Get immunised.

We can be immunised against the flu but we need to get a different vaccine each year. This is necessary because the flu viruses mutate rapidly, so the strain of flu we are most likely to catch changes too. This is why flu vaccines are mainly recommended for older people, people with ongoing illnesses and people taking medication that weakens their immune system.

Figure 17.3 Nothing can cure the flu, but taking medications like these relieves some of the symptoms and makes us feel a little better.

> **DID YOU KNOW?**
>
> In the years 1918–19, there was a global flu pandemic. It particularly affected the young men who had been fighting for many years in the First World War. No one is sure exactly how many people died as a result of the pandemic – it is thought that about 100 000 people died in the Caribbean, and worldwide, the estimates range from 20–40 million.

Bronchitis

Bronchitis is caused by an infection of the bronchi, which are the large airways leading to your lungs. It is a fairly common illness. Most cases of bronchitis happen because an infection inflames the airways. They produce more mucus than usual, and your body tries to remove this by coughing. Acute bronchitis comes on and goes again within a few weeks. Bronchitis may become chronic, with a succession of infections and constantly inflamed airways. This is particularly common in people who have bad or poorly controlled asthma (see Chapter 15) or in people who smoke (see Table 17.2 and Chapter 7). Bronchitis often leads to another infectious disease of the respiratory system, **pneumonia**.

Table 17.2 The causes, effects and treatment of bronchitis

Cause	How is it spread?	Signs/Symptoms	Treatment	Prevention
A virus or bacteria	Droplet infection	<ul><li>A cough</li><li>Fever</li><li>Headaches</li><li>Tiredness</li><li>Aching</li></ul> If the infection does not clear up, spreading further into small airways of the lungs, it can lead to pneumonia.	<ul><li>Rest.</li><li>Drink fluids.</li><li>Use bronchodilators and steroids (as used in asthma treatment).</li><li>Antibiotics are rarely used, as bronchitis is usually viral.</li></ul>	<ul><li>Avoid dust, smoke and areas with high air pollution.</li><li>Practise good hygiene.</li><li>Stop smoking.</li></ul>

Infections of the respiratory system

Pneumonia

Pneumonia is an infection of the small airways in the lungs. It often sets in after a respiratory infection such as the flu or bronchitis. The person may seem to be getting better, but then takes a turn for the worse. Sometimes just one section (or lobe) of the lung is affected (called **lobar pneumonia)**, but often all parts of the lungs are involved.

When we have pneumonia, the linings of the small airways in the lungs swell, blocking air from getting into the alveoli and making breathing difficult. Also, more mucus is made, and this blocks the airways and fills the alveoli. This, in turn, makes gaseous exchange (see Chapter 7) very difficult.

A doctor will hear rattling sounds in the chest of a person suffering from pneumonia when listening to their breathing through a stethoscope. However, often an X-ray is needed to make a definite diagnosis. Once the person starts to recover, they will cough up lots of sputum (phlegm) as the mucus that is blocking the airways starts to move. See Table 17.3 and then look at Figure 17.4.

Table 17.3 The causes, effects and treatment of pneumonia

Cause	How is it spread?	Signs/Symptoms	Treatment	Prevention
A virus or bacteria	Droplet infection	<ul><li>A cough develops that hurts the chest.</li><li>Breathing will become rapid, difficult and may be painful.</li><li>Since breathing is difficult, the muscles in the chest wall and ribs will be tight so the ribs stand out.</li><li>A fever develops, which causes sweating and shivering very like the flu.</li><li>The head and muscles ache and are painful.</li><li>The sufferer feels nauseous and may vomit.</li><li>The appetite is completely lost.</li></ul>	<ul><li>Antibiotics will help even if the pneumonia is caused by a virus, as they will kill or stop any bacteria from growing that might start up another infection.</li><li>Rest.</li></ul>	<ul><li>Practise good hygiene</li><li>Rest and isolate if you have the flu, COVID-19 or bronchitis.</li><li>Stop smoking.</li><li>Take the pneumonia vaccine (older people).</li></ul>

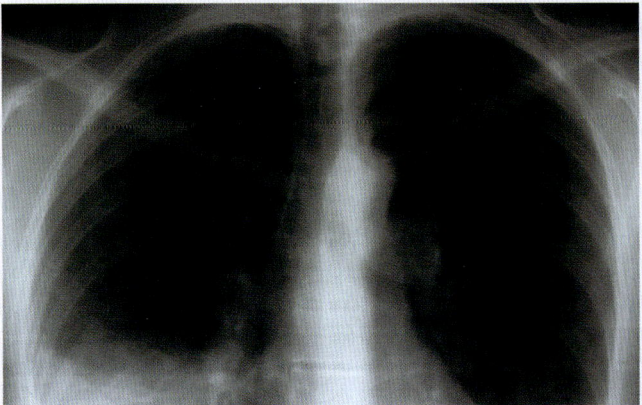 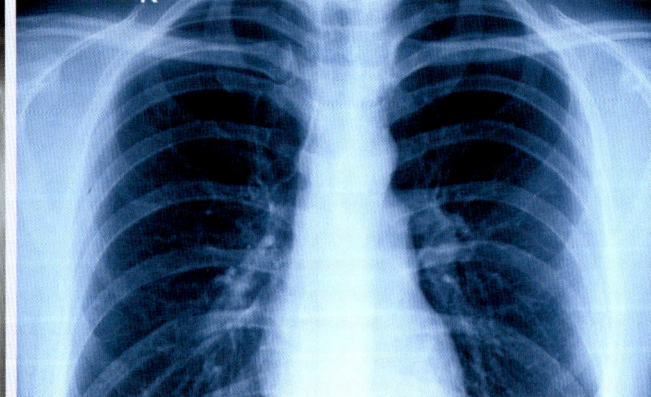

Figure 17.4 An X-ray photo of the lungs shows up pneumonia clearly. Look for the difference between the diseased and healthy lungs yourself – the lungs with pneumonia are in the image on the left.

Infectious diseases in action

> **Checkpoint question**
>
> 1. Explain why antibiotics are given to people with pneumonia, even though it is sometimes caused by a virus rather than a bacterium.

COVID-19

In 2019, a new virus emerged. **SARS-CoV-2** (see Figure 17.5) causes a respiratory disease, now known as **COVID-19**. This disease spread rapidly across the world, causing a global pandemic. It is very infectious – it passes easily from person to person. At the time of writing this book, the pandemic is two years old. It is estimated that around 520 million people around the world have been infected, and that at least 6.5 million people have died as a result. The virus affects the whole of the respiratory system, but especially the lungs themselves.

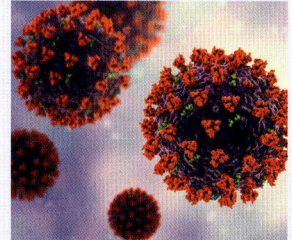

Figure 17.5 An illustration of SARS-CoV-2 – the new coronavirus that caused the COVID pandemic of the 2020s.

COVID-19 affects the respiratory system and, because it is a viral disease, antibiotics do not help. As you saw in Chapter 16, scientists everywhere worked to develop a number of different vaccines to immunise people. These vaccines limit the spread of the disease and reduce the risk of people becoming seriously ill or dying. Look at Table 17.4 to see the main features of COVID-19.

The virus causing COVID-19 mutates easily causing slightly different forms of the disease known as **variants**. Some variants are more infectious than others, and some seem to cause more serious disease than others. The Delta variant and the Omicron variant (see Figure 17.6) are currently the most widespread.

Around the world, countries have introduced testing so that people know if they are infected and can isolate to avoid passing on the disease. PCR testing allows scientists to identify the different variants. This helps them spot if another wave of the pandemic is coming and to change the vaccines if necessary.

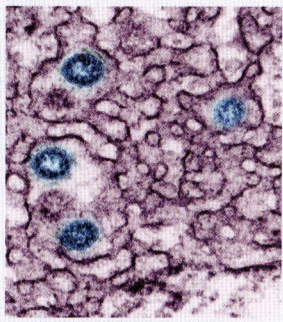

Figure 17.6 Transmission electron micrograph (TEM) of a cell (pink) infected with SARS-CoV-2 Omicron virus particles (blue). Omicron is a COVID-19 variant.

Table 17.4 The causes, effects and treatment of COVID-19

Cause	How is it spread?	Signs/symptoms	Treatment	Prevention
Virus SARS-CoV-2	<ul><li>Droplet infection</li><li>Direct contact</li><li>Contact with infected surfaces</li></ul>	<ul><li>High temperature</li><li>New, continuous cough</li><li>Loss or change to sense of taste and smell</li><li>Breathlessness</li><li>Tiredness</li><li>Headache</li><li>Sore throat</li><li>Feeling/being sick</li></ul>	At home: <ul><li>Rest.</li><li>Drink plenty of fluids.</li><li>Take painkillers for headache.</li><li>Sit upright in bed to help with breathing.</li><li>Take antiviral medicines (used to help those most at risk).</li></ul> In hospital: Take steroids to reduce inflammation. <ul><li>oxygen, ventilators.</li></ul>	<ul><li>Get immunised.</li><li>Keep rooms well-ventilated.</li><li>Practise good hygiene – hands and surfaces.</li><li>Wear face masks if you are infected to prevent spreading</li><li>Go into isolation.</li></ul>

Infections of the respiratory system

The biologist's toolkit: Using statistics carefully

The global COVID-19 pandemic has generated huge amounts of data. People use data to prove many things – but it is important as scientists to look at the data very carefully.

Here are three different graphs showing information about COVID-19 in five different Caribbean countries. The picture given in each set of data varies. Observe each one in detail; to get a realistic picture of what is happening, you need all of this data, and more.

Graph 1 shows you the number of confirmed cases of COVID-19 in five Caribbean countries as of May 2022.

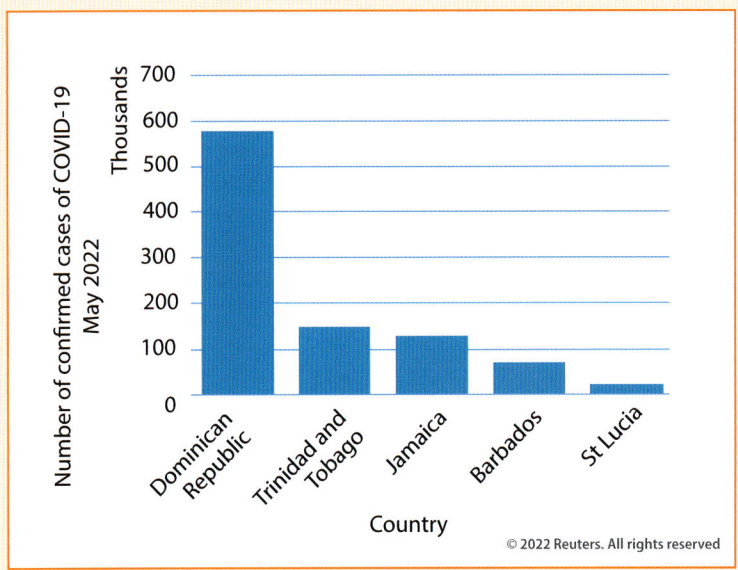

This information looks as if the Dominican Republic has been very badly affected, and that Barbados and St Lucia have had little problem. But look again …

Graph 2 shows the percentage of the total population of the five countries infected with COVID-19. By this measure, which gives us a lot more information, the Dominican Republic has done well and Barbados and St Lucia less so.

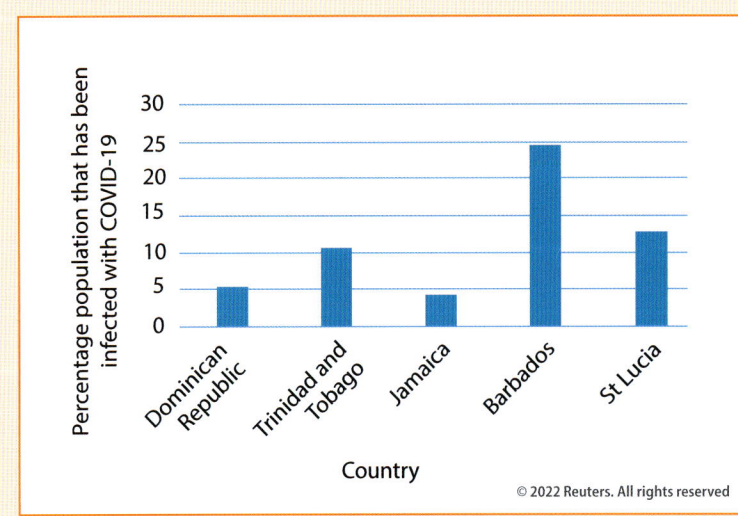

Infectious diseases in action

Graph 3 gives us different data again. It it shows the numbers of people who died of COVID-19 in each country.

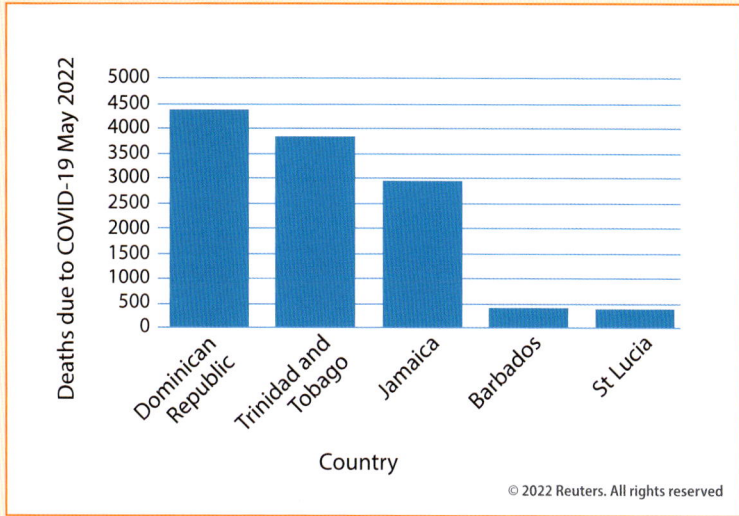

Now it looks as if Dominican Republic, Trinidad and Tobago and Jamaica have struggled. But let's think again – surely we need to know what percentage of the people who caught COVID-19 died of it to see how well the countries have been coping?

Graph 4 gives us this final piece of information.

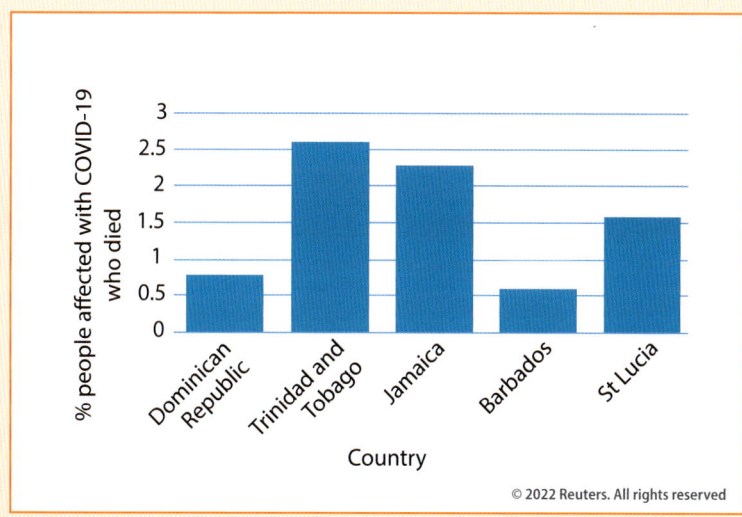

We can draw a number of different conclusions from this data – not least that the Dominican Republic and Barbados, with very different-sized populations, have both kept their death rates from COVID-19 low.

Look back to the graph of vaccination rates in Chapter 16 (Figure 16.14 on page 364) and you may find an explanation for this finding!

Remember: Whatever data you are presented with, make sure you have the full picture. Single sets of data, no matter how scientific they may look, can be very deceptive.

Infections of the digestive system

Our digestive system is well-protected from many invading microorganisms, because the acid produced in the stomach kills most of the microorganisms that we take in with our food or drink. But even that isn't a protection against everything. Here are two examples of infectious diseases that affect your gut.

Gastroenteritis

Gastroenteritis is an infection of the intestines that can be caused by viruses, bacteria and protozoa – it is very difficult to tell which is which.

In someone suffering from gastroenteritis, the linings of the stomach, small intestine and large intestine become inflamed and painful, and as a result, the body rejects and vomits out food. Also, water cannot be reabsorbed by the inflamed lining of the large intestine, resulting in liquid diarrhoea.

Whether the gastroenteritis is caused by viruses or bacteria, it is usually picked up from contact with someone who is already infected, or by taking in contaminated food or water. For example, if we eat shellfish that are grown in waters contaminated by sewage – particularly if, like oysters, we eat them raw, we may well become infected. In many cases, the gastroenteritis is passed on when someone with the virus prepares or handles food without washing their hands after using the toilet, and then they eat the food and the microorganisms together!

Gastroenteritis is most common in people who live in or travel to areas where the sanitation is poor, but it is a problem all over the world. See Table 17.5.

> **DID YOU KNOW?**
> Worldwide, more children die from vomiting and diarrhoea than any other disease. Around 370 000 children under five years old die of dehydration caused by diarrhoea every year.

Table 17.5 The causes, effects and treatments of gastroenteritis

Cause	How is it spread?	Signs/symptoms	Treatment	Prevention
Rotaviruses, the bacteria *Salmonella* and *Escherichia coli (E. coli)* and the protozoan *Giardia*	• Contact with an infected person • By ingesting contaminated food or water • When someone with the virus prepares or handles food without washing their hands after using the bathroom	• Violent abdominal cramps and pain • Nausea, vomiting or often both • Watery diarrhoea usually with no blood • A slight fever • General head and muscle aches • Dehydration	• Stop eating for 24 hours, with no drinks for a few hours to give your gut a complete rest. • Drink rehydration solutions that contain sugars and salts. • Slowly reintroduce small amounts of liquids into the body, graduating from water to clear soups or sweet drinks. • Rest. • Eat lightly as your appetite returns. Avoid rich, fatty and dairy foods until your gut feels back to normal.	• Always wash your hands thoroughly after using the toilet, and before preparing or eating food. • Make sure you know that the water you drink and use to wash salad, food and fruit is clean and safe. • Avoid eating meat, eggs, shellfish, etc. that are undercooked or raw. • If you know someone has gastroenteritis, stay away from them! • If you have to nurse them or visit, wash your hands thoroughly with soap and water afterwards.

Infectious diseases in action

Treatment

There is no effective treatment for either viral or bacterial gastroenteritis, so it is very important to prevent it wherever possible.

When we travel in other countries, we may catch gastroenteritis from contaminated food or water. Sometimes it is simply that the common gut bacteria are different to our own, and this can cause problems. To reduce the risk of problems when you go on holiday abroad, follow the same precautions you would at home, plus a few more! See Figure 17.7.

Figure 17.7 This food looks wonderful, but if the salad has been washed in dirty water, the shellfish or chicken is not cooked or the cooks did not wash their hands after visits to the toilet, it could be covered in the microorganisms that cause gastroenteritis.

- Drink only well-sealed, bottled or carbonated water and use it to brush your teeth.
- Beware of ice cubes and ice cream, which may be made with contaminated water.
- Avoid raw food – including peeled fruits, raw vegetables and salads, because they may have been washed in contaminated water or touched by unclean hands.

Common sense, good toilet, kitchen and food hygiene, and lots of hand washing should help to avoid gastroenteritis most of the time.

Gastroenteritis is common around the world. It is not usually serious in healthy people, but it may kill small children, the elderly or anyone who is already unwell. However, other gut infections such as cholera are a much greater threat to life. See Figure 17.8.

Figure 17.8 Rehydration drinks are not expensive and save lives.

Infections of the digestive system

Cholera

Cholera (Figure 17.9) is a bacterial infection that affects the intestinal tract. Cholera is spreading rapidly and extensively, at epidemic levels, in parts of Central and South America. A number of cases have also been reported in the Caribbean. People who may be at risk include those travelling to foreign countries where outbreaks are occurring, and people who consume raw or undercooked seafood from warm coastal waters that may be contaminated with sewage. In both instances, the risk is usually small. The most common way in which cholera is transmitted is through drinking water contaminated by the faeces of an infected person or people. This is why cholera is such a problem when there are major disasters and clean water becomes impossible to find. See Figure 17.10 and Table 17.6.

Figure 17.9 A scanning electron micrograph (SEM) of the bacteria that cause cholera. The bacteria infect the intestines, causing vomiting and diarrhoea. This results in extreme dehydration, which could cause death.

Table 17.6 The causes, effects and treatments of cholera

Cause	How is it spread?	Signs/symptoms	Treatment	Prevention
Bacteria called *Vibrio cholerae*	Eating or drinking food or water contaminated by the faecal waste of an infected person	• Mild to severe diarrhoea, vomiting and dehydration, but generally no fever • Pale, watery diarrhoea • Vomiting • Severe muscle cramps from the loss of salts • Dehydration and shock as the blood pressure plummets	• Rehydrate with rehydration fluids, but often also with an intravenous drip that puts the liquid straight back into the blood. • Take antibiotics, such as tetracycline, to treat the disease. They can reduce the time spent suffering from diarrhoea, and stop any more bacteria being shed in the faeces. • There is a cholera vaccine but it only gives partial protection (50%) and it only lasts 2–6 months.	• Drink from a supply of clean, uncontaminated water. • Ensure that sewage is disposed of correctly.

For travellers, the simplest way to avoid cholera is to avoid consuming uncooked foods or water in foreign countries where cholera occurs, unless they are known to be safe or have been properly treated.

Figure 17.10 When thousands of people lose their homes in a natural disaster or war, cholera is a major threat. Lack of sanitation and only dirty water to drink mean cholera often kills as many people or more than the original disaster, unless rehydration fluids and antibiotics are available.

Infectious diseases in action

Fungal infectious diseases

Not all diseases are deadly, or even serious. What's more, not all infectious diseases are caused by bacteria or viruses.

Ringworm

Ringworm is a skin infection caused by a **fungus** that can affect the scalp, skin, fingers, toenails or feet. Most people find it an inconvenience more than anything else. The organisms that cause these diseases are a group of fungi known as *Dermatophytes*. The fungi are all very infectious, and they particularly like warm moist areas of your body. The more sweat the better! See Figure 17.11.

The four main areas of the body that can be affected by ringworm are the:

- body
- scalp (where round bald patches occur)
- groin
- feet.

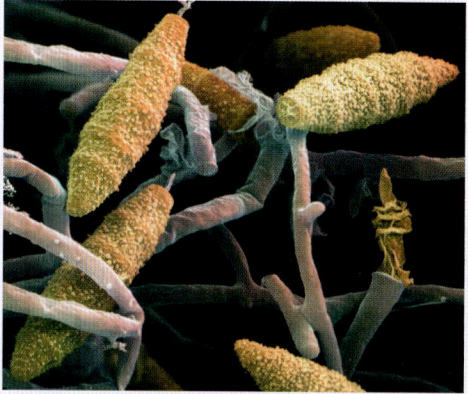

Figure 17.11 A scanning electron micrograph (SEM) of the fungus Microsporum gypseum which causes ringworm of the body shown in Figure 17.12.

Each of these areas can be attacked by a different but closely related fungus. When the fungus affects the feet it is called athlete's foot. Anyone can get ringworm, but children seem to be more susceptible than adults. The precise incubation period is unknown. For ringworm of the scalp and the body, it seems to be around 10–14 days after you have been in contact with someone or something that is affected by the disease.

Ringworm infections are spread relatively easily from person to person by direct skin contact. It can also be transmitted by contact with material that has been in contact with the skin of an infected person, such as towels, carpets, and so on. Ringworm of the feet and nails is only passed from person to person, but other types of ringworm also affect some animals, so it is possible to get ringworm from your family pet. See Figure 17.12.

Ringworm has nothing to do with worms, but the round red mark that can appear on your skin was originally thought to be a worm – hence the name!

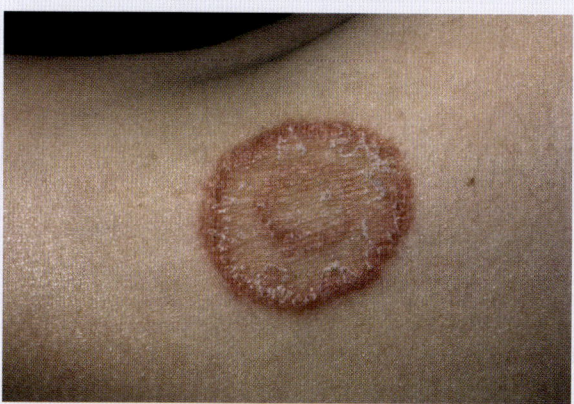

Figure 17.12 Ringworm of the skin

Symptoms

The symptoms of ringworm vary depending where on your body it is found:

- Ringworm on the scalp usually starts as a small pimple. As it grows and spreads, the area affected becomes larger. It produces scaly patches of temporary baldness, partly because the infected hairs become brittle and break off easily. Sometimes yellowish, oozing areas develop. See Figure 17.13.
- When ringworm fungi attack your nails, the affected nails become thicker, discoloured and brittle. On the other hand, they may become chalky and disintegrate completely. The infection can last for years because it is very difficult to get treatment to penetrate the nail.
- When ringworm appears on your body, you develop one or more flat, spreading ring-shaped areas. The edge of the area is reddish, and the circle may be either dry and scaly or moist and crusted. As the infection spreads spreads, the centre area recovers and the skin appears normal again.
- Ringworm of the foot (called athlete's foot) is very common, because the area between your toes is very warm and moist (sweaty feet!) and the skin is very soft. The fungus produces a scaling or cracking of the skin, especially between the toes. This can be itchy or – when the skin falls off to leave raw tissue beneath it – very sore.

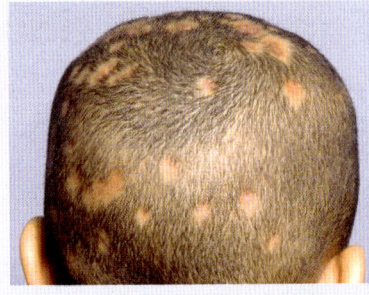

Figure 17.13 Ringworm of the scalp causes very typical bald patches.

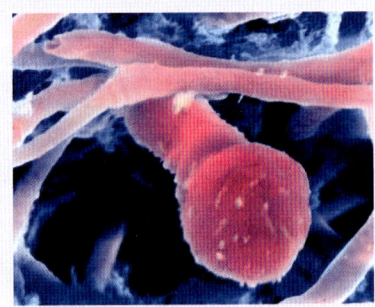

Figure 17.14 A scanning electron micrograph (SEM) of the fungus Trichophyton violaceum, which causes ringworm of the scalp (Figure 17.13).

Treatment

The most common way to treat all forms of ringworm is to use **antifungal** creams or powders. The creams can be rubbed into the infected skin or powders can be applied directly to the affected areas. You can easily buy these and treat yourself. If you have a really stubborn infection, your doctor may treat you with a prescribed fungicide that can be taken as tablets. As they sometimes have serious side effects, they are only used when it is really necessary.

To prevent the spread of ringworm, don't share towels, hats and clothing with an infected individual. When young children are affected, try to keep them away from other children as much as possible until they have been effectively treated. When many cases occur in the same area, seek advice from your local health department, as everyone many need to be treated at the same time.

Although ringworm is not at all serious, some fungal diseases attack the valves of your heart or get into your lungs or your brain. These diseases are very dangerous and can be fatal particularly in the elderly or people who are already ill with diseases such as cancer.

This chapter has given a brief glimpse of some of the infectious diseases that affect the health of millions of us every day. By building up a picture of how infectious diseases are spread and how they are treated, we discover how each of us as individuals and society as a whole can work to prevent the spread of many infectious diseases. In the next chapter, we will look at some diseases that are spread by parasites and vectors.

Possible SBAs based on infectious diseases caused by viruses and bacteria

→ Build a programme informing people about different infectious respiratory diseases with a step programme to try to avoid them.

→ Investigate the progression of the COVID-19 pandemic in your country and compare it to another country with a very different climate.

→ Write a history of the 1918–19 influenza epidemic including a timeline. Display as much of the data graphically as you can.

→ Organise an education campaign for local primary schools to teach children how gastroenteritis and other gut diseases are spread, and the basic hygiene steps needed to avoid it.

End-of-chapter summary

In this chapter you have learnt that:

- **influenza** is a viral infection of the lining of the respiratory tubes in the lungs; it comes on more rapidly than a cold; symptoms include high fever, headaches, shivering and feeling hot and cold; the spread can be prevented by washing hands every time you sneeze or cough into your hands; there is no cure but symptoms can be treated

- **bronchitis** is usually a viral infection of the bronchi – the airways leading to the lungs, although it is sometimes caused by bacteria; it is transmitted by droplet Infection through the air; it causes a cough, fever, aching and tiredness; it is usually treated by rest, fluids and, if it is severe, also with bronchodilators and steroids like those used to treat asthma

- **pneumonia** is an infection of the small airways in the lung; it may be caused by bacteria or by viruses; babies, children and older people who have chronic illnesses or disabilities are more likely to get pneumonia; usually sufferers are breathless, have a high fever, rapid breathing and are quite unwell; many would recover without treatment but do so much faster with antibiotics

- **COVID-19** is caused by the SARS-CoV-2virus; it affects the respiratory system, especially the lungs and caused the world pandemic of the 2020s; symptoms include coughing, fever and loss of or change in the senses of taste and smell; it may cause serious illness and death; there is no treatment but there are several very effective vaccines

- **gastroenteritis** can be caused by viruses or bacteria; people usually become infected with gastroenteritis by taking in contaminated food or water; the most effective method of treating this disorder is rehydration with lots of fluids containing electrolytes; good sensible hygiene practices minimise the spread

- **cholera** is a bacterial disease that affects the intestinal tract; the cholera germ is passed in the faeces; it is spread by eating or drinking contaminated food or water; infected people may experience mild to severe diarrhoea; it can kill very quickly if it is severe; it is a particular risk in areas of overcrowding with no proper sewage disposal

- some diseases – such as **ringworm** and **athlete's foot** – are caused by fungi; they are spread by contact with the infected skin and cured using antifungal medicines.

End-of-chapter questions

1. Identify the type of organism that causes influenza:
   - A Bacteria
   - B Viruses
   - C Fungi
   - D Protozoa (1)

2. Which of the following processes will help to avoid the spread of food-borne diseases?
   - i Use disposable cloths to wipe down surfaces.
   - ii Wash hands after visiting the toilet.
   - iii Wash hands three times a day.
   - iv Use disinfectant on kitchen surfacess.

   - A i, ii and iii
   - B ii, iii and iv
   - C i, ii and iv
   - D i, iii and iv (1)

3. Which **one** of the following diseases is only caused by bacteria?
   - A Influenza
   - B The common cold
   - C Gastroenteritis
   - D Cholera (1)

4. Which **one** of the following would you describe as an acute disease?
   - A Diabetes
   - B Bronchitis
   - C Heart disease
   - D Hypertension (1)

5.
   - a State **three** ways in which the flu can be caught. (3)
   - b Explain how the flu differs from a cold. (5)
   - c Describe how to avoid spreading the flu. (2)

6.
   - a State the organism that causes COVID-19. (1)
   - b State the parts of the body that are affected by COVID-19. (3)
   - c In the early 2020s, there was a COVID-19 pandemic.
     - i Describe a pandemic. (2)
     - ii Give **three** symptoms of COVID-19. (3)
     - iii Give **two** ways in which the pandemic was gradually controlled. (2)

7.
   - a Name the pathogens that cause influenza, bronchitis and pneumonia. (3)
   - b Describe the symptoms of influenza, bronchitis and pneumonia. (6)
   - c Describe how to treat influenza, bronchitis and pneumonia. (6)
   - d Suggest ways of preventing the spread of influenza, bronchitis and pneumonia. (6)

Infectious diseases in action

8 **a** Explain how cholera and gastroenteritis differ. (2)
   b Describe **three** ways of preventing both cholera and gastroenteritis, and explain why they are the same. (5)

9
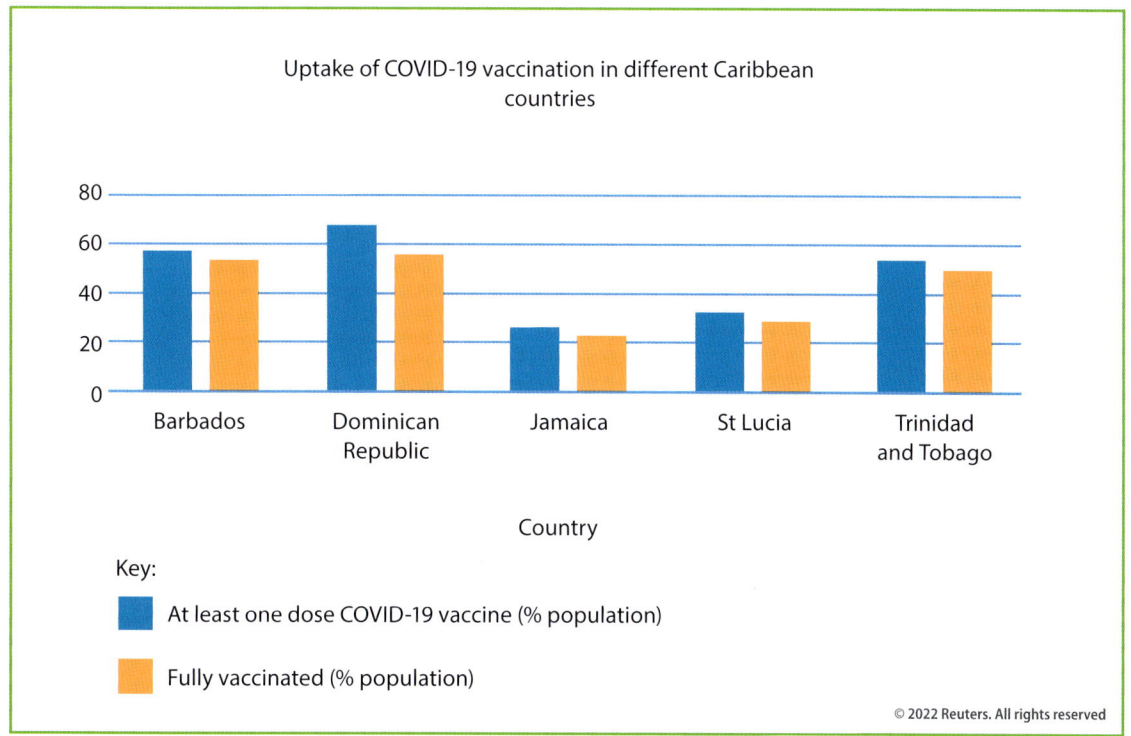

Look at the graphs showing the impact of COVID-19 in different Caribbean countries on pages 377–378. Using that data and the information on this graph from Chapter 16:
a Describe what the data tells you. (5)
b Suggest explanations for trends in deaths from COVID-19 shown in the data. (5)

Chapter 18 Lifestyle diseases

When you have completed this chapter, you will be able to:

- describe the way in which behaviour and lifestyle choices affect the spread of STIs
- discuss the causes, signs/symptoms, treatment and prevention of the STIs, gonorrhoea, syphilis, human papillomavirus (HPV), herpes and chlamydia
- explain the social and economic effects of drug misuse on the individual, family and community
- explain the effects of sexually transmitted infections on a pregnant mother
- explain the way in which sexually transmitted infections (STIs) are spread
- discuss the use and misuse of drugs by humans including prescription and non-prescription drugs
- discuss aspects of Human Immunodeficiency Virus/Acquired Immune Deficiency Syndrome (HIV/AIDS) including ethical considerations

What the examiners say

→ Candidates are generally unfamiliar with the methods of preventing the spread of sexually transmitted infections and are also confused about the symptoms of various sexually transmitted infections. Additionally, some candidates are unaware of the effects of sexually transmitted infections on a developing foetus.

→ There are several gaps in knowledge regarding the transmission of HIV/AIDS and its physiological effects on infected persons. Many candidates are also unable to recall the complete meaning of the acronyms HIV and AIDS.

Revision tip

If possible, revise in a place where you won't be distracted by the TV or other people. Try to find somewhere quiet where you can concentrate!

Figure 18.1 Mosquitoes can infect us with diseases such as malaria.

Many of the diseases that cause terrible health problems in the Caribbean, and around the world, are at least partly due to our lifestyle. If we don't keep our kitchen clean, we are more likely to suffer from gastroenteritis. If we eat a high-fat diet and do very little exercise, we increase our risk of heart disease. If we leave rubbish lying around, water will collect in it, mosquitoes (Figure 18.1) will breed and diseases such as dengue fever and malaria are more likely.

If we cough and sneeze without putting a hand in front of our mouth, other people are more likely to catch our cold or COVID-19. But in all of these examples, our own behaviour is only part of the story. However clean and careful we are, a fly will land in our kitchen carrying pathogens on its feet and in its saliva. We cannot avoid breathing in the viruses that other people cough out into the air, and how can we tell if a rat has urinated in the water we are wading in? See Figure 18.2.

Some diseases are very closely related to the personal choices we make and the lifestyle we choose to lead. In this chapter, we are going to look at a group of infectious diseases linked to sexual behaviour, and also at the potential impact of the misuse of drugs on our health. Both of these health issues are closely linked to the lifestyle we choose to lead.

Figure 18.2 Almost everything we do increases or reduces our risk of disease, but some of the choices we make have much more impact than others.

Sexually transmitted infections (STIs)

Sexually transmitted infections (STIs) are infectious diseases spread through sexual contact (see Figure 18.3). Many of the most common sexually transmitted infections are cured easily using antibiotics if caught early. Unfortunately, if they are not detected and treated in the early stages, these same infections cause great harm including infertility, brain damage and even death. What is more, they can be avoided altogether if you follow sensible sexual behaviour.

Remember – you do not have to have lots of sexual partners to get a sexually transmitted infection. It can happen the very first time you are sexually active.

The examples that we will look at in this section include gonorrhoea, syphilis, human papillomavirus (HPV), herpes and Acquired Immune Deficiency Syndrome (AIDS).

Figure 18.3 Young people may fall in love, but sexually transmitted infections are not very romantic. Fortunately, sensible choices will prevent them spoiling a relationship.

Checkpoint question

1　Define sexually transmitted infections.

Gonorrhoea

Gonorrhoea is an infection that is caused by the bacterium *Neisseria gonorrhoeae*. The gonorrhoea germs are found in the mucus-producing areas of the body (the vagina, penis, throat and rectum). Any sexually active person can be infected with gonorrhoea, but it is most often found in younger people (ages 15–30), and particularly in people who have multiple sex partners.

Gonorrhoea is spread through sexual contact, which could be vaginal, anal or oral sex. Having unprotected sex without using a **condom**, and particularly having many sexual partners, increases the risk of picking up the infection. A mother may also pass gonorrhoea to her child during birth. From the time someone is infected with gonorrhoea, they can spread the disease to any sexual partners they may have.

Symptoms of gonorrhoea

The symptoms of gonorrhoea differ between the early and late stages of the disease, depending on how quickly the person receives treatment. See Figure 18.4.

Early symptoms

The early symptoms of gonorrhoea are easier to identify in men than in women. Men infected with gonorrhoea will have:

- a burning sensation while urinating
- a yellowish-green discharge from the penis.

If women have any symptoms, they will have:

- a yellowish-green discharge from the vagina
- bleeding between periods
- possibly some burning while urinating.

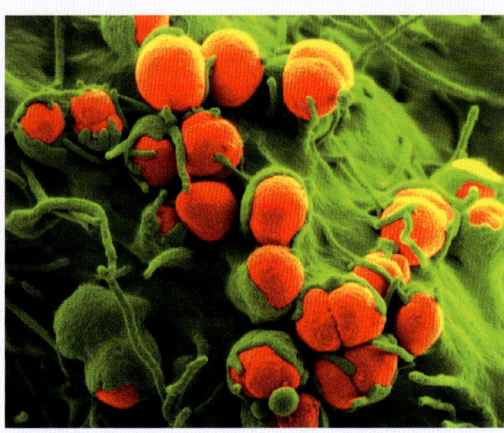

Figure 18.4 Gonorrhoea is caused by the bacterium *Neisseria gonorrhoeae*. This scanning electron micrograph (SEM) shows gonorrhoea bacteria (round, red) on a human epithelial cell.

Often, gonorrhoea causes no early symptoms; about 10–15% of men and around 50% of women have no indication at all that they are infected. Infections in the throat and rectum (from oral or anal sex) cause few symptoms. This is one of the great problems with gonorrhoea. People may spread the infection unknowingly, but also they do not get treatment and so put themselves at risk of developing more serious problems later.

Late symptoms

If an infected person is not treated quickly for gonorrhoea, there is a good chance that complications will occur.

> **Checkpoint question**
>
> 2 Suggest why some people do not realise for some time that they are infected with gonorrhoea.

- Women frequently get pelvic inflammatory disease (PID), a painful condition that occurs when the infection spreads throughout the reproductive organs. PID leads to infertility in women as the Fallopian tubes (see Chapter 12) become blocked and damaged.
- Men may suffer from swelling of the testicles and penis.
- Both sexes may suffer from arthritis, skin problems and other organ infections caused by the spread of gonorrhoea within the body.

Treatment of gonorrhoea

Gonorrhoea is caused by a bacterium, so it can be treated effectively using antibiotics, especially in the early stages. It is also cured in the later stages, but the damage caused cannot be undone. Unfortunately, some strains of the bacterium are becoming resistant to many antibiotics. People do not develop immunity to gonorrhoea so they may get repeated infections.

Lifestyle diseases

Gonorrhoea and pregnancy

If a pregnant woman has untreated gonorrhoea, she may pass the infection on to her baby as it passes out along the vagina when it is born. The infection will affect the baby's eyes and, if untreated, may result in blindness. If antibiotic treatment is given immediately, then this may save the sight of the baby. Babies born with gonorrhoea from their mothers are also at risk of developing sepsis or meningitis. These are both major infections that can kill rapidly even if treated. Testing pregnant women and treating anyone who has gonorrhoea will reduce or remove the risk to the unborn baby.

Prevention of gonorrhoea

The best ways to prevent the spread of gonorrhoea all involve a sensible and responsible approach to sexual relationships:

- Limit the number of sexual partners, as the fewer people you have sex with, the less likely you are to meet someone who is infected with gonorrhoea.
- Use a male or female condom because this prevents the bacteria passing from one person to the other.
- If you think you may be infected, avoid any sexual contact and visit your STI clinic, a hospital or a doctor.
- If you are infected, notify all your sexual contacts immediately so that they can be examined and treated, and do not have sex until your course of treatment is completed.

Syphilis

Syphilis is caused by the spiral-shaped bacterium *Treponema pallidum* (see Figure 18.5). Any sexually active person may be infected with this STI, although it is most common among young people between the ages of 15–30, perhaps because they are the group most likely to be very sexually active and to have a variety of partners. It is more common in towns than in the countryside, probably because there are more people available as sexual partners in urban areas.

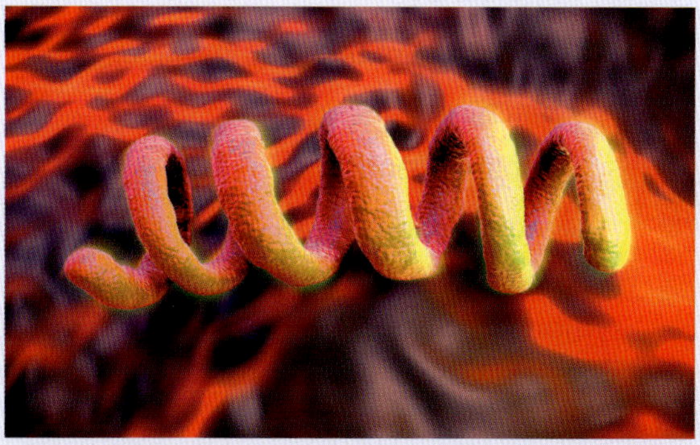

Figure 18.5 The bacteria that cause syphilis have a very distinctive spiral shape.

Like gonorrhoea, the most common way in which syphilis is spread is by sexual contact with someone already infected with the bacteria. However – again like gonorrhoea – the exception is **congenital syphilis**, which is spread from mother to foetus. The baby may have severe disabilities or die as a result.

For syphilis to be spread, a person must be in contact either with the mucus membranes of the infected person or with the moist sores caused by syphilis which may appear on the skin. The problem is that these lesions, which appear on or inside the genitals, may or may not be visible on the outside of the body. The more lesions someone has, the more infectious they are. An individual can pass on syphilis for up to two years or more after becoming infected.

However, syphilis cannot be spread by contact with toilet seats, baths, shared clothing or door knobs, no matter what your friends may tell you!

Symptoms of syphilis

Syphilis progresses in distinct stages. It can be treated in the earlier stages, but not in the later stages. The symptoms occur in stages called primary, secondary and tertiary (late) syphilis.

- **Primary syphilis:** The first or primary sign of syphilis is a painless sore or sores, which appear at the site of initial contact. They are commonly found on the penis, on the entrance to the vagina, in the mouth, in the rectum or inside the vagina itself, where they are invisible. There may be swollen glands, which usually develop within a week after the appearance of the initial sore. The sores may last from one to five weeks and may disappear by themselves even if no treatment is received. These sores are very infectious.

- **Secondary syphilis:** The second stage of the disease begins about six weeks after the sore first appears. The most common symptom during this stage is a rash. Others symptoms include:
    - tiredness
    - fever
    - sore throat
    - headaches
    - hoarseness
    - loss of appetite
    - patchy hair loss
    - swollen glands.

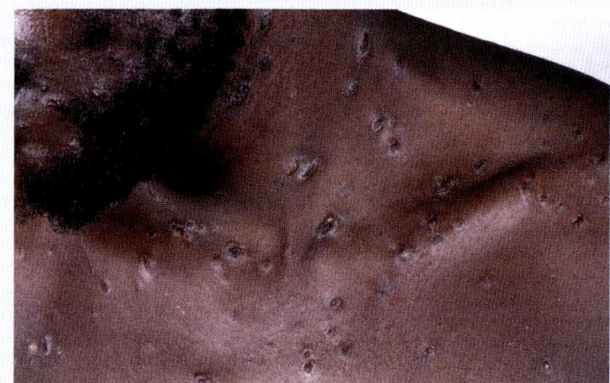

Figure 18.6 A typical secondary syphilis rash.

These signs and symptoms will last two to six weeks and generally disappear even without treatment. People often don't notice the symptoms or mistake them for other common minor illnesses such as heat rash, colds and flu, so they don't visit the doctor and the disease is not diagnosed or treated.

If left untreated, the disease goes into a long quiet phase. There are no obvious symptoms but the bacteria are still active in the body.

- **Tertiary syphilis:** The third stage, called tertiary or late syphilis (syphilis of over four years' duration), may involve illnesses of the skin, bones, central nervous system and heart. It causes severe and irreversible problems that cannot be treated successfully.

Treatment of syphilis

As syphilis is a bacterial disease, it is treated easily with antibiotics such as penicillin or tetracycline. The amount of treatment depends on the stage of syphilis the patient is in. At tertiary (third stage) syphilis, the damage already done to the tissues and organs cannot be undone. You do not develop natural immunity to syphilis, so a past infection offers you no protection. If you make poor lifestyle choices, you can catch syphilis again on the very day you finish your treatment.

Prevention of syphilis

As with gonorrhoea, lifestyle choices may prevent the spread of syphilis. The choices are very similar, as the diseases are spread in the same way:

- Limit your number of sexual partners.
- Use a male or female condom. Remember that the use of condoms may prevent the disease if the initial contact sore is on the penis or in the vaginal area. The bacteria cannot travel through the condom. However, transmission can still occur if there is a sore outside the areas covered by the condom – for example, in the mouth.
- If you think you are infected, avoid sexual contact and visit your local STI clinic, a hospital or doctor. See Figure 18.7.
- Notify all sexual contacts immediately so they can be examined and treated.
- All pregnant women should receive at least one prenatal blood test for syphilis so that they can be cured and any potential damage to their baby prevented.
- Do not have sex until your treatment for syphilis is completed.

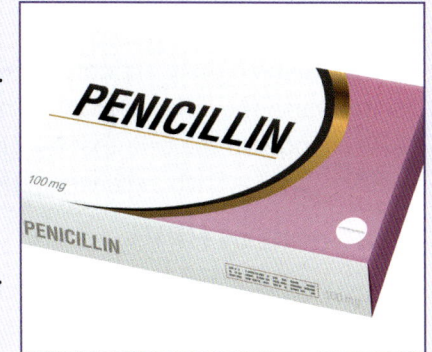

Figure 18.7 Penicillin is one of the antibiotics used to treat syphilis. Avoid any sexual contact until you have finished the treatment for syphilis.

Pregnancy and syphilis

Pregnant women with syphilis are treated with antibiotics to cure them and protect their baby. If the mother is not treated, the baby will be affected and about 40% of those babies will die before or just after birth. If a baby is born with syphilis from its mother, it needs daily antibiotic treatment for 10 days. Some babies will have suffered permanent damage from the infection before they are born. All babies born to mothers with syphilis needs regular testing to make sure the infection has gone and there are no long-term effects.

There is a simple blood test for syphilis. In many places around the world, including the Caribbean, pregnant women are tested to see if they have syphilis because if the infection is discovered and treated, any problems for the unborn child are completely prevented. By 2021, eight Caribbean states and territories had completely eliminated mother–child transmission of syphilis and HIV/AIDS – a major triumph of preventative medicine!

Checkpoint question

3 Compare and contrast gonorrhoea and syphilis.

Human papillomavirus (HPV)

You have already met HPV in Chapter 12 when you looked at diseases of the reproductive system. HPV causes genital warts in men and women. The viruses are very easy to catch from skin to skin contact in the genital area or from vaginal, anal or oral sex. See Figure 18.8.

Symptoms of HPV

Most of the time there are no symptoms of HPV infection. If you develop genital warts, they may or may not be visible. But as you have seen, HPV can cause changes in the cells that may lead to cervical cancer. HPV is also linked to other cancers, such as cancer of the penis, anal cancer and vaginal cancer.

Treatment of HPV

There is no treatment of HPV unless it causes symptoms such as genital warts or cervical cancer. Genital warts may be removed. The treatment for cervical cancer will depend on how extensively the cancer has spread – see Chapter 15 Table 15.1 *A summary of cancers* (page 344).

Prevention of HPV

There is one major lifestyle choice you can make make to prevent HPV and the cancers it is linked to. Get the HPV vaccine, which prevents you from catching or spreading HPV and makes sure that you don't develop the cancers linked to HPV (see Figure 18.9). You can also:

- use condoms whenever you have sex
- avoid having sex if you are being treated for genital warts.

Herpes II (genital herpes)

The next STD you are going to look at is, in some ways, much less serious than either gonorrhoea or syphilis. However, because it is caused by a virus, it cannot be treated, so once you have caught it from a partner, you will have the condition for the rest of your life.

Herpes simplex II is the virus that causes **genital herpes**, which is a sexually transmitted viral infection that often produces painful sores, usually in the genital area (head of penis, labia, anus or cervix). Once infected, a person will carry the virus throughout their life, although they will only have the symptoms occasionally when a new crop of sores breaks out.

The virus is spread during sexual contact with an infected person. It is secreted in the fluid produced from the sores (lesions) and the mucous membranes. In addition, herpes can be spread from an infected mother to her child during birth.

Symptoms of genital herpes

Some studies have shown that from one-half to two-thirds of people infected with the herpes II virus will have no symptoms. However, for those who do, typically the first sign of herpes appears about a week after someone becomes infected. The skin in the infected area becomes sensitive, tingling or sore, followed by a cluster of red, blister-like lesions in the genital area, anus or mouth. These are filled with a pale yellow fluid, which is released when they burst. It is filled with viruses and is highly infectious. They then form shallow ulcers that are very painful but crust and heal over within 4 to 15 days.

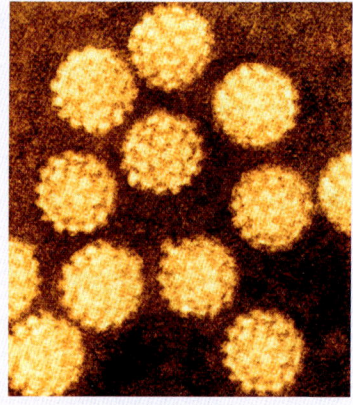

Figure 18.8 A TEM showing a group of papilloma viruses (yellow), which are linked to certain cancers, like cervical cancer.

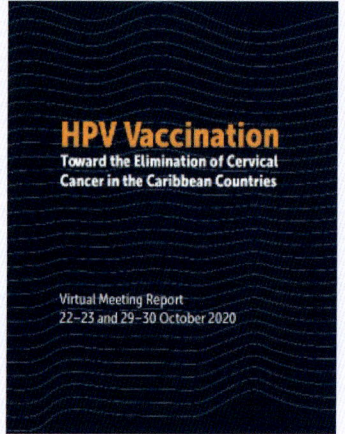

HPV Vaccination: Toward the Elimination of Cervical Cancer in the Caribbean Countries. Virtual Meeting Report, 22–23 and 29–30 October 2020. Washington, D.C.: Pan American Health Organization; 2021. License: CC BY-NC-SA 3.0 IGO.

Figure 18.9 Across the Caribbean, countries are working together to immunise students against HPV and protect them from cervical cancer.

Treatment of genital herpes

Since herpes is caused by a virus, there is no outright cure. Some new antiviral medicines:
- reduce the numbers of herpes II viruses in the blisters (making it less infectious)
- diminish the pain
- speed up the healing of primary herpes lesions.

When herpes occurs in the mouth from oral sex, this treatment seems to shorten the length of both primary and recurrent episodes (Figure 18.10). However, nothing can actually cure herpes or prevent it returning at a later date.

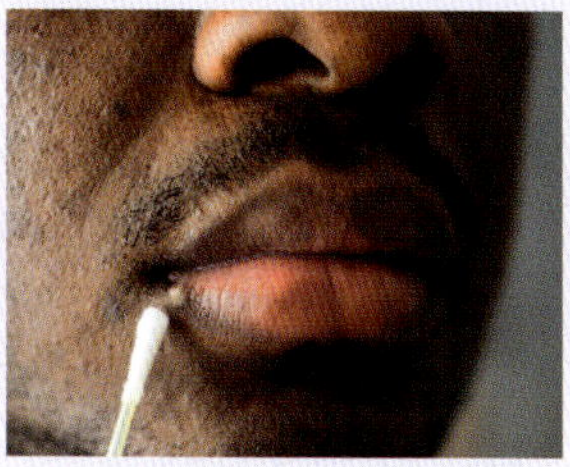

Herpes and pregnancy

If a woman is pregnant and knows she has genital herpes, she should make sure the doctors and nurses caring for her know as well. Then they can be prepared for any problems with the baby. Herpes is serious in a new-born infant – they become ill very quickly and if the virus spreads to their body organs, they will die. If someone has active herpes in late pregnancy, a Caesarean section is often recommended to try to avoid infecting the baby with the virus when it is being born.

Figure 18.10 The herpes viruses that cause cold sores affect the skin on the face, especially around the mouth.

> **Checkpoint question**
>
> 4 It is important to tell a new sexual partner that you have genital herpes, even if it is not active at the time. Explain why.

Cold sores are another type of herpes infection, which once present, remains in your body for life.

Chlamydia

Chlamydia is one of the most common STIs, but many people have never heard of it. It is also a major cause of female infertility.

Chlamydia is a bacterial infection. The bacteria (see Figure 18.11) are usually spread from one person to another through unprotected sex or contact with sexual fluids. You can pass on or catch chlamydia simply from contact between the genitals. It can also be passed on from infected fluids getting into the eye, and a pregnant woman can pass it on to her baby.

The danger of chlamydia is that most people who have the infection have no symptoms and so do not know that they are infected. This means they pass it on to others without knowing – and it can cause permanent damage inside their bodies.

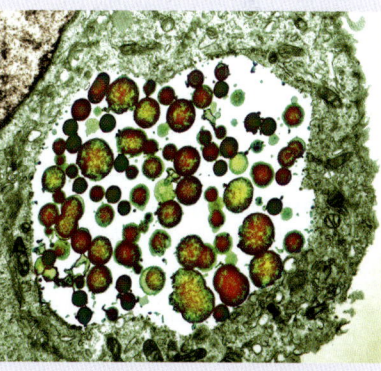

Figure 18.11 A transmission electron micrograph (TEM) of the chlamydia bacteria (red/yellow) inside a cell (green). Chlamydia is a major cause of female infertility.

Symptoms of chlamydia

Remember, most people don't get symptoms. If you do, they may include:
- pain when urinating
- an unusual discharge from the vagina, penis or rectum
- (women only) pain in the tummy, bleeding between periods and bleeding after sex
- (men only) swelling and pain in the testes.

Sexually transmitted infections (STIs)

Treatment of chlamydia

Chlamydia is easily treated in the early stages with a short course of antibiotics. Both partners must be treated at the same time. If chlamydia is not treated, it causes pelvic inflammatory disease in women, which often results in infertility. In men, it may cause swollen and painful testes. It can also cause a form of arthritis in both men and women.

Prevention of chlamydia

All the usual life choices apply, especially using a condom every time you have sex.

There is a simple test for chlamydia. Doctors advise everyone who is under 25, and sexually active, to have a chlamydia test at least once a year, and when you have sex with a new partner, make sure that any infection is treated and cured early. (See Figure 18.12.)

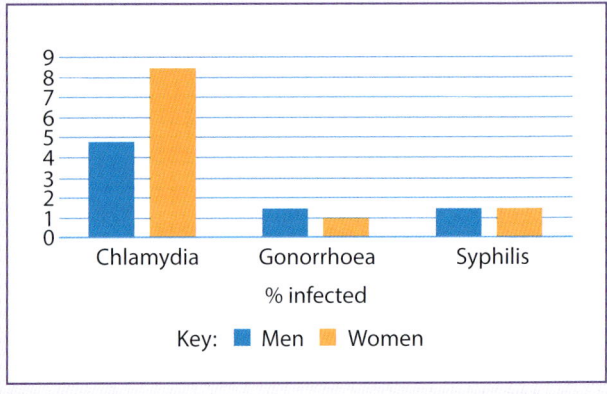

Figure 18.12 Chlamydia is one of the least well-known but most common STIs in men and women across Latin America and the Caribbean.

Human immunodeficiency virus and acquired immunodeficiency syndrome (HIV/AIDS)

Acquired immunodeficiency syndrome (AIDS) is the medical term for a combination of illnesses that result when the immune system is weakened or destroyed. It is the advanced form of an infection caused by the **human immunodeficiency virus (HIV)**, a virus that attacks the immune system, making the sufferer vulnerable (at risk) to other diseases. See Figure 18.13.

The human immune system is made up of special cells that protect the body from infection (see Chapter 15). Usually, when a virus enters the body, the cells that make up the immune system in the body all begin to work at once. The white blood cells produce antibodies that attack and destroy the pathogen, therefore helping the person to get better (see Chapter 16). The immune system of people affected by HIV/AIDS do not function in this way because the virus attacks and affects the immune cells themselves.

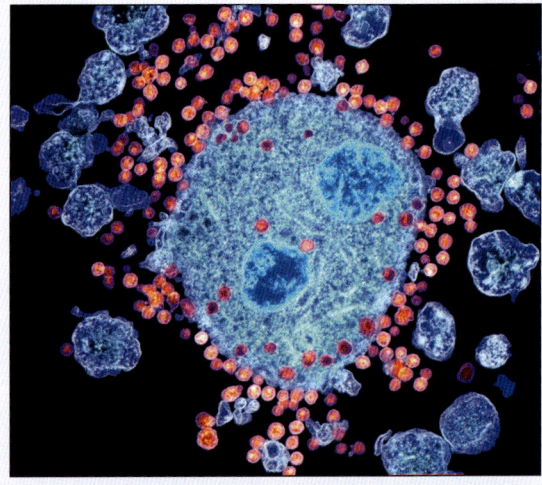

Figure 18.13 A transmission electron micrograph (TEM) of human immunodeficiency virus (HIV) particles (pink) attacking the white blood cells (blue). The particles enter the cell and use the cell to make copies of themselves. The new virus particles then burst out of the cell, killing it.

How is HIV spread?

HIV, the virus that leads to AIDS, can be spread through blood, semen, vaginal secretions and breast milk.

HIV is very fragile, so it cannot survive outside the human body for long. It is spread through:

- unprotected anal or vaginal sex (the most common way of spreading)
- an HIV-infected mother infecting her baby during pregnancy, at birth or through breastfeeding

Lifestyle diseases

- sharing contaminated needles, syringes or other equipment, such as for injecting illegal drugs or getting a tattoo
- blood–blood infection, which is possible, for example, if infected blood is accidentally transfused to an uninfected patient. This is very rare in the Caribbean because of the rigorous testing carried out on donated blood.

An individual can become infected with HIV from only one exposure; we only need to take a risk once, but once we are infected with HIV we can infect others.

> **Checkpoint question**
>
> 5 a List the main ways that HIV/AIDS is spread from one person to another.
>
>   b Explain why HIV cannot be caught from toilet seats, door handles or cutlery.

Signs and symptoms of HIV/AIDS

People infected with HIV/AIDS will experience the following signs and symptoms:

- For the first 2–12 weeks after infection there are no symptoms, and it may not be possible to detect the infection with a test because the body has not produced enough antibodies to be detected.
- Most people then have a short, flu-like illness but not everyone infected with HIV feels ill. At about three months after infection, HIV antibodies appear in the blood and the person becomes **HIV positive**.
- After this, the symptoms disappear for months or even years. During this time, the virus replicates rapidly, infecting the cells of the immune system and destroying them.
- Eventually, as the immune system weakens, an infected person may develop persistent signs and symptoms of AIDS. As the disease progresses, the symptoms become more noticeable and more severe, because the immune system no longer deals with infections. The clinical signs and symptoms of the final stages of AIDS include:
    - extreme fatigue
    - rapid weight loss
    - appearance of swollen or tender glands in the neck, armpits or groin
    - unexplained shortness of breath, frequently accompanied by a dry cough
    - infections such as tuberculosis (TB) and pneumonia
    - persistent diarrhoea
    - high fever, especially at night
    - appearance of one or more purple spots on the surface of the skin, and inside the mouth, anus or nasal passages caused by a rare cancer, Karposi's sarcoma
    - whitish coating on the tongue, throat or vagina as fungal infections take hold
    - forgetfulness, confusion and other signs of mental confusion
    - death.

Treatment for HIV/AIDS

There is no cure for HIV/AIDS, nor is there as yet an effective vaccine. Some of the early symptoms of HIV/AIDS can be treated:

- The secondary infections can be treated with antibiotics.
- The fungal infections can be treated with anti-fungal drugs.
- Any anaemia can be treated with iron and blood transfusions.

Figure 18.14 AIDS education is vital to prevent and eliminate this disease – this advert is for an AIDS awareness campaign in Soufriere, St Lucia.

The good news is that antiretroviral drugs can now control the reproduction of the virus and slow the progression of HIV-related diseases. Anti-HIV medications do not cure HIV infection and individuals taking these medications can still transmit HIV to others. A combination of antiretroviral drugs taken every day gives people who are infected with HIV a similar life expectancy as that of a healthy individual. The earlier the HIV infection is identified and the antiretroviral drugs are started, the more effective they are at lengthening the lifespan of an HIV-positive individual.

Preventing the spread of HIV/AIDS

The most effective ways of reducing or halting the spread of HIV/AIDS involve changing the way people, and young people in particular, behave. It is vitally important to be aware of the risk of HIV/AIDS in any sexual relationship. Ways of reducing the risk of HIV infection include the following:

- Use condoms in sexual relationships.
- Reduce the number of sexual partners, for example, by being faithful within marriage or a relationship.
- Have regular tests if you are at risk of getting HIV.
- Use PEP and PrEP, which are medicines taken before and after sexual contact with someone with HIV.
- NEVER share needles or other injecting equipment if you use drugs.
- Set up needle exchanges where intravenous drug users can get clean, sterile needles to help prevent the spread of HIV/AIDS.
- Educate people around the world about the risks of HIV/AIDS, how it is passed on and how to avoid it. This is one of the most important ways to prevent the spread of HIV/AIDS.

HIV/AIDS and pregnancy

Pregnant women need to be screened, and any found to be at risk should be given antiretroviral drugs to reduce the risk of infecting their unborn child. Caesarean deliveries are advised, to reduce the risk of transmission through blood. HIV-positive mothers should bottle-feed their babies, instead of breastfeeding, as HIV is also passed on to babies in breast milk.

The biologist's toolkit: Greater than, less than and equal to

Biologists often need to say that one result or piece of data is greater than or less than another. When things are the same, they are equal, so we use the = symbol.

Here are some more useful symbols. Look out for them in any data you see, for example, in the *Did you know?* on this page.

Symbol	Words	Example
=	equals	$1 + 1 = 2$
>	greater/more than	$5 > 2$
<	less than	$2 < 5$
≥	greater than or equal to	Percentage is ≥ 10
≤	less than or equal to	Percentage is ≤ 10

DID YOU KNOW?

If an HIV-positive mother is given antiretroviral treatment during pregnancy, has a Caesarian delivery and bottle-feeds her baby, the risk of HIV being transmitted to the baby drops from 40% to < 0.1%.

Checkpoint question

6 Discuss the main ways to avoid the spread of HIV/AIDS.

HIV/AIDS in the Caribbean

For many years, the Caribbean had a very high level of HIV/AIDS, second only to sub-Saharan Africa – at least 1.6% of the population was infected. But a report published by the WHO in 2020 showed that AIDS-related deaths and HIV infections had dropped by >30% across the Caribbean in the 10 years since 2010.

Many countries across the Caribbean have worked hard to improve testing and access to both antiretroviral and protective medicines. In 2017, 57% of all the people living with HIV were getting antiretroviral treatments, and in the best performing countries this was 80%.

One of the biggest problems is overcoming stigma and discrimination. Many people do not understand how HIV/AIDS is spread and this makes them treat those living with HIV with suspicion. For example, in one survey many people said they would not buy vegetables from a shopkeeper living with HIV – even though the scientific fact is that there is NO risk of catching HIV from doing so (see Figure 18.15).

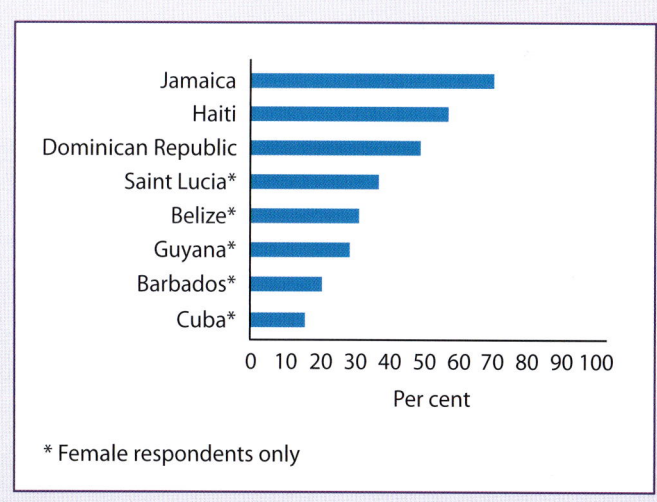

* Female respondents only

Copyright © 2022 UNAIDS https://www.unaids.org/sites/default/files/media_asset/miles-to-go_caribbean_en.pdf

Figure 18.15 People in some Caribbean countries do not understand how HIV/AIDS is spread, which fuels stigma and in turn leads to more infections, as people are afraid to get a diagnosis and treatment.

One reason why people with HIV have not gone for treatment in the past is that many do not understand how HIV is spread and are afraid of it. If people living with HIV fear discrimination like that shown in Figure 18.15, they do not come forward for treatment – so more people are infected. Education is key in overcoming these misconceptions. If people understand more, discrimination is reduced. In turn, more people get treated and infection rates fall. See Figure 18.16.

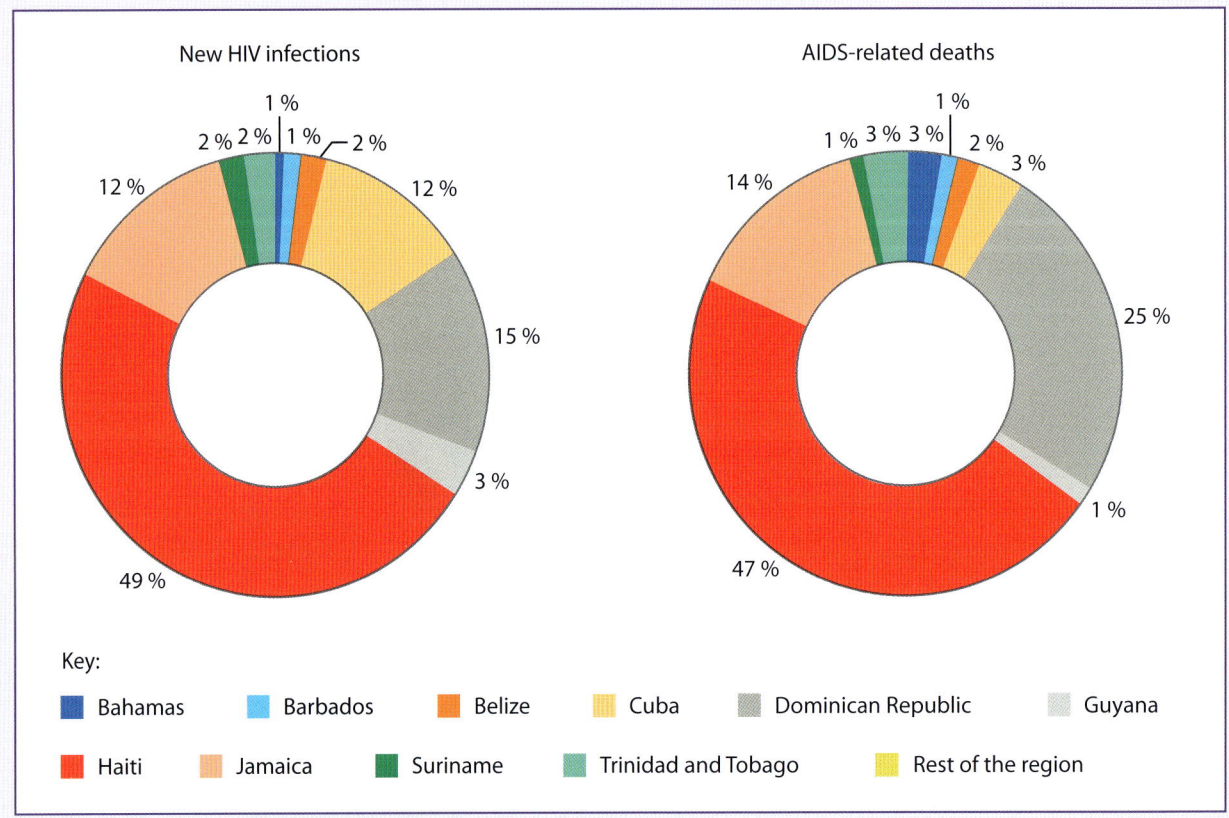

Copyright © 2022 UNAIDS https://www.unaids.org/sites/default/files/media_asset/miles-to-go_caribbean_en.pdf

Figure 18.16 New HIV infections are falling in most countries across the Caribbean, as education and improved treatments allow people to live with HIV within their communities and without spreading the disease.

Sexual behaviour is not the only lifestyle choice that can have an enormous impact on our health and well-being. Another growing problem in the Caribbean, and in many other areas of the world, is the use and abuse of drugs. In the next part of this chapter, we will look at some of the drugs most commonly used and misused, and the impact their use can have on individuals, families and society as a whole.

Lifestyle diseases

Drug use and misuse

Almost every person reading this book will have taken a **drug**, and not just medicine from the doctor. Most of us will have had a cup of coffee, a cup of tea or a drink of cola at some point in our life, and so will have taken in caffeine, a legal stimulant drug.

What do we mean by a drug? A drug is a substance that alters the way in which the mind, the body, or both, works. In every society there are certain drugs that are used for medicine and others that are used for pleasure. Usually some of these substances are socially acceptable and others are illegal. In the Caribbean, caffeine, nicotine and alcohol are the legal recreational drugs. In other parts of the world, such as the Arab states, alcohol is illegal.

Legal and illegal drugs

> **Checkpoint question**
>
> 7 List the most commonly used legal drugs in the Caribbean.

Both legal and illegal drugs are classified by the effect they have on the body, as shown in Table 18.1.

Table 18.1 Effects of drugs on the human body

Classification	Effect on the body	Examples
Stimulants	Speed up the heart and breathing rate, raise blood pressure and speed up reactions	Legal: Nicotine Illegal: Cocaine, methamphetamine
Depressants	Slow down the heart rate, and all the reactions of the body	Legal: Alcohol, prescription medicines such as anti-anxiety drugs Illegal: Cannabis, heroin
Hallucinogens	Affect the brain, so that the user sees things differently than they really are	Legal: Illegal: Ecstasy, LSD
Narcotics	Relieve pain and produce mood changes	Legal: Prescription medicines such as codeine and morphine Illegal: Opium, heroin

The status of a drug may be related to its effect on people, or simply down to the history of its use.

- **Prescription drugs** are used in medicine. Some prescription drugs, such as **antibiotics** (see Chapter 17) are never misused for pleasure because they do not affect human cells and so do not affect us. Other prescription drugs that are used in healthcare but also misused are, for example, **opioids**, **sedatives**, **painkillers** and **medicinal marijuana**.
- The drugs we use for pleasure tend to have a distinct effect on our mind. They are known as **non-prescription drugs**. Some of them are legal and some are not. These include **nicotine**, **alcohol**, **cocaine**, **methamphetamines**, **heroin**, **ecstasy** and **marijuana**.

Drug (or substance) use is when we choose to take a substance that affects our brain and/or body function and mental activity. We use drugs for a variety of reasons:

- Doctors give us prescription drugs to help us feel better – for example, they reduce anxiety, help us sleep or relieve pain.
- Legal non-prescription drugs are used for the mild pleasure they bring, to be sociable (Figure 18.17) and because using them becomes a habit.
- People start to use illegal non-prescription drugs for much the same reasons, because other people do and because they like the effect. However, as many of these drugs are highly addictive, they can soon lead to many problems.

Drug misuse is when we use a substance to the point of excess and/or dependence. When we take too much of a drug, we risk serious side effects and even death. **Drug dependence** is when we use a drug again and again, and become addicted to it. Drugs change the chemical processes in our body and this is why we can become addicted to them (dependent on them). If we are addicted to a drug, we cannot manage or function properly without it. Drug dependence may be:

- **psychological** – the need to keep using the drug becomes a craving or compulsion
- **physiological** – we develop a physical dependence, where our body no longer works properly without the added drug.

Once we are addicted to a drug, we cannot manage without it, and we need more and more of it to keep us feeling normal. When addicts try to stop using their drug, they will feel very unwell, often experiencing a combination of aches and pains, shaking, sweating, headache, cravings for the drug and even fever as the body reacts. These are called **withdrawal symptoms**.

Drug use, both legal and illegal, is a major public health problem that has a serious negative social and economic impact on Caribbean countries. School surveys have shown that alcohol, tobacco and cannabis (marijuana) are the substances most widely used by young Caribbean people. Alcohol and tobacco are legal drugs – they can be bought easily in shops and bars, but others are illegal across the Caribbean. These include cocaine, methamphetamine, heroin and ecstasy.

Figure 18.17 We use legal drugs such as nicotine and alcohol as part of our normal social life. No one at this party will be thinking of themselves as drug users – but that is what they are doing.

Checkpoint question

8 Define drug addiction.

The excessive use of alcohol and tobacco are linked to a wide range of health problems, many of which we have already met in earlier chapters. The health issues linked to these legal drugs are mainly the result of their effect on the systems of the body.

Illegal drugs also affect health in two quite different ways.
- Like legal drugs, they cause changes in the body and damage vital systems.
- Since non-prescription drugs are both illegal and addictive, drug abusers need considerable sums of money to feed their addiction. They may spend all of their cash on drugs, not feed themselves properly because they are buying drugs with their money, turn to crime or prostitution to raise the money they need, and take part in risky activities such as sharing needles, which increases the risk of becoming infected with HIV/AIDS or hepatitis (a liver infection that can be fatal).

Some of the most commonly used recreational drugs in the Caribbean all have the following features in common:
- They are addictive.
- They affect brain functioning and alter behaviour.
- They damage health, resulting in lower productivity and absence from school or work.

They adversely affect the individual, families, the community and the country.

Here is a summary of some of the most commonly misused drugs, both in the Caribbean and around the world.

Misused recreational drugs

Alcohol

Alcohol is one of the drugs most commonly used by young people in the Caribbean (see Figure 18.18). In one study conducted in Jamaica, 70% of school children admitted to having used alcohol and 29% admitted to its use in the previous 30 days. For many people, alcohol is part of their social life. They like to share a rum with friends or enjoy a glass of beer or wine with a meal. They probably don't think of themselves as drug users.

Figure 18.18 You can buy drinks legally in the Caribbean once you are 18. But you have to be over 21 in many places in the USA, and alcoholic drinks are completely illegal in several other countries.

In small amounts, alcohol makes people feel relaxed and cheerful. It makes them less inhibited. Shy people can feel more confident when they've had an alcoholic drink. But alcohol has a powerful effect on the body. It is very addictive and it is also very poisonous. In fact, alcohol is one of the most widely used drugs across the Caribbean.

Checkpoint question

9   Explain why alcohol is described as a drug.

How does alcohol affect your body?

Alcohol acts quickly because it is readily absorbed from the stomach into the bloodstream. From the blood it passes easily into nearly every tissue of the body. It dilates blood capillaries near the skin's surface, producing a feeling of warmth and well-being. It increases the heart rate and can increase feelings of hunger.

Alcohol gets into the nervous system and brain. This slows down the reactions. It can make the drinker lose their self-control. It contributes to poor muscular coordination, resulting in slurred speech and a lack of balance. Alcohol is a **diuretic**, which means that it makes the body lose water through increased urination. As the effects of the alcohol wear off, it can cause headaches due to dehydration and nausea.

Alcohol is poisonous but the liver can usually break it down. It gets rid of the alcohol before it causes permanent damage and death. If people drink large amounts of alcohol, such as a whole bottle of spirits, their liver may not be able to cope. They may suffer from alcohol poisoning. This can quickly lead to unconsciousness, coma and death.

Some people drink heavily for many years, becoming addicted to the drug. They are alcoholics. Their liver and brain suffer long-term damage and eventually the alcohol may kill them. They may develop cirrhosis of the liver. This disease destroys the liver tissue. They can also get liver cancer, which spreads quickly and can be fatal. In some heavy drinkers, the brain is so damaged (it becomes soft and pulpy) that it can't work any longer. This causes death. Even short bouts of very heavy drinking can cause the same symptoms to develop quite quickly.

DID YOU KNOW?
It takes your liver one hour to break down the alcohol in a single measure of rum or half a pint of beer (one unit of alcohol).

Checkpoint questions

10 State which organs are most affected by heavy drinking.

11 Give an example of a poor decision that someone under the influence of alcohol might make.

The effects of drinking on society

Alcohol puts people at risk because of the way it affects their behaviour. Since alcohol slows down their reactions, they are much more likely to have an accident. This is very dangerous if they drive after drinking. Alcohol is a factor in a high percentage of all fatal road accidents in the Caribbean. In fact, a survey in Trinidad a number of years ago showed that almost 50% of the men admitted to hospital following road accidents had alcohol-related conditions. See Figure 18.19.

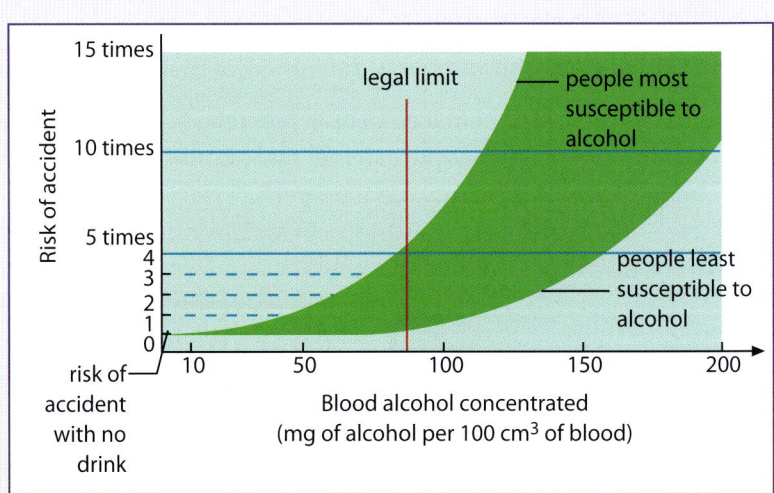

Figure 18.19 Blood alcohol content and the risk of road accidents.

Alcohol abuse affects personal lives as well. Domestic violence is often linked to patterns of heavy drinking. Many crimes take place when people are under the influence of alcohol, often mixed with other drugs.

Prescription drugs

The misuse of prescription drugs is another health problem in the Caribbean and elsewhere. For example, sedatives are drugs that slow down the response of the brain, making the user feel calm and sleepy. They are prescribed by doctors for people who are very anxious or find it hard to sleep. One group of sedative drugs, barbiturates, are very addictive. Doctors are very careful about prescribing sedatives, but there is still a big black market in barbiturates, which people use illegally to make themselves feel calm and happy.

Marijuana

Medicinal marijuana cultivation is now legal but regularised in Jamaica since 2015. Marijuana has also been decriminalised in Trinidad and Tobago and a few other Caribbean territories. The positive medicinal use of marijuana in treating glaucoma, pain in cancer patients and in treating mentally ill patients, is now documented by many leading universities including Harvard. The legal recreational use of marijuana in many US states, Canada and a few other first-world countries also needs to be addressed.

The effect of marijuana varies. Even the same person can react very differently depending on the following factors:

- **The potency of the drug:** Marijuana is a natural product and some plants produce more of the drug in their leaves than others. See Figure 18.20.
- **Smoking technique:** Some people draw the drug into their lungs to be absorbed much more effectively than others.
- **Dose:** A few puffs of a cannabis cigarette (spliff or joint) gives a very different dose to smoking several joints on one's own.
- **Setting:** If you are in a happy, comfortable setting with friends you are more likely to have a pleasant experience than if you are in a stressful setting or are nervous.
- **The biological vulnerability of the user:** Scientists are finding more and more evidence that marijuana use can tip vulnerable people into mental illness; illness that may well affect them for the rest of their lives.

It is not just the nervous system that is affected by marijuana. Some of the other major organs and systems are affected by cannabis on a long-term basis. They include the following:

- **The respiratory tract:** A joint of marijuana produces five times the amount of cancer-producing agent, benzopyrene, as a cigarette. Chronic coughs and bronchitis are not uncommon in marijuana smokers.
- **The reproductive system:** Marijuana is associated with early foetal death, decreased foetal weight and premature birth. It can also have adverse affects on fertility in males and the menstrual cycle in females.
- **The motor system:** Marijuana affects the reaction time, so it has a direct effect on the ability to drive and operate machinery. It can lead to road accidents and work-related accidents.

Figure 18.20 Different types of marijuana plants produce different amounts of the hallucinogenic drug. Even the same type of plant, grown in different conditions, will produce more or less of the drug.

> **Checkpoint question**
>
> 12 Make a table to show reasons why people use marijuana and why it can be dangerous and damaging.

It is difficult to see a way forward to reduce the use of this drug. Better education, so that people understand the health risks of using the drug, can only help.

Cocaine

When someone takes cocaine they get a rush of energy and a high that makes them feel very powerful. The downside is that they can end up feeling paranoid and depressed afterwards. Cocaine is an extremely addictive drug – the body quickly craves more and more of it. As a result, people become addicted very quickly. It is also quite expensive, so people end up spending all their money to get enough fixes to satisfy their craving.

Cocaine raises the blood pressure, causes the heartbeat to become fast and irregular, and increases the body temperature. It can kill people the first time they use it, as some people have complete heart failure as a reaction to the drug. Every time someone uses the drug, they are putting their mental and physical health at risk. What is more, most people snort the drug up through their nose. Eventually the chemicals eat away and destroy the tissue that divides the two sides of the nose, so they are left with a single opening, which only plastic surgery can repair. It is said that several famous actors and models have needed this operation after using too much cocaine.

Methamphetamine

This drug is closely related to amphetamine and it is a stimulant. It is smoked or injected, and it gives people an instant high. They feel exhilarated and aroused, but it also makes them paranoid, confused and aggressive. It is addictive, and it often increases risk taking. This in turn affects the risk of – for example – getting STIs through high-risk sexual behaviour. Methamphetamine may lead to severe psychosis, heart attacks and strokes.

Opium

The gum of the opium poppy plant is used to produce opium, which is further processed to make codeine and morphine, which are painkillers used in hospitals to ease the suffering of patients after surgery and with cancer.

Heroin

Heroin is usually found as a white or brownish powder that is dissolved in water and injected either under the skin or directly into a muscle or vein. See Figure 18.21. However, the drug can also be 'snorted' into the nose, smoked, eaten or inserted as a suppository into the rectum.

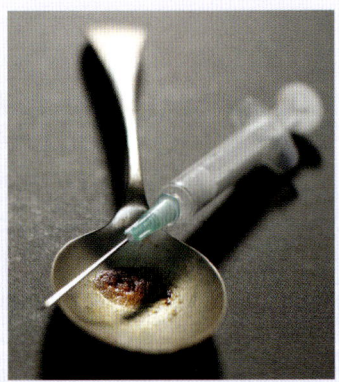

Figure 18.21 Injecting heroin into the veins gives users an immediate high – but sharing needles increases the risk of HIV infection as well as the dangers of the drug itself.

Immediately after heroin is injected into a vein, the user feels a strong surge of pleasure, known as a 'rush' followed by calm well-being. Most users do not feel hunger, pain or sexual feelings when they are under the influence of the drug.

Many of the risks are side effects of the lifestyle that often goes with heroin abuse, rather than from the drug itself.

- Many heroin addicts who need a fix of the drug are not too fussy about the sterility of the needles they use, or about preparing their skin first. For example, infection of the heart lining and the valves can result from dirty needles. Intravenous drug users who share needles run a high risk of getting HIV/AIDS. They also risk abscesses, liver diseases such as hepatitis and even brain damage.
- Since heroin addicts tend to spend most of their money on their drug, and because heroin use also suppresses the appetite, users tend to have a very unhealthy lifestyle with very poor nutrition. This leaves them vulnerable to diseases.
- With regular use, users develop a tolerance to the effects of heroin and so have to take higher doses to achieve the same desired intensity of effect. They become both physically and psychologically addicted to the drug.

Ecstasy

Another illegal drug that has become increasingly popular with young people is ecstasy (E). This is a synthetic drug that has both hallucinogenic and stimulant properties. It is used at large music and dance parties known as 'raves', where hundreds of young people get together, largely to dance for a very long time to very loud music. Ecstasy produces feelings of well-being and general warmth towards other human beings. It also gives increased energy and stamina, allowing users to dance for longer and with more enthusiasm without becoming exhausted (Figure 18.22).

Ecstasy has several physiological effects. It increases sweating, but also reduces the sensitivity of the body to a lack of water (see Chapter 10). This means that eventually the user stops sweating due to a lack of water, and the body begins to overheat. The ecstasy user does not feel thirsty. Some users have collapsed and died from dehydration and overheating. Sadly, other young people, in a desperate attempt not to overheat, drink so much water that they go into a coma and die from excess water in the system.

Ecstasy use also leads to psychological problems such as confusion, depression, sleep problems, cravings, severe anxiety, paranoia and psychotic episodes.

Figure 18.22 Ecstasy is popular at raves – it is easy to take and allows users to dance for longer without becoming exhausted.

Checkpoint question

13 Describe the main health risks of ecstasy use.

The social effects of drug misuse

The excessive and wrong use of drugs is called drug misuse. Often this arises because an initial liking becomes an addiction. Most drugs tend to be habit-forming, if not physically addictive. The sensation that accompanies the use of drugs is often very appealing and satisfying, which makes this 'high' sought after. However, the high is usually followed by depression and an extreme craving for another 'hit' just to feel better again.

Many people spend a lifetime chasing that high and it becomes very difficult to break the cycle. The effects of drug abuse are many and varied. The excessive use of drugs significantly affects the individual user, the family and the wider community.

Social and economic effects of drug misuse on individuals

Since most drugs affect the nervous system, someone who abuses drugs typically displays an unusual sense of relaxation on the one hand, and a superhuman energy and strength on the other. Drug abuse is sometimes accompanied by aggressive and abusive behaviour, especially when someone wants a fix of the drug and cannot get one.

Some drugs lead to an increase in appetite and consequently weight gain, while others cause a lack of appetite and weight loss. Many people who misuse drugs lose their job as they no longer perform well, so they turn to crime to fund their drug habit. Often this involves prostitution, so drug users are particularly prone to sexually transmitted infections. They often share needles, which leads to a greatly increased risk of catching blood-borne diseases such as hepatitis and HIV/AIDS. All these changes have a big impact on their work as well as their social and family life. People who misuse drugs often end up socially isolated, ill and struggling for money as a result of their habit.

Social and economic effects of drug misuse on family and friends

Very often, relationships at home are affected and friendships are seriously damaged. See Figure 18.23. Drug users may lose their job due to poor quality work, regular absences or the inability to function. Loss of employment means a loss of income, creating serious financial problems for their family. Drugs often impair sexual function, which will affect the relationship between drug users and their partners. Many drug users crave attention and affection, which often leads to an increasingly promiscuous lifestyle. Divorce is often the outcome of continued drug use.

In many cases, drug users will steal and sell items from home to buy their drugs, creating further distress to the family. They may borrow money from family and friends that they cannot pay back. Even when the drug user does not live in the family home, there is the emotional and psychological pain from being uncertain of his or her whereabouts and condition.

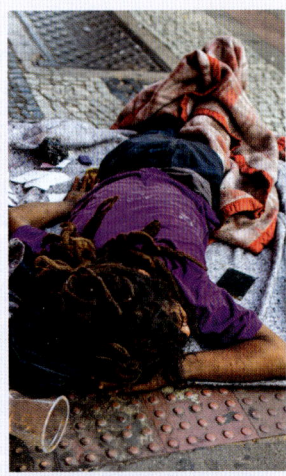

Figure 18.23 Drug misuse affects not only the affected individuals, but also their families and friends, and the wider community. These effects are both personal and economic, and the associated socials issues can make the community feel challenged and overwhelmed.

Children living in a family with a drug user can become lonely, depressed, aggressive and disruptive, and often become drug users themselves to manage their pain, and so the cycle continues.

Social and economic effects of drug misuse on the community

The wider community is severely challenged by the presence of those who have become abusers of drugs. Social issues including violence, theft, sexual abuse and assault, murder, damage to property and violent crimes affect many communities where drug misuse is high. Many accidents are the result of drug misuse, and accidents leave communities gripped by fear, anxiety and grief. People cannot work effectively and families break up. The impact of drug use on a community can be huge, in both economic and personal terms.

What is more, the community has to pay for the healthcare of addicts, for education programmes to prevent drug misuse, for hostels to house homeless addicts, and to help support the families left in despair. The cost of drug misuse to the community is a very high one indeed.

Possible SBAs based on lifestyle diseases

- Research the performance of your country in reducing the number of/treating/reducing maternal transmission to babies of any STI in this chapter.
- Compare the performance of different countries in preventing a range of STIs and display the data clearly to use to convince governments to look at how such issues are tackled in their countries.
- Find out as much as you can about any one of the STIs covered in this chapter and produce a booklet, podcast, blog series, TV or radio news item to inform young people about the risks of the disease you have chosen, how it is spread and how to avoid it. Make sure you encourage people to seek medical help if they are concerned and emphasise the successful treatments available for many STIs.
- Plan an awareness campaign to make young people across the Caribbean more aware of chlamydia and the potential fertility risks it holds.
- Produce a series of posters on the major sexually transmitted infections that affect people in the Caribbean.
- Write a play to highlight the social and economic price of drug misuse on an individual.
- Plan an education campaign to highlight the problems of drug misuse across the Caribbean/in your own country.

End-of-chapter summary

In this chapter, you have learnt that:

- sexually transmitted infections (STIs) are infectious diseases spread through sexual contact; many of the most common sexually transmitted infections are cured easily using antibiotics if caught early

- sexually transmitted infections are avoided by sensible sexual behaviour

- gonorrhoea is a bacterial infection spread by sexual contact; infected men will experience burning while urinating and yellowish-green discharge from the penis; women with symptoms will have a vaginal discharge and burning while urinating; antibiotic treatment is prescribed; responsible sexual practices are also recommended

- syphilis is a bacterial infection; any sexually active person can be infected with syphilis; the symptoms of syphilis occur in stages called primary, secondary and tertiary (late); it is treated with antibiotics; if untreated, syphilis leads to the destruction of soft tissue and bone, heart failure, blindness and a variety of other conditions; it can be prevented by healthy sexual practices

- the human papillomavirus (HPV) causes genital warts but is linked to the development of cervical cancer and other cancers of the reproductive organs; it is best prevented by the HPV vaccine and by sensible sexual behaviour

- herpes II is a viral sexually transmitted infection that produces painful sores, usually in the genital area; it is treated orally, it can be avoided through sensible sexual behaviour

- with HIV/AIDS, HIV is the virus that causes AIDS; it is spread mainly through blood, semen, vaginal secretions and breast milk; patients are initially asymptomatic but deteriorate and eventually show many signs and symptoms of a weakened immune system; treatment options include the use of antiretroviral drugs, healthy lifestyle practices and a strong support system

- most STIs will affect a foetus developing in the uterus and newborn babies; it is important to manage pregnancy and birth of mothers affected by STIs to reduce the risk of damage or death for the baby

- drug misuse is when you use a substance to the point of excess and/or dependence; when you take an excess of a drug, you risk serious side-effects and even death

- drug dependence is when you use a drug again and again and become addicted

- drugs change the chemical processes in your body, so you can become dependent on them; this means you cannot manage or function properly without the drug; this may be psychological – the need to keep using it becomes a craving or compulsion – or it may be a physiological, physical dependence where your body no longer works properly without the drug

- prescription drugs are used in medicine; some prescription drugs, such as antibiotics, are never misused for pleasure; other prescription drugs are used in healthcare but are also misused, for example, opioids, sedatives, painkillers and medicinal marijuana

- non-prescription drugs that are misused in the Caribbean for pleasure may be legal or illegal; they include nicotine, alcohol, cocaine, methamphetamine, heroin, ecstasy and marijuana

- alcohol, tobacco and marijuana are the most widely misused substances among schoolchildren

- drug misuse and dependence have serious social and economic effects on the individual user, their family and the entire community.

Lifestyle diseases

End-of-chapter questions

1. Which one of the following infections is not sexually transmitted?
   A   HIV/AIDS
   B   Gonorrhoea
   C   Diarrhoea
   D   Syphilis   (1)

2. Which of these STIs are caused by viruses?
   i   HIV/AIDS
   ii   Gonorrhoea
   iii Genital herpes
   iv   Syphilis

   A   i, ii and iii
   B   i, ii and iv
   C   ii, iii and iv
   D   i and iii only   (1)

3. Which **one** of the following is not a way to help prevent the spread of HIV/AIDS?
   A   Washing your hands after using the toilet
   B   Using a condom when you have sex
   C   Having only one sexual partner
   D   Not sharing needles for intravenous drug use   (1)

4. Which **one** of the following drugs is not legal in the Caribbean?
   A   Nicotine
   B   Alcohol
   C   Marijuana
   D   Caffeine   (1)

5. Which drug is described here? It produces feelings of well-being and general warmth towards other human beings. It also gives increased energy and stamina, so it can be very attractive to young people.
   A   Heroin
   B   Ecstasy
   C   Alcohol
   D   Cocaine   (1)

6. Smoking has a direct relationship with all but one of the following. Which **one**?
   A   Coronary heart disease
   B   Stroke
   C   Cancer
   D   Cholera   (1)

7 Which of the following organ systems is NOT damaged by using marijuana?
  A   Respiratory system
  B   Reproductive system
  C   Motor nervous system
  D   Skeletal system (1)

8 Which of the following STIs is treated with antiretroviral drugs?
  A   Syphilis
  B   HIV/AIDS
  C   HPV
  D   Gonorrhoea (1)

9 a   Sexually transmitted infections (STIs) are a problem in the Caribbean. Explain what is meant by the term 'sexually transmitted infection'. (3)
   b   Syphilis is an example of a sexually transmitted infection caused by a bacterium.
       i   Describe the signs and symptoms of syphilis and explain how it is transmitted from one person to another. (6)
       ii   Describe how syphilis is treated/cured. (2)
       iii Give **four** ways in which syphilis can be prevented. (4)

10 HIV/AIDS is a sexually transmitted infection that is still a major problem in the Caribbean.
   a   Describe the main signs and symptoms of HIV/AIDS. (10)
   b   State **four** ways of controlling or preventing the disease. (4)
   c   Discuss reasons why HIV/AIDS is such a problem in the Caribbean. (5)

11 a   Define the terms 'prescription drug', 'non-prescription drug' and 'drug dependence'. (4)
   b   State **two** features of drugs that are:
       i   Stimulants
       ii   Depressants
       iii Hallucinogens
       iv   Narcotics (4)
   c   Chose **one** prescription drug and **one** non-prescription and for each, describe the effects of misuse. (6)

12 The graph below shows the incidence of three STIs in the Caribbean.

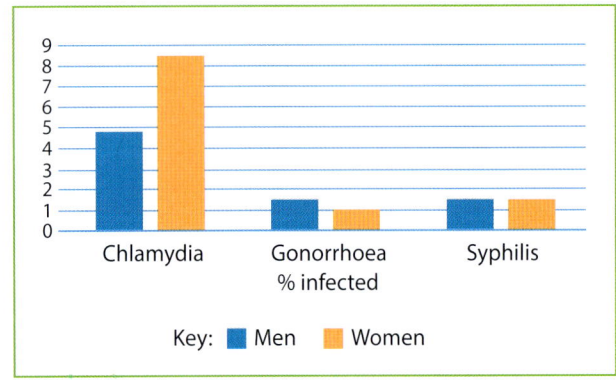

   a   Describe what is shown by this graph.
   b   Make a table to compare the **three** STIs shown. Compare the organism that causes the disease, the symptoms of the disease, the way it is spread, how it is treated and how it may be prevented. (24)

Lifestyle diseases

Chapter 19: Parasites and vectors

When you have completed this chapter, you will be able to:

- identify vectors that are common in the Caribbean
- explain the importance and methods of controlling vectors that affect human health
- describe the life cycle of the housefly and the mosquito
- explain the effects of vectors on the health of human beings
- define the term 'vector'
- discuss diseases including malaria, leptospirosis, dengue, zika and chikungunya

What the examiners say

- Candidates are often unable to demonstrate comprehensive understanding of the infectious diseases and how they are spread. In too many instances, responses are vague.
- Candidates need to specify that vectors are carriers of the organism that results in disease, rather than indicating that vectors are carriers of the disease.

Revision tip

Never study in bed! Find a well-lit area where you have plenty of fresh air and can sit upright. This avoids putting strain on your eyes or your spine!

Not all pathogens are spread by droplet infection or direct contact. The pathogens that cause some of the most damaging diseases in the world are spread by other living things – for example, mosquitoes and rats (Figure 19.1) carry pathogens from one person to another.

Figure 19.1 Most rats carry a bacterial disease called Leptospirosis, which can be fatal to humans. You will learn more about this later in the chapter.

The role of vectors in disease

A **vector** is an organism that transmits disease-causing microorganisms from one host to another. Some vectors simply transport pathogens from one host to another on their body – often the feet or mouthparts. A **housefly** is a good example, carrying bacteria from the faeces or rotting food in which it breeds onto the surface of food you are about to eat! Some animals are **biological vectors**. They are needed as part of the life cycle of an infective organism. Well-known examples of biological vectors are the *Anopheles* mosquito (Figure 19.2), which carries **malaria** and the *Aedes aegypti* mosquito (Figure 19.3), which spreads **dengue fever**.

Animal vectors are a nuisance – they have an impact on our life in one way or another. But more importantly, they play a large role in the spread of some of the fatal communicable diseases responsible for untold human suffering. We are going to look at a number of diseases in which vectors play an important role. In each case we have to think about ways of both treating the disease and controlling the vectors. Common vectors in the Caribbean include:

- houseflies
- mosquitoes – *Aedes aegypti* and *Anopheles spp*
- rats.

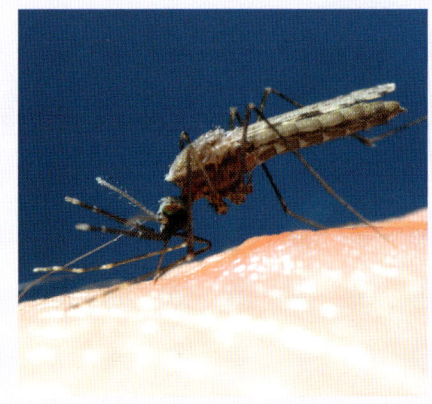

Figure 19.2 The *Anopheles spp* mosquito carries malaria.

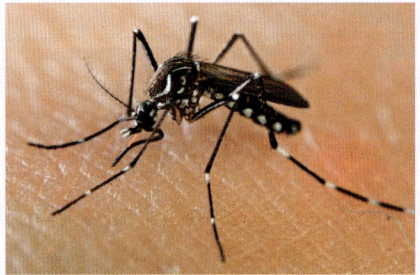

Figure 19.3 The *Aedes aegypti* mosquito

Checkpoint question

1 Define the term 'vector'.

The housefly as a vector of disease

The housefly is found almost everywhere, carrying disease wherever it goes. Houseflies are an example of a simple **mechanical vector**, as the pathogen is passed on without multiplying in the fly. See Figure 19.4.

Wherever houseflies go, bacteria and viruses go too. The average lifespan of an adult housefly is around one month. During that time, they feed and breed on garbage, sewage, rotting food and other sources of filth. They can only take in liquid food, so they vomit up digestive juices onto their food and suck up the liquid that is produced by the digestive enzymes. They suck up lots of bacteria and viruses too. The pathogens that cover their food are carried away on their mouthparts, feet and other body parts. Then, through their vomit, faeces and contaminated external body parts, the flies transfer bacteria and viruses to human and animal food or directly onto human skin.

As far as people are concerned, flies act as carriers of diseases. They are wholly undesirable from a hygienic viewpoint. On the other hand, the larvae of flies, known as maggots, play a vital role in the decomposition of waste and dead bodies.

Figure 19.4 Houseflies like this one are vectors of many diseases.

Over 100 pathogens that cause disease in humans and animals are associated with the housefly. These include typhoid, cholera (see Chapter 17), bacterial dysentery, tuberculosis, anthrax, ophthalmia (eye infections) and diarrhoea in children, as well as parasitic worms. Gastroenteritis (see Chapter 17) is one example of a fly-borne disease that you have already studied in some detail. What is more, not only do flies carry disease but they are a real nuisance, both to people and to animals.

The life cycle of the fly

During the life cycle of a housefly, it undergoes a complete metamorphosis. This means it changes form completely in one big step – in this case from a maggot to a fly. The life cycle has the following distinct stages (see Figure 19.5):

- egg
- larva or maggot
- pupa adult.

The number of eggs laid by a female fly depends largely on how well fed she is. However, a single, well-fed female can produce more than 500 eggs at a time, and she will lay five or six batches over a 3–4 day period.

Maggots emerge from the eggs in warm conditions only 8–20 hours after they are laid. They feed on the manure, dead body or garbage on which the eggs were laid. They finally crawl some distance away from their food source to find a dry, cool place. There they are transformed to the final pupal stage, which is when metamorphosis takes place.

DID YOU KNOW?

It has been calculated that if a pair of flies started breeding in April and all their offspring survived and bred, they would have produced 191 010 000 000 000 000 000 flies by August. Fortunately for us, conditions are never ideal and they don't all survive!

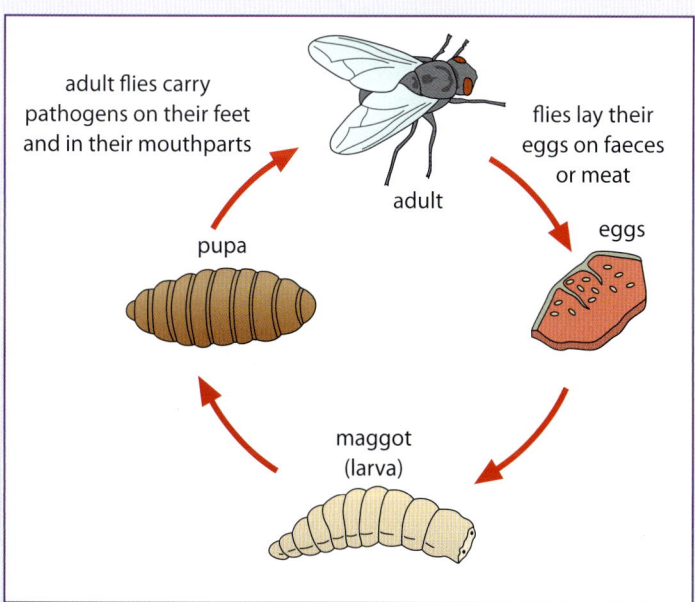

Figure 19.5 The timescale of the life cycle of the housefly depends on the temperature of the surroundings.

DID YOU KNOW?

Within an hour of death, flies smell a body and land on it to lay their eggs. If a body is discovered, forensic scientists use the numbers and ages of the larvae, pupae and empty pupal cases on the body to help them work out how long it has been dead. They also have to take into account the temperature and other local conditions. This is called **forensic entomology**.

The pupae are dark brown and about 8 mm long. The tough outer case is made from the last larval skin, which varies in colour as the pupa ages through yellow, red, brown and black. It takes 3–6 days for the metamorphosis of a larva into an adult fly to be completed.

Finally, the adult flies emerge from the pupae. The female is usually larger than the male. On average they are 6–7 mm long. Houseflies have reddish eyes and a spongy mouth, which is ideally suited for digesting their food, and for carrying pathogens onto ours!

Preventing the spread of fly-borne diseases

The best way to prevent the spread of the many diseases carried by flies is to control the numbers of the vectors themselves. If we can remove the materials that flies need to feed on and to lay their eggs, we will reduce the number of flies and reduce the spread of disease. We can use the following methods of control:

- **Chemical control** – fly traps and insecticides kill these pests if they get into our home. Flies are so common we almost ignore them – yet they spread more diseases than almost any other vector species.
- **Sanitary control** – flies love faeces! Careful removal and treatment of sewage reduces the housefly population.
- **Mechanical control** – keep food covered in the kitchen, so that flies cannot land and spread pathogens onto our food before we eat it.

> **Checkpoint question**
>
> 2   State the main ways in which we can control housefly numbers.

Mosquitoes as vectors of disease

Mosquitoes are vectors of many different diseases. Over one million people around the world die of diseases where the pathogens are carried by mosquitoes every year. Here in the Caribbean, we certainly do not escape the pathogens that the mosquitoes carry.

Mosquitoes and malaria

Malaria is disease in which *Anopheles* mosquitoes are the vector. Malaria is caused by the single-celled **parasite** *Plasmodium falciparum*, a protozoan with a very complicated life cycle. The parasite spends part of its life cycle inside a mosquito, part inside human red blood cells, and part inside the human liver!

In the Caribbean, very few places still have malaria. Haiti and the Dominican Republic still have endemic malaria at the time of writing this book. Other countries in the region no longer have this vector disease. But globally, malaria is still a major killer and has a big effect on societies and economies.

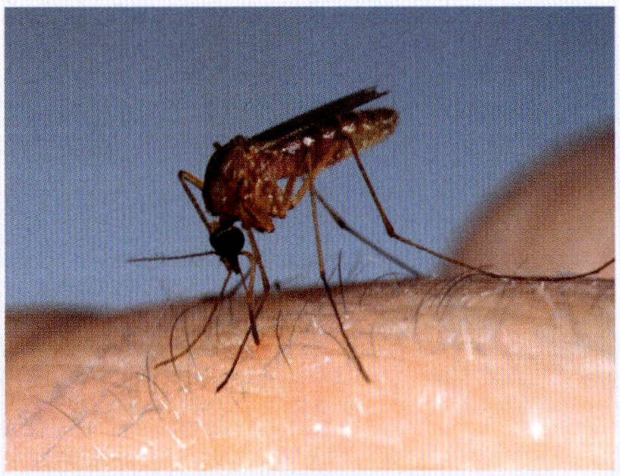

Figure 19.6 It seems amazing that such a small insect has such a devastating effect on human lives. Yet mosquitoes not only carry dengue, they also spread malaria – between them, two diseases that affect millions of people around the world.

How is malaria spread?

Plasmodium spends part of its life cycle in a mosquito and part in the human body. The *Anopheles* mosquito is a good example of a **biological vector**, in which the parasite multiplies inside the vector before being passed on. The life cycle of the mosquito means that the female *Anopheles* mosquito needs two meals of human blood to provide protein for her developing eggs before laying them. This is when the parasite that causes malaria passes from the mosquito to the next human host, spreading the disease (see Figure 19.7).

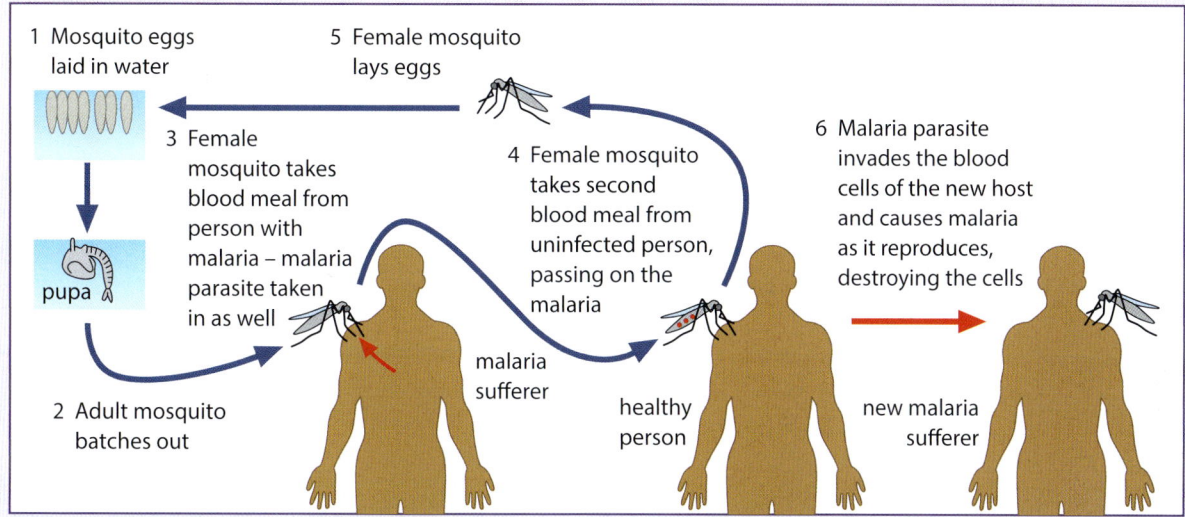

Figure 19.7 The life cycle of the *Anopheles* mosquito showing how the mosquito passes the malaria parasite from person to person, with the parasite spending time in the body of the mosquito.

Signs and symptoms of malaria

The symptoms of malaria come and go as the parasites causing the disease burst out of the red blood cells (see Figure 19.8) or stay inside reproducing. They include:

- fevers, sweats and chills
- headaches, feeling confused and uncertain
- exhaustion
- feeling sick, being sick, diarrhoea, stomach pain
- muscle pains
- difficulty breathing
- damage to the blood and liver means that people affected cannot work effectively.

Treatment of malaria

It is difficult to attack a parasite hidden in the red blood cells or the liver. There are a number of treatments but prevention is still the best approach. The only treatments are antimalarial drugs like quinine, which destroy the malaria parasite in the human body.

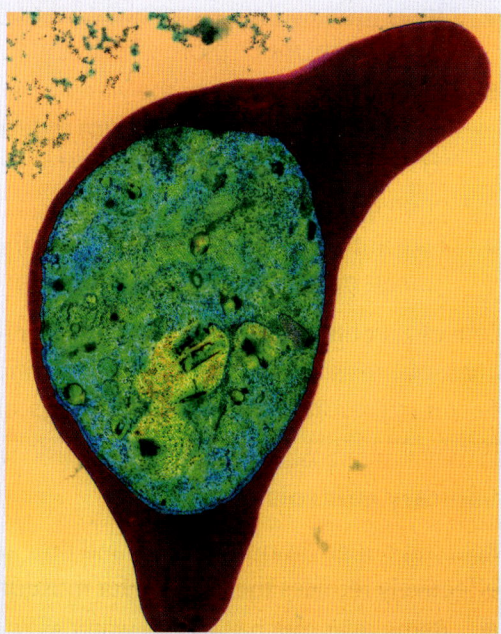

Figure 19.8 A transmission electron micrograph (TEM) of a red blood cell infected with a plasmodium malaria parasite. The parasite digests the haemoglobin in the cell, replicates and bursts out of the cell to invade other cells.

Mosquitoes as vectors of disease

Prevention of malaria

The best way to prevent malaria is to control the vectors. This is true of all vector-borne diseases. The four main ways of controlling the mosquitoes that carry malaria and other vector-borne diseases including dengue, zika and chikungunya (see more about these diseases from page 419) are:

- mechanical controls
- sanitary controls
- chemical controls
- biological controls.

Mechanical controls

Mechanical control methods help you to avoid contact with mosquitoes. They are all effective measures that protect you against mosquito bites and so against malaria. They are cheap and easy for us to use. Here are some examples:

- Have screens on doors and windows to prevent mosquitoes from getting in.
- Wear clothes that protect the skin against mosquitoes, for example, long sleeves and trousers.
- Sleep under a mosquito net (Figure 19.9); if possible, one treated with insecticide.

Figure 19.9 Mosquito nets protect us from mosquitoes and all the diseases they carry, while we are sleep.

Chemical controls

These control methods use chemicals to either keep mosquitoes away or to kill them. They are also effective measures against mosquito bites and so against malaria. We can use some controls as individuals. Others need the involvement of governments at local or national level and are relatively expensive:

- Use mosquito repellents.
- Spray insecticides on mosquito nets or buy insecticide treated nets; this makes them much more effective at keeping you safe at night.
- Spray insecticides onto the water where the mosquitoes breed to kill the eggs and the larvae. This in turn reduces the numbers of mosquitoes and so lowers the malaria infection rate. This can be small scale (in your yard) or on big rivers and lakes.
- Take antimalarial medicines that kill the malaria parasites inside the human body.

One big problem with chemical control methods is **resistance** (see antibiotic resistance – this is the same principle). Malarial parasites are becoming resistant to antimalarial drugs, and mosquitoes are becoming resistant to most of the widely used insecticides.

Sanitary controls

Keeping our environment clean and tidy, and dealing with human waste effectively reduces the chances for mosquitoes to breed. See Figure 19.10.

- Remove as much standing water as possible, as these are mosquito breeding grounds. Mosquitoes will lay eggs in any standing water – in a garden pond, old tyres, flower pots, old drink cans, and so on. Make sure you store rubbish out of the rain, and dispose of it properly.
- Dispose of sewage correctly. Managing human waste so that foul water is not left around will reduce the breeding places for the mosquitoes.

Figure 19.10 Conditions like these are ideal breeding grounds for mosquitoes, but with just a little effort we can get rid of the water and reduce the number of mosquitoes by stopping them from breeding.

Parasites and vectors

Biological controls

Scientists are developing biological weapons against mosquitoes to destroy them. These are expensive, and in many cases experimental, but they may eventually be the way we get rid of malaria and other mosquito-borne diseases. Methods of biological control include:
- developing fungi and bacteria that will attack and destroy mosquito as larvae in the water (see Figure 19.11) or as adults
- introducing fish that eat mosquito larvae in very large numbers
- genetic modification of mosquitoes to make them infertile so they cannot breed (see Chapter 14 for details of genetic modification).

Figure 19.11 Coloured scanning electron micrograph (SEM) of the *Metarhizium anisopliae* fungus covering the body of a mosquito *(Anopheles spp.)* larva, and killing it.

In addition, towards the end of 2021, a new anti-malaria vaccine was used in sub-Saharan Africa to protect children from malaria, alongside traditional measures such as bed nets. If this is successful, it may eventually be used in the Caribbean as needed.

Also, as you know from Chapter 14, people who are heterozygous for sickle cell anaemia are protected from the worst effects of malaria.

Mosquitoes and dengue

Dengue is another mosquito-borne disease caused by a virus. In fact there are four slightly different types of virus that can cause dengue, so in theory you can get slightly different versions of it four times – Dengue strain I, strain II, strain III and strain IV! The mosquito vector that takes the disease from host to host is *Aedes aegypti*. The viruses it carries have very simple names: DENV-1, DENV-2, DENV-3 and DENV-4.

The dengue virus does NOT spend any of its life cycle in the mosquito. It is simply a passenger in the mouthparts between one human and the next (see Figure 19.12 and compare it with Figure 19.7).

Dengue can develop into a more severe form called dengue haemorrhagic fever, which causes bleeding from the nose, gums and under the skin.

The disease occurs mainly in tropical Asia, China, India, the Middle East, Australia, the South and Central Pacific, Central and Southern America, and in the Caribbean, usually during the rainy seasons in areas with high numbers of infected *Aedes aegypti* mosquitoes. Epidemic levels have recently been reported in parts of the Caribbean and Central America.

How is dengue spread?

The life cycle of the mosquito means that the female needs two meals of human blood to provide protein for her developing eggs, and this is when she passes on her load of dengue viruses. If the first feed the mosquito takes is from someone infected with dengue viruses, the viruses remain in the mouthparts of the mosquito. When a mosquito feeds, she passes saliva containing an anticoagulant into the blood. This stops the blood from clotting and blocking up her delicate mouthparts. So the next time she feeds, the viruses pass into the blood of the victim along with the saliva, and someone else is infected with dengue.

> **DID YOU KNOW?**
>
> The term 'dengue' seems to have come from the Caribbean. It first came into use during a Caribbean outbreak of the disease in 1827–28. It comes from an attempt by Spanish speakers to pronounce the Swahili phrase 'kidenga pepo' which means 'a cramp-like seizure caused by an evil spirit'.

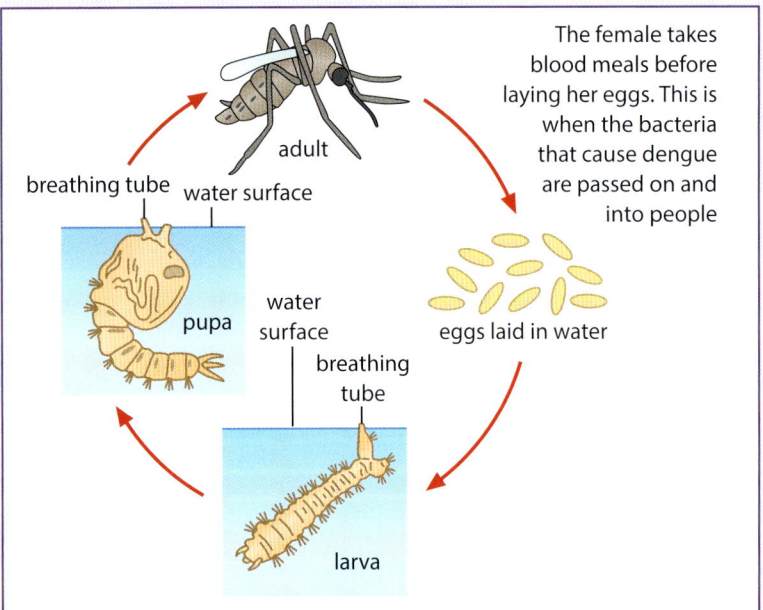

Figure 19.12 The life cycle of *Aedes aegypti*, the mosquito that carries the dengue virus from person to person

> **DID YOU KNOW?**
>
> One hundred million cases of dengue are reported every year by the World Health Organization (WHO), which makes it one of the most important viral diseases in the world.

Signs and symptoms of dengue

Dengue fever is characterised by:
- the rapid development of a fever that may last from 5–7 days
- intense headache
- joint and muscle pain
- a rash (Figure 19.13) that develops on the feet or legs 3–4 days after the beginning of the fever.

The haemorrhagic form of dengue is more severe. It is most common on a second or third infection. Symptoms include:
- loss of appetite
- vomiting
- high fever
- headache
- abdominal pain
- weak fast pulse
- bleeding gums or bleeding under the skin and fluid leaks from the blood vessels.
- shock and circulatory failure may occur.

Untreated, Dengue haemorrhagic fever results in death in up to 50% of cases.

Dengue may occur from 3–14 days after an infected mosquito has bitten you, but commonly symptoms appear within 4–7 days. Infection with one of the four strains of dengue virus usually produces immunity to that strain, but not the other strains. Most people recover from dengue, but they can be more seriously affected if they are infected again by one of the other strains.

The pain from dengue may be intense – its old name was 'break-bone fever'.

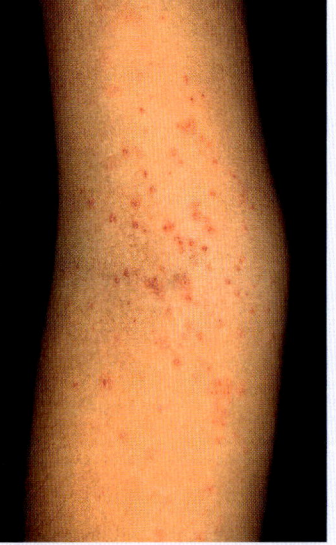

Figure 19.13 Dengue fever causes a rash that develops on the feet or legs after the fever starts.

Parasites and vectors

Treatment of dengue

As it is a viral disease, there is no specific treatment available for dengue fever.

- Bed rest and plenty of fluids are important.
- Paracetamol can be taken as a painkiller. Do not give aspirin or ibuprofen to patients with the haemorrhagic form as this will make the bleeding worse, because it has a thinning effect on the blood.
- Intravenous fluids and oxygen therapy are often used for patients who experience shock during their illness.

Prevention of dengue

The main methods of controlling and preventing dengue involve preventing ourselves from getting bitten. The same chemical, mechanical, sanitary and biological methods used to control *Anopheles* mosquitoes and prevent malaria are used to control the *Aedes aegypti* mosquitoes to prevent dengue. Simple measures such as using insect repellent, mosquito netting and minimising standing water are still the most effective and cheapest ways of overcoming dengue for most of us.

Some parts of the Caribbean are experiencing epidemic levels of dengue (see Figure 19.14), so we need to implement as many control methods as possible to protect ourselves. Unfortunately, these are the only options available as there is currently no vaccine to protect us against dengue.

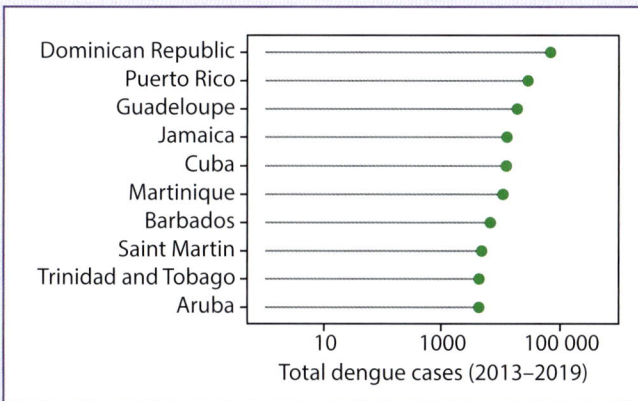

© Centre on Climate Change and Planetary Health, London School of Hygiene & Tropical Medicine, London, United Kingdom, Centre for Mathematical Modelling of Infectious Diseases, London School of Hygiene & Tropical Medicine, London, United Kingdom, Department of Geography, University of Florida, Gainesville, Florida, United States of America, Emerging Pathogens Institute, University of Florida, Gainesville, Florida, United States of America School of Life Sciences, University of KwaZulu-Natal, Durban, South Africa, The Caribbean Institute for Meteorology and Hydrology, St. James, Barbados, Caribbean Public Health Agency, Port of Spain, Trinidad & Tobago, Facultad de Ingeniería Marítima y Ciencias del Mar, Escuela Superior Politécnica del Litoral (ESPOL), Guayaquil, Ecuador, 9 Inter-American Institute for Global Change Research, Montevideo, Department of Montevideo, Uruguay, University of York, UNITED KINGDOM. Copyright © 2020 Lowe et al.

Figure 19.14 Dengue affects many lives across the Caribbean, so it is important to prevent as many cases as possible.

Checkpoint questions

3. Explain why dengue fever was known as 'break-bone fever'.
4. Explain how mosquitoes act as vectors for diseases such as dengue.
5. Summarise the main ways of controlling and reducing diseases spread by mosquitoes.

Mosquitoes and zika

Zika is another disease transmitted from person to person by a mosquito vector.

It has spread to the Caribbean from South America. The biggest problem with the zika virus is not for the person who is infected. The zika virus causes problems for a developing foetus. It affects the brain and the baby may be born with an unusually small head, known as **microcephaly** (see Figure 19.15), and have learning difficulties throughout its life. It is very important that pregnant women do not get bitten by infected mosquitoes. The virus remains in the body for some time after you recover from the symptoms of the disease, so it is also important to avoid mosquito bites when you are planning to have a baby.

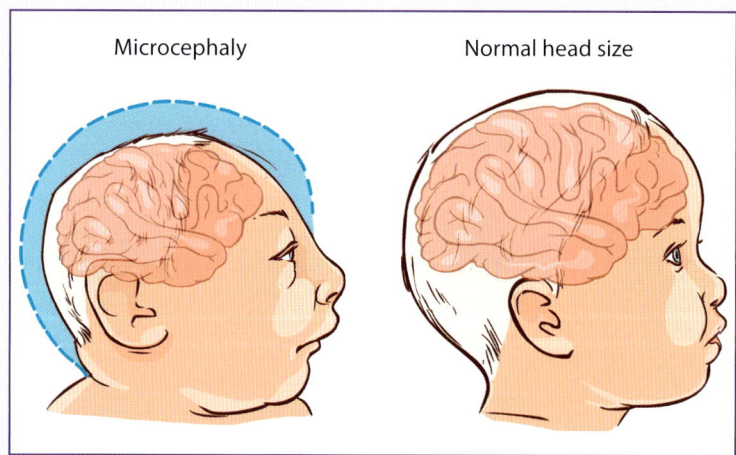

Figure 19.15 The impact of the zika virus on the brain and head size of during pregnancy is plain to see in these diagrams.

How is zika spread?

As with dengue, if a mosquito takes a blood meal from a person infected with a zika virus, it will pass the virus on in its saliva to the next person it bites. The virus does not complete any of its life cycle in the mosquito – it is simply transmitted from person to person.

Signs and symptoms of zika

The effect of zika on healthy adults is relatively small. Some people have no symptoms at all. Others have relatively mild symptoms, which last from 2–7 days. These symptoms may include:

- a high temperature
- sore, red eyes
- headache
- joint and muscle pain
- a rash and itchy skin.

It is easy to ignore the symptoms, but if you are pregnant, or get pregnant soon after infection, it could be life or death for your developing baby.

Treatment of zika

Zika is caused by a virus so there are no specific medicines to treat it. Follow these guidelines to help your body recover as fast as possible:

- Rest.
- Drink lots of fluids.
- Take simple pain relief such as paracetamol.

If you are pregnant, it is helpful to see medical staff and have a scan to see if your baby is developing normally.

Prevention of zika

Methods of controlling and preventing zika involves controlling mosquitoes, using the same biological, chemical, mechanical and sanitary controls that work against dengue and malaria. For most of us, using insect repellents, sleeping under mosquito netting (Figure 19.17) and minimising standing water will help us to avoid this disease and protect our unborn children.

Figure 19.16 shows the data on zika cases in the Caribbean from 2013–2019. There is no vaccine to protect us against the zika virus.

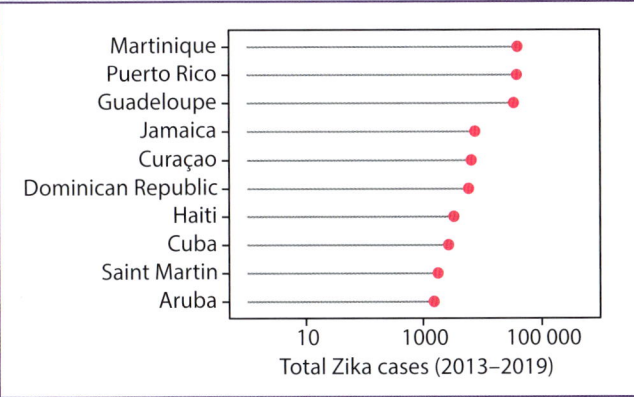

Figure 19.16 Data on zika cases across the Caribbean from 2013–2019. This does not show us how many unborn babies were affected.

Figure 19.17 Sleeping under mosquito nets and using insect repellent are good protection measures to prevent being infected with this disease.

Mosquitoes and chikungunya

Chikungunya is yet another viral disease where mosquitoes act as vectors, carrying the virus from one person to another.

Chikungunya has spread over a large area of the tropical world since it was first seen less than 100 years ago – see Figure 19.18. It only reached us in the Caribbean in 2013 but it has since spread rapidly across our islands and the Americas, carried by mosquitoes that travel freely from country to country.

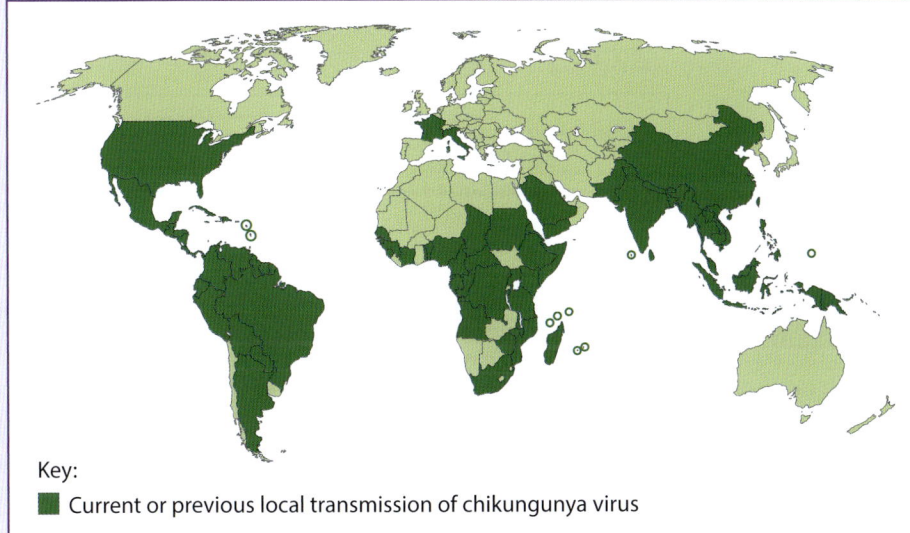

Figure 19.18 Countries or territories where chikungunya has been reported, as of October 2020

Mosquitoes as vectors of disease

How is chikungunya spread?

It is carried by the same mosquito species that spreads dengue – *Aedes aegypti* – and another species of mosquito, *Aedes albopictus*. The mosquito acts as a vector in the same way. The female needs at least two blood meals before laying eggs. If she bites a person infected with the chikungunya virus (Figure 19.19) in her first blood meal, she will transfer the virus in her saliva to the second person she bites.

Signs and symptoms of chikungunya

Almost everyone infected with the chikungunya virus develops symptoms that normally develop 3–7 days after infection. The main symptoms are:

- fever
- joint pain.

Less common symptoms are:

- joint swelling
- headaches
- rash
- muscle pain.

Most people infected with chikungunya feel better within a week, but some people suffer severe joint pain for several months. It is particularly serious for very young babies, older people and people with health problems such as diabetes, high blood pressure or heart disease – unfortunately that is a high proportion of many Caribbean populations. Chikungunya rarely causes death.

Treatment of chikungunya

As with most viral diseases, there is no specific treatment.

Prevention of chikungunya

Methods of controlling and preventing chikungunya involve using the same biological, chemical, mechanical and sanitary controls that work against malaria, dengue and zika.

Figure 19.20 shows the data on chikungunya cases in the Caribbean from 2013–2019. There is no vaccine to protect us against chikungunya.

The vectors we have looked at so far have all been insects, but almost all types of organism can act as vectors. The next disease we will be studying is carried by mammals, the group of animals to which we ourselves belong.

Figure 19.19 A scanning electron micrograph (SEM) of cells infected with the chikungunya virus. The *Aedes* mosquito carries the virus and infects humans when it feeds.

© Centre on Climate Change and Planetary Health, London School of Hygiene & Tropical Medicine, London, United Kingdom, Centre for Mathematical Modelling of Infectious Diseases, London School of Hygiene & Tropical Medicine, London, United Kingdom, Department of Geography, University of Florida, Gainesville, Florida, United States of America, Emerging Pathogens Institute, University of Florida, Gainesville, Florida, United States of America School of Life Sciences, University of KwaZulu-Natal, Durban, South Africa, The Caribbean Institute for Meteorology and Hydrology, St. James, Barbados, Caribbean Public Health Agency, Port of Spain, Trinidad & Tobago, Facultad de Ingeniería Marítima y Ciencias del Mar, Escuela Superior Politécnica del Litoral (ESPOL), Guayaquil, Ecuador, 9 Inter-American Institute for Global Change Research, Montevideo, Department of Montevideo, Uruguay, University of York, UNITED KINGDOM. Copyright © 2020 Lowe et al.

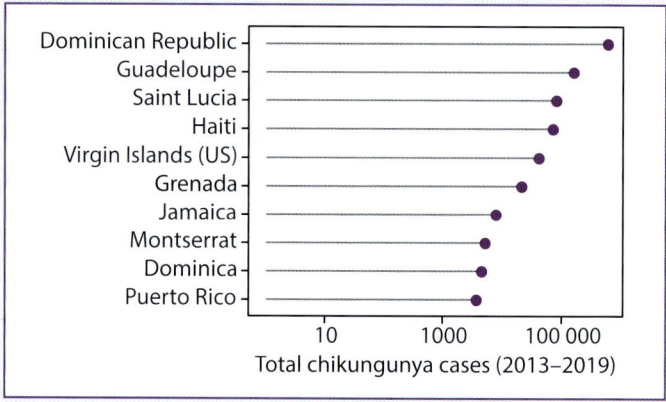

Figure 19.20 Data on chikungunya cases across the Caribbean from 2013–2019; it was almost unheard of in the Caribbean before 2013.

Rats and leptospirosis

Leptospirosis is a bacterial disease that ranges from a collection of flu-like symptoms to a fatal infection causing kidney and liver failure with internal bleeding. The bacterium is passed onto humans by infected animal vectors, the most common being rats. (see Figure 19.21)

How is leptospirosis spread?

The bacteria that cause leptospirosis infect rats and other animals. The bacteria are passed out in the urine and get into water. Rats urinate a lot to mark their territories, so any water – rivers, streams, ponds, puddles – is potentially contaminated with rat urine and most rats are infected with leptospirosis. If water containing the bacteria gets into someone's mouth, nose or eyes, or into a cut on the skin, the bacteria can enter their body and infect them.

Figure 19.21 Rats spread many different diseases – leptospirosis is just one of them.

Wherever humans settle, rats soon seem to arrive. They feed on human waste – bodily waste or food waste – and they breed very quickly.

Almost anyone is at risk from leptospirosis. Some people are at risk through their jobs, such as farmers, veterinarians and builders. Others put themselves at risk through leisure activities such as water sports and camping. Some of us are at risk simply as part of our everyday activities – keeping pets or other domestic animals, back-yard gardening, washing clothes in streams, or living in a rodent-infested area. Children are particularly at risk because they love to play and splash around in any little bit of water they can find (Figure 19.22). In fact, in some outbreaks, up to 40% of cases of leptospirosis have been in children under 15 years of age. So almost anyone is at risk of infection from a number of different sources.

Figure 19.22 Children playing in water are just having fun – but they are at risk of infection by leptospirosis, passed into the water in the urine of rats.

Symptoms of leptospirosis

Symptoms of leptospirosis range from relatively mild to very severe. About 10% of all the people infected with leptospirosis will die. Mild symptoms include:
- fever
- muscle aches
- chills
- nausea and vomiting
- inflamed, reddened eyes.

As you see, these could be the symptoms of many different diseases. Severe symptoms include:
- liver damage and jaundice (the skin and the whites of the eyes turn yellow)
- kidney failure
- internal bleeding.

Treatment of leptospirosis

Since leptospirosis is caused by bacteria (see Figure 19.23), it is treated using antibiotics. An early diagnosis is important, followed by immediate antibiotic treatment. The diagnosis can be confirmed by a blood test, but it is a good idea to start antibiotic treatment before the blood test results arrive.

Simple painkillers also help you to feel better. If you become seriously ill you will need hospital treatment.

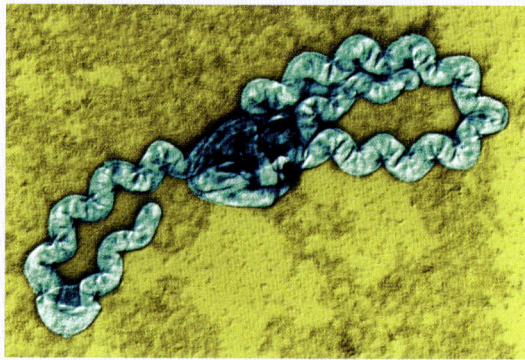

Figure 19.23 A transmission electron micrograph (TEM) of the *Leptospira* bacterium, which causes leptospirosis. Antibiotics are used to treat an infected person.

Prevention of leptospirosis

The spread of leptospirosis can be prevented by:
- implementing tight rodent control; if we get rid of the rats, we have removed the main source of human infection
- vaccinating domestic animals against the infection
- carefully disinfecting contaminated work areas
- prohibiting swimming in contaminated waters
- minimising or avoiding contact with potential sources of infection by wearing protective garments, including waterproof boots, gloves and eye protection when at risk of contact
- educating the public, which is vital; if mothers realise that their children are at risk if they play in water, then they can reduce the risk and look out for early symptoms of leptospirosis.

Checkpoint question

6 List the ways in which leptospirosis can be contracted.

Possible SBAs based on lifestyle diseases
NB these are largely project- or research-based

→ Make a large model of the life cycle of a housefly and indicate the different ways it spreads diseases (research some of the diseases known to be carried by houseflies).

→ Investigate and write a long article on forensic entomology and how it is used in solving crimes.

→ Malaria – either produce a podcast or programme about malaria – how it is spread, how it impacts lives globally and how it is prevented/controlled; OR investigate the Caribbean programmes that have eliminated malaria/are working to eliminate malaria.

→ Find out about and produce educational resources on zika. Target your material at young people who might be planning a family.

→ Develop an educational programme for your local primary school to encourage children to remove potential breeding grounds for mosquitoes around their school and community.

→ Either produce a display on the main methods of control of vector-borne diseases explaining how each works; OR investigate one method of vector control and identify data showing its effectiveness.

→ Investigate and write an account of an outbreak of zika virus and the consequences in terms of babies born with microcephaly.

→ Investigate and write an account of the spread of chikungunya across the Caribbean since 2013.

End-of-chapter summary

In this chapter, you have learnt that:

- a vector is an organism that transmits disease-causing microorganisms from one host to another
- houseflies are vectors of around 100 different diseases including gastroenteritis, cholera and dysentery; they breed and feed on human waste of all sorts, dead animals and garbage; to control their diseases carried by flies, we use chemical, sanitary and mechanical control methods
- mosquitoes are vectors of many different diseases including malaria, dengue, zika and chikungunya
- each disease has recognisable signs and symptoms
- malaria is caused by a protozoan; dengue types I–IV, zika and chikungunya are caused by viruses
- antimalarial drugs help to treat malaria but there are no effective treatments for the other diseases
- there is a vaccine that is helping to prevent the spread of malaria but there are no vaccines against dengue I–IV, zika or chikungunya
- ways of preventing mosquito-borne and other vector-borne diseases include biological controls, chemical controls, mechanical controls and sanitary controls
- rats are vectors of many diseases including leptospirosis; this is a bacterial disease that is spread through water contaminated with rat urine; it can be cured by antibiotics if it is caught early, but the severe form means it is fatal in 10% of cases.

End-of-chapter questions

1 Which of the following diseases is not transmitted by a vector?
   A   Leptospirosis
   B   Influenza
   C   Malaria
   D   Chikungunya (1)

2 One of these methods is not widely used in the control of leptospirosis. Which **one**?
   A   Biological control
   B   Chemical control
   C   Mechanical control
   D   Sanitary control (1)

3 Which **one** of the following is not a symptom of dengue fever?
   A   Reddened eyes
   B   Headache
   C   Fever
   D   Rash (1)

4 Which **one** of the following animals is the vector for malaria?
   A   *Aedes aegypti* mosquito
   B   Housefly
   C   *Anopheles* mosquito
   D   Flea (1)

5 Approximately how many diseases are carried by a housefly?
   A   50
   B   100
   C   150
   D   200 (1)

6 Which of the following measures can be used to reduce the spread of leptospirosis?
   i   Stop children playing in standing water.
   ii   Making sure at-risk workers wear protective clothing.
   iii Set up a rat extermination programme.
   iv   Use insecticide in the home.

   A   i, ii and iii
   B   i, ii and iv
   C   i, iii and iv
   D   ii, iii and iv (1)

7 Which **one** of the following diseases can be treated using antibiotics?
   A   Chikungunya
   B   Leptospirosis
   C   Dengue fever
   D   Gastroenteritis (1)

Parasites and vectors

8 Rats are the main source of infection of which of the following?
- A Malaria
- B AIDS
- C Leptospirosis
- D Rabies (1)

9 In one of these diseases, the pathogen spends part of its life cycle in the vector. Which **one**?
- A Zika
- B Malaria
- C Dengue
- D Chikungunya (1)

10
- a Define the term 'vector'. (1)
- b Distinguish between biological and mechanical vectors. (4)
- c Name **five** Caribbean vectors, giving an example of **one** disease in each case that they are responsible for transmitting. (10)

11 Describe the stages of the life cycle of the housefly. Explain how it spreads disease and how it may be controlled. (20)

12
- a Name a mosquito-borne disease you have studied. (1)
- b What is the causative agent? (1)
- c What is the name of the mosquito that spreads the disease? (1)
- d Describe the symptoms of the disease. (5)
- e How is the spread of this disease controlled? (2)

13
- a Name **four** diseases carried by mosquitoes. (4)
- b Make a table to compare the symptoms of these **four** diseases. (10)
- c In vector-borne diseases we try to prevent the disease by controlling the vector. Define and describe the following methods of controlling mosquitoes as vectors of disease. Give examples where possible. (12)
    - i Chemical control
    - ii Mechanical control
    - iii Sanitary control
    - iv Biological control

End-of-chapter questions 429

Chapter 20: Preventing disease

When you have completed this chapter, you will be able to:
- describe the impact of the Caribbean lifestyle on health
- explain the changes that can lower the risk of suffering non-communicable diseases
- understand the choices that lead to a reduction in infectious diseases
- summarise the importance of food hygiene in the prevention of food-related diseases
- explain the importance of personal hygiene in maintaining health
- explain how personal hygiene is maintained

What the examiners say

→ Candidates seem to lack the depth of understanding required for this topic. Most responses appear superficial, or include unscientific and ineffective home remedies.

Revision tip

Be unavailable! Turn off your phone and other devices so that you are not tempted to chat with your friends when you should be revising!

The preceding chapters have shown us that non-communicable diseases have overtaken infectious diseases as the main causes of death in the Caribbean. Nutrition-related chronic diseases such as obesity, diabetes and hypertension are responsible for much of the disability, illness and death in our region.

Throughout this book, we have discovered a number of ways in which sensible lifestyle choices (see Figure 20.1) can reduce the risk of developing many different types of diseases. This chapter draws together much of this advice and discusses the adoption of a healthy lifestyle and diet as methods of preventing a wide range of diseases.

Figure 20.1 A healthy lifestyle and diet, with exercise, can prevent a wide range of diseases, and be fun too!

Healthy diet

One of the large factors that affects our health in many ways is our nutrition. There are many cultural practices and popular myths that are important to us in the Caribbean. Have you ever heard any of the following?

- Cold water makes you fat.
- One meal a day is a good diet.
- Margarine is not fattening.
- Pregnant women must eat for two.
- You only put on weight if you eat sugary and fatty foods.

None of these statements is true. Nutrition was discussed in Chapter 5, so we know that the energy requirements of individuals vary throughout their lives and depend on a number of factors, including genetic factors, the balance of muscle and fat tissue in the body, and activity levels. If anyone consistently takes in more food than they need, the excess will be stored as fat. A certain amount of body fat is necessary – it cushions the organs of the body and provides an energy store for times of ill health, but the levels of fat can rapidly become a health problem in themselves.

Checkpoint question

1. Look at each of the bulleted statements above and use the knowledge you have built up through your course to explain why each of them is not true.

The best way of dealing with diseases such as hypertension and heart disease is to treat the cause (for example, obesity) and not just the effect. To do this we must address the quantity and quality of the food we eat. See Figure 20.2.

Figure 20.2 Enjoying our food and drink is an important part of our Caribbean lifestyle, but by making a few changes, we can continue to enjoy ourselves while living longer and more healthily.

Quality and quantity of food

Too much food and too little exercise can lead to the blood being overloaded with too many food substances such as glucose (sugar), other carbohydrates, fats and salt at any one time. Such behaviour is dangerous. Diabetics, for example, have a problem with the regulation of their blood sugar and overloading their systems can mean that they go into a hyperglycaemic coma. They have too much sugar and not enough insulin to deal with it.

It is often better to have smaller meals, more often, than to eat a large meal at each sitting spaced some time apart. Eating smaller meals more often allows a steadier absorption of the food substances and their nutrients. The less we eat, the less digested food will be absorbed into our bloodstream and carried to the cells of our body. In turn, this results in having less excess food being stored as fat.

Reducing our portions means that after a while our stomach tends to get smaller in size. This is very helpful in controlling our weight because it takes less and less food to satisfy us (for our 'belly to be full'). This eventually results in weight reduction, without us feeling hungry all the time. Reducing our body weight will in turn result in:

- reduced blood pressure
- less risk of heart attacks
- improved insulin action
- better diabetes control.

Figure 20.3 Food like this is delicious, but we need to cut down on the amount we eat for the sake of our long-term health.

The problem is not just the quantity of food eaten. The quality of our diet is also a major health concern. In the Caribbean we eat a lot of fried, fatty foods, such as stews, 'run down', thick gravy and sauces. We also have a culture of eating fast foods like hamburgers (Figure 20.3), patties, fried chicken and chips. **Saturated fats** and grease contribute to raised blood cholesterol levels and blocking of the arteries, as described earlier in Chapter 8 and Chapter 15. These in turn can lead to an increased risk of heart disease, heart attacks and strokes, so it is very important to control the amount of these fats in our diet. Not only do fats increase our risk of heart disease, but because they have a very high energy content, they also contribute to obesity levels. They contain many calories/kilojoules and are easily stored as body fat.

The good news is that reducing and eventually removing these high-calorie/kilojoule foods from our diet will eventually force our body to utilise the excess fats that have been stored, thereby:

- reducing our body weight
- lowering our blood cholesterol level
- improving blood pressure
- improving cardiovascular health
- helping with blood sugar control.

Checkpoint question

2. Explain why it is important to look at both the amount of food and the type of food we eat to be as healthy as possible.

Preventing disease

Food preparation

Much of Caribbean cuisine is highly seasoned. From the famous Jamaican jerk in the north to the curries of Trinidad and Guyana in the south, combined with the Chinese influence, Caribbean people use a great variety of seasoning salts. These salts contain sodium. For about 30% of the population, high salt levels in the diet tend to raise blood pressure and make hypertension worse, and none of us know if we are in that 30% until our blood pressure is too high. Good advice would be to use a lot more of our seasoning herbs while reducing the salt content. See Figure 20.4.

Figure 20.4 If we re-educate our taste buds to enjoy a little less salt, our jerk chicken will still taste good and we'll also be keeping our body healthy.

To promote healthy eating, we need to adapt some of our cultural methods of food preparation. We need to substitute the brown stews, curries, fricassee, fried food, and so on with steaming, baking, boiling and grilling. With a bit of ingenuity, and by resisting the lure of fast food as much as possible, we can continue to enjoy all the best of Caribbean cooking, but eat more healthily as well. That way our risk of obesity, and the diseases linked to it, will be reduced, and what is more, our immune system will be healthier to enable us to fight off infectious diseases.

If someone has a problem with their weight, it is not enough simply for them to cut down on what they are eating. It is very important to eat a balanced diet in sensible-sized portions, but to keep our weight under control and develop a healthy cardiovascular system we also need regular physical activity.

Benefits of physical exercise

Exercise does not necessarily reduce body weight but it does reduce body fat and increase the rate at which we break down food and use energy – our basal metabolic rate. One of the immediate benefits of exercise is to increase blood flow through the muscles, which removes the carbon dioxide produced as our muscles work. It also increases heartbeat, improves circulation and ventilation of the lungs. (See Chapters 7 and 8 for details of the impact of exercise on the heart, lungs and circulation.) Exercise doesn't have to mean lots of sport. Everything from housework and gardening to swimming, brisk walking, jogging, dancing, cycling and playing games such as football and cricket have similar benefits. Joining friends and family to exercise is a good idea – it is much easier to stick to regular exercise if we are enjoying it with people we know. It becomes a social occasion. See Figure 20.5.

As a way of tackling diseases such as hypertension, heart disease and diabetes, the type, duration and frequency of exercise should be tailored to the ability and health status of the person. A very overweight man with heart problems and high blood pressure would probably die of a heart attack if he tried to play squash, but would get a lot of benefit from regular short walks.

Figure 20.5 Exercise is for all ages and there are lots of ways to enjoy it.

Exercise is an extremely enjoyable leisure pursuit for many people and we are more likely to exercise if the activity is enjoyable. So it is important to find a form of exercise that we enjoy and which suits us and our lifestyle.

Exercise improves the fitness of our heart (see Figure 20.6) and the capacity of our lungs, and it burns up energy so it helps us to lose weight. This in turn improves our blood pressure and can reduce the effects of Type II diabetes if we are very overweight. In fact, by increasing our exercise, losing weight and watching our diet, we may get rid of Type II diabetes symptoms completely.

Exercise also has important effects on our musculoskeletal system. Using our joints helps to keep them flexible, and weight-bearing exercise helps to reduce the risk of developing osteoporosis. This is a disease that affects people, particularly women, from middle-age onwards, where the bones become weak and brittle. The more we exercise, the less likely we are to lose bone mass.

Another important area that we have considered in this section is the control of infectious diseases. This is very important in improving the health of the population of the Caribbean. A great deal has been done in recent years, but there is still more we can do to tackle the scourge of some of the infections that reduce life expectancy and make life a misery for thousands of people in the region.

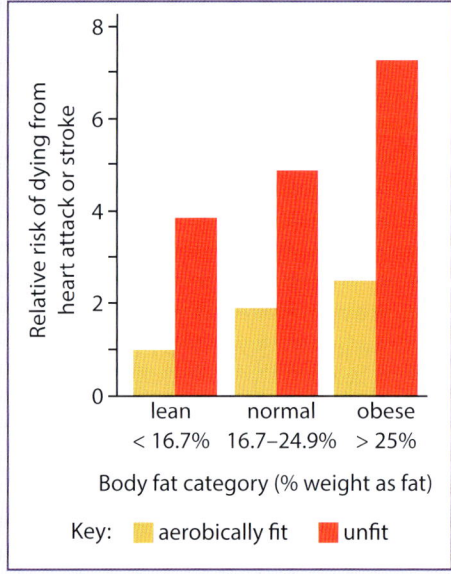

Figure 20.6 The risk of dying of a heart attack or stroke is closely linked to your percentage body fat. The amount of exercise you take – or how fit you are – also has an effect. Fit, obese people are less likely to have a heart attack than thin, unfit people. This gives you two ways of improving your health – watch your weight and get fitter!

Checkpoint question

3 List the ways in which exercise improves our general health.

Controlling infectious diseases

In this section, we have looked at many different infectious diseases and the steps we must take to prevent their spread.

Food hygiene – taking care in storing and preparing food (Figure 20.7) – can greatly reduce the incidence of diseases such as gastroenteritis, as you saw in Chapter 17. Clean water supplies and good sewage disposal are vital too, and in the next section of this book, we will look in more detail at how we deal with these issues in the Caribbean.

Alongside all of these measures, which are used to help combat infectious diseases, it is very important that we take good care of our body. We have already seen the importance of good food, exercise and sensible lifestyle choices in maintaining good health. But we also need to keep our body clean and fresh, to make it as difficult as possible for pathogens to invade our body and make us ill.

Figure 20.7 Washing fresh fruit and vegetables carefully can remove bacteria and reduce incidences of diseases such as gastroenteritis.

Preventing disease

Personal hygiene

Personal hygiene, such as bathing, is very important in Caribbean culture (Figure 20.8). It is expected that we will wash our body at least every day and use deodorants to stop body smells. Good personal hygiene is important socially. No one likes to be too close to someone who smells unpleasant, but it is also one of the most effective ways we have to protect ourselves and others from illness.

Small children tend not to smell even if they do not wash, as long as they keep clean after using the toilet. However, once we have gone through puberty in our early teenage years we find that our body quickly starts to smell if we don't keep it clean. Not all of the smells are automatically unpleasant, but left alone with the bacteria on our skin for too long they quickly become nasty.

Body smells are caused by a number of factors working in combination, including:

- chemicals in sweat, including pheromones, which are made by the body and sexually attract (or repel) other people
- wastes excreted through the skin, such as metabolised alcohol and certain spices
- the actions of bacteria that live on the skin and feed on dead skin cells and the protein that is in our sweat once we are sexually mature – the products of the excretion of these bacteria smell bad
- unwashed clothes, such as underwear and socks, which contain all of the above.

Figure 20.8 The easiest way to keep clean and stay smelling fresh is to shower or bathe once a day.

Bathing

The bacteria on our skin feed on pheromones and other proteins from our sweat, oil from our skin and dead skin cells. The waste products they produce smell unpleasant and they are not good for the skin. If they are left for too long, they can cause irritation and sores may develop. The areas where the sweat is particularly protein rich, and so contain particularly large colonies of bacteria, are the armpits, genital area and feet, so these are generally the smelliest bits of our body.

Regular washing of these areas, at least once a day, is very important both for personal hygiene and general health. We can use chemical deodorants under our arms, which help to prevent our sweat from smelling, but they are an addition, not an alternative to washing.

> **DID YOU KNOW?**
> Some of the bacteria that grow on our feet and make them smelly are from the same group of bacteria that give ripe cheeses their distinctive smell. No wonder our socks and trainers smell bad sometimes!

Female hygiene

Once a girl has gone through puberty it becomes very important that she keeps her genital area clean. The natural body secretions from the vagina are clean but bacteria on the external genitals will feed on the proteins. So no special care of the vagina is needed, other than washing the external genitals. Douches or anything similar should not be put into the vagina as the delicate skin can be damaged.

During menstruation, girls should wash their body, including the genital area, in the same way as they always do. It is important for girls to change their sanitary protection regularly, at least every 4 to 6 hours. This is because the bacteria from the skin will act on the menstrual flow in the same way as they do on sweat, and it will quickly become unpleasant smelling. It is particularly important to change tampons regularly. If they are left for too long, bacteria breeding on them can lead to toxic shock syndrome – a massive infection that spreads through the blood and can kill in hours. Girls should always wash their hands before and after handling tampons or pads.

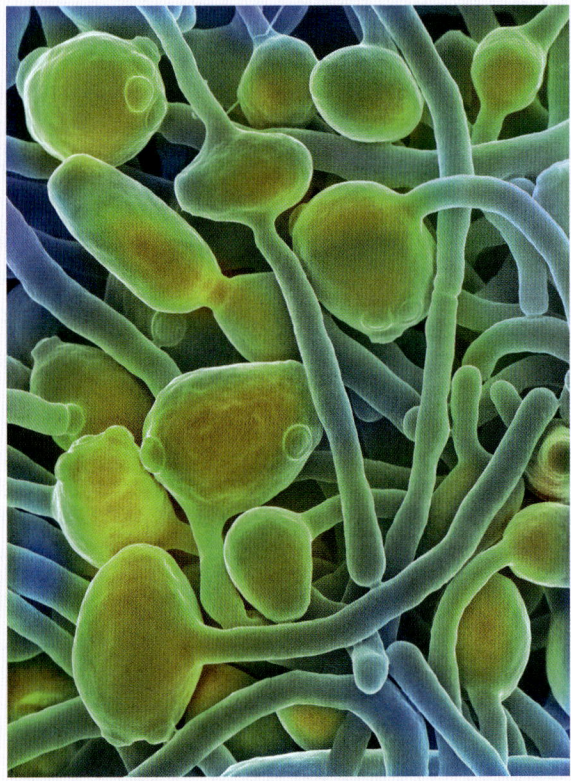

Figure 20.9 A scanning electron micrograph (SEM) of the yeast fungus Candida.

Keeping the genital area clean will also help girls to avoid **cystitis**, a bacterial infection of the bladder. This is a common condition for sexually active young women because sexual intercourse can push bacteria up from the outside the body into the urethra (see Chapter 10 for details of the anatomy of the bladder). Urinating after sexual intercourse can help to flush out any bacteria that may be in the urethra and bladder. However, cystitis is not confined to sexually active people, so keeping the skin clean helps everyone to avoid it.

Another common infection of the female genitals is **thrush**. This is caused by the yeast fungus Candida (Figure 20.9), which is there on healthy skin all the time.

Although it is important to keep the external genitalia clean, some soaps and detergents can irritate the delicate skin of this area and make thrush infections more likely. Mild, unscented soap or just plain water is recommended when washing, and unperfumed toilet paper for toilet hygiene.

Some women find that they get thrush when they use antibiotics. This is because the antibiotics destroy the 'friendly' bacteria in the vagina, which normally maintains a natural balance. This changes the pH of the area and creates the right environment for the yeast infection to develop. Wearing cotton underwear (Figure 20.10) in preference to tight, synthetic underwear is thought to help if thrush develops. The cotton absorbs the sweat and keeps the area dry, making it difficult for thrush to thrive. Thrush is easily treated using anti-fungal medicines in the form of tablets and ointments, which we get from a doctor or pharmacist.

Figure 20.10 Cotton underwear is the healthier option as it helps to prevent infections.

Preventing disease

Male hygiene

Men also need to make sure their genitals are washed regularly. A build-up of smegma, made up of secretions of the foreskin and dead cells, can form under the foreskin of uncircumcised men. This is very useful as it acts as a lubricant, but eventually bacteria will act on it and it will become unpleasant smelling. Uncircumcised males should gently pull back the foreskin when they have a shower and clean the area with water. Soap can be used, but must be rinsed off well. Male circumcision, when the foreskin is removed from the penis, can make personal hygiene easier. Often this is carried out when a boy is still a baby, for religious reasons or as a matter of family choice or tradition.

Washing the hands regularly is also very important (Figure 20.11). Many infections, such as colds and gastroenteritis, are caught when we put our unwashed hands with germs on them to our mouth. Some infections are caught when other people's dirty hands touch the food we eat. Hands and wrists should be washed with clean soap and water, using a brush if the fingernails are dirty. Dry the hands with something clean, such as paper towels or hot air dryers. We should always wash our hands after using the toilet, before making or eating food, after handling dogs or other animals (Figure 20.12) and if we have been around someone who is coughing or has a cold.

Figure 20.11 Keeping our hands clean is a big part of personal hygiene.

Checkpoint question

4 Explain why personal hygiene becomes more important after puberty.

It isn't enough just to wash our hands. We should also keep our nails short and clean. It helps to prevent the spread of diseases such as threadworms. All sorts of pathogens can get into the dirt under our nails – material from our visits to the toilet, scraps of raw meat from food preparation, dead skin cells and mucous from our nose, so it is very important that we clean our nails regularly rather than putting the bacterial cocktail into our mouth!

Keeping our body and hands clean is important, but it is not the whole story. We need to wash our hair regularly with an effective shampoo (Figure 20.13). Oil from our skin, dead cells and dust can build up in our hair and make it smell. What is more, clean hair and skin mean we are less likely to pick up skin infections.

Figure 20.12 Animals are often our friends and companions, but they still carry many pathogens and we should always wash our hands thoroughly after handling them.

Figure 20.13 Washing hair regularly reduces the likelihood of picking up skin infections.

> **Activity 20.1**
>
> ## Investigating the effect of pH on enzyme activity
>
> Produce a big poster to educate people in your community about the importance of good personal hygiene.
>
> Think carefully about the message or messages you want to get across, and then design or make a poster to display in school or at the local doctor's rooms or pharmacy to encourage people in your area to take personal hygiene more seriously.

Dental hygiene

As you saw in Chapter 6, it is very important to take good care of our teeth. Good dental hygiene includes regular brushing and flossing (Figure 20.14). If we fail to brush and floss regularly, we put ourselves at risk of tooth decay and infections of the gums and mouth. These not only cause us pain and may lose us our teeth, they also cause bad breath. Most people have bad breath first thing in the morning because saliva is not made while we sleep. Saliva helps to cleanse the mouth of the microorganisms that cause bad breath. Mouthwashes, mouth sprays and flavoured chewing gum can make the breath smell better for a while, but if we have a health problem in our mouth, we need to see our dentist.

Our battle against disease – both infectious and non-communicable – takes many forms. It involves a lot of input from our government and community, doctors, nurses and dentists. But in many ways, we can take at least some responsibility for our own health, by making sensible lifestyle choices about our:

- eating habits
- exercise levels
- sexual relationships.

We should have our children vaccinated, avoid drugs that we know put our health at risk and follow some simple, basic hygiene rules. In these ways, we can give our body the best possible chance to remain healthy and deal successfully with any diseases we may still be unfortunate enough to contract.

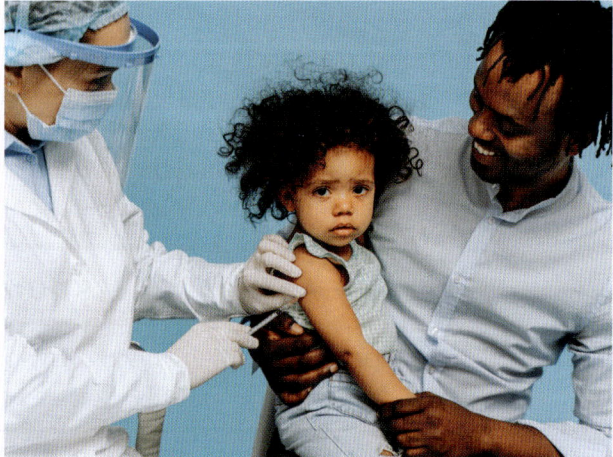

Figure 20.14 There are many different ways to protect ourselves and our children against disease – in many ways, our health is in our own hands.

Preventing disease

Possible SBAs based on preventing disease

→ Develop and produce a game – either a board game or a computer game – on how to stay healthy. It should be fun but, to succeed in the game, players must show that they understand healthy lifestyle choices and how to control infectious diseases.

→ Cooking challenge – try to develop some recipes, based on some of your favourite Caribbean foods, that are healthier options, using relatively little fat or salt in their preparation. Try out the recipes on family and friends, keep a photographic record of the dishes, and so on.

→ How clean are we? Design an anonymous survey to be done on paper or online to find out how often people in your school/community/family really wash their hands/bathe or shower/clean their teeth. Use the results of your survey as the basis of a report explaining the importance of good hygiene and highlighting EITHER on how good we are at maintaining personal hygiene OR how bad we are, depending on your findings.

End-of-chapter summary

In this chapter, you have learnt that:

- people can reduce their risk of both infectious and non-communicable diseases by changing their behaviour
- changes in both the quantity and quality of the food we eat will reduce our risk of obesity and the diseases associated with it, including hypertension, Type II diabetes and heart disease
- increasing the amount of exercise we take will help us to control our weight; it also helps to reduce blood pressure, lower the risk of heart disease and reduce the symptoms of Type II diabetes
- careful food hygiene prevents the spread of many diseases, including gastroenteritis
- personal hygiene is an important part of avoiding disease, and being socially acceptable
- after puberty, the sweat in our armpits, genital area and feet contains more protein, which is acted on by bacteria; the waste products of the bacteria smell unpleasant, so it is important to wash, bathe or shower every day
- genital hygiene is important in females to help avoid infections such as thrush and cystitis, and in males to prevent the build-up of smegma
- dental hygiene is important to maintain healthy teeth and gums, and to avoid bad-smelling breath.

End-of-chapter questions

1. Which of the following steps will not help to prevent heart disease?
   A   Taking more exercise
   B   Eating less food
   C   Eating food that is lower in fat
   D   Buying an exercise bike (1)

2. Which of the following are important in reducing the incidence and spread of infectious diseases?
   i   Vaccination programmes for children
   ii   Vector control programmes in the Caribbean
   iii Washing hands regularly
   iv   Sharing needles between drug users

   A   i, ii and iii
   B   i, ii and iv
   C   ii, iii and iv
   D   i, iii and iv (1)

3. Which **one** of the following is the correct explanation for body odour?
   A   The body produces pheromones that smell unpleasant.
   B   Bacteria break down the protein in sweat and the waste products smell unpleasant.
   C   Sweat smells unpleasant.
   D   None of the above. (1)

4. Why is it important for good health to keep the nails short and clean?
   A   They look nicer.
   B   Dirt and bacteria can be trapped under longer nails and carried to the mouth or passed to other people.
   C   Long nails sap your strength.
   D   It helps to suppress the appetite. (1)

5. a   Explain why it is important to wash your hands before preparing food. (2)
   b   Explain why you should always wash your hands after playing with your pet. (2)
   c   Explain why we are advised to clean our teeth at least twice a day. (2)

6. Explain what is meant by good personal hygiene and describe its role in personal and communal health. (15)

Preventing disease

End-of-section questions

1. a Define the following terms: 'health', 'signs', 'symptoms' and 'disease'. (4)
    b Describe **three** ways of maintaining high standards of personal hygiene. (6)
    c State what an antibiotic is and describe what antibiotics are used for. (3)
    d Explain how artificial passive immunity differs from natural passive immunity. (6)

2. a Some diseases are spread by vectors. Name **four** vectors. (4)
    b Name **one** disease that is transmitted by a vector. Explain how this disease affects the human body. (4)
    c Give **three** ways in which vectors may be controlled. (3)
    d Fill in the labels A–D in the diagram of the life cycle of the *Aedes aegypti* mosquito. (4)

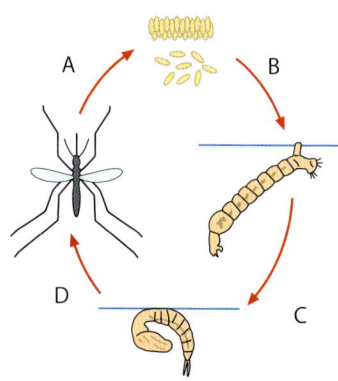

3. a Name **two** legal and **two** illegal drugs that people misuse in the Caribbean. (4)
    b i Describe drug addiction and give **two** different types of addiction (5)
       ii Name **one** drug and explain the negative physical, mental and emotional effect of this drug on users, especially users who become addicted. (8)
    c Name **two** social effects of drug misuse that are prevalent in the Caribbean. (2)

4. a Name **four** sexually transmitted infections that are common in Caribbean countries. (4)
    b i Name and describe the symptoms of a sexually transmitted infection. (4)
       ii Describe **four** methods that can be used to prevent the STI you named in part i. (4)
    c HIV/AIDS is a leading cause of death in the Caribbean. What do you think is the main reason for this? Explain your answer. (3)

5. a Give **five** ways in which infectious diseases are spread between people. (5)
    b Ringworm is an irritating but not deadly disease. Give the cause and main symptom of ringworm. (2)
    c Describe the cause, signs and symptoms, treatment and prevention of influenza. (10)

6. Prevention is better than cure of both infectious and non-communicable diseases. Give **five** lifestyle changes a person can make to ensure better health, and explain how each change helps to keep them healthy. (10)

Data analysis

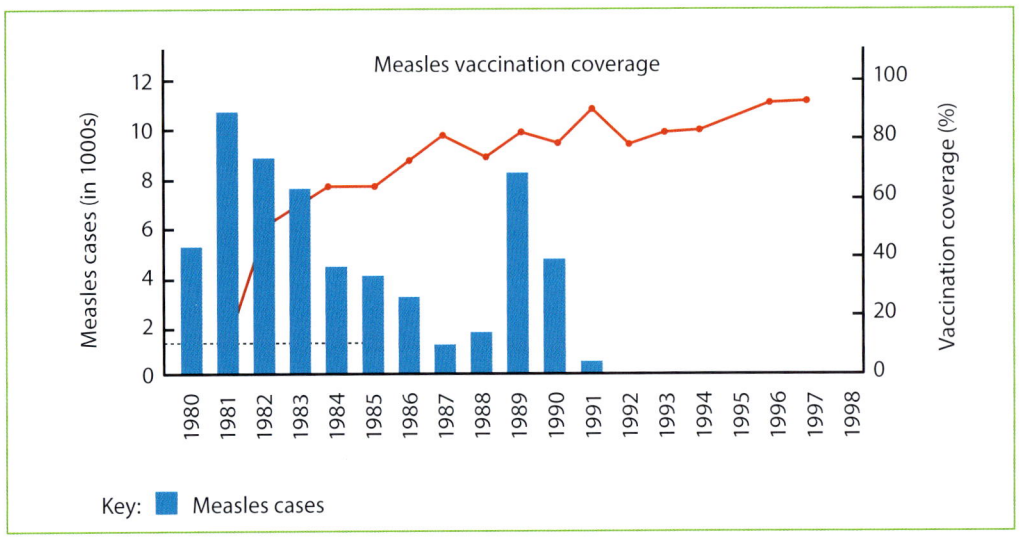

1. Describe what the data shows about the frequency of measles cases from 1981 to 1988. (3)

2. Describe the relationship between the frequency of measles cases and vaccination coverage from 1981 to 1988. (3)

3. Suggest a reason for the rise in the frequency of measles cases in 1989. (3)

4. Explain the effect of the measles vaccine coverage from 1991 onwards on cases of measles in the Caribbean. (4)

Case study

Mrs M was quite slim as a teenager, but loved to eat chocolates, fast food and ice cream. Since she was quite active and in good health, she justified eating these foods by saying 'I will burn them off because I work so hard'. When she was 40 years old, Mrs M started to notice drastic changes. She put on weight and would hear whispers of 'roly poly' when she walked past people on the street. She also felt the need to drink a lot more water than usual, she felt very tired when walking up long flights of stairs and anxious when she did not eat her meals regularly. Her heart also started to beat very quickly at times when doing the mild strenuous tasks she had been used to doing since her 20s. She is now unhappy with how she feels and looks.

1. Suggest causes for the changes that Mrs M has started experiencing. (4)

2. State **four** ways that Mrs M could improve her health. Explain why you have made these suggestions. (8)

3. Describe and explain the possible consequences if Mrs M does not make the changes you have suggested. (6)

Revision tool kit 4

While it is good to know and understand each topic individually, you will have realised by now that the topics do not exist in isolation. Therefore, your revision strategy should take into account that the topics are related to one another.

Also, the examination does not test individual topics, but instead tests your understanding of how you can apply principles to various contexts, what conclusions you can draw, as well as the recommendations you would make based on your knowledge and analysis.

For this reason, it makes sense to work through as many exam-type questions as possible. Not only will you become familiar with the structure of the exam before you actually take the paper, but you will also practise using knowledge, on which much emphasis is placed. Being able to use knowledge shows how well you can make the necessary connections across topics. This is where your analytical and critical thinking skills will be useful.

If you find that a particular question is challenging – do not give up. Ask your teacher or a peer for help, or do some more reading/research, but do not give up.

Section E: The impact of health practices on the environment

Learning outcomes

At the end of Section E, you will:

- appreciate the nature of the relationship between human beings and their environment
- understand that the environment is fragile and there is need to preserve it
- appreciate the contribution of modern technology to the maintenance of good health.

There are some things that we cannot live without, and water is one of them. Fortunately, in the Caribbean we have a plentiful supply of water, in the form of rain, which we can save and use as we need it. Water is constantly evaporating from the surface of the Earth, but is returned in the water cycle. You will be learning about the water cycle later in this section.

As you found out in Section D, many diseases can be carried in water, so it is important that the water we drink is purified. As you will see in the following chapters, the natural water in rivers, lakes and reservoirs can be contaminated in a variety of ways. We need to avoid polluting the water as much as possible, as well as purifying it later.

We humans not only take things from our environment, we put things into it as well. Unfortunately these are not always good things. It is natural for us to produce bodily waste – all animals do – and the waste produced is broken down by other animals, bacteria and fungi over time. But people tend to live close together in towns and cities, and bodily waste can carry disease. We cannot leave our waste lying around to decay naturally, so we need to find effective ways of dealing with our own waste materials. In this final section of the book, you will learn about how we deal with this problem in the Caribbean.

Figure E1 Clean water is one of the most basic needs of life, but just how clean is the water we drink?

Figure E2 Everyone produces rubbish. The challenge is how to deal with it to make the smallest possible impact on our environment.

It isn't only bodily waste that people produce. We produce waste from our factories, from our cars and we also produce huge quantities of household waste. All of this needs to be managed and dealt with, otherwise it can have serious effects on our environment, which in turn can affect our health and well-being. Worldwide, the levels of pollutants from burning fossil fuels have reached record levels, and most scientists believe we are seeing the impact of this in the process called 'global warming'. In this section of the book, you will be looking at some of these major issues, their impact on the Caribbean and how we as individuals and as a society can help.

Chapter 21 Clean water

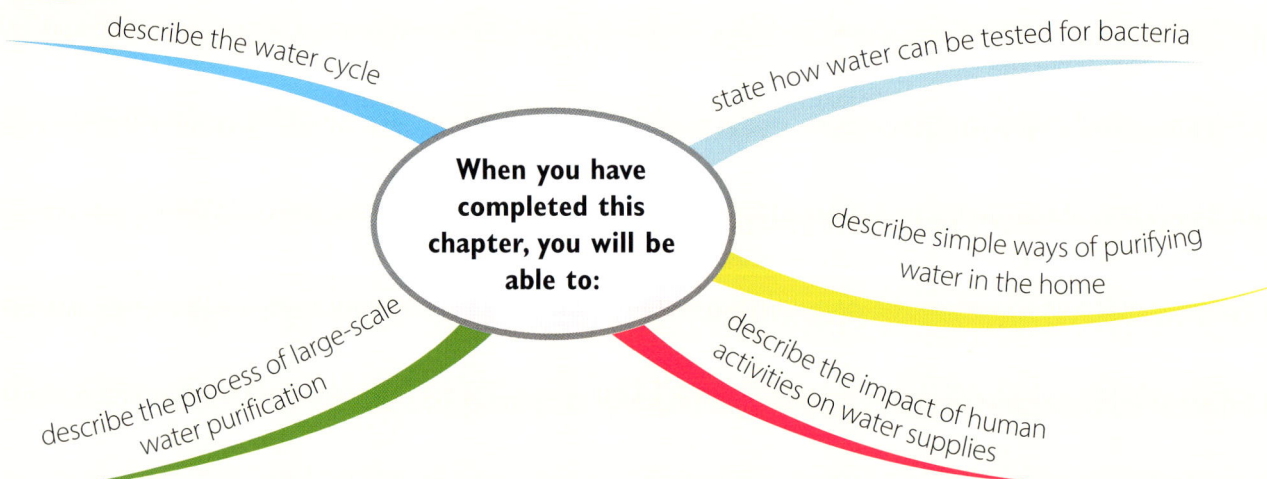

When you have completed this chapter, you will be able to:
- describe the water cycle
- state how water can be tested for bacteria
- describe simple ways of purifying water in the home
- describe the impact of human activities on water supplies
- describe the process of large-scale water purification

What the examiners say
→ To a large extent, candidates provide adequate responses to questions on this topic. However, there are some areas of weakness such as the methods of water purification and their effectiveness.
→ Additionally, candidates often confuse terms because they do not pay enough attention to the context of the question.
→ Candidates are unable to state the differences between sterilisation and disinfection.

Revision tip
Keep track of your progress. Make yourself a revision timetable – and make a record of the topics you have covered.

Water is vital for human life. As you know from earlier in the book, our body is made up of 60–70% water. It is vitally important that the concentration of our cell content, and of our blood and other body fluids, is kept as constant as possible. So, water is always needed. What is more, the water we drink must be clean and free of pathogens if it is to help keep us disease-free. Water that is fit to drink is called **potable**, and providing potable water for all can be a major challenge in the Caribbean, both for governments and for individual families.

There is a very close link between the Caribbean and water. Not only are the islands surrounded by the sea, but even the names of some of the islands and their towns and cities refer to water. For example, Jamaica, is described as the 'land of wood and water', while in Trinidad, Arima means 'plenty water' and Naparima means 'little or no water'. Yet many Caribbean countries are very short of drinking water and have suffered some severe shortages over the years. On average, 25% of us still do not have piped water to our home. In this chapter, we will look at where the water we need comes from and how it is stored, treated and supplied. See Figure 21.1.

Figure 21.1 If clean water is available whenever we turn on a faucet, it is easy to forget how lucky we are. In many parts of the world, a lack of clean water still causes millions of deaths every year. Even in the Caribbean, getting clean water piped to every home is still a challenge for the future.

Where does our water come from?

The water we drink goes through a constant process of recycling between the Earth and the atmosphere, sometimes passing through living organisms on the way. This process is known as the **water cycle**. See Figure 21.2.

- The water in the rivers, lakes and oceans evaporates in the heat of the Sun, forming water vapour that rises into the atmosphere.
- As the water vapour rises, the air is cooler.
- As the water vapour cools, it condenses into tiny droplets of liquid water, which forms clouds.
- The clouds cool more as they rise further, so more droplets form and join together.
- Eventually the droplets get too heavy and the clouds produce rain, replacing the water in the rivers, lakes and oceans.
- Some of this rain filters through soil, rock and layers of coral limestone structures until it reaches a body of rock or sediment that holds the water – these water-filled spaces are called **aquifers** and the water stored in them is called **groundwater**. The process is called **infiltration**.
- Some of the rain simply runs over the surface of the ground down to the seas and rivers. This is called **run-off**.

Living organisms are also involved in the water cycle. Water moves into plant roots by osmosis (see Chapter 2) from the soil and is lost by evaporation from the surface of their leaves through the process of **transpiration** (see Figure 21.3). Animals and plants release water vapour into the air through the process of respiration.

Animals drink water and then pass water out of their body in their urine and faeces. Water evaporates from this in the heat of the Sun. It's strange to think that the water we drink today may have passed through a dinosaur millions of years ago as part of the water cycle then and now.

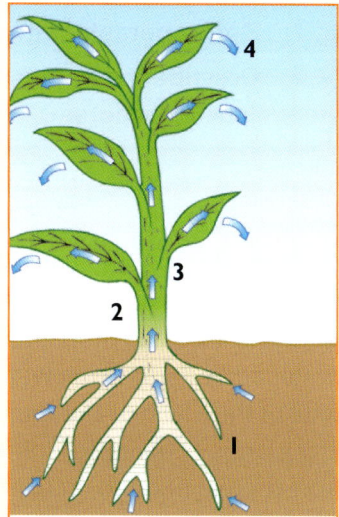

1. Water moves into the roots from the soil by osmosis. It replaces the water that is constantly moving up the stem.
2. Water moves up from the roots into the stem.
3. Water moves up through the stem and into the leaves to replace the water lost by evaporation.
4. Water is lost from the the leaves by evaporation through open stomata.

Figure 21.3 The transpiration stream: This is how plants move water from the soil to their leaves, where it evaporates into the atmosphere.

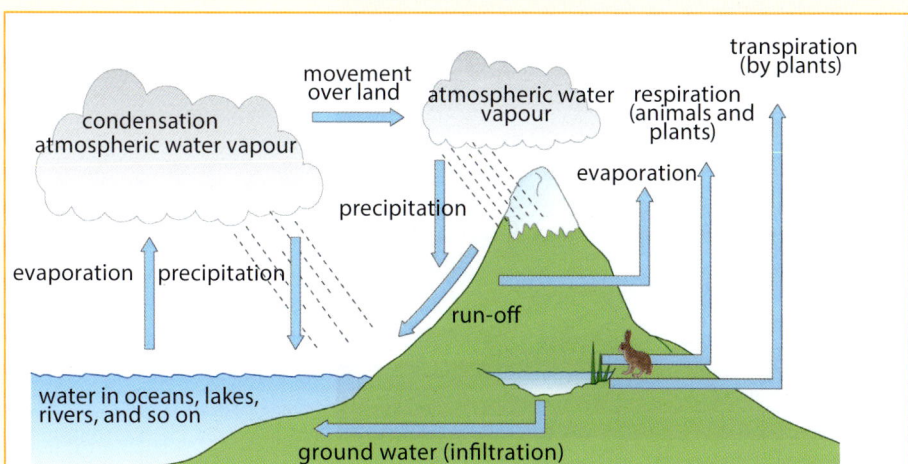

Figure 21.2 The water cycle in nature maintains the levels of water on the surface of the Earth and in the atmosphere.

Where does our water come from? **447**

> **Checkpoint questions**
>
> 1 Define potable water.
> 2 Describe the water cycle in nature through a large, clear and labelled diagram.

Water storage and water supply

The availability of water varies across the countries of the Caribbean depending on the season. Big problems for the region include the fact that most of the water arrives during the wet season, and most of this rain falls on the windward side of the islands – but the water is most needed all year round and all over the islands. Different countries have different ways of storing water and making sure there is enough for everyone all year round. These range from using groundwater, storing surface water in reservoirs, harvesting rainwater in tanks and **desalination** of the sea water all around us – see Figure 21.4.

Contaminated water and its problems

The water that falls as rain isn't always perfectly clean. It falls into lakes, rivers and reservoirs, which may be **contaminated** with bacteria and viruses from the waste products of the animals that live in them and from rotting vegetation. What is more, we humans are good at polluting water with our industries, with our body wastes and with the chemicals we use on our farms (see Chapter 22 for more details of this type of **pollution**).

DID YOU KNOW?

We will take in about 16 000 gallons (72 737 litres) of water in a lifetime – that's about 128 000 pints – if we drink the recommended 5 pints (2.8 litres) of water a day!

Figure 21.4 Where large reservoirs like the Mona reservoir in Jamaica are possible in the landscape, they play an important role in the drinking water supply. But it's important to remember that the water in a reservoir isn't safe to drink without treatment!

The water from a deep well should be relatively easy to make safe to drink, but care always has to be taken that wells are not contaminated by run-off water – the water that drains from agricultural fields or from housing. Sensible positioning and design of the well should reduce the risk of contamination. You also have to be careful with the positioning of **pit latrines** – you certainly do not want contamination from them in your water supply. You will find out more about this in the next chapter.

Drinking contaminated or polluted water may cause serious health problems as a result of infectious diseases such as gastroenteritis or cholera (see Section D), so as far as possible, the water that we drink needs to be treated to make sure it is free from disease-causing organisms. How do we set about turning water from a lake, river or reservoir into potable water?

On a large scale, this is done in water treatment plants, but it can also be done on a small scale in the home. However the water is treated, it is important to remember that it is clean but not pure. All the water we drink has many minerals and chemicals dissolved in it. The only way to get completely pure water is to make distilled water – when water is boiled and the vapour produced is collected and then cooled to condense it. This is completely unnecessary for drinking, but it is needed for many chemical reactions.

Checkpoint question

3 Explain why water should be treated before we drink it.

Home purification methods

There are many places in the Caribbean, particularly in rural areas, where no piped water is available. In these areas, water may have to be taken from a well. River water is commonly used for washing and bathing in these communities (Figure 21.5), so it is certainly not safe to drink without treatment. The water for other domestic use also needs to be safe.

Anywhere in the Caribbean, we are vulnerable to our water supplies being affected by extreme weather conditions such as hurricanes. At times like this, the water in our rivers and lakes can be so full of mud and sediment that the water treatment plants cannot cope. Even areas that usually have piped drinking water have to rely on water from wells. Yet we still need to drink clean water.

Figure 21.5 When people use river water directly in their home or for washing and bathing, it needs careful treatment to make sure it is safe to drink.

Contaminated water and its problems

Wherever we live, it makes sense to know how to make our water safe for drinking if it is contaminated in any way. Here are some **sterilisation** methods that can be practised at home, which will provide a family with water that is safe for drinking.

- **Add bleach:** A cap-full of bleach can be safely added to four litres of drinking water. Bleach is a powerful disinfectant. This will usually kill all the pathogens that are harmful to human health, making it safe to drink. The water will not necessarily be completely sterile BUT all the harmful microorganisms will have been destroyed. It is cheap and readily available, but has a typical smell.

- **Add purification tablets:** Water purification tablets contain a combination of compounds – often iodine, chlorine and chlorine dioxide – that kill most of the common water-borne disease-causing microorganisms. They are light and easy to carry anywhere but they are a bit more expensive than bleach and many types give the water a typical taste.

- **Boil:** Water can be boiled at room temperature on an open flame on a kitchen stove or fireside. This kills all harmful microorganisms – the water is sterile. Boiling is the safest and most effective way to make sure the water you and your family drink is safe and will not cause digestive system or other problems. See Figure 21.6.

Figure 21.6 Boiling water kills all harmful microorganisms, making the water sterile and safe to drink.

If you have access to a laboratory, test water for the presence of bacteria by inoculating agar plates with a sample of the water, incubating them no higher than 25 °C and waiting a few days to see if any bacterial cultures grow (see Chapter 16 for the technique).

Checkpoint questions

4 Explain why you are more likely to need to purify water at home during the hurricane season.

5 Compare and contrast the three main ways of home water purification.

Water treatment plants

Much of the potable water in the Caribbean is derived from rainfall. Some islands, such as Barbados, remove the salt from seawater (**desalination**) as an alternative source of potential drinking water.

Our rainy season is from June 1 to November 30 each year – hurricane season. Water that comes from boreholes in the ground is usually fairly clean. It is filtered as it passes through the limestone layers and other rocks. The water we take from rivers, lakes and reservoirs usually needs considerably more treatment. This involves a number of physical and chemical processes.

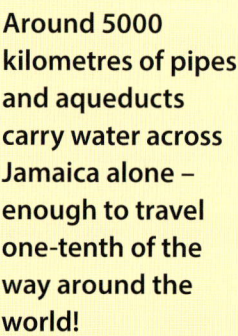

DID YOU KNOW?

Around 5000 kilometres of pipes and aqueducts carry water across Jamaica alone – enough to travel one-tenth of the way around the world!

As the water is drawn into the treatment plant, it goes through a number of stages, which are explained below (see Figure 21.7).

1. **Screening:** Large items such as logs and plants are screened out as water is drawn into the treatment plant. The screen is made of metal bars placed close together. Water from the ground is screened naturally as it passes through limestone rocks or coral underground.

2. **Sedimentation:** Chemicals, such as alum or lime, are added to the water to help remove impurities and destroy any bad taste or odour. Sometimes other chemicals are included to remove excess minerals that make the water 'hard' or cause rust to form on pipes. The water and chemicals are mixed and sent into a large tank where the chemicals and impurities stick together, forming larger, heavier particles that settle to the bottom of the tank and form a layer of sludge. This is **sedimentation**. When the water moves on through the treatment plant, the sludge is left behind so that chemicals and impurities are removed from the water at the same time.

3. **Filtration:** From the sedimentation tank, the water continues on its journey through the filters. Layers of fine sand and gravel are used to remove any other tiny impurities that are left in the water. The water now looks very clean but it can still contain harmful bacteria. The next stage is designed to kill these microorganisms.

4. **Chlorination:** A small amount of chlorine is added. This kills any bacteria that are present and prevents any bacteria growing as the water is distributed to the customer.

The clean, potable water is now stored in a reservoir or tank until it is needed. It then travels through large pipes to the consumer.

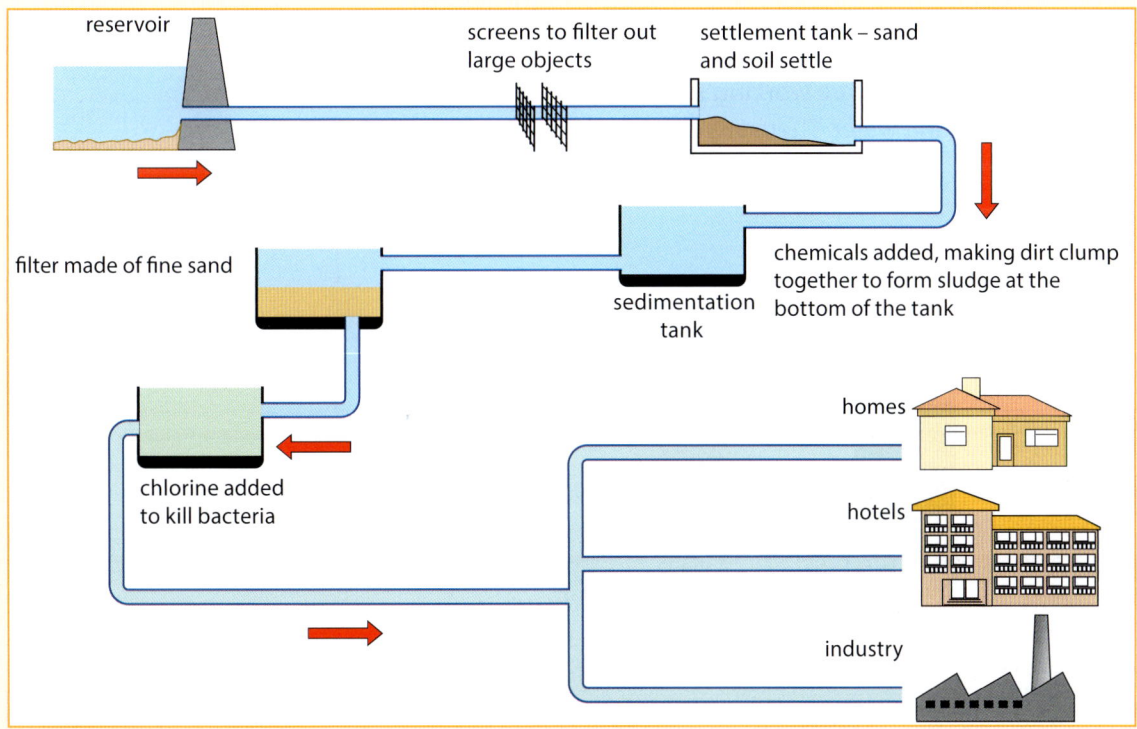

Figure 21.7 Water treatment plants like this provide safe, clean drinking water for all.

Human impact on the water supply

Humans have an enormous impact on the water supply. We take water for our needs, but as our lifestyles change, we are using more and more water. Regular showers and baths are excellent for personal hygiene and health (see section D Chapter 20) but they increase the amount of water we use. Washing clothes regularly is the same – good for us but increasing water usage – particularly as more homes have washing machines. Flush toilets also use a lot of water every time someone uses the toilet. As more people change from pit latrines to flushing toilets, the demand for water grows. The problem is that the amount of water available does not increase in line with the increase in demand.

In fact, because of the changing weather patterns we are seeing in the Caribbean, the amount of potable water available is actually falling and this lack of water has affected many people all across the Caribbean over the last few years. Many already have a water supply problem. These changes are brought about by climate change as a result of global warming resulting from burning fossil fuels and cutting down rainforests, and so on. Scientists think climate change will bring even more damaging effects in the future and that water stress will be a growing problem across the Caribbean:

- Rising temperatures mean more water will evaporate from the land surface and from plants as they transpire more.
- Levels of rainfall are predicted to drop across most of the Caribbean by around 30% in the next 50 years or so.
- More extreme weather events such as category 4 and 5 hurricanes and severe rainfall will cause structural damage and overwhelm the systems.

One way many Caribbean countries are working to cope with climate stress is to increase their ways of storing water. There are different ways of doing this – see Figure 21.8.

DID YOU KNOW?

Flush toilets use a lot of water – between 2–4 gallons (10–20 litres) for every flush! However, if you put a brick in the cistern of your toilet, it will still work perfectly, but less water will be flushed away every time you use the loo.

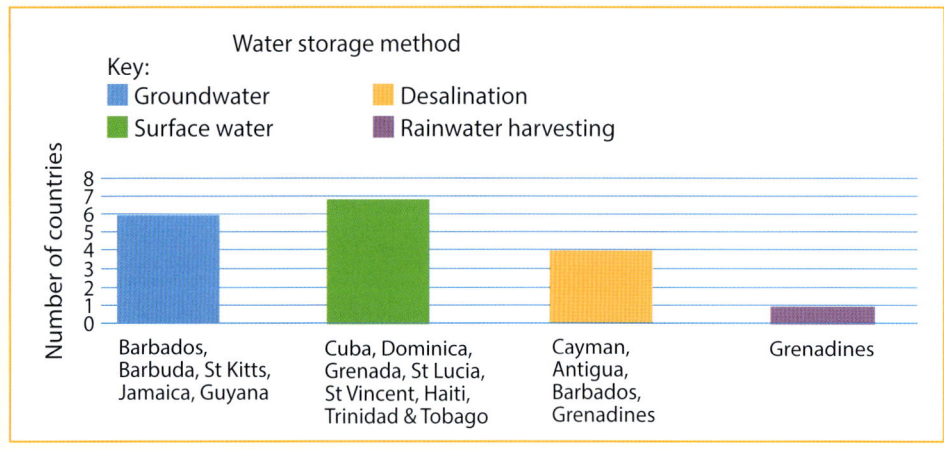

© Global Water Partnership-Caribbean (GWP-C) Secretariat. www.gwp-caribbean.org.

Figure 21.8 Different Caribbean countries rely on different ways of smoothing out the water supply across the seasons.

Simple things like turning off the faucet while we are brushing our teeth will help to save water. Having plants that don't need a lot of water is also a good idea. In most Caribbean territories, it is a crime to use water sprinklers during the dry season when water is in short supply. Does this happen where you live?

Other approaches include finding ways to reuse our 'grey' water. This is the waste water from washing up, baths and showers. If you **recycle** this water – even if only for watering your plants – then this also helps to save the levels of our cleanest, potable water.

If everyone took care not to waste water all through the year, then the water shortages we suffer would be reduced – but in most Caribbean countries we need to find more ways to both collect and store water for the future.

We also have a major impact on our water supplies in the way we dispose of our waste materials (sewage) and in the ways we pollute our waterways (see Figure 21.9). These are very important aspects of water management, which we will be considering in detail in the next chapter of this book.

Figure 21.9 How we dispose of sewage has major impact on our water supplies.

Possible SBAs based on the water cycle and safe water resources

- Develop an animation or interactive model of the water cycle.
- Investigate and produce a report into the different ways of water storage in different Caribbean countries and the factors that affect the methods that work in different places.
- Produce a short film explaining to people how best to make their water safe to drink after an extreme weather event such as a major hurricane.
- Develop an animation or interactive model of a water treatment plant.
- Carry out a practical investigation into the microorganisms in water from different sources, including your own school or home, and a local river or pond. Remember to carry out all appropriate health and safety measures and never take the lids off the Petri dishes once you have grown microorganisms.
- Produce a leaflet or podcast on the impact of human activities on water availability in the Caribbean and produce EITHER a water-saving manual OR a plan for ways to protect against water shortages for the future.

End-of-chapter summary

In this chapter, you have learnt that:

- the details of the water cycle involves evaporation of water from lakes, rivers and seas, the condensation of water vapour as it cools to form clouds, run-off of rain from the land, filtration through layers of limestone and other rock into aquifers, and the role of living organisms including transpiration, drinking and urination
- water can be purified in the home in different ways – by adding bleach, by adding water purification tablets and by boiling
- water goes through different processes as it passes through a typical Caribbean water treatment plant in order to deliver clean potable water to homes; the stages include screening, sedimentation, filtration and chlorination
- a simple test for bacteria in water samples is to inoculate agar plates and incubate them to see if bacterial colonies appear
- water is used and wasted in the Caribbean in different ways – and how global warming is creating potential challenges to our water supply
- conserving water around the home is important to reduce the demand on limited resources.

End-of-chapter questions

1. Which **one** of the following is the correct term used to describe water that is safe to drink?
    A   Portable
    B   Potable
    C   Percolate
    D   Pure (1)

2. Which **three** processes are parts of the water cycle in nature?
    A   Evaporation, condensation and crystallisation
    B   Evaporation, sublimation and condensation
    C   Evaporation, condensation and transpiration
    D   Transpiration, evaporation and diffusion (1)

3. Which **one** of the following is not a common source of drinking water in the Caribbean?
    A   The sea
    B   Rivers
    C   Reservoirs
    D   Lakes (1)

4. Which **one** of the following is not a reason for water treatment in the Caribbean?
    A   To get rid of bacteria
    B   To get rid of large debris such as tree branches
    C   To remove dirt
    D   To improve the taste (1)

5 Which **one** of the following stages of the water purification process is the point at which bacteria are killed or removed from the drinking water?
- A Chlorination
- B Filtration
- C Sedimentation
- D Screening (1)

6 Which of the following is NOT a method of purifying water in the home?
- A Adding chlorine
- B Boiling
- C Adding salt
- D Adding water treatment tablets (1)

7 Which **one** of the following is not an effective way to reduce the use of water in your home?
- A Don't leave faucets dripping.
- B Don't leave the water running as you clean your teeth.
- C Put a brick in your toilet cistern if you have one.
- D Run deep hot baths daily. (1)

8 Describe the various stages of water treatment from the point of the catchment area to delivery to your household, explaining why none of the stages can be left out if the water is to be safe to drink. (A flowchart can be very helpful.) (20)

9
- a Name **two** sources of water that can be used where no piped supply is available. (2)
- b Describe **two** household methods of water purification. (6)
- c Describe a simple test that can be conducted to show the presence of bacteria in drinking water. (8)

10
- a Name **two** obvious ways in which humans have an impact on the water supply and explain how each occurs. (4)
- b Describe **two** examples of ways in which you can reduce water wastage in the home. (2)

Chapter 22: Pollution and waste management

When you have completed this chapter, you will be able to:

- distinguish between the terms 'biodegradable' and 'non-biodegradable'
- identify pollutants in the environment and discuss the causes of water and air pollution
- describe the effects of pollutants on human beings and the environment, and explain methods of controlling pollution
- describe measures used to control solid waste volume, including reduce, reuse, recycle
- distinguish between proper and improper sewage disposal tactics, and discuss the impact of improper sewage disposal
- evaluate the impact of solid waste on the environment
- differentiate between a dump and a landfill, describe the operations at a landfill and discuss the importance of landfills in the Caribbean
- outline the methods for the treatment of sewage

What the examiners say

→ Candidates tend to respond fairly well to questions on this topic, but the weaker students are unable to accurately define or explain terms such as 'non-biodegradable', 'biodegradable', 'degradable' and 'recyclable'; are generally unable to adequately discuss the social effects of pollutants; cannot explain how landfills are created; are unable to state the advantages of processes such as incineration, recycling and composting; and possess a superficial understanding of the methods used in sewage treatment.

Revision tip

Reward yourself when you meet your study targets – a drink, a mango, a chat with friends, watching a TV show you enjoy … be kind to yourself to help you to meet your revision targets and succeed in your exams!

Figure 22.1 Caribbean ecotourists can take part in coral reef restoration projects while on holiday.

Environmental issues have been occupying the minds of conscientious West Indians for a long time, but recently the environment has moved up higher up the agenda around the world. On the Caribbean islands, this heightened awareness is mainly because our economy depends so heavily on tourism. Tourism in the Caribbean is expanding fast. There is always the attraction of 'sun, sand and sea', but tourism is extending beyond the beauties of our wonderful coasts and climate. More and more people are coming to our islands to enjoy the ecological variety we can offer. See Figure 22.1.

Some Caribbean countries are building up increasing levels of 'health tourism', when people visit to have health procedures such as infertility treatments, plastic surgery and cosmetic dentistry carried out away from home. They have their medical treatment and then, as they recover, enjoy the beautiful and relaxing atmosphere of the Caribbean.

However, having more tourists places more demands on our islands – tourists want clean drinking water, modern toilets and good waste disposal (see Figure 22.2). At the same time, our own population also wants higher standards. As you have seen in the previous chapters, increased personal hygiene, more flush toilets and washing machines mean people are healthier, but more clean water is needed. More disposable goods available in our shops mean more rubbish. More vehicles on the roads mean more exhaust gases going into the air. The impact of pollution and management of the waste we produce are becoming increasingly important issues in the Caribbean.

Figure 22.2 When tourists visit the Caribbean from the USA or Europe, they come to relax and enjoy themselves, but their presence raises some real challenges for Caribbean countries in terms of water supply, water standards and pollution.

What is pollution?

Pollution is the contamination of the natural environment by harmful substances as a result of human activities. Anything that contaminates the air, soil or water is known as a **pollutant**. Pollution may be small scale in your own home environment or very large scale, affecting whole countries. **Acid rain** and **eutrophication** are two examples of large-scale pollution you will be learning about in this chapter.

Different types of pollution include the following types.

- **Domestic pollution:** This is pollution on a very small, local scale. For example, every time you drop litter or a dog fouls the sidewalk, the local environment is polluted. The litter and the dog faeces are the pollutants.
- **Industrial pollution:** This is pollution, on a large scale by industries. There are many types of industrial pollution, including **thermal pollution**, when an industry raises the temperature of the environment locally, **heavy metal** pollution when industries pollute an area with heavy metal ions that are often toxic to plants and animals, and **chemical pollution** such as the dumping of oils at sea, or the release of toxic compounds into rivers.

- **Agricultural pollution:** This is pollution that results from farming practices, for example, fertilisers or pesticides ending up in rivers or lakes, or pollution from animal slurry.
- **Noise pollution:** People make a lot of noise. Some of these sounds affect our environment. For example, noise pollution in our towns and cities affects songbirds and disturbs human sleep patterns. Noise pollution from ships' engines in the water interferes with the communication of sea mammals such as dolphins and whales, and may cause them to strand themselves on the land. Preventing noise pollution is difficult. For these examples mentioned above, it involves developing quieter engines for both land and water vehicles.

> **Checkpoint question**
>
> 1 Describe the difference between pollution and a pollutant.

Air pollution

Clean air is essential for life (see Chapter 8) as it supplies the oxygen for cellular respiration. We breathe air in and out of our lungs from our birth to our death. Unfortunately, some of our other activities release substances that pollute the air and are harmful to humans, other animals and plants. **Air pollution** comes in various forms, which all have potentially serious effects on our health and well-being. Here are some of the main examples of air pollution.

Fossil fuel pollutants

Burning fossil fuels (coal, oil or gas, or electricity produced using them) in vehicles, factories, power plants and homes produces many different types of air pollution. Pollutants from the **combustion** of fossil fuels include smoke and carbon dioxide.

Smoke

Tiny particles of unburnt fuel are released into the air from diesel smoke, charcoal BBQ smoke (Figure 22.3) and home fires. The particles are very small, but huge quantities are released around the world. Scientists have shown that smoke particles in the atmosphere are blocking out some of the light from the Sun and this is known as **global dimming**.

Carbon dioxide

Carbon dioxide is produced by living organisms as a waste product of respiration. It is used by plants in the process of photosynthesis (see Chapter 3). Carbon dioxide is also produced as a result of burning wood and fossil fuels. Why is carbon dioxide becoming a problem? Remind yourself by looking back to Chapter 4.

Figure 22.3 Caribbean festivals are not a major cause of air pollution, but they certainly add just a little more smoke to the air.

- For millions of years, the levels of carbon dioxide released by living things into the atmosphere has been matched by the plants taking it out and the gas dissolving in the seas. As a result, the level in the air stayed about the same from year to year. You learnt about the carbon cycle in Chapter 4.

- Over the last centuries, the amount of carbon dioxide produced has increased quickly as a result of human activities. We burn huge amounts of fossil fuels in our cars, our planes and also in power stations to generate electricity. The natural carbon sinks (processes that use up carbon dioxide) cannot cope with this increase, so the levels of carbon dioxide are building up (see Figure 22.3).

- The situation is made worse because, all around the world, large-scale deforestation is taking place. We are cutting down trees over vast areas of land for timber and to clear the land for farming. In this case, the trees are felled and burned in what is called 'slash-and-burn' farming. No trees are planted to replace those cut down. These trees are natural carbon sinks that soak up carbon dioxide. We are removing them just when we need them most.

This build-up of carbon dioxide gas in the atmosphere is generally believed to contribute to the **greenhouse effect**, also referred to as **global warming** (see Figure 22.4). Carbon dioxide is known as a **greenhouse gas.**

Methane

Methane is another greenhouse gas that causes air pollution, and the levels of this gas are also rising. It has two major sources. As rice grows in swampy conditions, called paddy fields, methane is released. Rice is the staple diet of many countries, so as the population of the world has grown, so has the farming of rice.

DID YOU KNOW?

Cows produce methane all through the day from both ends, but especially burps! A single cow releases from 100–400 litres of methane per day. That's a lot of greenhouse gas.

Cows produce methane during their digestive processes and release it at regular intervals. In recent years, the number of cattle raised to produce cheap meat for fast food like burgers has grown enormously, so the levels of methane in the atmosphere are rising. Many of these cattle are raised on farms produced by deforestation.

The greenhouse effect is the result of gases like carbon dioxide and methane in the atmosphere. They act like a blanket around the Earth, trapping some of the heat energy from the Sun. This is vital for life on Earth, because the greenhouse effect keeps the temperature warm enough for life to exist.

Figure 22.4 The scientific evidence clearly shows that global warming is the result of a build-up of air pollutants such as carbon dioxide from the combustion of fossil fuels. The pollution is produced all over the world, but we are already feeling the effects here in the Caribbean with extreme weather events and rising sea levels.

Air pollution

However, as a result of human activities, the amount of carbon dioxide (and methane) in the air is increasing. This traps more heat close to the surface of the Earth, causing the temperature at the surface of the Earth to rise. Global warming is driving climate change, with temperatures rising, the land ice in the South Pole melting is driving rising sea levels and an increase in extreme weather events such as hurricanes.

Carbon monoxide

Carbon monoxide is another gas generated by the burning of fossil fuels. It is produced by cars and fires as well as by home water heaters if they are not functioning properly. Carbon monoxide is very dangerous because it combines irreversibly with haemoglobin in the red blood cells, reducing their oxygen-carrying capacity (see Chapter 8). There is carbon monoxide in cigarette smoke, which is why it is so dangerous for a pregnant woman to smoke, because it can deprive the unborn baby of oxygen. Carbon monoxide poisoning can eventually lead to death. Since the gas has no colour or smell, there is no way of knowing if it is leaking into your home from a faulty water heater.

Sulfur and nitrogen oxides

Fossil fuels often contain sulfur impurities. When these burn, they react with oxygen to form sulfur dioxide gas. At high temperatures, for example, in car engines, nitrogen oxides are also released into the atmosphere. Sulfur dioxide and nitrogen oxides pollute the air and can cause serious breathing problems for people if the concentration is too high. They form a haze of pollution called **smog** (Figure 22.5), which can be a real problem in big cities where there are millions of motor vehicles. Sulfur dioxide and nitrogen oxides are also involved in the formation of acid rain. This pollutes land and water over a wide area.

Figure 22.5 Smog, which is visible air pollution from burning fossil fuels in car engines, as seen in Shanghai

Acid rain

The sulfur dioxide and nitrogen oxides formed when fossil fuels are burned dissolve in raindrops and react with oxygen in the air to form dilute sulfuric acid and nitric acid. This makes the rain more acidic, which is why it is called acid rain (see Figure 22.6). Not surprisingly, acid rain has a damaging effect on the environment. If it falls onto trees, the acid rain can cause direct damage. It may kill the leaves and, as it soaks into the soil, even the roots of the tree may be destroyed. In some parts of the world, huge areas of woodland are dying as a result of acid rain. Acid rain is a result of air pollution but it causes **water pollution**. Acid rain falling into lakes and ponds may make them so acidic that nothing lives in them. Acid rain also destroys limestone buildings – and limestone coral reefs.

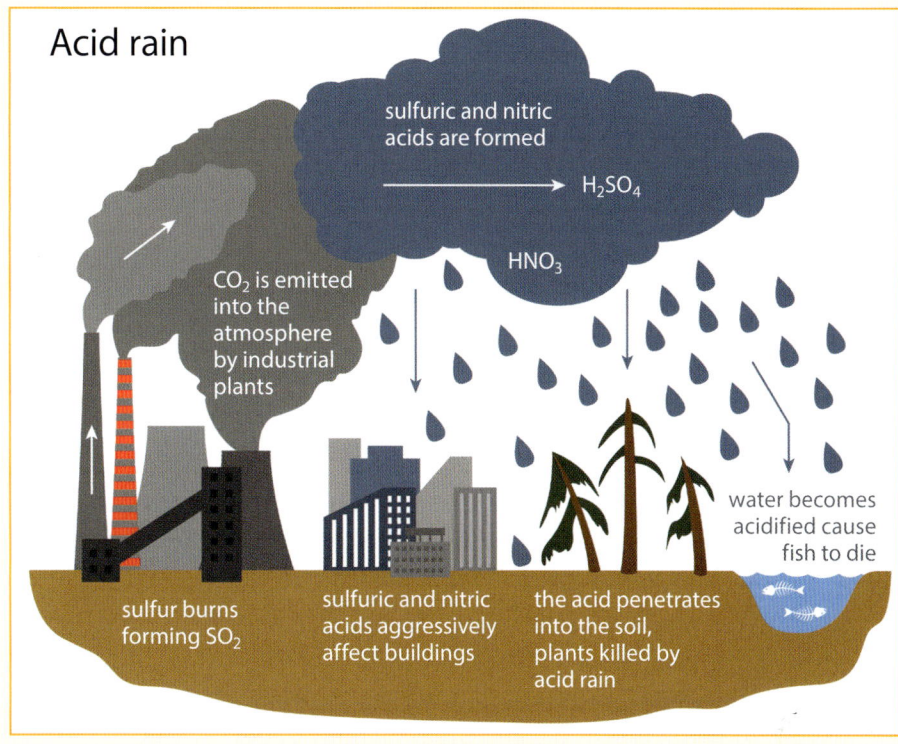

Figure 22.6 The formation and effects of acid rain

The worst effects of acid rain are often not felt by the country that produced the pollution in the first place (see Figure 22.7). The sulfur and nitrogen oxides are carried high in the air by prevailing winds. As a result, it is often the relatively 'clean' countries that receive the pollution and the acid rain from their dirtier neighbours. Their own clean air goes on to benefit someone else!

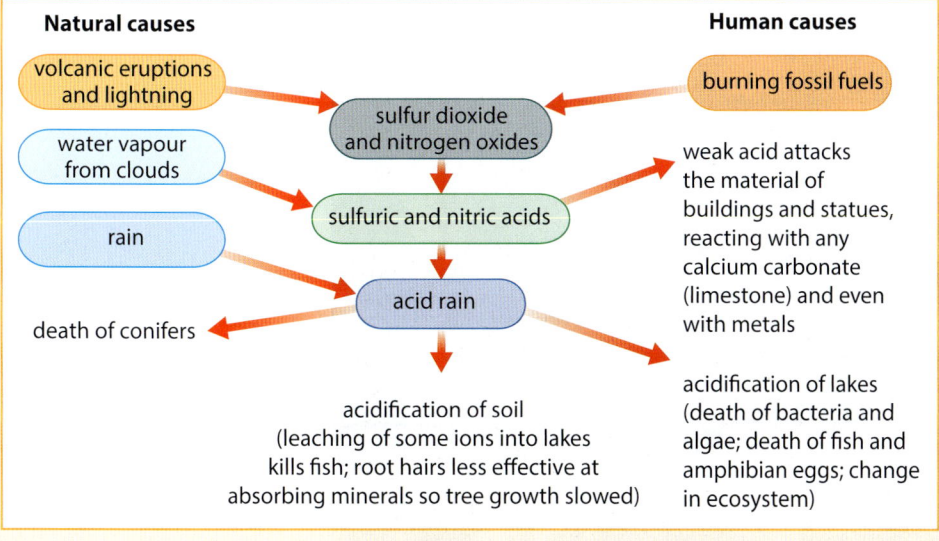

Figure 22.7 Air pollution in one place may cause acid rain and serious pollution problems somewhere else entirely. Depending on the prevailing winds, it may even be in another country!

> **DID YOU KNOW?**
>
> Normal rain has a pH of around 5. It is slightly acidic because of the carbon dioxide that is dissolved in the atmosphere. Acid rain has been measured with a pH of 2.0 – more acidic than vinegar!

Air pollution 461

> **Checkpoint questions**
>
> 2 State the main gases involved in the formation of acid rain.
> 3 Explain how acid rain kills trees.

Impact of air pollution on people and the environment

In some ways it is impossible to separate the effects of air pollution on the environment and on ourselves, as we are so closely linked.

One of the ways in which air pollution is affecting the Caribbean environment is by killing life in the seas that surround our islands. Our beautiful coral reefs (see Figure 22.8) are very vulnerable to air pollution in two ways:

- First, the temperature of the seas is increasing as a result of global warming, which is the result of air pollution in the form of rising greenhouse gas levels. Just a small rise is enough to kill the algae on which the coral polyps feed. In turn, the polyps die and the coral reefs begin to bleach and die as well.
- Second, acid rain falls into the sea and the change in pH affects the calcium salts dissolved in the water and breaks down the coral skeleton itself.

Without the coral, much of our unique reef ecosystems will be destroyed forever, and one of the big attractions to the tourists we depend on will be lost, so controlling air pollution is very important indeed here in the Caribbean. What is more, as you will see later in the chapter, the coral reefs are also vulnerable to water pollution, so there is a lot of work to be done if we are to protect this vital part of our ecology.

Figure 22.8 The coral reefs of the Caribbean are one of the most spectacular ecosystems in the world. It is very important to preserve them and the biodiversity they support.

Air pollution can affect our health in many ways.

- Examples of short-term effects include irritation to the eyes, nose and throat, and upper respiratory infections such as bronchitis and pneumonia. This has been a serious cause for concern in the bauxite industry in Jamaica. Other symptoms can include headaches, nausea, allergic reactions and severe attacks of asthma.
- Long-term health effects of air pollution can include chronic respiratory disease, lung cancer, heart disease (particularly from cigarette smoke), and damage to the brain, nerves, liver or kidneys.

> **Checkpoint question**
>
> 4 State the main pollutants found in car exhaust gases.

Controlling air pollution

People have become more aware of the problems caused by air pollution. In many countries in the world – as in the Caribbean – steps are being taken to stop the damage to our environment from air pollution. Scientific groups at the University of the West Indies are studying the damaging effects of air pollution on plant, animal and human life. As we have seen, one of the biggest causes of air pollution are cars and other vehicles (Figure 22.9). The fumes from car exhausts contain carbon dioxide, carbon monoxide, sulfur dioxide and oxides of nitrogen oxides. All of these gases have both a direct and an indirect effect on human health. We are working hard to reduce the levels of sulfur dioxide and nitrogen oxides in car exhaust fumes. More and more cars are being fitted with **catalytic converters**. Once hot, these remove the acidic gases before they are released into the air.

There is a move towards electric cars that are charged using electricity generated by clean, **renewable** technology such as solar power and wind power. These add no carbon dioxide to the atmosphere.

Renewable energy supplies in general are increasingly being used, with wind, solar, nuclear and wave power replacing burning fossil fuels to generate electricity.

The governments of the Caribbean region are drafting laws to control emissions from cars and from factories. Experience from other areas of the world suggests that this type of legislation can be very effective.

In addition, it is possible to prevent many types of air pollution that are not regulated, through personal, careful, attention to our interactions with the environment.

Figure 22.9 We all want to use vehicles like this because they are so convenient – but they are producing almost every type of air pollution you can think of!

Air pollution

You as an individual can make a definite difference:
- Reduce the amount of electricity you use by switching off lights and other electrical goods when you don't need them.
- Reduce the level of the air conditioning.
- Walk or cycle sometimes instead of using cars or buses.
- Buy local produce with as little packaging as possible – that reduces the fuel used to get your food to you, and the chemicals processed to make the packaging.

All of these things can make a real difference in the long term, and the more people who help the better.

Water pollution

Air pollution is a problem at both a local and a global level. Water pollution is much more of a local problem, but no less serious for that. When water pollution affects rivers and lakes, it can have a drastic effect on the environment as animals and plants are damaged or killed. If water pollution affects the drinking water, it can have very serious effects on the health of the local population. Since drinking water comes from a variety of sources, we have to be very careful to avoid pollution in many different ways.

Water pollution may comprise any of the following types.
- **Organic/biological:** Our water supply may be contaminated by disease-causing organisms or human waste.
- **Inorganic:** Compounds of various types may be released into the water. This includes acid/base pollution, which is usually the result of acid rain falling into rivers and lakes.
- **Radioactive pollutants:** This is very unlikely, but if it happens through waste from nuclear power stations or radioactive waste from hospitals going astray, it can be extremely dangerous, as we take the radioactivity into our body.

Table 22.1 describes the two main causes of water pollution.

Table 22.1 Causes of water pollution

Indirect	Direct
Contaminants enter our water supply from soils or groundwater systems and from the atmosphere via rainwater. • **Agricultural:** Fertilisers, pesticides and manure slurry are washed away by the rain and flow into rivers and other water sources. The chemicals can affect other animals and plants. • **Industrial:** Industrial wastes that have not been disposed of properly can also contaminate the soil and then, in turn, cause water pollution.	• **Domestic:** This involves pouring pollution directly into the water. This includes untreated sewage, effluent (waste) from factories and refineries, as well as waste and liquids from sewage treatment plants.

> **Checkpoint question**
>
> 5 Describe the main types of water pollution.

What are the effects of water pollution?

Water pollution affects the ecosystem on various levels:

- Pesticides washed into the water from farmland may poison animals and, what is more, it can accumulate in food chains. If the smallest animals in a food chain pick up some pollution from the water, they may not be affected themselves. But at each stage up the chain, more and more organisms are eaten, so the levels of pollutants build up until finally the largest animals in the food chain start to die. (See Chapter 4 to remind yourself of food chains and bioaccumulation.)
- Chemical pollution in the water such as heavy metal contamination may affect humans directly when we drink the water, or indirectly when we eat the animals that have accumulated toxins from the environment over time.
- Water pollution can leave ecosystems such as rivers and lakes unbalanced, so that they can no longer support full biological diversity (the maximum range of animals and plants). This may simply be that certain types of animals or plants are killed by a particular pesticide or herbicide. As you will remember from your work on feeding relationships in Chapter 4, many different animals and plants are interconnected in food chains and webs. So if one or two organisms are killed by water pollution, it can have a big effect on the whole ecosystem. On the other hand, if the water pollution is from acid rain, the change in pH can kill most of the plant and animal life. Look at Figure 22.10.

DID YOU KNOW?

Some years ago, a factory in Japan dumped huge amounts of mercury waste into Minamata Bay (an example of heavy metal pollution). Soon afterwards, people in the city of Minamata began to get ill, give birth to deformed children and even die. What's more, their pet cats suffered the same symptoms. Eventually people realised that the fish in the bay had taken in low levels of mercury, but when people and cats ate lots of fish, the levels of mercury built up until they were high enough to cause disease and death.

Figure 22.10: Sometimes, when an oil tanker is damaged at sea or sinks, there is massive oil pollution of the surrounding seawater and coastline. This has a direct and distressing effect on wildlife, and a huge, indirect economic effect on the local human population, as fishing and tourism are badly affected.

- The nitrates and minerals that are leached into the waterways from agricultural fertiliser (Figure 22.11), slurry or untreated sewage may lead to the rapid growth of water plants in a process called **eutrophication**. The pollutants provides nitrates for the water plants, causing a surge in their growth followed by death. The bodies are then broken down by decomposers, greatly increasing the use of oxygen in the water by microorganisms and eventually leading to the death of the animals and plants that live there.

Eutrophication

Eutrophication consists of the following stages:

- Minerals and nitrates are washed into water bodies, which increases the natural levels of these substances.
- They act as fertilisers so water plants and algae grow rapidly.
- The algae block the light from the plants, which die. Animals feed on the plants and excrete.
- There is a big increase in the number of microorganisms feeding on and decomposing these dead plants.
- The process of decomposition uses up a lot of oxygen, so the increase of microorganisms breaking down the plants lowers the levels of oxygen in the water.
- There is not enough oxygen in the water to support some of the fish and other animals. They die and are decomposed by microbes, using up yet more of the oxygen.
- The oxygen levels in the water become so low that all the fish and other aquatic animals die of suffocation due to lack of oxygen.

Figure 22.11 Fertilisers are an important part of modern farming but they must be used with care to protect our waterways and all the life in them.

Impact of water pollution on people and the environment

At a local level, much of the water pollution in the Caribbean is linked to the problem of dealing with human waste (sewage) or **improper sewage disposal practices**. Currently almost 85% of the wastewater in the Caribbean region goes untreated into our oceans and in many areas about 52% of households do not have sewer connections. This makes it much harder to deal with sewage properly, but it is possible.

Improper sewage disposal may involve the release of large amounts of untreated sewage from a town, or a hotel, into rivers or the sea. Poorly sited pit latrines are also a problem. Pit latrines are a relatively cheap and safe way of disposing of human waste, but only if they are correctly sited to prevent the waste contents from contaminating the drinking water.

The consequences of improper sewage disposal practices include the following:

- Eutrophication and death of organisms in waterways (see Figure 22.12)
- Algae growth covering the coral reefs, causing death of the coral polyps and damage to and death of the reef
- Contamination of groundwater, leading to the spread of gastrointestinal diseases in the water
- Breeding grounds created for disease-carrying insects
- Pollution of streams and rivers, creating dreadful odours.

Water pollution from untreated sewage and agricultural run-off is having a serious impact on ecotourism in the Caribbean. No one wants to visit dead waterways, particularly if they are smelly from untreated sewage or decomposing organisms.

Figure 22.12 Improper sewage disposal practices may lead to the complete eutrophication of our waterways. This green pond is completely dead.

Even more importantly, the sea itself is becoming heavily polluted around some of our islands. Eutrophication happens in the sea as well. High levels of nitrates cause algae to grow and smother the coral reefs. The coral reefs are vital for ecotourism, to protect our islands from the sea, and for fishing. So this major threat to them is a big concern. For example, the seas around Jamaica are heavily polluted with sewage. The levels have become so high that they are seriously threatening the coral reefs all around the coast. Scientific research shows that effective sewage treatment should be implemented all over Jamaica as soon as possible.

Controlling water pollution

Proper sewage disposal practices avoid all of the problems highlighted above. Sewage, which consists of the waste water and human waste from our home along with waste water collected from commercial and industrial sources, is disposed of properly when it is carried away in sewers, or in some cases collected from our cesspits, to undergo further treatment.

Ideally, sewage is treated in a sewage treatment plant, where there are a number of stages in the process.

These involve:

- removal of solid items (screening)
- microorganisms removing organic substances
- release of water that is fairly unpolluted.

There are two main types of sewage treatment plants:

- the filter method of sewage treatment
- activated sludge method of sewage treatment.

Biological filter method of sewage treatment

The **biological filter method** of sewage treatment is used in many countries around the world. See Figure 22.13.

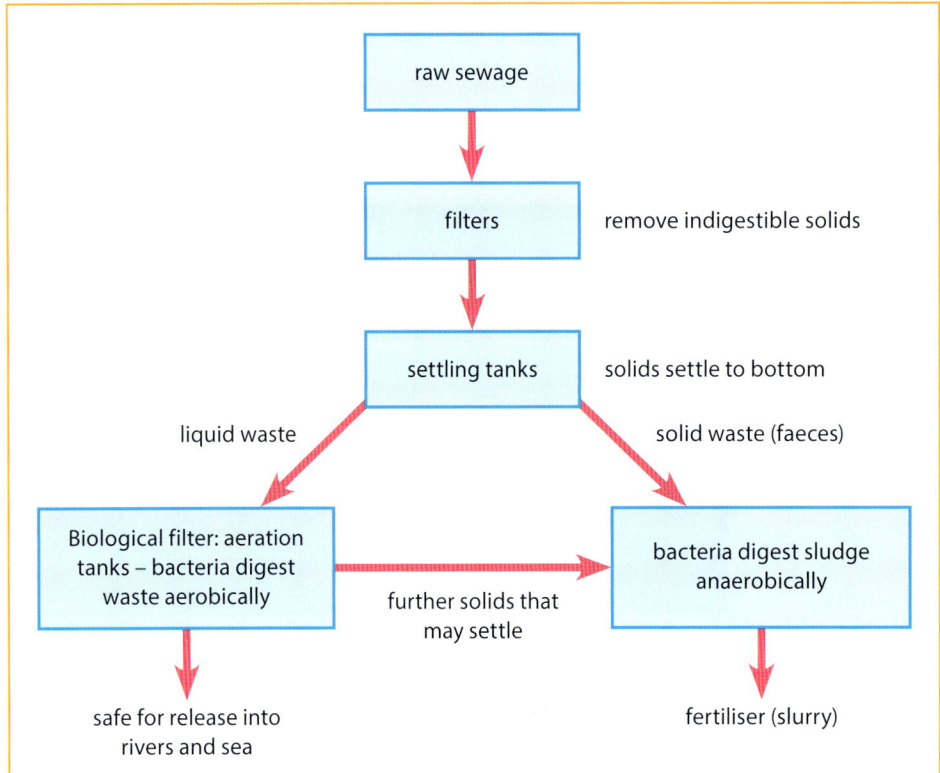

Figure 22.13 A sewage treatment plant based on biological filters

The main stages of the biological filter method include the following stages.

1 **Screening:** Rubbish is removed. The solid waste is raked off, broken down into small pieces and then mixed back into the waste water.

2 **Grit settling tank**: The mixture moves slowly along channels that allow the grit and sand to settle to the bottom. This sand and grit is dredged out, washed, and used to fill in holes in the ground.

3 **Primary sedimentation**: The waste water flows into large tanks (called primary sedimentation tanks). The solids, called crude sludge, settle to the bottom and are pumped to the **sludge digestion tank**.

4 **Secondary sedimentation**: The remaining liquid (called **primary effluent**) is sprayed over a **biological filter**. This is made up of tanks full of sand or clinker, and each grain is covered with a thin layer of bacteria and other microorganisms called a **biofilm**. They feed on the human waste matter in the effluent, digesting it and breaking it down, leaving only harmless gases and water, acting as a **biological filter.** The microorganisms are aerobic – they need oxygen to break down the human waste. A common way of providing the microorganisms with the oxygen they need is to spray the waste water onto the biological filter so that it absorbs oxygen from the air (see Figure 22.14).

Figure 22.14 The primary effluent is oxygenated as it is sprayed over the biological filters in this secondary treatment plant.

Pollution and waste management

5 **Final sedimentation**: The microorganisms are separated from the water in final sedimentation tanks so that they can be reused. The water that is left is clean enough to flow into a river or the sea without seriously upsetting the balance of organisms living there.

6 **Sludge digestion tank:** Meanwhile, the crude sludge is pumped to the sludge digestion tanks where anaerobic microorganisms destroy the unpleasant and smelly materials in the solid waste and change them into a gas that contains methane. The sludge settles to give water (liquor) and a thicker sludge. The sludge (now called digested sludge) can be used on land as a soil fertiliser after some of its water has been removed. It is a very useful fertiliser, but care must be taken to make sure that the nitrates contained in it do not leach out into waterways and cause eutrophication.

Activated sludge method of sewage treatment

In this type of sewage treatment plant there are no biological filter tanks (see Figure 22.15). It is a very useful way of treating sewage in areas where there is a relatively large population but relatively little land – such as the Caribbean nations.

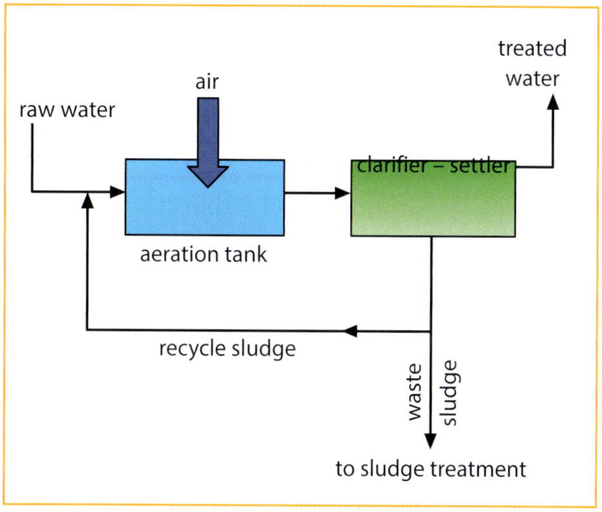

The main stages of the activated sludge method include the following stages.

1 **Screening:** Rubbish is removed (not shown on diagram).

2 **Activated sludge tank:** Raw sewage is pumped into the tank along with recycled activated sludge full of microorganisms. Air or oxygen is blown into the tank to smash the solids and produce a biological soup in which the microorganisms from the activated sludge use oxygen to break down the human waste. See Figure 22.16.

3 **Clarifier/settling tank:** The excess sewage liquor is passed into a settling tank. Live bacteria sink to the bottom, and dead bacteria float on the top and form a crust. The water between the two layers is now clean enough to be discharged into a waterway. The activated sludge at the bottom of the tank is pumped back to the first tank and the processs is repeated.

Figure 22.15 An activated sludge sewage treatment plant.

Figure 22.16 Activated sludge tanks in action.

Once sewage passes through a sewage treatment works, it is no longer a threat to people or the environment. However, many people in the Caribbean still do not have their sewage collected and taken away to a sewage treatment plant. In fact, on some Caribbean islands, nearly 40% of the population still use pit latrines. Sited and used properly, these are also a safe and effective way of dealing with human waste.

Water pollution

The pit latrine

The pit latrine is one of the simplest and cheapest means of disposing of human wastes. If well designed and built, correctly sited and well maintained, it contributes significantly to the prevention of diseases (see Section D).

How do pit latrines work? At their simplest, a pit latrine is a deep hole partially covered by a slab where people go to pass their bodily waste, which is then broken down by microorganisms over time. At the other end of the pit latrine spectrum, the latrine is still a pit with a slab but it has a seat for people to sit on, a simple building to enclose the latrine, a vent for gases and flies and even a water-filled pan to wash away the urine and faeces into the pit.

It is important to cover the hole when it is not in use to keep smells from coming out and flies from going in. See Figure 22.17.

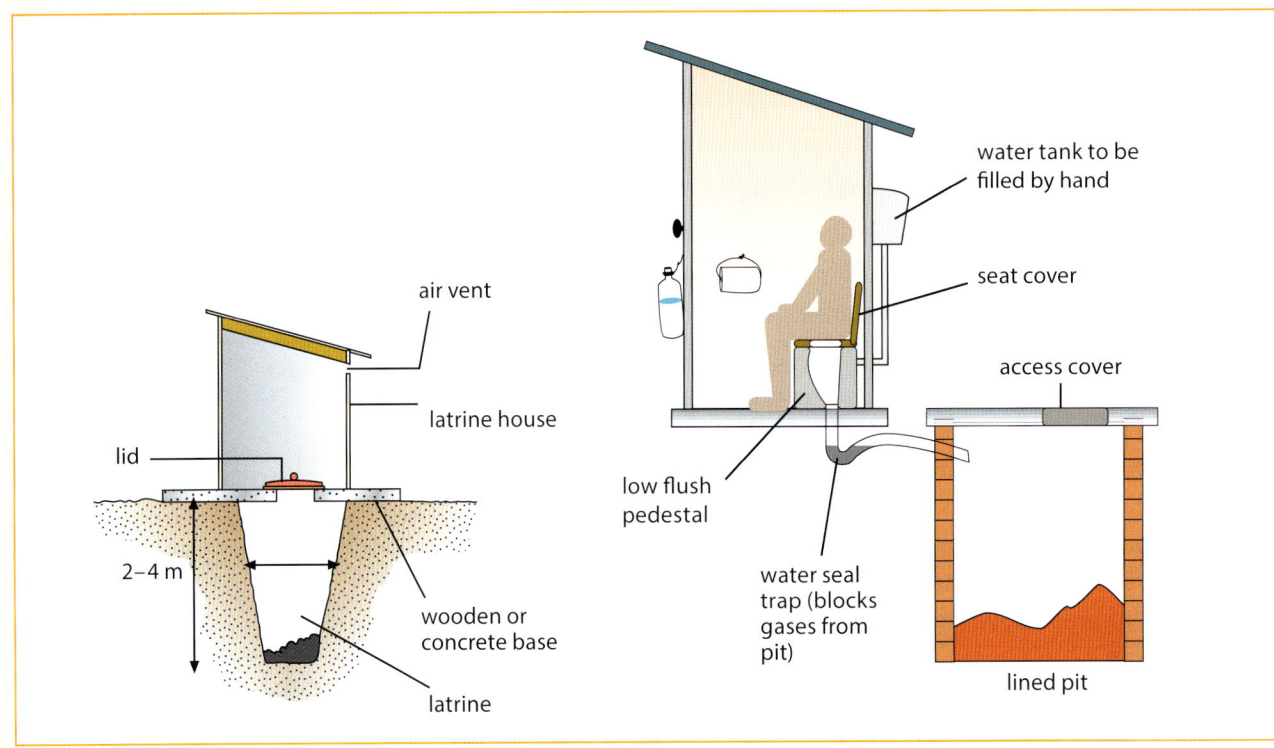

Figure 22.17 Pit latrines are a basic but effective way of dealing with human waste, as long as they are designed, built and sited correctly.

Checkpoint questions

6 Describe how improper sewage disposal methods contribute to water pollution in the Caribbean.

7 Describe the key features of proper sewage disposal practices, without going into details of any one method.

Pollution and waste management

Here in the Caribbean, we have the ability to deal effectively with human waste, but we do not always manage to do so. The pit latrines that have worked well for centuries are strained by the size of modern populations, and if they are not built and sited correctly they can cause pollution problems. We have not yet built enough sewage treatment plants to deal with the amount of human waste now produced on our islands. This is the challenge for the future, because we can no longer simply discharge raw sewage into the sea. The damage to our environment by the levels of pollution we have produced is no longer acceptable. So the phasing out of pit latrines and effective sewage treatment for all has to be high on our agenda for the next few years.

Dealing with household and industrial waste

People don't just produce bodily waste. We also produce huge amounts of waste of all sorts – and as we use more and more disposable goods and wrappings, the amount of waste we produce increases. Look at the data in Figure 22.18.

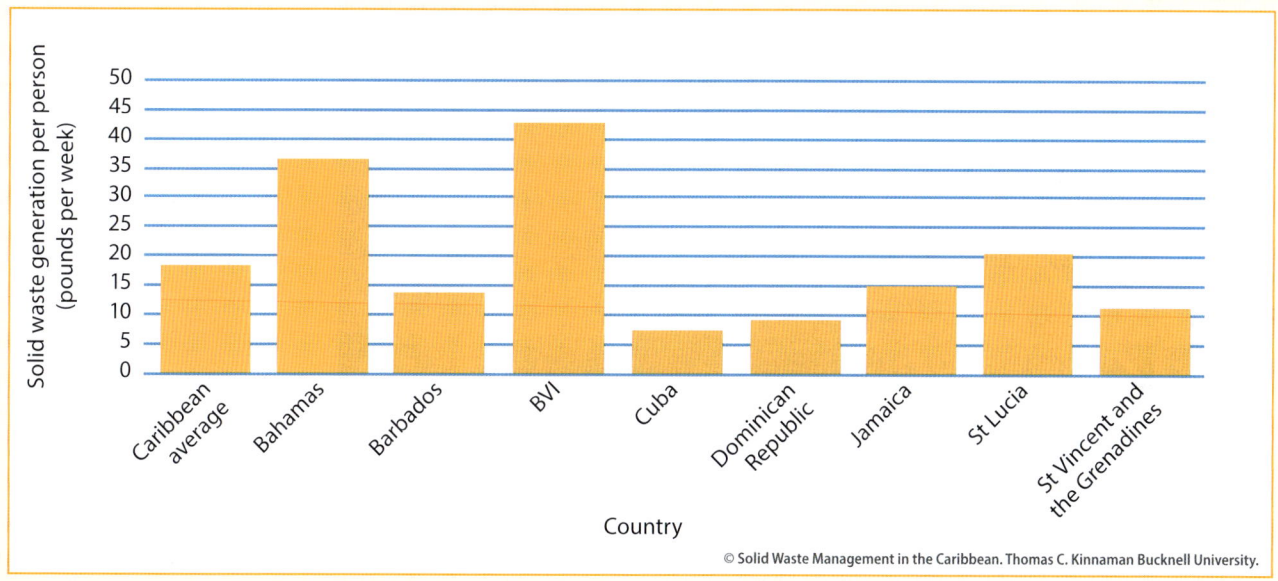

Figure 22.18 The mass of solid waste produced per head of population varies considerably across the Caribbean, as the data on this graph clearly shows.

The solid waste that needs to be dealt with falls into a number of different categories. Household or municipal waste is the one we are most familiar with. This is the waste we throw out from our home every week. It includes:

- **organic waste** – fruit and vegetable peelings, and leftover food
- **toxic waste** – old paints, batteries and household chemicals we throw away
- **recyclable materials** – paper, glass, metal and plastic.

However, it isn't just our household waste that has to be dealt with. We must also safely get rid of **industrial waste** and **hospital waste**. All of these different types of solid waste must be dealt with effectively.

Dealing with household and industrial waste

Why does managing waste matter?

If waste disposal is not carried out properly, it can have serious impacts on our health and cause major problems to the surrounding environment. It also has important economic impacts.

The tourist industry in the Caribbean produces huge amounts of waste (see Figure 22.19) – far more per head than the local population – but we need the tourists to support the economy of our islands. Yet if we do not manage our waste problems effectively, it will help to prevent tourists from visiting.

Figure 22.19 Tourists produce more waste than locals – but don't want to see it on the beaches!

Here are some of the problems that result from waste that is not disposed of properly:

- Piles of waste obstructs storm/hurricane water run-off, resulting in the formation of stagnant pools that become breeding grounds for disease. Mosquitoes and other vectors of disease take advantage of the water to lay their eggs, so their numbers increase rapidly.
- Waste dumped near a water source causes contamination of the water body or the groundwater source. Direct dumping of untreated waste into rivers, seas and lakes results in the accumulation of toxic substances in the food chain, through the plants and animals that feed on it. We have already seen the problems that this sort of pollution can bring.
- Burning waste results in air pollution from the smoke, and carbon dioxide is released into the atmosphere along with poisonous chemicals often produced when plastics are burnt. See Figure 22.20.

Pollution and waste management

- The unhygienic use and disposal of plastics has a number of bad effects on human health. Coloured plastics are harmful as their pigments contain heavy metals that are highly toxic. Some of the harmful metals found in plastics are copper, lead, chromium, cobalt, selenium and cadmium. Interestingly, research at the University of the West Indies in Jamaica is revealing that some schools are situated in areas of heavy metal contamination. These institutions were founded on sites where **non-biodegradable** materials, such as car batteries, were buried. These do not break down into simpler, less harmful end products with time. Plastics also remain in the environment indefinitely unless they are special **biodegradable** plastics. What's more, as we have already seen, they produce toxic gases when they are burnt.

Figure 22.20 Burning rubbish results in several different types of air pollution, from smoke to CO_2.

- Industrial and hospital waste can be **hazardous**. For example, hospital waste in the form of disposable syringes, swabs, bandages and body fluids is highly infectious and can be a serious threat to human health if not managed responsibly.
- Old batteries, shoe polish, paint tins, old medicines and medicine bottles are also hazardous. These wastes can be highly toxic to humans, animals and plants. They can be corrosive, highly inflammable or explosive, and they could react when exposed to other chemicals.

Managing solid waste

Have you ever wondered where the solid waste that you dispose of goes to after it is picked up by the rubbish truck?

- Much of it goes to landfill or dumps (see Figure 22.21).
- Some is **reused** or used again. For example, drinks companies may ask that glass bottles are preserved and returned. These are sterilised and then returned to the bottling lines to be filled again.
- Some waste gets **recycled**. It is processed and then converted into other useful articles. This method is usually employed for paper, glass and cans. Some countries – such as Belgium in Europe – recycle up to 50% of their waste.
- Some is burnt to produce energy – it is used to generate electricity. This is known as 'waste to energy'.
- Some is simply burned.

What happens to our waste depends partly on what we are throwing away.

Figure 22.21 Lots of Caribbean countries are working hard to tackle household waste disposal.

Dealing with household and industrial waste

Biodegradable or not?

Some of our solid waste is **biodegradable** – microorganisms break it down or decompose it. Examples include all our food waste, paper, card, wood, compostable plastics and some fabrics such as cotton or wool. We can compost our biodegradable waste to produce fertiliser for our garden (see Chapter 4) – or on a big scale use it to produce methane to burn for electricity generation.

The rest of our solid waste is **non-biodegradable.** It cannot be broken down by microorganisms. Examples include glass, plastics, metals, batteries, and so on.

In some ways, it is much easier to deal with biodegradable waste than non-biodegradable waste. Biodegradable waste will eventually be broken down by microorganisms to produce gases such as methane and carbon dioxide, although these in turn contribute to global warming! On the other hand, much of our non-biodegradable waste can be recycled and reused.

Dump or landfill?

Dumps and landfills are the two main ways in which we dispose of our solid waste.

A **dump** is just an open hole in the ground where waste is literally dumped and sometimes buried. (Dumps are often confused with landfills, which are much less polluting).

- A dump has various animals (rats, mice, birds) as well as many flies and other insects swarming around. All of these animals can act as vectors of disease.
- Water gathers in dumps, acting as a breeding ground for mosquitoes, the vectors of malaria, dengue, chikungunya and zika.
- The foul water that collects in a dump, or drains through it when it rains, can pass through and pollute the groundwater very easily.

A landfill, on the other hand, is a carefully designed structure (see Figure 22.22), built into or on top of the ground, in which waste is isolated from the surrounding environment (groundwater, air, rain):

- A landfill has a bottom liner, which prevents liquid from the waste from draining out into the surrounding soil and/or groundwater, isolating the waste from the environment.
- The waste is covered daily with a layer of soil, which reduces the numbers scavenging animals that feed there. This also reduces the numbers of potentially disease-carrying vectors. In a **sanitary landfill**, a clay liner is used to isolate the waste from the environment, while a **municipal solid waste (MSW) landfill** uses a synthetic (plastic) liner to do the same job.

Figure 22.22 On Antigua, several new landfill sites have been developed to cope with the growing levels of waste produced on the island.

Pollution and waste management

A Caribbean perspective

In the Caribbean, not all of the waste we produce gets collected and transported to well-managed landfill sites. In Jamaica, around 36% of households do not have waste collection services, while in the Dominican Republic that figure is 25%. Yet on many other islands, almost 100% of homes have a waste collection service. Why is there so much variability?

- **Limited space:** There is a limited amount of land on each island country. Developing landfill sites is not always top priority, as far more money can be made from tourist facilities and local people want land to build homes and grow food.
- **Lack of resources:** Sometimes we cannot afford collection services and management of landfill sites. If money is short, other priorities such as healthcare often get the money rather than refuse management.
- **The infrastructure of our islands is not always as efficient as we would like it to be:** Sometimes waste is not collected because the vehicles break down, the roads are inadequate or there is no more room at the dump or landfill.

How can we help?

As the Caribbean, along with many other countries in the world, struggles with a mountain of waste, is there anything that we can do to help? If everyone made an effort, we could have a big impact on the waste problem and really help to solve it. Up to 80% of the rubbish in our oceans – and so on our beaches (see the graph in Figure 22.23) – is plastic. Plastic does not break down and decompose – but a lot of it can be reused or recycled, even if we don't have a waste collection service.

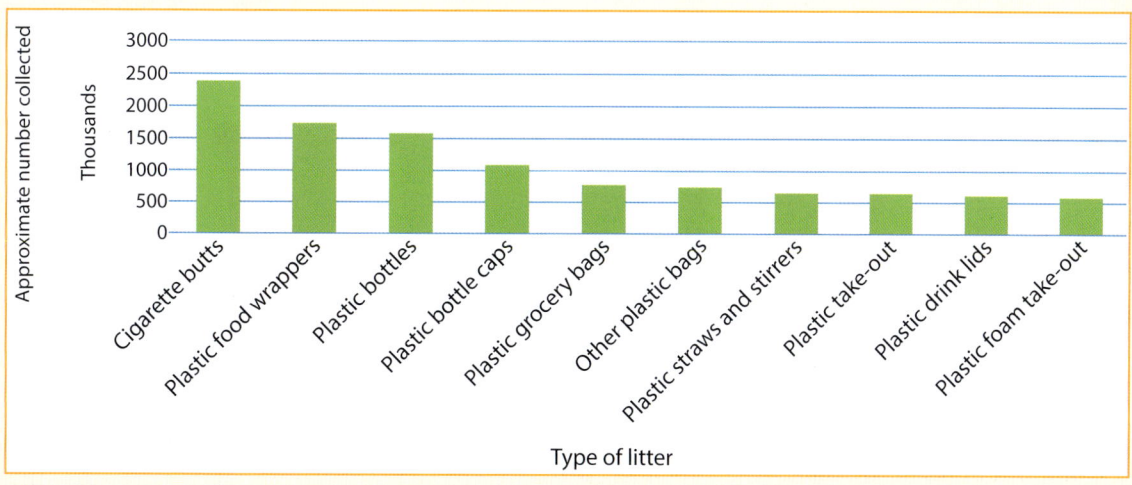

Figure 22.23 This data from global seaside litter collections shows that most of this litter is plastic. Much of this litter could be recycled or reused, instead of being dropped to spoil beautiful beaches.

We need to reduce the volume of solid waste produced, so that landfill sites last longer and our rubbish takes up less space. In our own home, we can contribute to waste reduction in many different ways. Some of them seem very small, but they all help. Often the waste is actually physically compacted and crushed, so that it takes less space in the landfill sites. This uses energy, and also slows down the process of decomposition as air cannot reach the material in the centre of the crushed material very easily.

Separation of waste

Household waste should be separated daily into different bags for the different categories of waste. We should have one bag for easily biodegradable waste such as leftover foodstuff, vegetable peels, etc. This will break down over time by the action of bacteria, so can be put in a compost pit. Even better, use a compost bin (see Chapter 4), which will keep out the rats that might otherwise feed on our leftovers. We can use the compost we produce as a fertiliser in our garden. It will both feed our plants and improve the texture of the soil. Where possible buy biodegradable plastics. They take quite a long time to break down, so won't really work on a compost heap, but they will ultimately reduce the volume of material in our landfills.

It is a good idea to have a box or bin for toxic wastes such as medicines, batteries, dried paint, old bulbs and dried shoe polish. These need to be disposed of carefully. See Figure 22.24.

Figure 22.24 Separating household waste will reduce the amount of waste in our landfills.

Our key actions to solve the problems of solid waste are (see Figure 22.25):
- **reduce** the amount of waste we make
- **recycle** and **reuse** as much as we can.

Pollution and waste management

Reduce

We need to reduce the volume of waste we produce as much as possible. To reduce the amount of waste material, we need to look carefully at what we buy. We should buy less and choose goods with as little packaging as possible, as much of our solid waste is actually packaging material.

Reuse

One of the biggest ways to reduce the waste we produce is to look for every possible opportunity to reuse items. So, for example, we can use old calendars or magazines to cover and protect books. This uses up paper and extends the life of the books. Old greeting cards can also be reused to make new ones with just a little effort. We can reuse plastic bags and paper carriers. Ice cream containers become storage boxes. Old clothes can be given to charity or reused for pet bedding. Paper can be shredded and used for packaging, or added to a compost heap or recycled to make more paper. Modern disposable nappies make life very easy for those of us with a baby, but they take centuries to break down and add greatly to the volume of solid waste in our landfills. Cloth nappies are reusable and so are a much more ecologically aware choice.

Figure 22.25 Reducing, reusing and recycling is good for the Caribbean and good for the planet.

Recycle

Dry or non-biodegradable waste consisting of cans, aluminium foil, plastics, metal, glass and paper can be put aside to be recycled. It takes a lot less energy, and therefore creates a lot less pollution, to recycle materials like aluminium, glass and paper than it does to extract them or make them. The material is processed, often melted down, and then reused to form new items. So recycling not only reduces the volume of solid waste going into landfill sites, it also produces less global warming. The number of materials that can be recycled is increasing all the time. See Figure 22.26.

In recycling centres, the different materials are separated. Some materials are sent to local businesses to be recycled and reused. Others are shipped as far away as China where they are again recycled and reused. This is big business.

Figure 22.26 Recycling resources is a very positive way to use less energy and produce less pollution. The Sustainable Barbados Recycling Centre sorts through 50 truckloads of waste every day.

How can we help?

Another exciting approach to recycling, which is growing in Caribbean countries is the production of **biodiesel** from waste vegetable oil. Schools, hotels and restaurants use huge amounts of vegetable oils in cooking. This oil is regularly replaced. This waste oil is filtered to remove bits of food, collected and converted into biodiesel. This in turn is used as a fuel in agricultural vehicles, lorries, and so on. Not only does making biodiesel from waste oil reduce waste – it also reduces the use of fossil fuels, so it benefits the environment in two ways.

Sorting and managing our waste like this will greatly reduce the volume of solid waste we send off to the landfill sites or dumps each week. If everyone does the same, just imagine the effect it would have.

There are many benefits of effective management of solid wastes across the Caribbean – but there are also some costs. To be successful in managing our ever-growing waste problems, governments and populations have to be convinced that the benefits outweigh the costs. See Table 22.2.

DID YOU KNOW?

At the Sustainable Barbados Recycling Centre, about 1000 tons (50 truckloads) of waste is delivered every day – and about 70% of that is sorted and recycled. That's a huge amount of waste that doesn't go to landfill and a lot of resources that are used sustainably.

Table 22.2 Benefits and costs of effective waste management systems

Benefits	Costs
• Reduces many public health problems, such as diseases spreads by vectors such as mosquitoes, rats • Prevents contamination of water sources by pathogens or chemical pollutants, and heavy metals • Provides direct economic benefits to individuals and community in recycling and reusing materials • Preserves the beauty of our environments for local people and tourists	• Initial investments needed to establish landfills, recycling centres, waste to energy processing, and so on • Relatively high annual operating costs for national waste management • Land used for landfill sites cannot be used for profit in building hotels/homes and so on

In our everyday life, we can do a great deal to minimise pollution if we take care to recycle materials and act responsibly with household chemicals and their disposal. There are choices we make each day that affect what we introduce into our environment. See Figure 22.27.

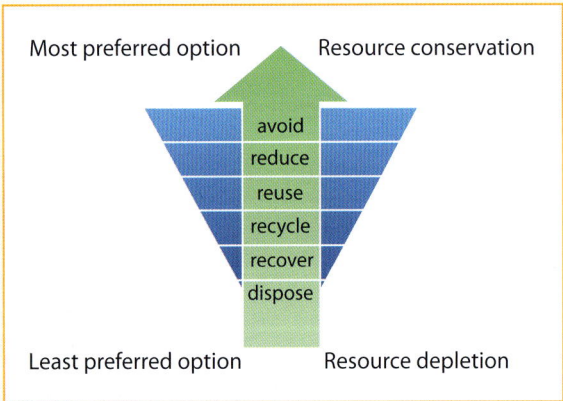

Figure 22.27 We make choices as individuals, schools, communities and nations about how we deal with our waste materials. For the health of our countries, we must all try to make the right choices.

Pollution and waste management

Environmental issues: The human impact

We humans are affecting our environment in many different ways. Air pollution, water pollution and land pollution are all creating environmental issues that are challenging ecosystems around the world. As we have seen through this chapter, the environmental issues we create also have an effect on us. In this final section, we will look at the impact of environmental issues on three important areas of human life. The main focus of these impacts is from global warming. Here in the Caribbean, we are particularly vulnerable to many of the impacts of global warming, especially rising sea levels and extreme weather events.

Food security

Food security means that everyone, all the time, has access to sufficient safe and nutritious food to meet their dietary needs and preferences to allow them to lead an active and healthy life.

There has never been a time when the whole world has experienced food security, but a growing world population and a range of environmental issues, which are the result of human activities, are making things worse rather than better. In the Caribbean we are often dependent on imported foods. Climate change as a result of global warming is one big factor threatening food security both in the Caribbean and globally.

- Temperatures are rising and there is less rainfall in areas of the world that grow much of our food, so there is less food to go round.
- In the Caribbean we are very vulnerable to losing crops to extreme weather events, especially flooding and hurricane damage. See Figure 22.28. This both reduces the amount of available food for us, and also damages our economy, as we have less to export.
- Extreme weather events in other parts of the world risk reducing global supplies of basic foods such as rice and wheat.
- As the sea temperatures increase, fish move to cooler areas – for example, yellow fin tuna and flying fish are moving northwards to cooler waters. This removes the livelihood and food supply of those who depend on them.

Another impact of global warming, which also has an impact on food security, is land security.

Figure 22.28 Global warming causes more extreme weather events like flooding, which threatens the Caribbean's food security and its economy.

Land security

Land security is threatened when sea levels rise, or when land becomes so damaged, it can no longer sustain life. As the climate of the Earth warms, land ice is melting faster and faster. As a result, sea levels are rising. Rising sea levels threaten small island nations, some of which will lose substantial amounts of land or even disappear completely. This is a real threat in the Caribbean.

In other areas of the world fertile land will become desert. This loss of land will mean many people will migrate to regions where there is still some fertile soil. This will also threaten global food security – see the previous page.

Health

Environmental issues impact the health of people everywhere. For example:

- Waste that is dumped becomes a breeding ground for vectors of diseases such as malaria, dengue and leptospirosis.
- Pollution of water sources with sewage or other waste increases the risk of diseases such as gastroenteritis OR of poisoning from heavy metals. See Figure 22.29.
- Increasing temperatures around the world mean the distribution of diseases will change – for example, tropical diseases such as malaria, dengue and so on, will spread to areas such as Europe.

As you have used this book, you have learnt a great deal about the human body and how it works in health and disease. You have seen the different ways in which our actions affect our health, from the lifestyle choices we make about our diet and our sexual behaviour to the way we manage our waste and wash our hands before preparing food. If everyone pulls together, we can reduce pollution, lower the spread of infectious diseases and see deaths from lifestyle diseases fall dramatically. Life in the Caribbean is good, but by acting on what you have learnt, you can make it even better.

The best tool we have in the fight against disease and pollution is education, enabling young people like you to make the right choices as you grow up and influence the country in which you live.

Figure 22.29 Polluting water sources with sewage increases the risk of diseases and reduces access to clean water.

Pollution and waste management

Possible SBAs based on feeding relationships and the carbon cycle in nature

→ Put together a demonstration/display of different types of pollution found in the Caribbean with labelled exhibits and explanations of the problems that result from each.

→ Put together a presentation on air pollution in your area – look at types of pollution, local sources, how this compares to the global picture and put together a plan of action to reduce air pollution locally.

→ Produce an animated/interactive model of global warming that demonstrates the effect of increasing or lowering the levels of greenhouse gases in the atmosphere.

→ Carry out pH testing on water from a number of sources near you and see if there is any evidence of acid rain.

→ Find pieces of coral on the beach and use them to investigate the impact of water at different pHs on the structure of coral – include explanations of acid rain and information on the threat of pH changes on Caribbean ecosystems.

→ Using agar plates, investigate local drinking water and other water sources to see if they are contaminated with bacteria. Observe all health and safety precautions when doing this.

→ Plan an investigation to compare the effectiveness of home water treatments.

→ Produce a leaflet or an online broadcast to educate people about the risks of eutrophication.

→ Investigate the way human waste is treated in your country – types and capacity of sewage treatment works, numbers of people with pit latrines, and so on, and write a report on sewage treatment in your country with as much data as possible.

→ Analyse the waste produced by your classmates/your school people in your family/your local community. Use the information you gather to develop a scheme to reduce waste production. If possible, evaluate the success of your ideas.

End-of-chapter summary

In this chapter, you have learnt that:

- there are pollutants of air, land and water
- the effects of pollutants on the environment are severe
- the main air pollutants include smoke, carbon dioxide, carbon monoxide, sulfur dioxide and nitrogen oxides
- a lot of air pollution comes from the burning of fossil fuels
- the increase in carbon dioxide and methane in the atmosphere is leading to global warming and climate change
- car exhausts are a major cause of air pollution
- using renewable energy resources such as wind and solar power to generate electricity reduces the carbon dioxide being emitted (going) into the air
- sulfur dioxide and nitrogen oxides produced by burning fossil fuels dissolve in the water droplets in the clouds to form acid rain; acid rain damages plants and buildings, and causes the pH of lakes, rivers and the sea to fall, killing animals and plants
- acid rain is a threat to the Caribbean coral reefs
- water pollution is a major problem in the Caribbean
- large-scale water purification plants ensure that most people have clean, safe water to drink
- water is made safe to drink at home by adding bleach or water purification tablets, or by boiling
- human effluent (sewage) is one of the main causes of water pollution
- there are two main ways of dealing with human sewage properly – sewage treatment plants and pit latrines
- untreated sewage causes serious damage to the environment, including the risk of diseases and eutrophication
- water is also polluted by run-off from farms – pesticides, herbicides and fertilisers
- solid waste needs to be properly managed as the impact of solid waste is far-reaching in our home and society at large
- the best and least polluting method of disposing of solid waste is to use landfill sites
- biodegradable waste can be broken down by microorganisms and should be composted; for example, food waste, paper
- waste cooking oil is used to produce biodiesel
- non-biodegradable waste cannot be broken down by microorganisms; for example, plastics, metal, glass
- there are a number of effective ways to cut down the volume of solid waste produced – reduce, reuse and recycle your waste.
- environmental issues impact on human food security, land security and health.

End-of-chapter questions

1. Which **one** of the following chemicals is not involved in air pollution?
   - A Carbon monoxide
   - B Carbon dioxide
   - C Sulfur dioxide
   - D Nitrogen (1)

2. Which **one** of these statements about the greenhouse effect and global warming is not true?
   - A Carbon dioxide and methane are the two gases that contribute most to the greenhouse effect.
   - B Smoke particles are an important feature of global warming.
   - C Most of the carbon dioxide comes from burning fossil fuels.
   - D Deforestation adds to the greenhouse effect, as there are fewer plants to absorb carbon dioxide during photosynthesis. (1)

3. Which is the correct order for these statements describing the formation of acid rain?
   - i The rain that falls containing the dissolved gases can have a pH as low as 2.
   - ii When fossil fuels are burned, sulfur dioxide and nitrogen oxides can be produced.
   - iii The gases dissolve in the water droplets in the clouds.
   - iv Acid rain lowers the pH of the surface layers of the sea and can damage coral.

   - A i, ii, iii, iv
   - B i, ii, iv, iii
   - C ii, iii, i, iv
   - D ii, iv, iii, i (1)

4. Which of the following is NOT a way of dealing with human body waste?
   - A Activated sludge treatment method
   - B Pit latrine
   - C Biological filter treatment method
   - D Sanitary landfill (1)

5. Which **one** of the following do bacteria and microorganisms need to live and be able to purify sewage?
   - A Adequate oxygen
   - B Adequate food
   - C A suitable temperature
   - D All of the above (1)

6. Which of the following examples of solid domestic waste is not biodegradable?
   - A Vegetable peelings
   - B Aluminium foil
   - C Meat trimmings
   - D Leaves (1)

7. Which **one** of the following is non-biodegradable?
   - A Car batteries
   - B Medicine bottles
   - C Plastics
   - D All of the above (1)

8. a Define the terms 'pollution' and 'pollutant', being careful to distinguish between the two. (2)
   b Name **three** ways in which air pollution can affect the environment. (3)
   c Explain how smoke can be a pollutant. (3)

9. Describe the formation and effect of acid rain. (10)

10. Cars and other motor vehicles are a major source of air pollution.
    a List the main pollutant gases produced in car exhausts. (10)
    b Explain how the increasing use of cars seems to be contributing to global warming. (8)
    c What are the main impacts of global warming on the Caribbean so far? (4)

11	a	Giving **three** named examples for each, explain the short- and long-term effects of air pollution on health.	(10)
	b	Explain how we can minimise, if not prevent, the damaging effects of air pollution.	(5)
12	a	What are the main sources of water pollution in the Caribbean?	(5)
	b	Give a brief account of the effects of water pollution.	(5)
	c	What advice would you give to help to decrease the problems of water pollution?	(5)
13	a	Draw an annotated flow diagram to show how an activated sludge sewage treatment plant works.	(10)
	b	Discuss the advantages and disadvantages of a sewage treatment plant and a pit latrine as ways of dealing with human sewage.	(5)
14	a	Define the term 'sewage'.	(1)
	b	Compare and contrast biological treatment in the secondary sedimentation tanks of a biological filter sewage works with the activated sludge tanks of an activated sludge sewage works.	(8)
15	a	Draw a suitably labelled diagram of a simple pit latrine.	(10)
	b	Name **two** diseases that may occur because of contamination of the water supply by human sewage.	(2)
	c	Give **two** advantages and **two** disadvantages of pit latrines over large sewage treatment plants.	(4)
16	a	Define the terms 'reduce', 'recycle' and 'reuse'.	
	b	Distinguish between a dump and a landfill site.	(6)
	c	What is the purpose of a landfill?	(1)
	d	Suggest **three** ways in which the volume of solid waste in the Caribbean might be reduced.	(3)

17

Kinnaman, Thomas C.. "Solid Waste Management in the Caribbean." Journal of Eastern Caribbean Studies (2010) : 38-60.

	a	Using data from the graph, state which countries in the Caribbean produce more than the average amount of solid waste and which produce less.	(9)
	b	Suggest **three** ways in which solid waste can be disposed of.	(3)
	c	Explain why it is so important to dispose of solid waste effectively.	(3)

End-of-section questions

1. a. Air pollution can negatively affect people who have respiratory illnesses such as asthma. What are the major sources of air pollution in the Caribbean? (3)
    b. What can individuals and people as a whole do to reduce air pollution? (4)
    c. Students in schools are told to follow the three Rs to help manage waste. What does this mean? (3)
    d. Give **five** examples of each of the methods in part **c** that you could follow. (5)

2. a. Draw a diagram of the water cycle and explain how it works. (6)
    b. Drinking water that is polluted can cause many illnesses.
        i. List **three** major sources of water pollution in the Caribbean. (3)
        ii. Briefly explain **two** different methods that you can use to purify water at home. (2)
    c. Tourism is being affected negatively by water pollution in the Caribbean. Explain what is meant by this statement. (4)

3. a. Explain the meaning of the terms: 'global warming', 'deforestation', 'composting' and 'sewage'. (4)
    b. Name **one** cycle in nature (not the water cycle) in which biodegradable waste matter is reduced. Explain briefly how it works. (5)
    c. Compare the advantages and disadvantages of the flush toilet and the pit latrine in the Caribbean. (6)

4. a. Methane and carbon monoxide are air pollutants. How are they produced? (2)
    b. What can we do to decrease the impact of methane and carbon dioxide? Give **two** suggestions for each gas. (4)
    c. Explain why the process of photosynthesis in plants is so important in controlling air pollution and global warming. (4)
    d. List the major causes of acid rain and explain briefly the major effects seen in non-Caribbean countries. (3)

5. a. Label the simple drawing of a water treatment plan below. Write the answers in your book next to the corresponding letters in the diagram. (4)

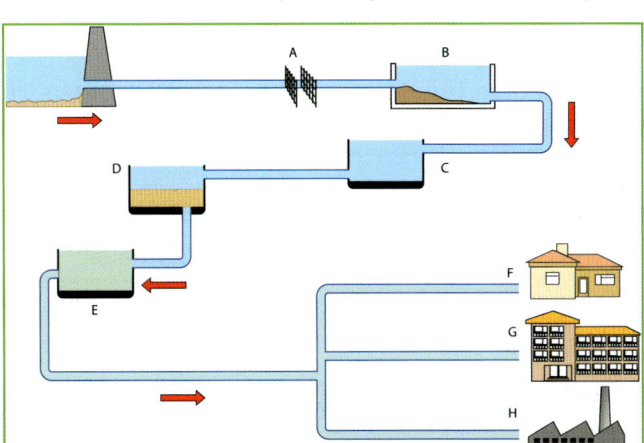

    b   Explain the activities in these stages of water purification in a water treatment plant.
- i Screening (2)
- ii Sedimentation (4)
- iii Chlorination (2)

    c   Name **three** illnesses caused by the long-term effects of air pollution. (3)

Data analysis

1 Using the data in the table below, construct a pie chart showing the amount of air pollutants in an air sample. (6)

Air pollutant	Concentration in air sample (p.p.m)
Carbon monoxide	560
Methane	450
Sulfur dioxide	350
Nitrous oxide	200
Hydrogen sulphide	100

2 Do you think that the air in this sample came from polluted air? Why? (3)

Case study

A new eco-resort is being planned in a Caribbean country that has been having decreasing amounts of rain. This eco-resort is being built in a forested hillside area approximately 10 kilometres from a very busy industrial estate that releases high levels of waste gases.

1 Which **three** pollution issues must be considered? Explain your answer. (6)

2 What advice would you give to the owner of this resort on how to reduce the effects of the pollution? (6)

3 What part will the surrounding forested hillside play in reducing the greenhouse effect on the eco-resort? (3)

Revision tool kit 5

Study groups

Study groups can be very effective for some people, but do not work very well for others. How well they work will depend on the members of the group! So, before you join a study group, you need to ask yourself a number of questions:

- How many people are in the group? Can you work well as a group of that size?
- Are there people in the group whom you find distracting? Will you distract others?
- Does the group meet at convenient times and places?

Here are some pros and cons of study groups.

If, after weighing the pros and cons, you decide that it is one of the revision strategies for you, the following may help you to make a study group work for you.

1. Look for people who motivate and inspire you. This is your future you are working on; you cannot involve people who are going to waste your time when you have worked hard to create a plan.

2. Keep it small. You might be tempted to have a 'more the merrier' approach to a study group, but that is really a party! A small group is easier to control; the discussions are less likely to get out of hand, and it is easier to arrange group meetings and make other decisions.

3. Select a group leader who is responsible and organised. This is especially important if organisation does not come naturally to you. It is good to have someone who can help to create a schedule of when you are going to meet, what you are going to study, who is responsible for preparing which aspect of the topic, and so on.

4. Be prepared. When you become part of a study group, you have an obligation to your group members in the same way that they have an obligation to yourself. When it is your turn to prepare a topic for discussion, you need to ensure that you turn up to that group meeting armed with all the information to share with your group members. Remember that they have done it for you and will continue to do it for you until the exam.

5. Develop a procedure. To make the study group meaningful and cut down on the possibility of off-topic discussions, it may be useful to develop a procedure. For example, when you meet, whoever is responsible for leading the topic discussion begins, whoever is responsible for supplying the examples to illustrate each point will insert them at the relevant points, and at the end, there is a summary of the main points and a working through of at least one exam-type question.

6. Everyone's views matter. Some people, if given the opportunity, will dominate the group. In order to avoid this, create discussion guidelines that allow each person to make at least one point.

7. Have fun. While you should never forget the purpose of the study group, you should never be so tense and serious that you completely forget to relax and have fun in each other's company.

8. Tick-tock. Study sessions should not last for several hours at a time. It is better to plan a shorter and more effective session than a long, non-productive session.

9. Avoid junk food. If you are going to snack during the study group session, eat something healthy that will not deplete your energy. By now, you should know that some foods benefit the body and others do not.

Section F: The School-Based Assessment (SBA) and the Alternative Paper

Learning outcomes

At the end of Section F you will:

- understand how to carry out a School-Based Assessment research project
- be able to produce your School-Based Assessment project report
- understand the assessment criteria used to mark your School-Based Assessment
- understand how to approach a Case study question in the Alternative Paper.

As you would have realised by now, Human and Social Biology (HSB) focuses on the structure and functioning of the human body and the interaction of human beings with their environment, both living and non-living. As importantly, from studying it you also learn how to adopt a healthy lifestyle, which is important to your enjoyment and quality of life.

For students in schools, completion of the **CSEC**® HSB course includes doing the SBA project, which is worth 20% of your final examination marks. Your SBA report should be presented electronically. For private students, 20% of your final examination marks is obtained from writing an Alternative Paper instead of doing an SBA project. In doing the SBA and Alternative Paper, you must apply your knowledge of research techniques, deep thinking and problem-solving skills. This enables you to further develop a critical understanding of how all living organisms work together in our world environment to sustain life.

HSB is an interdisciplinary course with a basis in subjects such as biology, chemistry, physics, geography, social studies and environmental studies. It provides a foundation for further study or specialised training in fields such as nursing, physiotherapy, dietetics, healthcare, some related environmental fields, public health and certain medical technician fields.

Chapter 23: Planning, conducting and reporting on your School-Based Assessment (SBA) project

The SBA project

You and your classmates should follow the SBA guidelines in the **CSEC®** Human and Social Biology syllabus. You can work either individually or in groups for the research part of the SBA project. The recommended group size is three to six students. Each student's SBA report should have their own presentation of data, analysis and interpretation of data, conclusion, recommendation, reflection and bibliography.

Planning your SBA project

You should plan your SBA project work as early as possible, so that you can meet the deadline for submission of the SBA project report to **CXC®**. This is important because your SBA project is a continuous or formative assessment exercise. If you present your teacher with a report without showing your work from planning and execution, it is very likely your teacher will ask for supporting evidence that you have carried out all stages of the project. This may also result in you having to do an oral evaluation.

Discuss realistic deadline dates with your teacher for doing your actual research, collating and analysing your findings and submitting written components of your project, including the final version. Certain project tasks may have to be done as field work outside of normal school hours, whereas others can take place during normal class time. Continuous assessment of draft components of your project and feedback by your teacher will help you to improve your drafts until the final version is ready for submission to **CXC®**. Your teacher plays a very important part in helping you to obtain the best possible score.

Keep a folder with your working documents, including your planning ones, notes, ideas and observations from start to end of your SBA project. These items will be useful in helping you to write your project report. Draw up a personal planning calendar to help you keep to deadlines. Add any time adjustment(s) as your project progresses. Highlight the final submission deadline for your SBA project report so that it's always visible!

Choosing an appropriate School-Based Assessment (SBA) topic

The issue you select should be related to the syllabus and must be on a current health-related or environmental problem or issue. Write a shortlist of issues you are interested in or feel strongly about. Then carry out some background research on each one. For example, read the relevant chapter in your textbook, look at a YouTube video or do some internet research. Write a problem statement for each issue you have on your shortlist.

Discuss your problem statements with your teacher who will advise on their suitability and help you to select the most appropriate one. The following guidelines are also helpful:

* The problem you select should be very interesting and meaningful to you. This will inspire you to do the research.
* The problem statement should be specific, feasible and realistic.
* Choose a problem for which the research and project report writing can be completed within the allocated timeframe and for which you can meet deadlines for the project stages. Remember that you will most likely have to do SBAs for other subjects and must allocate your time wisely.
* Plan your research project so that it can provide data on your analysis and interpretation, which can help you to make useful recommendations to relevant persons within the school system, the local or wider community, to help alleviate the problem.

> **Remember**
> Aim to have 1000 words in your project report. If your project report is longer than 1150 words, you will lose 10% of your project score.

Some ideas for SBA project topics

In addition to suggested SBA project topics provided throughout the textbook, here are a few more:

* The prevalence of hypertension in different age groups in the community
* Organising a programme to control an insect-borne disease such as dengue fever or malaria
* Classification of waste in the school/classroom over a defined period of time
* Can vertical farming make growing plants more efficient for small homeowners?
* Investigating the benefits of certain foods using food labels
* Is there an increase in the use of genetically modified (GM) foods by families in the community?
* Changes from unhealthy to positive lifestyle choices can improve health and quality of life over a defined time period
* The impact of rat infestation at a particular site
* Prevention of eutrophication in named waterways
* Investigating different types of pollution in the neighbourhood
* Food tests experiments to determine the food nutrients in foods regularly eaten

Support from your teacher

* **Choice of issue or topic:** Your teacher can help you to select an issue that is related to your own experiences and is meaningful to you. Discuss this with your teacher.

* **Problem statement:** Your teacher can help you refine this.

* **Setting of stages:** Your teacher will help you to decide on stages in your SBA project and deadlines for each of these.

* **Timely and constructive feedback:** Your teacher can provide you with feedback as you move from stage to stage of your research project. This will help you to grow in confidence and competence. Remember to make required changes to your project work based on your teacher's feedback.

* **Requirements of the project, the assessment criteria and the mark scheme:** Discuss all of these with your teacher to ensure that you align your project with what is required by **CXC®**.

* **Anti-plagiarism check:** As your teacher monitors and evaluates your work, they will impress upon you and remind you that your work is done by you and is not copied from the work of others.

* **In class and out of class work:** Your teacher will schedule in-class time for you and your classmates to work on your SBAs. Some work like field research would most likely have to be done outside of class time, but your teacher can help you plan this so that you carry this out efficiently.

* **Marking of SBA reports and submission to CXC®:** Your teacher will collect and mark all SBA projects, keep records and submit these to **CXC®**.

* **Moderation of the SBAs in your school:** Your teacher will advise you on the moderation procedure and let you know if you will be expected to participate in this and, if so, in what manner.

Format of your SBA report, assessment criteria and mark scheme

Once you have chosen your topic, written your problem statement and have decided on the deadlines for the stages in your project, then you can start your research and writing parts of your first draft of your project report. Your SBA report should have the following components in the order shown in the next few pages. You can use Table 23.1 as a checklist for coverage of items and marks allocation, to help you achieve maximum coverage and the highest mark attainable.

Table 23.1 SBA report format, assessment criteria and mark scheme

No.	Components	Assessment criteria and elaborations	Marks
1	Cover page	Your full name as on your registration form, your **CXC®** candidate number, name of subject and date of SBA submission should be on the cover page.	(See Component 12 on page 498)
2	Table of contents	This should have the list of components of the report in the correct order (numbers 2–11) with the corresponding page numbers.	
3	Background/ Overview of the issue	*Write a comprehensive, detailed description of the health-related or environmental issue you have chosen for your research project.* Background information should indicate the root of the problem being studied. (Use questions to guide you, for example: *What is the issue? Where is its impact felt? Who is affected by/has caused this issue? When? How does it affect humans and/or the environment? Why did you choose this issue?*) Demonstrate that you have thoroughly researched your topic; show this by discussing the breadth and depth of prior work in this area. Provide the foundation to support your research question. It should not be ambiguous, disorganised and have unrelated information.	2 (Total = 2)
4	Problem statement	*Write a clear and precise research problem, in one sentence.* *Write the research problem as an observation or question.* A problem statement is a short, clear sentence defining the the part of the issue you are researching. It should logically lead to the research objective.	1 1 (Total = 2)
5	Research objective	*When writing your research objective, ensure that it is linked to the issue being researched.* Reread your problem statement and your research objective. They should show a clear correlation to each other. *Your research objective should be specific, measurable, achievable, realistic and time bound.* The research objective should state what the purpose of your study is. It can be written in one sentence starting with 'To' and a verb, for example: *To investigate …*	1 1 (Total = 2)

No.	Components	Assessment criteria and elaborations	Marks
6	Methodology	The methodology describes how you collected data. It should be written in the past tense. It should describe the sample used to collect the data, the method used to collect the data, the data collection instrument and at least one limitation of the data collection method.	
		Sample – Identify and describe.	2
		Who is participating in the research, are they in the school or the community? How many participants? The larger number sample may give better results but may be too time-consuming. Use of a consent form may be needed stating the research objective; participants or guardians by signing this are confirming their willingness to honestly take part in the research, with the understanding that you will treat their contributions with confidentiality and protect their anonymity.	
		Data collection instrument – Identify and describe.	2
		What is the data collection instrument you have used? Give a clear description of how it was used.	
		Data collection method – Identify and describe.	2
		Name the data collection method and clearly describe how it was applied. You can write the steps on how it was done as part of the description.	
		Deciding on the instruments and methods to collect data should be done after careful thought, preliminary research and consulting your teacher or group members. Data can be collected by observation checklist, questionnaires, interviews, surveys, focus groups and experiment/investigation.	
		Data collection method – Justify or give your reason for using this method.	1
		Based on your thoughts, discussions with your teachers and group members, write your reason(s) for selecting the method of data collection you used.	
		One limitation of the data collection method	1
		The limitations of the data collection method can impact or influence the interpretation of the findings from the research. Some limitations can be sample selection, unavailability of resources, small sample size, limited access to data and time constraints.	(Total = 8)

No.	Components	Assessment criteria and elaborations	Marks
7	Presentation of Data	*Appropriate presentation using tables, graphs, charts or pictures*	1
		Select the best format(s) so that all of your data can be utilised and displayed optimally. When your data is displayed properly it is easy to discover trends or findings.	
		Data is presented in at least two formats	2
		Such forms can be the use of bar graphs, pie charts, line graphs, histograms and tables. Pictures may also be appropriate for certain issues. Using two formats should assist you in better analysing your data.	
		Data presentation formats are labelled accurately	1
		All items should have an accurate title and be numbered in logical sequence. Graphs and charts should have labelled axes and appropriate units. Any essential rubrics should be included. Tables should have explicit headings and be easy to understand.	
		Data used is accurate	1
		The data you present should be accurate. Check your data collation before you put it into your presentation formats.	
		Re-check your data, your collation of your data and the formats that you have created. Data presented in the various formats should match the original data collected.	(Total = 5)
8	Analyse and Interpret Data	*Data is analysed using appropriate methods*	1
		Data analysis and interpretation refers to obtaining meaning from the data you collect. From analysis, you can propose a conclusion, and support the importance and implications of findings. Data can be analysed using calculations such as fractions, percentages, using statistical methods, for example, mean, median, mode and grouping of discrete or whole items under themes or headings.	
		Data is summarised accurately and relate back to the issue	1
		When analysing data you can use terms like majority, minority, less than, more than and use percentages if appropriate. You should identify patterns and recognise relationships from the data your research has revealed. Data must be seen to relate to your chosen issue and problem statement.	
		Write at least two statements of findings.	2
		What have you noticed from your data and its presentation in specific formats? You should write at least two findings but do not limit yourself to two, if there are other important findings.	
		Findings are aligned with analyses	1
		Check that your two (or more) statements of findings are substantiated by your analyses of the data.	(Total = 5)

No.	Components	Assessment criteria and elaborations	Marks
9	Conclusion	*Comprehensive summary of the project*	1
		Your conclusion must relate back to the research objective. It must be concise and straightforward with no descriptions, opinions, speculations or interpretations.	
		Your summary must make sense and be based ONLY on the findings of your research.	1
		Read your conclusion aloud. Does it make sense? Is it based solely on your research findings?	
		Ask a group member or another classmate to read your conclusion and then explain what they understand from it.	(Total = 2)
10	Recommendations	*Write at least two recommendations based on your findings.*	2
		Recommendations are specific solutions to the problem investigated in your research project. As you were working through your project, some solutions to the problem should have become visible. Write down all of these and then select the two that seem easiest to implement.	
		Your recommendations should be realistic.	1
		They should be concrete, specific, realistic and achievable. They should not create new questions or problems; they should be achievable with resources that are available or easily accessible.	
		Your recommendations should be based on your findings.	1
		After you have written your recommendations, check them against your findings to ensure that they are based on your findings.	(Total = 4)

No.	Components	Assessment criteria and elaborations	Marks
11	Reflection	*Write two lessons you have learnt from working on the research project.*	2
		Your reflections are personal and subjective. Even if you worked in a group to collect data, each group member will have different reflections. How has this project changed your perception and thoughts on how to do research? How has it changed your opinion on any of your beliefs or behaviours?	
		What is one way in which you can apply the lessons you have learnt to your personal life?	1
		What positive action can you take or what skill can you commit to or continue to improve based on your learning during the project?	
		What is one way in which the project could be improved?	1
		Refer to any problems or limitations you had while carrying out the project. Look at your notes in your project folder to make sure you have correctly identified the problem. Suggest one way to improve your project.	
		One social impact that the issue could have on the school/community	1
		This refers to the effect of the issue on the health and well-being of the wider community and the school, if it is not handled properly.	
		One economic impact that the issue could have on the school/community	1
		This refers to the effect of the issue on the finances and availability of goods and services for use by the community or school, if it is not handled properly.	(Total = 6)

No.	Components	Assessment criteria and elaborations	Marks
12	Overall presentation	*The layout of the report follows the set format* This is shown in the Components numbered 1 to 11 on the previous pages. It also includes the bibliography and the appendices. Appendices contain raw data or supplementary materials, which are expansive, such as questionnaires, interview questions, surveys, checklist or extensive raw data. Each piece of information must be in a separate Appendix. *Bibliography contains a complete list of sources used* The bibliography refers to a list of sources (for example, books, articles, websites) used to write the project. It usually includes all the sources consulted even if they are not directly cited in the research. The bibliography should contain names of authors, publishers and dates of publication. *Spelling and grammar* You must ensure that your spelling and grammar throughout the submitted report are correct. Use spellcheck on your computer as well as any grammar check.	2 1 1 (Total = 4)

Conducting your research

Continue to use your working document folder to keep copies of all of your plans, ideas, notes to yourself and observations as you conduct your research. Your teacher can have a look at this as evidence that you are doing authentic research. Write a chronological list of the steps you have taken as your research progresses. This will help you to manage your project and make timely changes.

Data collection methods and instruments

Data can be collected from speaking to people, conducting experiments, surveys, observations of real-life situations or the environment. Primary data is collected directly by you from the source of the data. Secondary data can be extracted from existing or published data, such as hard and electronic copies of reports by reputable organisations and statistical offices, newspaper articles. Data collection instruments should be placed in the Appendices.

Some data collection methods are the following:

* **Observation checklist**

    An observation checklist is a list of questions that you want answers for on the specific issue being investigated. This checklist should have details of what, when and where you will be carrying out your observations.

- **Investigation**

    An investigation is observing or studying the natural world, without interference or manipulation. An investigation in its simplest form involves collecting information. Some science investigations can be experiments if they involve testing a hypothesis by changing one variable while keeping the other factors constant.

- **Experiments**

    Experiments should have controls and variables. Measurements should be linked to the identified problem in the research. Experiments chosen should be suitable for the health-related or environmental issue. An experiment is a tightly controlled investigation that involves variables (independent/manipulated and dependent/outcome) and establishes cause-and-effect relationships.

- **Interviews and focus groups**

    Interviews are a method of data collection that involves two or more people exchanging information through a series of questions and answers in order for the interviewer to collect data. You as the interviewer should prepare your questions before the interview so you can keep your focus on the topic of the interview. You can use a mix of closed and open-ended questions.

    - **Focus group interview** is a similar approach where a group of respondents are **interviewed** together. Focus group encourages more discussions. All members must respect one another. However everyone must be given a chance to express their opinion. The student can use either way in order to gain an in-depth understanding of the issue. Interviews allow the student to take control of the discussions and to add questions if needed.

    - **Surveys**

        Surveys are research methods that use questionnaires to gather information from a specific set of respondents, evaluate this data and draw conclusions. A survey may contain questionnaires and other data collection methods. Students can use a survey to collect, analyse and interpret data from a wide range of individuals. However, it is costly and time-consuming.

    - **Questionnaire**

        A questionnaire is a research instrument consisting of a series of questions for the purpose of gathering information from an individual (respondent) or groups of persons. The questions should be precise and should elicit concise answers. It should be user-friendly. It is a relatively inexpensive, quick, and efficient way of collecting large amounts of data even when the researcher is not present to collect those responses first-hand. Writing questions for a questionnaire takes time, thought and practice!

Types of questions you can write for a questionnaire

1. **INFORMATION OR QUANTITY** – These require open-ended answers in words or numbers. Two examples are:

    a What is your favourite subject ? _____
       (Write name of subject, e.g. Human and Social Biology)

    b How many subjects are you currently studying? _____
       (Write number, e.g. 7)

2. **CATEGORY** – This requires choosing of an answer from a group of answers, e.g.

    Have you completed your homework on time over the last term?

    Yes _____   No _____   Sometimes _____

3. **SELECT OR MULTIPLE CHOICE** – Chose one or more answers in response to a question, e.g.

    Circle all of the subjects you have studied since you entered secondary school

    English   Language   Mathematics   Social Studies   Integrated Science

    Spanish   French   History   Geography

4. **SCALE** – Answers to the question are given on various scales from low to high values either in words or numerically.

    a **Attitude** – The question asks you to choose an opinion from a set scale, e.g. How important is HSB to you in preparing for a career? Tick the box.

Unimportant ☐	Not very important ☐
Somewhat important ☐	Important ☐
Very important ☐	

    b **Ranking** – The question ask you to rank each answer from high to low or vice versa, e.g.
       How interesting are the following subjects to you?
       Rank these with 1 being the most interesting and 6 being the least.

Mathematics ☐	Biology ☐
Spanish ☐	Physical Education ☐
Physics ☐	Technical Drawing ☐

    c **Likert scale** – A series of related questions are given and you are asked to select the most appropriate answer from those given, e.g.

Tick one answer for each question.

Q	Question	Agree	Agree somewhat	Neutral	Disagree somewhat	Disagree
1	I love attending school.					
2	School is important in my holistic growth.					
3	All subjects at school are equally important.					
4	Male students are better at Physical Education than female students.					
5	Homework is necessary to do well in all subjects.					

You can research further information on these and other types of questions. Be sure to consult your teacher as you prepare your questionnaire.

Presentation of data

You can use a variety of different techniques to present data. Your ultimate aim is to make the data, patterns and relationships easy to see, not to create confusion for the reader. If a method does not show everything, it may not be the right one to use. Using one type of data presentation over another can have its own benefits. For example, a pie chart shows the relative size of each of the categories compared to each other by using percentages. A good data presentation will also make the reading of the results more interesting to the reader. You would want your teacher and moderator (if your report is selected) to understand your presentations, as well as find them interesting.

Pie chart

This is a visual representation of percentages as a relative proportion of the total of a circle. A pie chart is a circle divided into sectors (think of them as the slices of a cake). 100% represents the whole complete circle, 50% represents a half circle, 25% is a quarter circle, and so on. It can be drawn manually or be using Microsoft Word or Excel.

Calculate the percentages of each category as follows:

$$\text{Percentage of each data category (\%)} = \frac{\text{value of the specific data}}{\text{total value of data}} \times 100$$

Calculate the angle at the centre of the circle as follows:

$$\text{Angle of sector (}^\circ\text{)} = \frac{\text{percentage value of category}}{100} \times 360^\circ$$

The SBA project

Example

The table below shows the number of items disposed of by 25 students in appropriate bins in a classroom over one week.

Type of Item	Total number of items in one week	Percentages of each item type %	Calculations of the angles for each sector °
Plastic	172	$\frac{172}{435} \times 100 = 39.54$	$\frac{40}{100} \times 360 = 144$
Glass	33	$\frac{33}{435} \times 100 = 7.59$	$\frac{8}{100} \times 360 = 29$
Metal	76	$\frac{76}{435} \times 100 = 17.47$	$\frac{17}{100} \times 360 = 61$
Paper	154	$\frac{154}{435} \times 100 = 35.40$	$\frac{35}{100} \times 360 = 126$
	TOTAL 435		

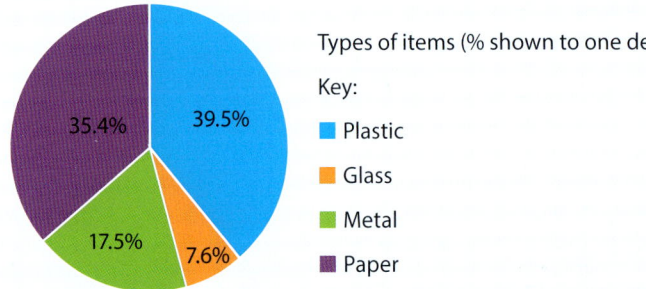

Types of items (% shown to one decimal place)

Key:
- Plastic
- Glass
- Metal
- Paper

Figure 1 Pie chart showing the number of items disposed of by 25 students in appropriate bins in a classroom over one week.

Line graphs

Line graphs are used when the dependent variable changes with respect to the independent variables. The data from both dependent and independent variables are measured in numbers. The graph must have a clear, descriptive title that outlines the relationship between the dependent and independent variable.

A variable is any factor that can be controlled, changed, or measured in an experiment. Variables are an important part of science. A variable is a factor or a quantity whose value may vary depending on the values of the other variables.

Independent or manipulated variable: The independent variable is changed or controlled in an experiment. It is sometimes referred to as a predictor variable because it helps to 'predict' and influence changes in response or the dependent variable.

Dependent or responding variable: The dependent variable is the variable that you measure or observe since it is dependent on the manipulated variable.

Features of line graphs

* The independent variable is placed on the *x*-axis and the dependent variable is placed on the *y*-axis.

* The scale must remain the SAME along the entire axis and use easy intervals such as 10s, 20s, 50s, and not intervals such as 7s, 14s, and so on, which makes it difficult to read information off the graph.

* Each axis must be labelled with what is shown on the axis and must include the appropriate units in brackets, for example: Temperature (°C), Time (days), Height (cm).

Example

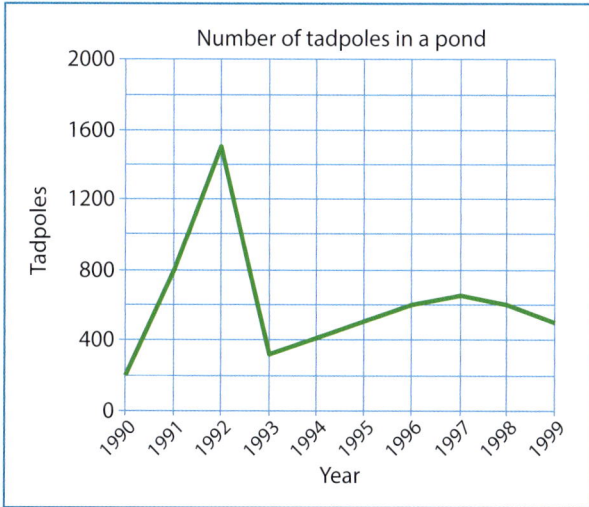

Figure 2 Graph showing the number of tadpoles in a pond each year between 1990 and 1999

Bar graphs

* These are used when the independent variables represented on the *x*-axis are not numerical. For example, when examining the protein content of various food types, the order of the food types along the horizontal axis is irrelevant.

* The data are plotted as columns or bars that do not touch each other since they represent discrete items.

* The bars must be the same width and be the same distance apart from each other.

* The height of the bars varies according to the data plotted as the dependent variable.

* A bar graph can be displayed vertically or horizontally.

* A bar graph must have a clear, descriptive title, which is written beneath the graph.

The SBA project

Example

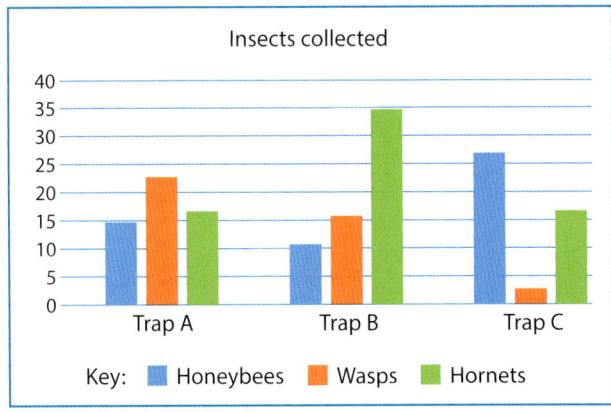

Figure 3 Bar graph showing the different types and number of insects collected in three traps over a week

Histograms

* In a histogram, the data are plotted as columns or bars that touch each other. Histograms are used when the independent variable (*x*-axis) represents information that is continuous, such as numerical ranges, that is, 0–9, 10–19, 20–29, etc.

* The numerical ranges *must not* overlap and must be exclusive so that there is no doubt as to where to put a reading, for example: 0–9, 10–19, 20–29, and so on.

* The bars can be vertically or horizontally drawn.

* A histogram must have a descriptive heading, which is written below it.

* The *x*- and *y*-axes must be labelled.

Example

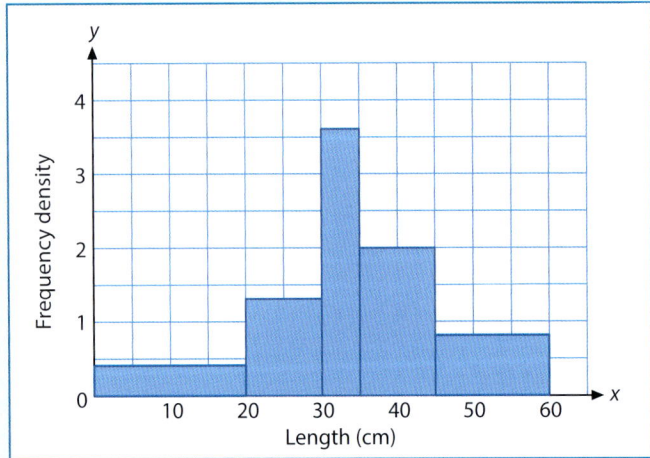

Figure 4 Histogram showing the number of plants with length/height after one month of seedling growth

Planning, conducting and reporting on your School-Based Assessment (SBA) project

Tables

Tables are used to summarise information and to compare related things or aspects. The results of an experiment can be recorded on a table. A table can illustrate patterns and trends. Tables are used to record the data that will be used to construct graphs.

* A table is a summary of data, using as few words as possible.
* It is a grid made up of rows and columns.
* The title is placed above the table and must include both independent and dependent variables.
* The independent variable is placed in the first column.
* For numerical tables, the column headings should mention physical quantity measured and the units used for example: Time (min), Weight (kg), Length (cm) and Temperature (°C)
* Be consistent with using two decimal places for numerical data, unless told otherwise.
* For non-numerical tables, the correct headings must be given and all data must be recorded.

Example

Plant growth in soils with different pH values

Plant Group	pH of soil	Average plant growth (cm)
1	6.0	25.4
2	6.2	33.0
3	6.4	50.8
4	6.6	53.3
5	6.8	53.3
6	7.0	30.5
7	7.2	22.9

Writing a bibliography

A common format for writing the bibliography is the APA style created by the American Psychological Association.

* Alphabetise the entries in your list by the author's last name. Only the initials of the first and middle names are given. If the author's name is unknown, alphabetise by the title, ignoring any 'A', 'An', or 'The'.
* If there is more than one author, use an ampersand (&) before the name of the last author. If there are more than six authors, list only the first one and use *et al.* for the rest.

* For dates, use either the day-month-year style (22 July 1999) or the month-day-year style (July 22, 1999) and be consistent. Place the date of publication in brackets immediately after the name of the author.

* You should capitalise only the first word of a title and subtitle and proper names in a title. Publication names should be in italics.

Formats and examples

1. Books

Author's last name, first initial. (Publication date). Book title. Additional information. City of publication: Publishing company.

Example

Nicol, A. M., & Pexman, P. M. (1999). *Presenting your findings: A practical guide for creating tables*. Washington, DC: American Psychological Association.

2. Dictionary/Encyclopaedia

Author's last name, first initial. (Date). Title of Article. Title of Encyclopedia (Volume, pages). City of publication: Publishing company.

Example

Merriam-Webster's collegiate dictionary (10th ed.). (1993). Springfield, MA: Merriam-Webster.

3. Magazine and Newspaper Articles

Author's last name, first initial. (Publication date). Article title. Periodical title, volume number (issue number if available), inclusive pages.

Example

Harlow, H. F. (1983). *Fundamentals for preparing psychology journal articles*. Journal of Comparative and Physiological Psychology, *55*, 893–896.

4. Website or Webpage

* When citing internet sources, refer to the specific website document. If a document is undated, use 'n.d.' (for no date) immediately after the document title.

* Break a lengthy URL that goes to another line after a slash or before a period.

* There is no period following a URL.

* If you cannot find some of this information, cite what is available.

Online periodical

Author's name. (Date of publication). Title of article. Title of Periodical, volume number, Retrieved month day, year, from full URL.

Online document

Author's name. (Date of publication). Title of work. Retrieved month day, year, from full URL.

Health Canada. (2002, February). *The safety of genetically modified food crops*. Retrieved March 22, 2005, from http://www.hc-sc.gc.ca/english/protection/biologics_genetics/gen_mod_foods/genmodebk.html

Hilts, P. J. (1999, February 16). *In forecasting their emotions, most people flunk out*. New York Times. Retrieved November 21, 2000, from http://www.nytimes.com

End-of-chapter summary

- Plan your project research as early as possible. Do a project plan with deadline dates for the stages and keep to this.
- Your SBA report should **not have more than 1000 words**. This excludes bibliography, charts, graphs, tables, pictures, references and appendices, since these are not part of the word count.
- 10% of your project score will be deducted if your report exceeds the maximum length of 1000 words, by more than 150 words.
- Your project must be **authentic**, that is, done by you. **Plagiarism or copying** of other persons' reports or publications can result in your report being rejected.
- The **role of your teacher** is critical in carrying out a successful SBA project, so ask your teacher for guidance and tips as you work through your project.
- Follow the **format, assessment criteria and mark scheme** prescribed by CXC® for the SBA for this subject to achieve the highest mark possible.
- There are several research methods and instruments. Select those best suited to your research topic so that you can obtain meaningful data.
- There are various ways to present data. Use the ones which best present your project data.

Chapter 24: A Sample School-Based Assessment (SBA) report and a worked Case study

Why look at samples?

Even though you may understand the theoretical approach to doing a School-Based Assessment report, it is useful to look at samples of SBA reports so that you can see how these were actually done. This can help guide you in writing up your own SBA report. You should remember, however, that your approach should be appropriate to the topic you have chosen and that you should not copy the reports of other students.

If you have signed up as a private candidate, you will be writing an Alternative Paper 032, which consists of questions based on a Case study. These questions relate to the same knowledge and skills that are being assessed in the SBA research project in Paper 031. Looking at samples of Case studies and questions based on these will help you to understand the types of questions that can be asked and how to approach answering these questions.

A sample School-Based Assessment Report

Topic
The impact of the COVID-19 pandemic on my community.

Background
The COVID-19 epidemic started in late November 2019 and quickly developed into a pandemic in early 2020. This was due to the high transmission rate of the virus by airborne droplets by both symptomatic and asymptomatic carriers. In order to avoid large numbers of persons contracting and transmitting the disease, countries asked their populations to stay at home, exercise social or more precisely physical distancing, wash and sanitise hands frequently and wear masks.

The COVID-19 pandemic has had far-reaching social and economic consequences. Sectors that rely heavily on physical presence, including passenger transportation, the arts, entertainment, cinemas, restaurants, tourism, education and sports were closed.

The result of the restrictions resulted in the closure of many companies, a reduction in staff and decreased spending capacity, particularly for lower-income families. Many individuals temporarily or permanently lost employment. Reductions in social gatherings such as family get-togethers, weddings, attending religious events, funerals and other functions resulted in social isolation due to social-distancing guidelines. Students were unable to attend physical classes but were given classes online. Some students were unable to attend online classes since they had neither electronic devices such as personal computers, smartphones or tablets, nor reliable internet connectivity.

Problem statement
How did the COVID-19 pandemic impact families in my community in relation to health and finances?

Research objective
To investigate the impact of the COVID-19 pandemic on the physical and mental health, and financial situation of families in my community

Methodology
I did a survey using a questionnaire on how COVID-19 affected the health and financial status of 30 families. I asked a few of my classmates to fill it out before I gave it to one member of each of 30 families in my community. The families were randomly selected and they agreed verbally to take part in the research. Through a combined letter and questionnaire (see Appendix A), the families were informed formally of the research project and given assurances that their anonymity and confidentiality of responses would be kept. They were asked to complete the questionnaire honestly.

The questionnaire consisted of 16 questions, as seen in Appendix A. Questions 1–5 asked about the physical health of members in the families. Questions 6–11 dealt with stress as one mental health issue caused by COVID-19. Questions 12–16 asked about the impact of COVID-19 on the family finances. I collected the completed the questionnaires after one week.

I chose a questionnaire to collect data because it offered a fast, efficient and affordable way of gathering large amounts of information from the community sample. Because they are anonymous, I expected families to put forward truthful answers so that I would obtain reliable data.

One limitation of using the questionnaire I designed is that it did not allow for collection of data, which family members may not want to write down but would probably have been more willing to speak about.

Presentation of data

A tally chart was used to collate the responses provided by all 30 families. The tally chart is shown in Appendix B. The data collected is in Table 1, parts (a) and (b).

Table 24.1 (a) and (b) show the responses from 30 families in my community on how COVID-19 impacted their lives in the first two years of the pandemic.

Table 24.1 Table showing responses from 30 families on how COVID-19 impacted on their lives in the first two years of the pandemic

a

Question number	Question content	No change	Less	More
1	Consumption of junk and fast foods	5	0	25
2	Alcohol consumption in persons over 18 years.	23	7	0
3	Physical exercise	4	26	0

b

Question number	Question content	Yes	No
4	Did any members of your family become ill with COVID-19? If yes, how many?	25	5
5	Death of a family member or very close friend as a result of COVID-19	24	6
6	Being worried that a family member will catch COVID-19 because they have an underlying medical condition	20	10
7	Any family member suffered with depression, loneliness or anxiety	17	13
8	Stress of children/grandchildren losing education over closure of school	17	13
9	Worrying that you may become infected with COVID-19 and then infect other people	17	13
10	Worried about attending place of worship, weddings, funerals, social events	30	
11	Feeling stressed due to not being able to meet with family and friends	30	
12	Experiencing financial hardship due to job loss or loss of earnings	24	6
12	Having difficulty getting supplies when needed, e.g. medicines, food, drinks or other essentials because of lack of income	20	10
14	Inability of household to meet financial obligations (loan repayments, household bills, etc.)	20	10
15	Any financial support from the government, friends, family or community	13	17
16	Concern about the future with respect to financial security, stability	30	0

Questions 1–16 above were placed into three categories of impact by COVID-19 families: physical health, stress and finances. The physical health and finances categories consisted of five questions. The stress category consisted of six questions.

Table 24. 2 shows the number of families affected in each category. A bar chart was used to illustrate the responses.

Table 24. 2 Table showing the number of families affected by COVID-19 with respect to health, stress and finances

Question	Physical health	Question	Stress	Question	Finances
1	25	6	10	12	24
2	7	7	13	13	20
3	26	8	17	14	20
4	5	9	17	15	13
5	6	10	30	16	30
		11	30		
Total responses per category	69		117		107

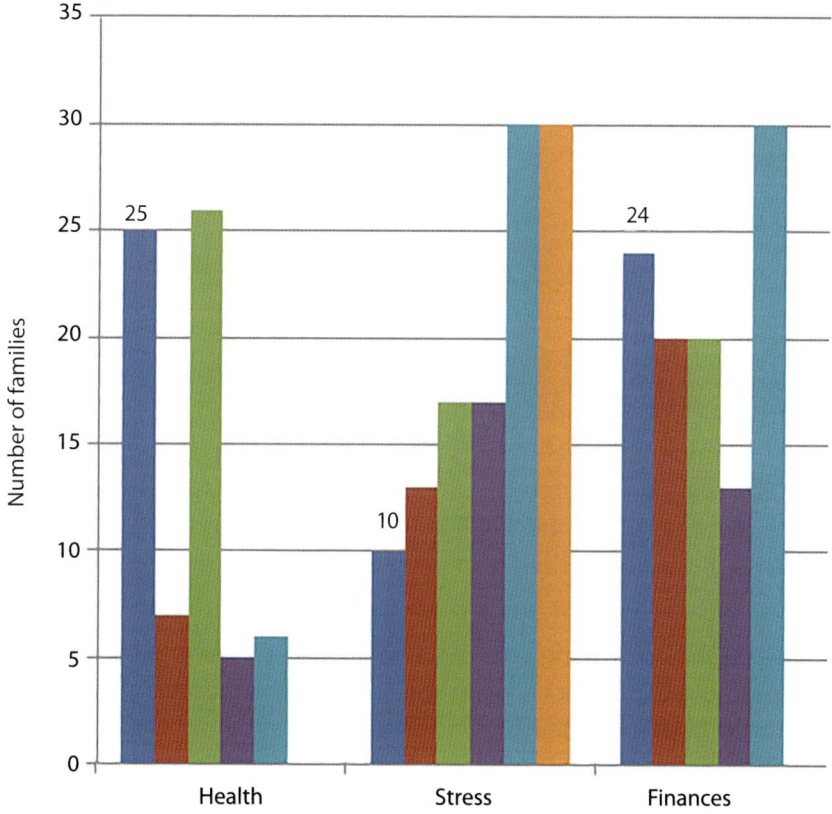

Figure 24.1 Bar chart showing the number of families affected by COVID-19 in terms of physical health, stress and financial issues

Table 24.3 was constructed to show another representation of data to facilitate the comparison of the negative impact of COVID-19 on the physical health, stress and finances of the families.

Table 24.3 Table showing the total negative impact responses per categories of health, stress and finances using percentages

Factors	Total number of responses	% of total number of responses	Angle of sector for each category 0
Physical health	69	23.5	84.6
Stress	117	40	144
Finances	107	36.5	131.4

The above percentages are represented in the pie chart below.

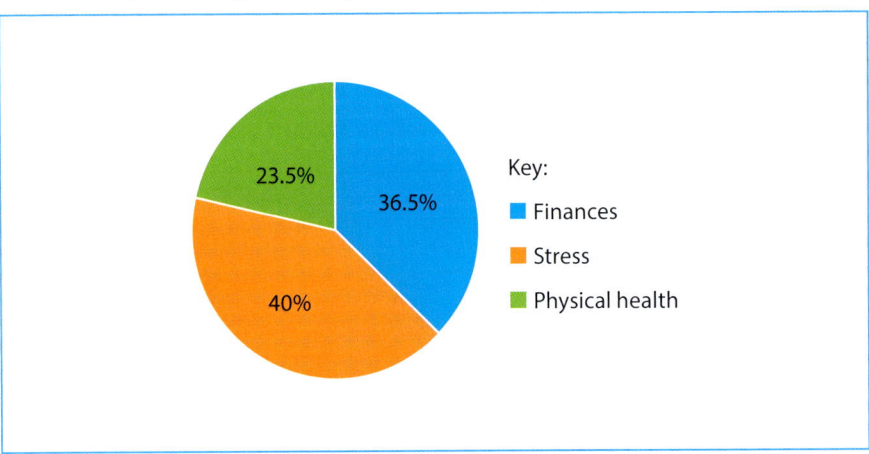

Figure 23.2 Pie chart comparing the negative impact of COVID-19 on the health, stress level and finances of families.

Analysis and interpretation of data

Seven (7) families recorded increases in the alcohol consumption by members while twenty-five (25) families recorded increases in the amount of sweets and junk food consumed. Together with this, twenty-six (26) families reduced their physical activities. These changes can cause the deterioration of physical health for many of the 30 families. Members of 5 families were infected with COVID-19 and 6 persons died from the disease. 10 people had other illnesses and were concerned about contracting COVID-19.

All 30 families were stressed because of not being able to interact with families and friends; 17 families worried about education and schooling for their children and also about contracting the disease. Thirteen (13) families suffered from loneliness, depression and anxiety; larger families were less likely to suffer from loneliness, which may be the reason for the figure recorded.

All families had concerns about their financial security in the future. Seventeen (17) families received financial assistance from the government, community, family and friends. This showed some level of care and concern for people in this time of need. Because of a lack of income, ten (10) families had difficulties in obtaining essential goods and services and paying household bills. Six (6) people lost their jobs due to COVID-19 and were affected financially. Financial stress can add to mental stress.

Two statements of findings based on the data presented are:

1. The data shows that 40% of families were affected by stress as a result of COVID-19. The impact on the stress level in families was larger than the impact on physical health or their financial situation.
2. The data showed that 36.5% of families suffered financial difficulties as a result of COVID-19, resulting in their diminished capacity to meet financial commitments.

> **Conclusion**
>
> COVID-19 has negatively impacted families. The effect on the community shows a clear increase in stress, poor physical health practices and financial difficulties.

RECOMMENDATIONS

The following recommendations are proposed, based on the findings:

1. Quick and easy access to hotlines for people to converse about their stresses caused by restrictions due to COVID-19. Professionals should be accessible to assist people in their mental state while maintaining anonymity and confidentiality.
2. Educate the community through newspaper, pamphlets, social media and television on the importance of eating healthy foods, demonstrating what healthy foods are and the importance of physical exercise. This should be projected regularly and be emphasised.

REFLECTIONS

Two lessons learnt from completing this project are:

1. Diseases can destroy the physical health of the body. By eating healthy foods and exercising the body can be more immune to deal with diseases.
2. It is important to maintain contact with families and friends. Such interactions can help to reduce mental stress.

Personally, I will take as many precautions as possible to prevent contracting the disease and listen to technical and professional persons.

One way to improve the project is to increase the sample size, which should give a fairer assessment and also to collect information on which families are low, middle or high income.

One negative social impact is the loss of opportunity for children to develop physical strength through not being more physically active by playing outdoors and at school. This can cause a setback in the normal development of children.

One negative economic impact is that the decrease in finances will decrease the spending capacity of families with respect to foodstuff, medicines and paying bills.

BIBLIOGRAPHY

Chrome Wikipedia. (2021, July). Retrieved April 26,2022, from
https://en.wikipedia.org/wiki/Social_impact_of_the_COVID-19_pandemic

Chrome Wikipedia. (2021, July). Retrieved April 27,2022, from
https://en.wikipedia.org/wiki/Economic_impact_of_the_COVID-19_pandemic

Effects of COVID-19 questionnaire.jcm-09-03481-s001.pdf (n.d.). Retrieved April 30, 2022 from https://www.google.com/search? =questionnaire+on+effect+of+covid-

Effects of COVID-19 questionnaire. Social_Impact_of_COVID-19_questionnaire.pdf (n.d.). Retrieved April 29,2022 from https://www.google.com/search?=questionnaire

APPENDIX A

Letter to Family

Dear Family

I am a Form 4 student of the Bright Life Secondary School. I am asking for your assistance with my research project for Human and Social Biology. My research is on the effects of COVID-19 on families during the two (2) years on lockdown.

There is no need for any names and all information will remain confidential. Please complete the questionnaire honestly. I wish to collect the completed questionnaire in one week from today.

Thank you for taking part and assisting me.

Sincerely

Jane Doe

QUESTIONNAIRE

This questionnaire aims to collect information on the effects of the COVID-19 pandemic on your family. Please answer all of the questions.

Circle your answers for questions 1 to 3.

1. How did your family members change their consumption of fast foods and snacks?

   No change   Ate less   Ate a lot more

2. How did your family members change their alcohol consumption? Circle your answer. This question is only to be answered for family members who are 18 years old and over.

   No change   Drank less   Drank more

3. How did your family change its physical exercising?

   No change   Exercised less   Exercised more

For questions 4 to 7, fill in the table below.

Question number	Question	Yes (If your answer is Yes, insert the number of persons in the last column)	No	Number of persons
4	Did any family member(s) become ill with COVID-19?			
5	Did any family member(s) die from COVID-19?			
6	Did any family member(s) worry that they would catch COVID-19 because they have an underlying medical condition?			
7	Did any family member(s) suffer with depression, loneliness or anxiety?			

For questions 8 to 16, tick (✓) Yes or No.

No	Question	Yes	No
8	My family experienced stress because of children/grandchildren losing education over closure of school.		
9	My family worried that we may become infected with COVID-19 and then infect other people.		
10	My family was worried about attending worship, weddings, funerals, social events.		
11	My family experienced stress because of not being able to meet with family and friends.		
12	My family experienced financial hardship due to job loss or loss of earnings.		
13	My family had difficulty getting supplies e.g. medicines, food, drinks or other essentials because of lack of income.		
14	My family could not meet household and financial obligations (loan repayments, household bills, etc.).		
15	My family received financial support from the government, friends, family or community.		
16	My family is concerned about the future with respect to financial security and stability.		

APPENDIX B

Tally charts to collate the various responses provided by the 30 families on how COVID-19 impacted on their lives.

Questions 1 to 3

No	Question	No change	Less	More
1	Families eating junk food and snacks	##		## ## ## ## ##
2	Families with persons over 18 years alcohol consumption change	## ## ## ## ///	## //	
3	Families' physical exercising	////	## ## ## ## ## /	

Questions 4 to 16

No	Question	Yes	No
4	Families with members who become ill with COVID-19	## ## ## ## ##	##
5	Families experiencing death of members	## ## ## ## ////	## /
6	Families who worried that a family member will catch COVID-19 because they have an underlying medical condition	## ## ## ##	## ##
7	Families with member suffering with depression, loneliness or anxiety	## ## ## //	## ## ///
8	Families experiencing stress because of children/grandchildren losing education over closure of school	## ## ## //	## ## ///
9	Families worrying they may become infected with COVID-19 and then infect other people	## ## ## //	## ## ///
10	Families worried about attending place of worship, weddings, funerals, social events	## ## ## ## ## ##	
11	Families with stress of not being able to meet with family and friends	## ## ## ## ## ##	
12	Families with financial hardship due to job loss or loss of earnings	## ## ## ## ////	## /
13	Families with difficulty getting supplies when needed, for example, medicines, food, drinks or other essentials because of lack of income	## ## ## ##	## ##
14	Families unable to meet financial obligations (loan repayments, household bills, and so on.)	## ## ## ##	## ##
15	Families receiving financial support from the government, friends, family or community	## ## ///	## ## ## //
16	Concern about the future with respect to financial security and stability	## ## ## ## ## ##	

ALTERNATIVE PAPER TO THE SBA (PAPER 032)

The Alternative Paper 032 to the School-Based Assessment, is for private candidates only. It takes the form of a written examination, which consists of a Case study with related questions involving a health-related issue in a named Caribbean territory.

The intention of this Alternative Paper is to examine the same skills as those tested by the School-Based Assessment. The Case study in this paper will test knowledge, skills, critical thinking and problem-solving skills that are related to the subject.

Guidelines for Paper 032

- Paper 032 is designed for private candidates who have no marks for Paper 031 from a recognised educational or private institution.
- All candidates writing the **CSEC®** Human and Social Biology at the January examination for the first time MUST write Paper 032.
- Candidates taking Paper 032 are NOT required to submit a SBA report.
- Paper 032 is one hour and fifteen minutes and consists of 12 questions.
- Paper 032 is worth 20% of the student's final examination marks. These marks contribute to the final marks and grades that are awarded to students for their performance in the examination. The 80% is acquired from Paper 01 and Paper 02.
- The questions in the Case study follow similar pattern as the report for Paper 031, the SBA report. They may be based on any of the following:
  - Statement of the problem
  - Objective of the investigation
  - Methods of data collection
  - Limitations of the data collection method
  - Presentation of data
  - Questions on analysis and interpretation
  - Social and economic implications
  - Conclusion
  - Recommendations
  - Reflection

A SAMPLE CASE STUDY

INSTRUCTIONS: The following case study contains background information for a research project. Read the case study carefully and then answer the questions after each section.

CASE STUDY

In the Caribbean, obesity is a serious health issue. 40% of the population in several Caribbean countries is classified as obese. Childhood obesity is a complex problem, with many contributing factors, including learnt behaviour, environment, genetics and culture. Obesity is caused by consuming more calories (particularly those in fatty/sugary foods and drinks) than a person can burn off through physical activity. As a result, excess energy is stored by our body as fat.

Child obesity can negatively affect children in many ways and it is important to note that it can lead to serious and potentially life-threatening illnesses. Some of these are: Type II diabetes, coronary heart disease, breast and bowel cancer and strokes.

The World Health Organization (2017) breaks down the contributing factors to:

- busy families cooking less and eating out more
- easy access to high-calorie fast food and junk food
- bigger food portions both in restaurants and at home
- children consuming huge amounts of sugar in sweetened drinks and snacks
- children spending less time actively playing outside and more time watching television, playing video games, and sitting at the computer.

As a student, you are to investigate the reasons for child obesity and make recommendations for 50 families in your community, each of which has more than one obese child.

1. Write a suitable statement of the problem.
   Obesity negatively affects the health of children in the Caribbean
   (Marking scheme: 1 mark for stating the problem as obesity and 1 mark for stating the target group as Caribbean children. Total = 2 marks)

2. What is the objective of the investigation?
   To determine the reasons for child obesity in 50 families in the community.
   (Marking scheme: 1 mark for stating the topic of investigation and 1 mark for identifying the target group. Total = 2 marks)

3. Methodology
   Give one method of collecting data for your investigation.
   Interviewing a parent, guardian or responsible adult family member using a semi-structured approach involving a questionnaire. (1 mark))

4. a For the method chosen in 3 above, describe the process by which the data will be collected.
   – *I produced a plan for my interview, which included using a brief questionnaire with closed and open questions.*
   – *I plan to speak with one adult member of each of 50 families in the community who have obese children.*
   – *I promise to safeguard the identity of the persons being interviewed and not refer to them using any other identifiers.*

- The questionnaire consists of 10 questions, which require simple answers or short responses.
- I am using closed questions to verify that I have correctly identified reasons for child obesity and open questions to allow for the collection of other reasons specific to the community. (5 marks)

   b Give one reason for choosing the method in 3 above to collect the data.
   A mixed methods approach allows for cross-checking that the information obtained is true via verbal responses being correlated back to the questionnaire responses. (1 mark)

5. State one limitation of the method of data collection used
   One limitation of using this data collection method is finding a place where the interview can be conducted confidentially. (1 mark)

Presentation of data

The total number of children in 50 families is 55. The results of the responses for five of the questions are presented in Table 1.

TABLE 1: Table showing the number of children in responses to eating habits and physical activity

Habits	Number of children
Eating out at least four days per week.	40
Having constant access to fast food and junk food	40
Daily consuming of sugar above the daily recommended amount for children, for example, in sweetened drinks and snacks	50
Playing video games and/or looking at television for at least three hours every day after school	45
Eating second serving at dinner time every day	35

6. Using the grid provided (graph page), plot a bar graph of the data for the number of children presented in Table 1. (4 marks)

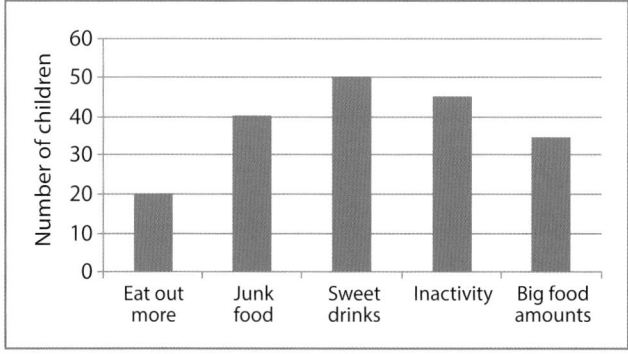

Bar graph showing the number of children with different eating habits and physical activity

Mark scheme:

1 mark for each axis labelled correctly	(2 marks)
2 marks for ALL 5 bars plotted accurately	(2 marks)
If 1 bar plotted incorrectly	(1 mark only)
If 2 or more bars plotted incorrectly	(0 marks)

7. Calculate the percentage of children who partake in sedentary or low-level physical activities.
   $\frac{45}{50} \times 100$ (1 mark) = 90% (1 mark) (2 marks)

8. Suggest one other way in which the data could be presented.
   The data could be presented using a pie chart. (1 mark)

Analysis and interpretation

The Body Mass Index (BMI) is a useful measure of overweight and obesity. It is calculated from your height and weight. Table 2 shows the relationship between BMI and weight.

TABLE 2: Table showing the relationship between BMI and weight

BMI	Weight status
Below 18.5	Underweight
18.5–24.9	Healthy
25.0–29.9	Overweight
30.0 and Above	Obese

9. a The BMI was recorded for one child from each of the five categories of eating habits taken from Table 1. Using the information from Table 1 and Table 2, complete Table 3 by indicating the weight status of each child.

   TABLE 3: Weight status based on the BMI of five children whose families eat outside of home regularly.

Name of child	BMI	Weight status
Sam	23.2	Healthy
Mary	32.5	Obese
Ann	30.1	Obese
David	25.0	Overweight
Lee	28.3	Overweight

   (5 marks)

   b. Identify the healthiest child and suggest one reason for this even though eating out regularly can lead to obesity.
   Sam is the healthiest. (1)

   Maybe this is because Sam is eating one portion of healthy foods like lean meats, salads and complex carbohydrates, as recommended for his age when he eats out. (1) (Total = 2 marks)

   c. Which child seems most predisposed to developing health issues related to obesity? State the evidence which suggests this.
   Ann show the highest predisposition to developing health issues since she is classified as obese and has the highest BMI. (2 marks)

   d. What are two diseases which obese children can develop?
   Type II diabetes, coronary heart disease, breast and bowel cancer, stroke.
   Any 2 of given answers (2 marks)

e. An increase in unhealthy eating habits can lead to more children becoming obese. This will lead to more children with diseases related to obesity. Suggest one social and one economic impact of obesity in the community.
Social impact: lack of confidence, bullying, social exclusion, stress. Any 1 of answer given (1 mark)
Economic impact: increase in healthcare cost (1 mark)

Recommendations

10. Give TWO recommendation to Ann's parents to help with reducing her obesity. Justify your recommendations.
I would tell Ann's parent to reduce her intake of sugary foods and drinks because the energy from these in excess of her dietary requirements become stored in her body as fats. (1) Her parents can get her involved in a physical activity she enjoys, for example, sports, aerobics or dance to help her burn calories. (1) (Total of 2 marks)

11. Give ONE recommendation to the principal of your school to help reduce obesity in the students. Justify your recommendation.
I would advise the principal to ensure that the curriculum of the school for all class levels include classes that encourage physical activity, for example, physical education, sports, dance, and so on. This would help with burning calories through increasing their activities in healthy ways. (2 marks)

12. Give two recommendations to the government that can assist in preventing obesity in children.
    1. *Educate the country through newspaper, pamphlets, social media and television (1) on the importance of eating healthy foods, demonstrating what healthy foods are and the importance of physical exercise. (1)* (2 marks)
    2. *Create a policy on the types of snacks and drinks that are made available in the school cafeteria and in school-supplied meals. Also ensure that the recommended portion sizes are served in the School-supplied meals.* (2 marks)

Total marks for the Alternative Paper = 40 marks

Glossary

abortion the termination of a pregnancy

absorption digested food moving from the gut into the blood

accommodation the way the lens of the eye changes to focus on different objects

acid rain rain with a low ph as a result of dissolved gases produced during the burning of fossil fuels

actin a protein found in muscle cells involved in muscle contraction

action potential the electrical event of the nerve impulse

activation energy the energy required for a chemical reaction to take place

active site the region of an enzyme that binds to the substrate molecule

active transport the movement of substances against a concentration gradient using energy from respiration

acute a sudden onset, severe disease

addicted physically or psychologically dependent on something

adrenaline a hormone that prepares the body for flight or fight

aerobic respiration breaking down food to release energy using oxygen

agar a jelly used to grow microorganisms

albinism a genetic condition in which no melanin pigment is formed in the cells

alcohol in this context: drinks containing ethanol

alleles different forms of the same gene

alveoli microscopic air sacs in the lungs with a large surface area

amino acid the building block of protein

amniotic fluid the fluid that surrounds the developing foetus in the uterus

amputation the surgical removal of a limb

anabolism reactions in the body that build small chemicals into larger ones

anaemia a deficiency disease from lack of iron

anaerobically without oxygen

anaerobic respiration breaking down food to release energy without oxygen

anorexia a mental illness where people have a distorted body image and deliberately stop eating

antagonistic pair muscles that work opposite each other

antibiotics drugs that kill bacteria

antibodies proteins made by white blood cells that inactivate pathogens

antidiuretic hormone (ADH) a hormone that reduces water loss from the kidneys

antigens proteins found on the outer surface of all cells

antiseptics chemicals that are used outside of the body to kill bacteria

aquatic habitat home in the water

asexual reproduction reproduction that only involves one parent, resulting in offspring that are identical to their parent

assimilation taking in and use of the digested food by the body

asthma the response of the respiratory system to one or more allergens – the tubes narrow and excess mucus is produced

astigmatism a common eye defect, causing irregular eye shape and affecting the way light is focused on the retina

atherosclerosis when fatty deposits form in the arteries, which supply the heart muscle with oxygen

ATP (adenosine triphosphate) the main energy-storing molecule in the cells

atria the top chambers of the heart

autoclaving sterilising in water at temperatures above boiling point to kill pathogens

autosomes all chromosomes other than the sex chromosomes (X, Y)

autotroph plant; organism that makes its own food

axon long process from nerve cell that carries the nerve impulse

bacteria microorganisms, members of the monera; microorganisms that cause disease

beri-beri deficiency disease from lack of vitamin B

biodegradable able to be broken down by microorganisms

biological vector a vector in which the disease-causing organism multiplies before being passed on

biomass the mass of biological material in an organism

biuret reagent a compound used to show the presence of protein

bleach sodium hypochlorite, a chemical that kills pathogens in water

bronchi (singular: bronchus) pair of air tubes leading to each lung

bronchioles branching air tubules inside lungs

bulimia an eating disorder in which the person binge eats then vomits and/or takes laxatives

calcified made rigid with calcium salts
carbohydrase an enzyme that breaks down carbohydrates
carbohydrate a major food group made up of carbon, hydrogen and oxygen
carbon cycle the cycling of carbon compounds between the living and the non-living world
carbon dioxide a widely produced greenhouse gas
carnivore an animal that eats other animals
carrier an individual with a normal phenotype who has one normal dominant allele and one recessive allele, which codes for a genetic condition such as albinism
cartilage a smooth, rubbery tissue that reduces friction and cushions joints
catabolism reactions in the body that break down large chemicals into smaller ones
catalase an enzyme that breaks down hydrogen peroxide
catalytic converters units that remove the main pollutants from car exhausts
cataracts a clouding of the lens of the eye
causative agent the pathogen responsible for a disease
cell membrane the outer layer of living cell that controls the movement of substances in and out
cellulose the complex carbohydrate that makes up plant cell walls
cell wall the outer layer in plant cells and bacteria that is freely permeable
central nervous system (CNS) the brain and spinal cord
cerebellum part of the brain that helps with balance and coordination
cerebrovascular relating to the blood supply of the brain
cerebrum the cerebral hemispheres, the site of intelligence and thinking in the brain
chlorination adding chlorine to drinking water to destroy pathogens and make it safe to drink
chlorophyll the green pigment that traps light energy in plants
chloroplast a plant cell organelle containing green chlorophyll for photosynthesis
cholesterol a compound produced by the body; raised blood levels indicate a high risk of heart disease
choroid a dark layer in the eye
chromatids two identical strands of a chromosome formed just before cell division
chromosome a strand of DNA-carrying genetic information
chronic long term or prolonged

ciliary muscles the muscles that change the shape of the lens
clone genetically identical individual
collagen protein fibres found in connective tissue and bone
coma a state of unconsciousness
combustion burning
composite a substance made of more than one material
compost product of the decay of edible food remains and garden rubbish
concentration gradient the difference between an area of high concentration and an area of low concentration
conception the fusing of an ovum and a sperm
condensation the process by which water vapour from the air becomes liquid water
cones light-sensitive cells on the retina that give colour vision
congenital cataracts clouding of the lens of the eye that is present at birth
congenital defect a problem that is present from birth
consumer an organism that feeds on other organisms
continuous variation the range of a genetic trait that can show any value between the two extremes
contraception a method or device of preventing pregnancy
converging coming together
cranial reflexes reflexes that involve the unconscious areas of the brain
curare a nerve poison
cuticle the waxy covering of epidermis of a leaf
cystic fibrosis a genetic condition particularly found in white European communities
cystitis a urinary tract infection
cytoplasm a jelly-like substance containing cell organelles
decomposer an organism that feeds on and breaks down dead organisms and biological material such as fallen leaves and faeces
defibrillate to restore the normal rhythm to the heart
degenerative gradual deterioration of an organ, tissue or lifestyle
dendrites finger-like processes that connect nerve cells and receive impulses
desalination the removal of salt from seawater
diabetes the condition when your body stops making or no longer responds to the hormone insulin
diabetes mellitus a condition that results from a lack of insulin produced by the pancreas, or from a lack of response by the body tissues to insulin

dialysis machine a machine that takes over the functioning of the kidneys

diastole the relaxed heart when the blood pressure is lowest

diffusion the movement of particles from an area of high concentration to an area of low concentration along a concentration gradient

disaccharide two simple sugar units joined together

discontinuous variation the range of a genetic trait that can show only specific values with no possibilities in between

disinfectant a chemical used to kill pathogens on surfaces

diuretic chemicals that increase the output of urine

diverging spreading apart

dominant an allele where the characteristic is expressed in the phenotype even if only one copy of the allele is present

drug dependence the state when you cannot function properly without a particular drug

duodenum the first part of the small intestine where bile and pancreatic juice enter

ecosystem all the animals and plants that live in an area along with the things that affect them such as soil and the weather; also the interaction between many different types of living organisms and the non-living features of their home

effector organ a muscle, organ that responds to a nervous signal

egestion the removal of undigested food from the body (faeces)

electron microscope a microscope that uses a beam of electrons to give very high magnification

embryo a developing baby in the uterus for the first 12 weeks of development

embryonic stem cells cells from the early embryo that have the potential to form almost any other type of cell

emulsify break down into smaller droplets

endocrine gland a ductless gland that secretes directly into the bloodstream

endoplasmic reticulum membrane stacks in the cytoplasm

environment an organism's home and its surroundings

enzyme a protein molecule that acts as a catalyst in cells

epidemic a disease that spreads rapidly and affects large numbers of the population

eutrophication the process by which waterways that are polluted by nitrate rich fertilisers become lifeless as a result of a mass of plant and animal growth followed by excessive decay, which uses up the oxygen in the water

evaporation the change from liquid water to water vapour – the opposite of condensation

excretion the removal of the waste products of cell metabolism

exocrine gland the gland where the secretion is passed along a duct or tube

exoskeleton a skeleton on the outside of the body

fermentation anaerobic respiration in yeast that produces ethanol

fertilisation the meeting of an ovum and a sperm

filtration filtering out large particles

flaccid floppy, limp

foetus the developing human from about 12 weeks of pregnancy onwards

food chain the links between plants and animals that feed on each other

food web a model of a habitat showing how many animals and plants are interconnected through their feeding habits

forensic entomology the study of insects to help solve crimes

fovea the area of the retina with the clearest and best focus

friction the force caused by two surfaces moving against each other

genes units of inheritance

genetic engineering changing the genetic material of an organism by adding a gene/genes from another organism

genotype a description of the alleles of an individual for a given genetic trait

glaucoma the term covering a number of eye conditions in which the optic nerve carrying impulses from the eye to the brain is damaged where it leaves the eye

global dimming particles in the atmosphere shield the Sun's rays and act to cool the temperature at the surface of the Earth

global warming high levels of greenhouse gases in the atmosphere prevent heat radiating away from the surface of the Earth and cause the temperature at the surface of the Earth to rise. this is also known as the greenhouse effect

glomerular filtrate the liquid produced in the Bowman's capsule of the kidney tubules

glycogen the carbohydrate store in animals
greenhouse effect see global warming
guard cells cells that control opening of stomata
habitat place where an animal or plant lives; its home
haemoglobin the red pigment in the red blood cells that carries oxygen around the body
haemophilia a genetic disease where at least one of the normal clotting factors for the blood is missing
hallucinogen a drug that affects the mind and produces images of things that do not exist
hazardous (waste) waste that is potentially poisonous or could cause disease, e.g. the waste produced by hospitals, chemical factories or nuclear power stations
herbivore an animal that eats only plants
heterotroph an organism that feeds on other organisms
heterozygous having two different alleles for a particular feature
homeostasis the maintenance of a constant internal environment
homologous pair a matching pair of chromosomes in a nucleus, one from each parent
homozygous having two identical alleles for a particular feature
housefly a vector of disease
human immunodeficiency virus (HIV) the virus that causes AIDS (acquired immunodeficiency syndrome), a disease that affects millions of people worldwide and for which we have so far no cure
hydrolysis breaking down using water
hydrostatic skeleton a skeleton made up of trapped fluid
hyperglycaemia high blood sugar
hypertension high blood pressure
ileum the main part of the small intestine where digestion is completed and nutrients are absorbed
immune system the system in the body which protects you against invading microorganisms and foreign proteins
implantation the process by which the early embryo embeds in the lining of the uterus
infectious disease (communicable or transmissible) a disease that can be passed from one person or animal to another
infertility an inability to become pregnant
ingestion taking in food
insertion the point where a muscle joins a bone furthest from the heart
internal fertilisation the process by which a sperm fuses with an ovum inside the body of the mother
IVF (*in vitro* fertilisation) the fertilisation of ova with sperm outside of the body of the mother in a laboratory; as the embryos begin to develop, some are replaced in the uterus
karyotype an arrangement of an individual's chromosomes into ordered pairs, for study
kwashiorkor a deficiency disease from lack of protein
labour the process by which a baby is delivered from its mother's uterus to the outside world
lactic acid the product of anaerobic respiration in animal cells
landfill a waste site in which the waste is kept isolated from the surrounding environment
laparoscopy a procedure that allows doctors to examine the Fallopian tubes for damage
legume member of the bean, pea and clover families
ligament connective tissue that attaches muscle to bone
limiting factor something that limits the rate of photosynthesis, e.g. light, temperature
lipase an enzyme that breaks down fats
lipids the general name for fats
lobar pneumonia a chest infection that affects the lobes of the lungs
lubricant fluid that reduces friction in a joint
lymphocytes white blood cells that are an important part of the immune system
lymph system a series of tubules that return tissue fluid to the blood
macronutrient food you need in large amounts
marasmus a protein-energy deficiency disease
mastication chewing
mechanical vector vector in which the disease-causing organism does not multiply before being passed on
medulla an area of the hind brain
meiosis cell division that reduces the chromosome numbers and forms the sex cells
menopause the point in a woman's life at which there are no eggs left in the ovary and the periods stop
menstruation the loss of blood by women in the monthly reproductive cycle
mesophyll photosynthetic cells between upper and lower epidermis of a leaf
metabolic rate the rate of all chemical reactions taking place in the body
metabolism the sum of anabolism and catabolism in the body

methane a greenhouse gas made in large quantities by animals such as cows and by rice plants
micronutrient food you need in only very small amounts
microscope an instrument for magnifying specimens
minerals a class of micronutrients
miscarriage a natural abortion when the developing embryo/foetus dies and is lost from the body
mitochondria organelles where cellular respiration takes place
monohybrid inheritance characteristics inherited through a single pair of genes
monosaccharide a single sugar unit
mutation a change in the dna
myelin sheath the fatty insulating sheath that grows around many nerves
myosin a protein found in muscle cells involved in muscle contraction
negative feedback a feedback mechanism that responds to a stimulus by reducing the stimulus
nephron a kidney tubule
neurons nerve cells
neurotransmitter the chemical that passes messages between neurones across a synapse
nitrogen cycle cycling of nitrogen compounds between the living and non-living world
non-biodegradable not able to be broken down by the action of microorganisms
non-identical twins twins who develop from two separate ova
nucleus found in all cells at some stage during their life cycle; contains chromosomes
obese very fat with a BMI over 30
obesity having a BMI of over 30 – being very overweight
omnivore an animal that eats both animals and plants
organelle a small body with a specific function in a cell
origin the point where a muscle is attached to a bone nearest the heart
osmoreceptors sensory organs that respond to the concentration of the body fluids
osmoregulation controlling the osmotic potential (concentration) of a fluid
osmosis the movement of water from an area of high concentration to an area of low concentration along a concentration gradient through a partially permeable membrane
osteoarthritis a disease of the joints often due to wear-and-tear

osteoporosis a disease of the skeleton where calcium is lost and the bones become brittle
ova (singular ovum) the female sex cells
ovary the female sex organ that produces ova
ovulation the release of one or more ova from the ovaries each month
pacemaker the area in the right atrium of the heart where the heartbeat starts
pancreas an organ that produces hormones and digestive enzymes
pandemic a disease that spreads rapidly and affects large numbers of people across a number of countries
parasite an organism that lives and feeds off another living organism
partially permeable allows the passage of some substances but not all
pathogens microorganisms that cause disease
penicillin an example of a common antibiotic
pepsin an enzyme that breaks down proteins
periodontal disease a disease of the gums
peripheral nervous system nerves that go to all parts of the body
peristalsis squeezing of the gut muscles, which moves food through the digestive system and mixes it with enzymes
personal hygiene keeping your body clean
phagocytes white blood cells that engulf bacteria
phenotype a description of the physical appearance of an individual relating to a given genetic trait
phloem living transport tissue in plants that transports food
photosynthesis building food molecules using light, used by plants
phytoplankton very tiny plant-like organisms living in water
pit latrine a simple toilet that disposes of human waste by decomposition
placenta the organ that supplies the developing foetus with nutrients and oxygen from the mother, and removes the waste products of the foetus
plasma the liquid part of the blood
plasmolysis shrinking of the cytoplasm of a plant cell away from the cell wall due to osmotic movement of water
platelets cell fragments in the blood needed for clotting
pollen the male sex cells of plants
pollutant a substance that causes pollution

pollution the contamination of the natural environment by harmful substances as a result of human activity

polysaccharide a long chain of single sugar units

population a group of organisms of the same species living in the same habitat at the same time

potable drinkable

primary consumer an animal that eats plants

producer a plant

protease a protein-digesting enzyme

puberty the physical process by which the body changes from a child to a sexually mature adult

recessive an allele where the characteristic is only expressed in the phenotype if two copies of the allele are present

recombinant DNA the DNA of an organism, e.g. bacteria to which a gene from another organism has been added in genetic engineering

recycle the process of returning materials such as glass, plastic, aluminium and paper to be reused in various ways rather than simply disposing of them as waste

red blood cell the type of blood cell that carries oxygen around the body

reducing sugar a sugar that reacts with Benedict's solution to form an orange precipitate

reflex arc a pathway from sensory to relay to motor neurone to create a rapid response without thought

relay neurons neurones in the CNS that connect sensory to motor neurones without further input, to cause reflexes

ribosome an organelle involved in protein synthesis

rickets a disease that results from a lack of vitamin D or calcium

rods light-sensitive cells on the retina that respond to light and dark and movement

saturated fat a fat with no double bonds in the structure

sclera white outer layer of the eyeball

scurvy a deficiency disease from lack of vitamin C

secondary consumer an animal that eats an animal that eats plants

sedative a drug that calms you down

semen the fluid produced by the testes and the associated glands released at ejaculation

sewage human bodily waste and waste water from homes, factories, etc.

sex-linked disease a genetic disease involving alleles that are inherited on the sex chromosomes

sexually transmitted infections (STIs) diseases that are mainly transmitted through sexual contact

sexual reproduction reproduction that involves two parents, introducing variety and resulting in offspring that are genetically different to their parents

sickle cell anaemia a genetic condition that in the homozygous form causes changes in the haemoglobin so the red blood cells become sickle shaped; they cannot carry oxygen properly and clog up the blood vessels; in the heterozygous form it gives protection against malaria

smog a fog that contains many pollutant particles

solute a chemical that is dissolved in a solvent

solvent a liquid in which a solute is dissolved

sperm the male sex cell

spermicide a chemical that kills sperm

sphincter a ring of muscle

stomata (singular: stoma) openings in epidermis of leaf for gaseous exchange

symptoms feelings we experience as a result of disease

synapse the junction between two neurones

synovial fluid fluid produced by the synovial membrane to lubricate the most active joints

synovial joints joints that have synovial membranes and fluid

systole the heart as it contracts, when the blood pressure is highest

tendon connective tissue that holds bones together at a joint

thrush a fungal infection of the mucous membranes

tissue fluid the liquid forced out of the capillaries; plasma without the large proteins

transpiration the process by which water moves through plants

trophic level the level in a food chain/pyramid to which an organism belongs

turgor a strong pressure of the cytoplasm of a plant cell against the cell wall when the vacuole is full due to osmosis

umbilical cord the cord that attaches the developing foetus to the placenta

uterus (womb) the organ in which a new individual forms

vaccination (immunisation) the use of dead or weakened strains of pathogens to produce immunity to dangerous diseases

vascular bundle veins in a plant

vasoconstriction when the blood vessels narrow

vasodilation when the blood vessels widen
vector an organism that transmits disease without being harmed by it
ventilation moving air into and out of the lungs
ventricles the large, lower chambers of the heart that pump the blood out to the lungs and the body
villi finger-like projections of the lining of the small intestine
virus a microscopic particle that causes disease
vitamin a class of micronutrients
water cycle the cycling of water through a process of evaporation, precipitation and condensation
withdrawal symptoms the symptoms suffered by drug addicts when they stop using their drug

X-rays a form of ionising radiation that penetrates soft tissues and shows up the bones
xylem dead transport tissue in plants; moves water and minerals
zooplankton very tiny animals living in water
zygote the cell formed when an ovum and sperm fuse

Index

abortion 252, 269, 270–271, 276, 522 see also miscarriage
absorption 18, 39, 41, 114, 129, 131, 132, 206, 210, 432, 522
accommodation 241–242, 522
acid rain 457, 460–462, 464, 465, 481, 482, 522
acrosome 19, 259
actin 20, 522
action potential 229, 235, 247, 522 see also nerve impulse
activation energy 117, 522 see also chemical reaction
active site 116–117, 120, 132, 522
active transport 24, 34, 39–40, 41, 129–130, 156, 522 see also concentration gradient and respiration
acute (bronchitis, disease, illness or infection) 153, 357, 374, 375, 522
addiction/addicted 151, 152, 264, 402–404, 406, 407, 410, 522
adenosine diphosphate (ADP) 155
adenosine triphosphate (ATP) 155–157, 522
adipose tissue 20, 216 see also lipids
adrenaline 175, 229, 246, 341, 348, 522 see also 'fight or flight'
aerobic respiration 155–157, 159, 161, 522
agar 366, 450, 454, 481, 522
agglutination 180
air pollution 458–461, 482 see also acid rain, carbon monoxide, methane and smog
  controlling 463–464
  impact of 462–463
albinism 300, 308, 309, 319, 522
alcohol consumption 339, 510, 512, 514, 516 see also depressants
algae 63, 67, 68, 461, 462, 466, 467 see also eutrophication
alimentary canal 123, 127
alveoli 29, 136–138, 142, 147, 150, 160, 340, 376, 522
  for gaseous exchange 143–144
amino acid(s) 46, 73, 90–92, 111, 115, 129–130, 166, 209, 220, 522
amniotic fluid 260, 262, 522
amphetamine 147, 406 see also methamphetamine
amputation 222, 336, 522
amylase 120, 122, 126, 129
anabolism 156, 522
anaemia 95, 96, 177, 185, 398, 419, 522 see also sickle cell anaemia
anaerobic respiration 135, 156–157, 159, 161 see also fermentation and lactic acid

anaphase 287, 288
angioplasty 338
anorexia 104, 522
antagonistic pairs 198, 199, 202, 522
antenatal/prenatal care 263–265
antibiotic-resistant gene 292 see also genetic variation
antibodies 91, 166, 186, 361–363, 370, 396, 522 see also birth process, blood groups, infectious diseases and plasma
  HIV 397
  Rhesus 181
anticoagulant 419 see also mosquito(es)
anti-depressants 351
antidiuretic hormone (ADH) 211, 213, 223, 246, 522
anti-fungal agents 367
anti-fungal medicine 436 see also thrush
antigens 179–181, 186, 361–362, 370, 522
antiseptics 356, 359, 365–366, 367, 368, 369, 370, 522
anxiety 145, 147, 161, 330, 346, 352, 353, 401, 402, 407
  symptoms and causes of 347
  treatment of 348
approximation 158
aquatic feeding relationship 66
asexual reproduction 253, 275, 286, 288, 290–291, 297, 302, 522
assimilation 130, 131, 132, 522
asthma 173, 331, 340–342, 352–353, 374–375, 385, 463, 522
athlete's foot 367, 383, 384, 385 see also fungal under infections
atria (singular atrium) 169, 170, 171, 174, 186, 522 see also pacemaker
autoclave 368, 522
autosomes 285, 311, 522
autotroph 44, 522
axon 19, 228–232, 235, 247, 522
  bacteria 15, 292, 316, 327, 366–367, 522
  asexual reproduction 253
  cholera 382 see also cholera
  gonorrhoea 390
  listeria 110
  respiratory system 137
  syphilis 391
  tooth decay 125
bacterium 316, 361, 377, 389–391, 425–426
balanced diet 98, 100–102, 106, 111, 266
barbiturates 405
bar charts 100, 141, 145, 333, 334, 511
bar graphs 100, 183, 495, 503–504, 519
Benedict's solution 88–89, 111, 145 see also reducing sugar

benzopyrene 405 see also marijuana
beri-beri 97, 98, 522 see also deficiency diseases
bibliography 490, 498, 505, 507, 514
bioaccumulation 71, 465
biodegradable 456, 473, 474, 476, 482, 522
biodiesel 478, 482
biodiversity 462
biofilm 468
biogas 159, 161 see also methane
biological catalysts 115
biological drawings 12
biological filter method 468–469
biomagnification 71
biomass 68–71, 76, 77, 522
birth control 252, 266–269, 271, 276
birth process 262–263, 276
Biuret reagent 91–92, 522
bleach 365, 368, 450, 454, 482, 522 see also water purification
blood capillaries 130, 147, 160, 182, 403
blood circulation system 166
blood clotting 178, 186, 209, 370 see also clotting
blood groups 165, 179–181, 186, 293, 296
blood pressure 150, 175, 183, 186, 267, 382, 401, 406, 432–434, 439
  high 95, 105, 106, 182, 184, 264, 334, 337, 339, 353, 424
  low 182
  measuring 171–172
BMI calculations 105
Body Mass Index (BMI) 104, 108–109, 183, 333, 336, 520
bolus 126, 137
bone marrow 177, 191, 193
Bowman's capsule 211, 214 see also glomerular filtrate
breathing rate 145–149, 156, 160, 208, 401
breeding grounds 418, 427, 467, 472, 474, 480
bronchi/bronchus 136–137, 150, 160, 375, 385, 523 see also bronchitis
bronchioles 137, 143, 160, 522
bronchitis 150, 153, 161, 331, 374–376, 385, 405, 463
bronchospasm 341
bulimia 104, 522
bypass surgery 338
Caesarean 265, 395, 398
calcium salts 192–194, 202, 462, 523
calculating
  magnification 10
  mean (average) of a set of data 108
  obesity 105

rate 27
surface area to volume ratio 28
cancer 342–346
   bowel 106, 342, 344, 518, 520
   breast 330, 343, 344, 352, 353
   cervical 267, 273–275, 276, 330, 343, 344, 352, 394, 410
   lung 150–153, 161, 330, 342, 352, 353, 463
   mouth 152
   ovarian 274, 276, 330, 343, 344, 352
   prostate 274, 276, 330, 334, 345, 352, 353
   reproductive system 272–275
   testicular 274
cannabis 153, 401, 402, 405, see also depressants
carbohydrase 126, 129, 523
carbohydrates 46, 51, 59, 85–87, 111 see also carbon cycle, macronutrients and obesity
   complex 87 see also polysaccharides
   refined 106
   starchy 339
   testing for 87
carbon cycle 73–76, 77, 481, 523
carbon dioxide 523 see also greenhouse gas
carbon monoxide 148, 149, 152, 161, 460, 463, 482 see also fossil fuels and smoking
carcinogens 150, 152, 343, 353
cardiac muscle 169, 174, 197, 202
cardiopulmonary resuscitation (CPR) 153–154, 160, 338
carnivore 60, 61, 63, 65, 69, 523 see also food chain
carrier 309, 310, 313–314, 319, 413–414, 523 see also genetic condition
cartilage 17, 189, 191, 193, 194, 196, 198, 200–202, 523 see also joints
cartilage rings 136, 137, 138
catabolism 156, 523 see also metabolism
catalase 117, 120–122, 523
catalyst(s) 115, 116, 120, 524
catalytic converters 463, 523
cell membrane 9, 12, 13, 15, 16, 21, 24, 28, 35, 46, 93, 245, 248, 523
   partially permeable 10, 30, 31, 32, 39, 41, 527 see also osmosis
cellulose 13, 15, 21, 46, 55, 130, 523
cell wall 13, 15, 16, 21, 34, 46, 53, 157, 367, 523 see also plasmolysis and turgor
central nervous system (CNS) 228, 229, 231–233, 236, 247, 330, 392, 523 see also relay under neuron(s)
cerebellum 232, 233, 523
cerebrovascular 185, 523 see also cardiac disease
cerebrum 232, 233, 523

chemical reaction(s) 116, 117, 155, 333, 449 see also activation energy and metabolic rate
chemotherapy 343–345, 353 see also cancer
chlamydia 395–396, 409 see also STIs
chlorination 451, 454, 523
chlorine 365, 450, 451, 523
chlorophyll 13, 14, 21, 45, 46, 54, 55, 68, 73, 141, 523
   the need for 51–52
chloroplast 13, 21, 45, 51–55, 87, 523
cholera 358, 360, 381, 382, 385, 415, 427, 449
cholesterol 92, 183, 334, 432, 523
choroid 239, 240, 523
chromatids 287–289, 523 see also mitosis and meiosis
chromatography 55
chronic obstructive pulmonary diseases (COPD) 150 see also smoking and tar
ciliary muscles 239–243, 248, 523
cirrhosis of the liver 404
clone 253, 259, 275, 288, 291, 297, 523
clotting 98, 166, 177, 181, 338, 419 see also blood groups, platelets and Vitamin K
clotting factors 176, 183, 209, 314, 319 see also haemophilia
cocaine 147, 401, 402, 406, 407, 410, 411 see also stimulants
codeine 401 see also narcotics
collagen 91, 192–193, 194, 202, 523
coma 221, 404, 407, 432, 523
combustion 73, 74, 458, 459, 523 see also fossil fuels
complication(s) 181, 183, 265, 339, 390
composite (material) 192, 202, 523
compost 72, 456, 474, 476, 477, 482, 523 see also active transport
concentration gradient 26–30, 34, 39–41, 129, 142–143, 168, 211, 261, 523
   (humans) 129, 142, 143, 168, 211, 261
   (living systems) 26, 27, 29, 30, 34, 39–40, 41, 522
conception 259, 266, 267, 268, 276, 523
condensation 447, 523 see also evaporation and water cycle
condensation reaction 86, 87, 94, 95, 115
condom 266–267, 389 see also birth control
cones 155, 227, 240, 248, 523
congenital defect 309, 523
congenital syphilis 391
constipation 99, 106, 131 see also fibre
consumer 61, 68, 73, 77, 451, 523, 527, 528
contamination 365, 366, 449, 457, 465, 467, 472, 473, 478 see also groundwater, heavy metal and pollution

continuous variation 284, 294, 296, 297, 523
contraceptive/contraception 266–268, 278, 523
converging (lenses) 243, 523
coral reef(s) 63, 447, 451, 456, 461–462, 481–482
coronary artery(ies) 169, 171, 183, 337–338
covalent bonds 93, 94
cranial reflexes 235, 523
curare 230, 523 see also synapses
cuticle 46, 53, 523 see also epidermis, mesophyll and stomata
cystitis 436, 439, 523
cytoplasm 5, 9, 21, 33–35, 46, 161, 523
   anaerobic respiration 156–157
   bacterial cell 15
   biological drawings 12
   chloroplasts 53
   dentine 124
   egg cell 19
   lipids 93
   proteins 91
   telophase 287
   unspecialised animal cells 10, 13
   yeast cells 16, 157
data collection 40, 494, 498, 499, 517, 519
daughter cells 286, 287, 288, 297 see also mitosis
decomposers 15, 16, 72, 73, 77, 466, 523
defibrillator 338, 523
dengue (fever) 76, 327, 388, 413, 414, 416, 418–424, 427, 474, 480
deoxyribose nucleic acid (DNA) 283, 285, 297
depressants 147, 401
depression 330, 346, 350–353, 407, 408
   causes and treatment of 351
   symptoms of 350
dermis 216
desalination 448, 450, 452, 523
detritus feeders 72
diabetes mellitus 328, 335, 339, 353, 354, 355
dialysis machine 214, 225, 524 see also kidneys
diarrhoea 99, 109, 131, 380, 382, 385, 397, 415, 417
diastole 165, 170, 171, 524 see also blood pressure
diffusion gradient 129
digestive enzyme(s) digestive 17, 19, 115, 124, 127, 234, 414 see also pancreas
diploid cells 286, 288, 296
disaccharide(s) 86–88, 111, 524
diseases
   blood-borne 408
   cardiovascular 337, 339
   chronic 328, 330, 340, 342, 352–353, 430 see also asthma

communicable/infectious/
  pathogenic 329–330, 356, 414, 525
defences against 360
deficiency 96–98, 103, 110, 111,
  193 see also anaemia, beri-beri,
  kwashiorkor, marasmus and scurvy
  nutritional 330, 352 see also goitre
  and rickets
fly-borne 415, 416
food-borne 109–110, 111
gastrointestinal 467
non-communicable 329–330, 332,
  351, 352, 357, 374, 430, 439
pelvic inflammatory disease
  (PID) 390, 396
sex-linked 300, 311–312, 528
social and economic impact 331–332
transmissible 356, 370, 525
vector-borne 330, 418, 427
water-borne 110, 450
disinfectant 110, 359, 365–70, 450, 524
diuretic 404
diverging 242, 243, 524 see also lenses
DNA (deoxyribose nucleic acid) 73, 96,
  282–285, 287, 296, 297, 307, 315–317,
  343 see also mutation
  bacterial 367
  damage 342
  pathogen 362
  recombinant 316, 319, 527
dominant (allele) 301, 303–305, 307–
  311, 318, 523 see also inheritance and
  inherited conditions
Down syndrome 291, 310, 312, 318, 330,
  352
drug dependence 402, 410, 524
drug misuse 347, 402, 408–409, 410
drugs
  anti-clotting 184
  anti-fungal 398
  antiretroviral 398, 399, 410
  illegal 147, 153, 264, 330, 397, 401,
    403
  legal 147, 151, 401, 402, 403
  non-prescription 401–403, 410
  prescription 401–402, 405, 410 see
    also depressants and narcotics
duodenum 123, 128, 524
dust mites 340, 341
ebola 360
ecosystem 39, 72, 461, 462, 465, 479,
  481, 524
ecotourism 467 see also coral reef(s)
ecstasy 401, 402, 407, 410 see also
  hallucinogens
effector organ 197, 229, 247, 524
effluent 464, 468, 482 see also waste
egestion 130–131, 132, 524 see also
  faeces
electrocardiogram (ECG) 184 see also
  heart disease

electron microscope 9, 193, 524
electrons 87, 93, 178, see also clotting
Elton, Charles 61 see also food chains
embryo 17, 19, 21, 254–256, 258–260,
  268–270, 276, 524
embryo development 174
embryonic stem cells 20, 524
embryo plant 92
emphysema see chronic obstructive
  pulmonary diseases (COPD)
emulsify (fats and lipids) 128, 132, 524
endocrine gland 208, 227, 232, 245, 248,
  524
endoplasmic reticulum 5, 9, 10, 13, 21,
  524 see also cytoplasm
energy for life 68–69
energy reduction 69–71 see also tropic
  level
enzyme activity 120–122, 438
epidemic 332, 353, 374, 382, 385, 419,
  421, 509, 524 see also pandemic
epidermis 14, 35–36, 53, 216 see also
  cuticle, mesophyll and stomata
equation 44–47, 55, 157, 159
estimation 158
ethanol 41, 47, 52, 94, 111, 157–159, 161,
  264, 522
eutrophication 457, 466–467, 469, 481,
  482, 491, 524 see also algae
evaporation 216–218, 447, 454, 524 see
  also thermoregulation and water cycle
excretion 8, 21, 130, 208, 209, 214, 222,
  435, 524
excretory organ(s) 209, 216
exercise
  benefits of 433–434
  breathing 349
  physical 172, 173, 513, 521
  regular 160, 185, 345, 352, 353, 433
exhalation 139
exocrine gland 245, 524
exoskeleton 189, 524
eye defects
  astigmatism 244, 248, 522
  cataract(s) 244, 247, 248, 523
  glaucoma 244, 248, 405, 524
  hyperopia/hypermetropia/long-
    sightedness 243, 248
  myopia/short-sightedness 242–243,
    248
faeces 69 see also egestion
fair test 48, 54, 64
farming 458, 459, 466, 491
feedback mechanisms 208–209, 213,
  223 see also negative feedback
fermentation 157, 159, 160, 524 see also
  anaerobic respiration
fertilisation 252, 256, 258–259, 276, 286,
  289, 296, 524
  internal 254
  IVF (in vitro) 271, 526

fibre 99, 100 see also constipation
'fight or flight' 175, 246 see also
  adrenaline
filtration 451, 454, 524 see also water
  treatment
flaccid 34, 524 see also turgor
flush toilet 452, 457 see also pit latrine
foetal alcohol syndrome 264
foetus 161, 168, 256, 258, 260–265, 269,
  270, 276, 524
food chain 5, 59–66, 68–72, 76, 77, 465,
  472, 524 see also trophic level
food security 479–480
food web 60, 65–67, 71, 72, 76, 77, 524
forensic entomology 415, 427, 524
fovea 239, 240, 524
friction 196, 524 see also cartilage and
  lubricant
gaseous exchange 138, 142–143, 145,
  153, 235 see also alveoli and stomata
gastroenteritis 358, 360, 380–381, 385,
  388, 427, 434, 437, 439, 449, 480 see
  also housefly
genetically modified organism
  (GMO) 316, 319
genetic condition 318, 319 see also
  albinism, carrier and cystic fibrosis
genetic diseases 269, 304, 309–310, 312,
  317, 318, 319 see also haemophilia
  curing 315
genetic disorders 309
genetic engineering 15, 284, 300,
  314–316, 318, 319, 524
  advantages and disadvantages of 317
genetic information 19, 253, 259, 284,
  285–286, 296, 297, 302, 327
genetic material 10, 15–17, 96, 253, 290,
  310, 315–316, 330, 343, 357 see also
  genetic engineering
genital herpes 358, 394–395
genitals 391, 395, 436, 437
genital warts 393–394, 410 see also
  human papillomavirus (HPV)
genotype 284, 303–306, 308–309, 311,
  313–314, 318, 524
global dimming 458, 524
global warming 74, 76–77, 445, 452, 454,
  459–460, 462, 477, 479, 481–482, 525
  see also carbon cycle, fossil fuels and
  greenhouse gas
glomerular filtrate 211, 525
glycogen 85, 87, 220–221, 223, 525
goitre 96 see also deficiency under
  diseases
gravity 83, 127, 139, 146, 189, 190, 191,
  197–198 see also muscle tone
greenhouse effect 459, 525
greenhouse gas 459, 462, 481, see also
  global warming and methane
groundwater 447, 448, 452, 464, 467, 472
guard cells 53–54, 525 see also stomata

habitat 62–63, 64–65, 525 *see also* food web *and* population
    aquatic 62, 522
haemoglobin 95–96, 111, 149, 176–177, 186, 310–311, 417, 460, 525 *see also* sickle cell anaemia
haemophilia 300, 313–315, 319, 327, 525
haemorrhoids 131 *see also* egestion
hallucinogen 401, 405, 407, 525
haploid cells 286, 288, 296, 297
hazardous (waste) 473, 525
heartbeat 170, 172, 433 *see also* cocaine *and* pacemaker
heart disease 7, 95, 105, 106, *see also* diabetes, smoking *and* obesity
    atherosclerosis 183–184, 186, 522
    coronary 184, 186, 337, 339, 351, 353, 518, 520
    heart attack 150, 153, 182–184, 186, 337–339, 406, 432–434
    hypertension 182–184, 186, 353, 430–431, 433, 439, 491, 525
    stroke 95, 150, 182, 183, 186 *see also* hypertension
heart rate 146, 148, 156, 173–175 *see also* alcohol consumption *and* depressants
hepatitis 358, 368, 403, 407, 408
herbivore 60, 61, 65, 69, 525 *see also* food chain
heroin 401, 402, 406–407 *see also* depressants *and* narcotics
herpes simplex II *see* genital herpes
heterotroph 44, 84, 525
heterozygous 303–305, 310–311, 313, 318, 419, 525
HIV/AIDS 327, 330, 356, 358, 367, 388, 393, 396–400
homeostasis 33, 83, 206, 207–210, 215, 216, 222, 232, 525
home purification methods 449–450 *see also* water treatment
homologous pair(s) 285, 289, 297, 525
homozygous 303–304, 307, 310, 312, 314, 318, 525
hookah smoking 152–153, 160, 161 *see also* vaping
hormone system 208, 246 *see also* homeostasis
housefly 414, 415–416, 427, 525
human immunodeficiency virus (HIV) 396, 525
human papillomavirus (HPV) 274, 343, 344, 389, 393–394, 410 *see also* genital warts
human skin 216, 414
human waste 72, 418, 425, 427, 464, 466–471, 481 *see also* pit latrine *and* sewage
hydrochloric acid 89, 116, 120, 127
hydrolysis 91, 115, 525

hydrostatic skeleton 189, 525
hygiene 359
    dental 438–439
    female 436
    food 109–110, 381, 434, 439
    personal 99, 435, 437–439, 452, 457, 527
hyperglycaemia 335, 525
hypothesis(es) 40, 52, 67, 100, 107, 499
ileum 123, 129–130, 132, 525
immune system 98, 131, 179, 341, 343, 361–363, 396–397, 525 *see also* anorexia *and* lymphocytes
immunity 266, 356, 364, 390, 392, 420 *see also* vaccination
    active 362, 363, 370
    artificial 362–363
    natural 361–362
    passive 362, 363, 369, 370
implantation 258, 259, 268, 525
indicators 119, 145
infection
    bacterial 331, 373, 382, 395, 410,
    droplet 358, 370, 374, 375, 376, 377, 385, 413
    fatal 137, 425
    food-borne 110
    fungal 367, 397–398 *see also* ringworm *and* thrush
    respiratory 185, 340
    sexually-transmitted (STIs) 267, 272, 330–331, 358, 388–399, 408–410, 528
    uterine 268
    virus 343, 363
infertility 268, 270, 271–272, 275, 389, 390, 395–396, 457, 515
influenza 330–331, 356, 358, 374, 385 *see also* communicable *under* disease
ingestion 123, 131, 132, 525
inhalation 139
    inheritance 181, 283, 285, 300–301, 302, 310, 312
    monohybrid 302–303, 307, 311, 318, 319, 526
inherited conditions 307–308, 319
insertion 198, 202, 525 *see also* skeletal muscles
insulin *see* diabetes mellitus
intercostal muscles 136, 138, 139, 160, 235 *see also* exhalation *and* inhalation
interphase 287
intracellular enzymes 116
IVF (*in vitro* fertilisation) *see* fertilisation
joints 105, 190, 194–200, 202, 374, 434, *see also* arthritis, cartilage, osteoarthritis *and* synovial fluid
junction 229, 230, 247 *see also* synapse(s)
karyotype 285, 291, 312, 318, 525
keratin 91
kwashiorkor 103, 199, 327, 525 *see also* deficiency diseases

labour 262–263, 525 *see also* birth process
lactic acid 156 –157, 161, 169, 525 *see also* anaerobic respiration
landfill 456, 473–478, 482, 525
land security 479, 480
laparoscopy 272, 525
legume 73, 535
lenses (eye) 227, 243, 244, 248,
lenses (microscope) 9, 10, 14
leptospirosis 413, 425–429, 480
life cycle 414, 415–417, 419, 420, 422, 427 *see also* nucleus
life processes 7–8, 15, 21, 82–281
ligament 20, 189, 191, 194, 196, 198, 202, 525 *see also* tendons
    suspensory 239–242, 248
light energy 57 *see also* chlorophyll
limiting factor 55, 525
line graph 49–50, 495, 502–503
lipase 117, 128–129, 525
lipids 85, 89, 92–95, 100, 111, 114, 183, 525 *see also* heart attacks, lipase *and* macronutrients
listeria 110 *see also* bacteria
litter concentration graphs 475
liver diseases *see* hepatitis
lobar pneumonia 376, 525
lubricant 196, 202, 437, 525
lymphocytes 177, 179, 182, 186, 525
lymph system 130, 132, 182, 186, 361, 370, 525
macronutrients 85, 95, 525
magnesium 46, 87, 96, 111
magnitude 158, 191
malaria *see* mosquito(es)
malnutrition 84, 100, 103–104, 111, 199, 284
marasmus 103, 199, 330, 352, 525 *see also* deficiency *under* diseases
marijuana 153, 161, 402, 405, 406, *411*, *412*
    medicinal 401, 405, 410
mastication 124, 525
matrix 20, 91, 192, 194, 202
medulla 174, 211, 232–233, 525
meiosis 253, 283–284, 288–291, 296, 297, 300, 312, 525
Mendel, Gregor 301, 318 *see also* inheritance
menopause 184, 258, 274, 344, 345, 525
menstrual cycle 246, 252, 254, 256–259, 266, 275, 276, 405
menstruation 256, 436, 526
mental health 104, 270 332, 352, 509
mental health problems 330, 346–351, 353
mesophyll 53, 526 *see also* cuticle, epidermis *and* stomata
metabolic rate 101, 217, 218, 223, 246, 526
    basal 433

metabolism 96, 103, 115, 156, 166, 232, 333, 526 *see also* anabolism, catabolism *and* excretion
metaphase 287, 288, 289
methamphetamine 401, 402, 406, 410
methane 159, 161, 459–460, 469, 474, 482, 526 *see also* biogas *and* greenhouse gas
microcephaly 422 *see also* mosquitoes *and* zika
micrograph 12, 230
    scanning electron (SEM)
        blood clot 178, 209
        breast cancer cells 343
        Candida 436
        chikungunya virus 424
        cholera bacteria 382
        flu viruses 356
        fungus 383, 384, 419
        gonorrhoea bacteria 390
        white blood cell antibody 361
    transmission electron (TEM)
        cell infected with SARS-CoV-2 Omicron virus 377
        chlamydia bacteria 395
        HIV virus 396
        Leptospira bacterium 426
        malaria-infected red blood cell 417
        MRSA bacteria 292
        virus particles infecting epithelial cells 373
micronutrient 85, 90, 95, 97, 111, 526
microvilli 28, 129–130, 132, 211 *see also* villi,
minerals 29, 46, 54, 72, 84, 85, 95–96, 100, 526 *see also* blood clotting, malnutrition, nutrients *and* vitamins
    storage 190, 191, 202
miscarriage 269, 276, 526
misconceptions 6, 141, 206, 400
mitochondria 5, 9–11, 13, 19, 155–157, 228, 255, 526 *see also* active transport, muscle cells *and* specialised cells
mitosis 253, 275, 282–283, 284, 286–288, 290, 291, 296, 297, 300 *see also* asexual reproduction
monosaccharides 86–89, 91, 111, 115, 526
morphine 401, 406 *see also* narcotics
mosquito(es) 310, 330, 358, 360, 388, 413, 418
Aedes aegyti 414, 419–421, 424
    Anopheles 414, 416–417, 419, 421
    and chikungunya 423–424
    and dengue 419–421
    and malaria 416–419
    and zika 422–423 *see also* microcephaly
muscle cells 10, 17, 20, 21, 155, 156, 174, 246 *see also* actin, aerobic respiration, anaerobic respiration *and* muscle tone 198, 202
mutation 290–291, 307–312, 314, 319, 353, 526 *see also* cancer *and* non-communicable *under* diseases
mycelium 16
myelin sheath 19, 229–230, 526 *see also* muscle cells *and* nerve cells
myocardial infarction 184 *see also* heart attacks
myosin 20, 526 *see also* muscle cells
narcotics 401
natural selection 291–292, 293, 294, 297, 310–311 *see also* sickle cell anaemeia
negative feedback 208, 213, 223, 526 *see also* osmoregulation
Neisseria gonorrhoeae 389, 390 *see also* gonorrhoea
nephron 210, 211, 214, 526
nerve cells 17, 19, 21, 208, 221, 228, 240, 526 *see also* dendrites *and* neurons
nerve impulse 19, 228–230, 233–235, 238, 246, 522 *see also* action potential
nervous system 93, 197, 208, 227, 230, 234–235, 238, 246, 347, 404–405, 408
    autonomic 228
    central (CNS) 228, 229, 231, 233, 236, 247, 330, 392, 523
    involuntary 197
    peripheral 228, 229, 231, 527
    neuron(s) 19, 228, 526
    motor 229, 230, 231, 235, 236, 247 *see also* reflex arc
    relay 231, 235, 236, 248, 527
    sensory 229, 230, 231, 233, 235, 236, 240, 247
neurotransmitter 229, 230, 347, 348 *see also* anxiety
nicotine 148 *see also* smoking *and* stimulants
nitrogen cycle 73, 526
non-biodegradable 456, 473, 474, 477, 482, 526
non-identical twins 259, 526
nucleus 526 *see also* unspecialised animal cells *and* unspecialised plant cells
obesity 90, 99, 104–106, 110, 111, 145, 148 *see also* atherosclerosis, hypertension *and* malnutrition
    childhood 107–109
omnivore 60, 61, 124, 526
opioids 401, 410
organelle(s) 10, 13, 15, 21, 51, 155, 228, 526 *see also* chloroplast, cytoplasm, mitochondria
origin 198, 202, 526
osmoreceptors 213, 526
osmoregulation 34, 208, 212, 213, 526 *see also* negative feedback
osmosis in animals 33–34
osmosis in plants 34–39

osteoarthritis 196, 200, 327, 330, 352, 526
osteoporosis 193, 196, 434, 526
ovulation 252, 256, 258, 266–267, 271, 276, 526
oxidation reactions 87
oxygen debt 146, 156, 157, 161
pacemaker 174, 175, 526
painkiller(s) 377, 401, 406, 410, 421, 426 *see also* leptospirosis
pancreas 17, 123, 128–129, 526 *see also* diabetes mellitus, obesity *and* smoking
pandemic 353, 362, 385, 509, 510, 514, 526
    COVID-19 217, 332, 347, 359, 364, 377, 378, 385
    influenza and 374–375
partially permeable membrane 10, 30–32, 39, 41, 526
PCR testing 377
penicillin 367, 392–393, 526 *see also* antibiotics
penis 255, 276, 410 *see also* birth control, chlamydia, fertilisation, genital herpes, gonorrhoea, HPV *and* male hygiene
pepsin 116–117, 120, 122, 127–129, 526 *see also* proteins
periodontal disease 125, 526
peristalsis 99, 127, 129, 132, 526
Petri dish 271, 366, 367, 453
phagocytes 177, 179, 182, 186, 361, 370, 526
phenotype 284, 303–308, 311, 314, 318, 526 *see also* carrier, dominant *and* recessive
phloem 54, 526
photosynthesis 45–55, 526 *see also* limiting factor
phytoplankton 63, 67, 69, 526
pie charts 102, 336, 501, 502, 512
pit latrine 449, 452, 466, 469, 470–471, 481, 482, 526 *see also* human waste *and* sewage
placenta 168, 259, 526 *see also* antenatal/prenatal care *and* birth process
    role of the 260–261
plasma 176–178, 186, 211, 526 *see also* blood groups
plasmolysis 34, 41, 527
platelets 183, 186, 209, 211, 527 *see also* clotting *and* clotting factors
pneumonia 331, 356, 375, 376, 385, 397, 463 *see also* lobar pneumonia
pollen 21, 136, 527 *see also* asthma
pollutant 445, 456, 457, 463–466, 482, 527 *see also* catalytic converter
    air 459
    chemical 478

fossil fuel 458
radioactive 464
pollution 457–458, 527 see also air pollution and water pollution
polysaccharide 86–87, 91, 111, 527
postnatal care 265–266, 275
potable 446, 448–452, 527
prediction 40
pregnancy 101, 149, 181, 252, 256–272, 275, 276, 396, 399, 410
    gonorrhoea and 391
    herpes and 395
    HIV/AIDS and 398
    syphilis and 393
    termination of 270, 522
    zika and 422
presentation 495, 498
presentation of data 490, 501, 510, 512, 517, 519
producer 59–60, 61, 62, 68, 77, 527 see also food chains
prophase 287–288
protease 127–129, 131, 527
puberty 19, 256, 258, 276, 312, 435, 439, 527 see also female hygiene
pulmonary circulation 167, 186
Punnett square 305, 307, 311, 313, 314, 318
rabies 363
radiation 77, 217, 218,
    ionising 260, 307, 319, 343, 353 see also X-rays
    therapy see also cancer 343
    ultraviolet 342
rats and leptospirosis 425–426
reabsorption 206, 210, 211
reactants 99, 116–117, 132,
recessive (allele) 300, 303–305, 307–313, 318, 319, 527 see also carrier and haemophilia
recombinant DNA 316, 319, 527
recycle 477–478, 527
reducing sugars 88–89, 111, 145, 527
reduction reactions 87
reflex arc 235–237, 527
relay neurons 231, 235–236, 248, 527
renal dialysis 214
reproductive system 18
    female 19, 252, 255–257, 276, 353
    male 254–255, 276, 353
retina 97–98, 235, 239–244, 248, 312 see also astigmatism
ribosome 9–11, 13, 18, 21, 527
rickets 95–98, 193, 199, 330, 527
ringworm 358, 367, 383–385
rods 155, 227, 240–241, 248, 527
saturated fat 432, 527
sclera 239–240, 248, 527

scurvy 97, 98, 327, 330, 352, 527 see also deficiency disease(s)
secondary consumer 61, 68, 527
sedative 148, 401, 405, 410, 527
semen 255, 258, 267–268, 272, 276, 396, 410, 527
sewage 358, 380, 382, 385, 414, 416, 418, 453, 466, 527
sewage treatment 72, 456, 464, 467–471, 481, 482
sexual reproduction 253–254, 258, 275, 284, 289–291, 297, 302, 527
sickle cell anaemia 185, 300, 310–311, 315, 318, 319, 330, 352, 419, 527
skeletal muscle(s) 182, 197, 198, 202
smallpox 363
smegma 437, 439
smog 460, 527
smoking (cigarettes) 150, 260, 273, 339
    passive 151
solute 30, 34, 527 see also solvent
solvent 30, 41, 93, 99, 527 see also solute
specialised cells 9, 17–20, 21
spermicide 267, 527
sphincter 127–128, 212, 255, 527
stent 338 see also heart disease
stimulants 401
STIs see under infections
stomata 51, 53–54, 447, 527 see also guard cells
synapse(s) 229–230, 231, 246, 247, 527 see also neurotransmitter
    spinal 233–234, 236
synovial fluid 196, 202, 527
systemic circulation 167, 186
systole 170, 171, 527
tar 150, 152, 161, 343, 353 see also smoking
telophase 287, 288 see also meiosis
tendon 17, 91, 189, 191, 194, 197–199, 202, 527
thermoregulation 216
thrush 436, 439, 527
tissue fluid 99, 181–182, 186, 207, 209, 527 see also excretion
tonsillitis 356, 367
toxic shock syndrome 436 see also female hygiene
transpiration 35, 447, 454, 527
trophic level 61, 62, 66, 69–71, 77, 527 see also food chains
tuberculosis 327, 358, 361, 397, 415
turgor 34, 41, 527
Type I diabetes 221, 335, 336
Type II diabetes 221, 222, 328, 333, 334, 336, 353, 434, 439, 518, 520 see also obesity
umbilical cord 260, 261, 528
unspecialised animal cells 9–11, 15, 21

unspecialised plant cells 9, 13, 15, 21
untreated sewage 464, 466–467, 482
uterus (womb) 18, 169, 209, 254–260, 268, 528 see also abortion, infertility and labour
vaccination (immunisation) 266, 360, 362–364, 369, 370, 379, 528
vaping 152–153, 161 see also hookah smoking
variation 310, 318, 523, 524
    discontinuous 284, 294, 296, 297, 524
    environmental 293, 295
    genetic 290–295, 297
vascular bundle 54, 528
vasoconstriction 218, 223, 528
vasodilation 217, 223, 528
vector 330, 414, 427, 474, 478, 480, 528
    animal 358, 370, 413, 416, 425 see also mosquito(es)
    biological 414, 417, 523
    destroying or controlling 360, 370
    mechanical 414, 526 see also housefly
ventilation 138, 139, 160, 433, 528
ventricle(s) 169–170, 171, 174, 186, 528
vertebrae 190, 194, 199, 201, 231, 247
villi 29, 129–130, 132, 261, 528
vital capacity 140, 145, 160
vitamin K 98, 111, 178, 179, 186 see also platelets
vitamins 85, 97–98, 528 see also micronutrients
water cycle 445, 447, 448, 453, 454, 528
water pollution 461–462, 464, 480, 482
    controlling 467–469
    effects of 465–467 see also eutrophication
water purification 446, 450, 454, 482
water supply 452–453
water treatment plants 449, 450–451, 453–454
well-being 100, 148, 330, 352, 400, 403, 407, 411, 445, 458 see also ecstasy and air pollution
    calm 148, 407 see also nicotine and heroin
    human 59, 76, 77, 79, 326
    overall 350 see also depression
    physical and mental 332
    physical, mental and social 329, 352
withdrawal symptoms 402, 528
X-ray(s) 260, 307, 329, 337, 343, 353, 376, 528
xylem 35, 54, 528
zooplankton 63, 69, 528
zygote 17, 254, 259, 528

Acknowledgements

The Publishers would like to thank the following for permission to reproduce copyright material. Every effort has been made to trace or contact all copyright holders, but if any have been inadvertently overlooked, the Publishers will be pleased to make the necessary arrangements at the first opportunity.

Text acknowledgements
pp. 488–507 CSEC® Human and Social Biology syllabus, © Caribbean Examinations Council 2008, 2020.

Photo acknowledgements
4 cl, **p. 23** cl © Ami Images/Science Photo Library; **4** br, **p. 7**cr © Paul/Adobe Stock Photo; **5** cl © Science History Images/Alamy Stock Photo; **5** cr © Dr Morley Read/Science Photo Library; **7** cc, **p. 8** tc, bc, **p.16** bl, bc, **p. 29** br, **p. 30** tr, **p. 39** br, **p. 40** cl, cr, **p. 45** tr, **p. 90** cl, cc, cr, **p. 97** br, **p. 210** tr, **p. 231** cr, **p. 254** tr © Anthony Short; **7** bc © All Canada Photos/Alamy Stock Photo; **8** lc © Monkey Business/Adobe Stock Photo; **8** lc © Wirestock/Adobe Stock Photo; **8** lc © Tankist 276/Adobe Stock Photo; **12** bc © Ed Reschke/Getty Images; **13** br © J C Revy Ism/Science Photo Library; **15** br © Dr Gary Gaugler/Science Photo Library; **16** cr © Dr Linda Stannard Uct/Science Photo Library; **16** cl © Power And Syred/Science Photo Library; **17** br © Ed Reschke/Getty Images; **19** br © Herve Conge Ism/Science Photo Library; **20** cr © M I (Spike) Walker/Alamy Stock Photo; **20** cl © Erebor Mountain/Shutterstock.com; **24** bc © Richard Carey/Adobe Stock Photo; **29** cr © Pikovit/Shutterstock.com; **31** cr © Marek Photo Design.com/Adobe Stock Photo; **31** br © Vrx 123/Adobe Stock Photo; **47** cr © Cordelia Molloy/Science Photo Library; **48** tr Contributor:Nigel Cattlin/Alamy Stock Photo; **54** © Eh Point/Adobe Stock Photo; **54** cl © Pipicato/Adobe Stock Photo; **59** cr © Chalermpong/Adobe Stock Photo; **60** © Gig Oliver/Shutterstock.com; **60** cl © J Dross 75/Adobe Stock Photo; **60** cr © Monkey Business/Adobe Stock Photo; **61** tr © Mario/Adobe Stock Photo; **61** cl, **p. 65** cl © Paul Sat/Adobe Stock Photo; **61** cc, **p. 65** cc © Russell/Adobe Stock Photo; **61** cc, **p. 65** cc © Lisa Strachan/Adobe Stock Photo; **61** cr, **p. 65** cc © Brizard H/Adobe Stock Photo; **63** © Solarisys/Adobe Stock Photo; **63** br © Оксана Лебедева/Adobe Stock Photo; **64** cr © San Hanat/Adobe Stock Photo; **64** br © Butus/Adobe Stock Photo; **69** br © Anca Enache/Adobe Stock Photo; **70** br © N Nerto/Adobe Stock Photo; **70** cl © Gary/Adobe Stock Photo; **71** cr © Ermess/Adobe Stock Photo; **71** br © Vector Mine/Shutterstock.com; **72** br © Pixavril/Adobe Stock Photo; **74** tr © Алексей Филатов/Adobe Stock Photo; **74** bl © Richard Carey/Adobe Stock Photo; **76** tr © Drew/Adobe Stock Photo; **82** cl © Science Photo Library; **82** br, **p. 101** cl © Michael Bradley/Getty Images; **83** br, **p. 207** cl © Mauritius Images Gmbh/Alamy Stock Photo; **83** cc © Scie Pro/Adobe Stock Photo; **84** cl © Koles Nikovserg/Adobe Stock Photo; **84** cr © Ruzz/Adobe Stock Photo; **84** cl © Николай Григорьев/Adobe Stock Photo; **84** br © Sean/Adobe Stock Photo; **85** tr © Hemis/Alamy Stock Photo; **86** tr © Steheap/Adobe Stock Photo; **86** cr © Evgenia Sh/Adobe Stock Photo; **87** tr © Davide Angelini/Adobe Stock Photo; **87** cr © Dennis Kunkel Microscopy/Science Photo Library; **87** cr © Dr Jeremy Burgess/Science Photo Library; **88** cr © Andrew Lambert Photography/Science Photo Library; **88** br © Andrew Lambert Photography/Science Photo Library; **89** br © Warning Signs/Adobe Stock Photo; **90** cr © Anaumenko/Adobe Stock Photo; **92** cl © Rana Photos/Alamy Stock Photo; **92** br © JPC Prod/Adobe Stock Photo; **94** br © Andrew Lambert Photography/Science Photo Library; **95** cr © Jeff Rotman/Alamy Stock Photo; **96** tr © Medicimage Education/Alamy Stock Photo; **97** tr © Gail Philpott/Alamy Stock Photo; **97** cc © Florida Stock/Shutterstock.com; **97** cr © Bit 24/Adobe Stock Photo; **98** br © Eva Fesenuk/Adobe Stock Photo; **99** tr © Fizkes/Adobe Stock Photo; **99** br, **p. 106** cr © Marilyn Barbone/Adobe Stock Photo; **100** cr © Paul Brighton/Adobe Stock Photo; **100** br © George Nazmi Bebawi/Shutterstock.com; **102** br © Kolonko/Adobe Stock Photo; **103** tr, **p. 433** br © Wavebreak Media Micro/Adobe Stock Photo; **104** br © Samuel B/Adobe Stock Photo; **105** br © Abhijeet Bhosale/Shutterstock.com; **109** cr © Jim West/Alamy Stock Photo; **110** tr © Dr Gary Gaugler/Science Photo Library; **110** cr © Africa Studio/Adobe Stock Photo; **119** bc © Blue Ring Media/Adobe Stock Photo; **124** © Peter Hermes Furian/Adobe Stock Photo; **125** cr © Sandor Kacso/Adobe Stock Photo; **137** cl © Aldona Griskeviciene/Shutterstock.com; **146** tr © Tyler Olson/Adobe Stock Photo; **147** br © Dz Mitrock 87/Adobe Stock Photo; **148** br © Dr Arthur Tucker/Science Photo Library; **149** bl © Pavel Chagochkin/Shutterstock.com; **150** cl © Astrid & Hanns-Frieder Michler/Science Photo Library; **150** cr © Carolina Biological Supply Company/Science Photo Library; **151** br © PSL Images/Alamy Stock Photo; **153** © Anastasia/Adobe Stock Photo; **155** © CNRI/Science Photo Library; **157** cr © Steve Gschmeissner/Science Photo Library; **159** tr © Eye Ubiquitous/Alamy Stock Photo; **166** cr © Maurizio De Angelis/Science Photo Library; **168** cl © Lolo Stock/Adobe Stock Photo; **168** cr © Olga/Adobe Stock Photo; **171** br © Stivog/Adobe Stock Photo; **172** tr © Andrey Popov/Adobe Stock Photo; **172** cr © Syda Productions/Adobe Stock Photo; **175** cl © Metamor Works/Adobe Stock Photo; **178** cr © Steve Gschmeissner/Science Photo Library; **180** br © Image Broker/Alamy Stock Photo; **181** cr © Chana Wit/Adobe Stock Photo; **184** tr © Bsip Vem/Science Photo Library; **184** cr © Michael Jung/Adobe Stock Photo; **189** br © John Serrao/Science Photo Library; **193** cr © Susumu Nishinaga/Science Photo Library; **200** cc © Pixel-Shot/Adobe Stock Photo; **200** cr © Chris Robbins/Alamy Stock Photo; **201** © Editable line icons/Adobe Stock Photo; **208** © Maxim Images Com/Alamy Stock Photo; **209** cr © Anne Weston Em Stp The Francis Crick Institute/Science Photo Library; **214** cr © Mailson Pignata/Adobe Stock Photo; **214** cc © Art Inspiring/Shutterstock.com; **215** tc © Evgenii/Adobe Stock Photo; **215** tr © Nastya Smirnova Rf/Shutterstock.com; **217** tr © Smikey Mikey 1/Adobe Stock Photo; **217** br © Andrey Popov/Adobe Stock Photo; **218** tr © Alexey Seafarer/Adobe Stock Photo; **221** br © Jpc Prod/Adobe Stock Photo; **228** tr © Michael Jung/Adobe Stock Photo; **228** br © Axel Kock/Adobe Stock Photo; **229** br © Yustyna Olha/Adobe Stock Photo; **230** br © Science Photo Library/Alamy Stock Photo; **231** © Jeniffer/Adobe Stock Photo; **232** cl © CNRI/Science Photo Library; **233** br © Paul Thompson Images/Alamy Stock Photo; **235** tr © Top Ten 22 Photo/Adobe Stock Photo; **235** cr © Bsip Chassenet/Science Photo Library; **235** cr © Bsip Chassenet/Science Photo Library; **238** cr © Natural History Museum London/Science Photo Library; **240** tr © Anthony/Adobe Stock Photo; **240** br © Science Photo Library/Alamy Stock Photo; **253** tr © Digital Skillet 1/Adobe Stock Photo; **253** cr © A Dowsett, National Infection Service/Science Photo Library; **253** br © Dennis Kunkel Microscopy/Science Photo Library; **258** cr © Digital Skillet 1/Adobe Stock Photo; **259** tr © D Phillips/Science Photo Library; **259** cr © Digital Skillet 1/Adobe Stock Photo; **262** tr © Prostock Studio/Adobe Stock Photo; **263** cr © Monkey Business/Adobe Stock Photo; **264** cr © Prostock Studio/Adobe Stock Photo; **265** tr © Darren Brode/Adobe Stock Photo; **265** cr © Piksel Stock/Adobe Stock Photo; **266** tr © Drazen Zigic/Shutterstock.com; **267** tl © Stock Photos Art/Adobe Stock Photo; **267** cl © Nito/Adobe Stock Photo; **267** cl © Windy Night/Adobe Stock Photo; **268** cl © Rfbsip/Adobe Stock Photo; **269** tr © George Rudy/Adobe Stock Photo; **271** cr © Andriy Bezuglov/Adobe Stock Photo; **272** cr © John Walsh/Science Photo Library; **282** cl © One Inch Punch/Adobe Stock Photo; **282** br © Thom Leach/Science Photo Library; **283** tr © Adimas/Adobe Stock Photo; **283** bl © Monkey Business/Adobe Stock Photo; **284** br © Alex Mit/Shutterstock.com; **285** br © CNRI/Science Photo Library; **287** tc © Akor 86/Shutterstock.com; **287** br © M I Walker/Science Photo Library; **288** cr © Akor 86/Shutterstock.com; **289** cl © Designua/Adobe Stock Photo; **290** cl © Dr Jeremy Burgess/Science Photo Library; **291** cl © Kate Ryan Kon/Science Photo Library; **291** cr © Karel Noppe/Adobe Stock Photo; **291** cr © Bios Photo/Alamy Stock Photo; **292** cl © Keith Chambers/Science Photo Library; **292** cr © Dr Kari Lounatmaa/Science Photo Library; **293** cl © Ton Koene/Alamy Stock Photo; **293** cr © Yvette Cardozo/Alamy Stock Photo; **294** tr © Nigel Cattlin/Alamy Stock Photo; **294** cr © Tim Gartside Travel/Alamy Stock Photo; **295** cl © Tetra Images Llc/Alamy Stock Photo; **295** cr © John Birdsall Social Issues Photo Library/Science Photo Library; **300** br © Natali Mis/Adobe Stock Photo; **301** tr © Science History Images/Alamy Stock Photo; **301** cr © Martin Shields/Alamy Stock Photo; **302** cr © Robin/Alamy Stock Photo; **303** cr © Design Pics/Alamy Stock Photo; **304** cl © Bro Creative/Shutterstock.com; **304** cr © Srinivasan Clicks/Adobe Stock Photo; **304** br © Blackday/Adobe Stock Photo; **307** cl © Graham Dunn/Alamy Stock Photo; **308** tr © Wanderluster/Alamy Stock Photo; **310** cr © Eye Of Science/Science Photo Library; **310** tr © Auu San AKul/Adobe Stock Photo; **310** cr © Kateryna Kon/Adobe Stock Photo; **311** cr © Svetlana/Adobe Stock Photo; **311** cr © Libin/Adobe Stock Photo; **312** tr © Monica Schroeder/Science Photo Library; **312** cr © Monica Schroeder/Science Photo Library; **312** br © The Image Zone/Alamy Stock Photo; **314** tr © World History Archive/Alamy Stock Photo; **315** bl © Luchschen F/Adobe Stock Photo; **316** tr © Sinclair Stammers/Science Photo Library; **317** © Wr Lili/Adobe Stock Photo; **326** cll © Egg Eeggjiew/Adobe Stock Photo; **326** br, **p. 414** © Wild Pictures/Alamy Stock Photo; **327** cl © Mediscan/Alamy Stock Photo; **328** bc © Greg Boiarsky/Adobe Stock Photo; **328** br © Poco Bw/Adobe Stock Photo; **329** cl © Pheelings Media/Adobe Stock Photo; **329** cr © Dr P Maraxxi/Science Photo Library; **330** cl © Douglas/Adobe Stock Photo; **330** cr © St Mary'sHospital Medical School/Science Photo Library; **330** cr © Rawpixel Com/Adobe Stock Photo; **330** br © Rohane/Adobe Stock Photo; **331** bl © Tommy Stock Project/Adobe Stock Photo; **332** tr ; **332** br © Senic Photo/Adobe Stock Photo; **334** cr © Steve Gschmeissner/Science Photo Library; **334** tr © Mood Board/Adobe Stock Photo; **335** br © Jpc Prod/Adobe Stock Photo; **336** tr © Science Photo Library/Alamy Stock Photo; **336** br © Cultura Creative Rf/Alamy Stock Photo; **337** br © Bsip Sa/Alamy Stock Photo; **338** br © Designua/Adobe Stock Photo; **339** br © Tetra Images Llc/Alamy Stock Photo; **340** br © Pixel Shot/Alamy Stock Photo; **341** tr © Clouds Hill Imaging Ltd/Science Photo Library; **341** br © Science Photo Library/Alamy Stock Photo; **342** tr © Allstar Picture Library Ltd/Alamy Stock Photo; **342** bl © Designua/Adobe Stock Photo; **343** tr © Steve Gschmeissner/Science Photo Library; **343** br © Mark Kostich/Adobe Stock Photo; **344** tr © Timolina/Adobe

Stock Photo; **347** *br* © Diego Cervo/Adobe Stock Photo; **348** *tr* © Seventy four/Adobe Stock Photo; **348** *br* © Douce Fleur/Adobe Stock Photo; **349** *tr* © Olesia Bilkei/Adobe Stock Photo; **350** *tr* © Юля Бурмистрова/Adobe Stock Photo; **351** *br* © Svyatoslav Lypynskyy/Adobe Stock Photo; **356** *br* © Lennart Nilsson Boehringer Ingelheim International Gmbh Tt/Science Photo Library; **358** *tr* © Robson Photo/Adobe Stock Photo; **358** *cr* © Jan Sochor/Alamy Stock Photo; **359** *tr* © Eugenio Marongiu/Adobe Stock Photo; **360** *tr* © Mariya Sokolova/Adobe Stock Photo; **360** *cr* © Weerajata/Adobe Stock Photo; **361** *cr* © Science Photo Library/Science Photo Library; **364** *cr* © Mark Thomas/Science Photo Library; **365** *cr* © Sn 040288/Shutterstock.com; **365** *br* © Sora Pop/Adobe Stock Photo; **366** *tr* © CNRI/Science Photo Library; **367** *tr* © Saiful 52/Adobe Stock Photo; **368** *cr* © Okrasiuk/Adobe Stock Photo; **368** *br* © Tsanan/Adobe Stock Photo; **369** *tr* © Warloka 79/Adobe Stock Photo; **369** *cr* © Matka Wariatka/Shutterstock.com; **373** *br* © National Institute Of Allergy And Infectious Diseases National Institutes Of Health/Science Photo Library; **374** *cr* © Nibsc/Science Photo Library; **375** *tl* © William/Adobe Stock Photo; **376** *bl* © Cnri/Science Photo Library; **376** *br* © Dmitry Rukhlenko/Shutterstock.com; **377** *cr* © Kateryna Kon/Science Photo Library; **377** *cr* © National Institutes Of Health/Science Photo Library; **381** *tr* © Aathan Allen/Adobe Stock Photo; **382** *tr* © Dennis Kunkel Microscopy/Science Photo Library; **382** *br* © Wesley Bocxe/Science Photo Library; **383** *tr* © Eye Of Science/Science Photo Library; **383** *bl* © Cid Ism/Science Photo Library; **384** *tr* © Mediscan/Alamy Stock Photo; **384** *br* © E Gueho/Science Photo Library; **388** *tr* © Banker Fotos/Adobe Stock Photo; **389** *tr* © Dave Stamboulis/Alamy Stock Photo; **389** *cr* © Michael Jung/Adobe Stock Photo; **390** *tr* © Science Photo Library/Science Photo Library; **391** *cr* © Prb Arts/Adobe Stock Photo; **392** *cr* © Dr M A Ansaty/Science Photo Library; **393** *tr* © Peter Hermes Furian/Adobe Stock Photo; **394** *tr* © Dr Linda Stannard Uct/Science Photo Library; **395** *tr* © Andrey Popov/Adobe Stock Photo; **395** *cr* © Biomedical Imaging Units Southampton General Hospital/Science Photo Library; **396** *cr* © Nibsc/Science Photo Library; **398** © Neil Bowman/Alamy Stock Photo; **402** *cr* © Nattakorn/Adobe Stock Photo; **403** *cr* © Image Professionals Gmbh/Alamy Stock Photo; **405** *cr* © PLG/Adobe Photo Stock; **406** *br* © Tek Image/Science Photo Library; **407** *cr* © Nomad Soul/Adobe Stock Photo; **408** *br* © Juliana Swenson/Alamy Stock Photo; **413** *tr* © Heiko Kiera/Shutterstock.com; **414** *tr* © Bios Photo/Alamy Stock Photo; **414** *cr* © João Burini/Alamy Stock Photo; **416** *br* © Slow Motion Gli/Adobe Stock Photo; **417** *tr* © Dr Tony Brain/Science Photo Library; **418** *tr* © Malajscy/Adobe Stock Photo; **418** *br* © A Crump Tdr Who/Science Photo Library; **419** *tr* © Eye Of Science/Science Photo Library; **420** *cr* © Zay Nyi Nyi/Alamy Stock Photo; **422** © Corbacserdar/Adobe Stock Photo; **423** *cr* © Riccardo Niels Mayer/Adobe Stock Photo; **424** *cr* © Science Photo Library/Science Photo Library; **425** *cr* © Heiko Kiera/Shutterstock.com; **425** *bl* © Csi Productions/Alamy Stock Photo; **426** *cr* © Eye Of Science/Science Photo Library; **430** *tr* © Wave Break 3/Adobe Stock Photo; **431** *bl* © John Cole/Science Photo Library; **432** *cr* © More Pixels/Adobe Stock Photo; **433** *tr* © Drew/Adobe Stock Photo; **433** *cr* © Afri Pics Com/Alamy Stock Photo; **434** *br* © Drazen/Adobe Stock Photo; **435** *cr* © Lightfield Studios/Adobe Stock Photo; **436** *tr* © Dennis Kunkel Microscopy/Science Photo Library; **436** *br* © Natali/Adobe Stock Photo; **437** *cr* © Mast 3 R/Adobe Stock Photo; **437** *cr* © Donovan/Adobe Stock Photo; **437** *br* © Suphaya/Adobe Stock Photo; **438** *cr* © One Inch Punch/Adobe Stock Photo; **438** *cr* © Fizkes/Adobe Stock Photo; **438** *br* © Karen Struthers/Adobe Stock Photo; **444** *cl* © Fazon/Adobe Stock Photo; **444** *br* © Lotus Studio/Adobe Stock Photo; **445** *cl* © Alter Photo/Adobe Stock Photo; **445** *cr* © Vera New Sib/Adobe Stock Photo; **446** *cr* © Victor Koldunov/Adobe Stock Photo; **448** *bl* © Mr Innis/Shutterstock.com; **449** *br* © Digi Shooter/Adobe Stock Photo; **450** *cr* © Jan Mach/Adobe Stock Photo; **453** *cl* © Wvd M Photography/Shutterstock.com; **456** *br* © Focused Adventures/Adobe Stock Photo; **457** *cl* © Kurhan/Adobe Stock Photo; **457** *cr* © Fazeful/Adobe Stock Photo; **458** *br* © Chris Pictures/Alamy Stock Photo; **459** *br* © Drew/Adobe Stock Photo; **461** *tr* © a7880ss/Adobe Stock Photo; **462** *bl* © Georgette Douwma/Science Photo Library; **463** *br* © Daniel Prudek/Shutterstock.com; **465** *bl* © Upi/Alamy Stock Photo; **466** *tr* © Art Directors & Trip/Alamy Stock Photo; **467** *cr* © Manishankar Patra/Shutterstock.com; **468** *tr* © Robert Brook/Science Photo Library; **469** *tr* © QQ 47182080/Adobe Stock Photo; **472** *cl* © Jeffrey Isaac Greenberg 19+/Alamy Stock Photo; **473** *tr* © Budimir Jevtic/Adobe Stock Photo; **474** *br* © David Nunuk/Science Photo Library; **476** *cc* © Tartila/Adobe Stock Photo; **477** *tr* © Rada Covalenco/Adobe Stock Photo; **479** *br* © Monika Mlynek/Alamy Stock Photo; **480** *br* © Freeman 83/Adobe Stock Photo; **488** *br* © Sylvie Bouchard/Alamy Stock Photo; **488** *cl* © Jan Sochor/Alamy Stock Photo.

t = top, *b* = bottom, *l* = left, *r* = right, *c* = centre